Fundamental Constants

Quantity	Symbol	Approximate Value	Current Best Value[†]
Speed of light in vacuum	c	3.00×10^8 m/s	2.99792458×10^8 m/s
Gravitational constant	G	6.67×10^{-11} N·m²/kg²	$6.67259(85) \times 10^{-11}$ N·m²/kg²
Avogadro's number	N_A	6.02×10^{23} mol⁻¹	$6.0221367(36) \times 10^{23}$ mol⁻¹
Gas constant	R	8.315 J/mol·K = 1.99 cal/mol·K = 0.082 atm·liter/mol·K	8.314510(70) J/mol·K
Boltzmann's constant	k	1.38×10^{-23} J/K	$1.380658(12) \times 10^{-23}$ J/K
Charge on electron	e	1.60×10^{-19} C	$1.60217733(49) \times 10^{-19}$ C
Stefan-Boltzmann constant	σ	5.67×10^{-8} W/m²·K⁴	$5.67051(19) \times 10^{-8}$ W/m²·K⁴
Permittivity of free space	$\epsilon_0 = (1/c^2\mu_0)$	8.85×10^{-12} C²/N·m²	$8.854187817... \times 10^{-12}$ C²/N·m²
Permeability of free space	μ_0	$4\pi \times 10^{-7}$ T·m/A	$1.2566370614... \times 10^{-6}$ T·m/A
Planck's constant	h	6.63×10^{-34} J·s	$6.6260755(40) \times 10^{-34}$ J·s
Electron rest mass	m_e	9.11×10^{-31} kg = 0.000549 u = 0.511 MeV/c^2	$9.1093897(54) \times 10^{-31}$ kg = $5.48579903(13) \times 10^{-4}$ u
Proton rest mass	m_p	1.6726×10^{-27} kg = 1.00728 u = 938.3 MeV/c^2	$1.6726231(10) \times 10^{-27}$ kg = 1.007276470(12) u
Neutron rest mass	m_n	1.6749×10^{-27} kg = 1.008665 u = 939.6 MeV/c^2	$1.6749286(10) \times 10^{-27}$ kg = 1.008664904(14) u
Atomic mass unit (1 u)		1.6605×10^{-27} kg = 931.5 MeV/c^2	$1.6605402(10) \times 10^{-27}$ kg = 931.49432(28) MeV/c^2

[†]Reviewed 1993 by B. N. Taylor, National Institute of Standards and Technology. Numbers in parentheses indicate one standard deviation experimental uncertainties in final digits. Values without parentheses are exact (i.e., defined quantities).

Other Useful Data

Joule equivalent (1 cal)	4.186 J
Absolute zero (0 K)	-273.15°C
Earth: Mass	5.97×10^{24} kg
Radius (mean)	6.38×10^3 km
Moon: Mass	7.35×10^{22} kg
Radius (mean)	1.74×10^3 km
Sun: Mass	1.99×10^{30} kg
Radius (mean)	6.96×10^5 km
Earth-sun distance (mean)	149.6×10^6 km
Earth-moon distance (mean)	384×10^3 km

The Greek Alphabet

Alpha	A	α	Nu	N	ν
Beta	B	β	Xi	Ξ	ξ
Gamma	Γ	γ	Omicron	O	o
Delta	Δ	δ	Pi	Π	π
Epsilon	E	ε	Rho	P	ρ
Zeta	Z	ζ	Sigma	Σ	σ
Eta	H	η	Tau	T	τ
Theta	Θ	θ	Upsilon	Y	υ
Iota	I	ι	Phi	Φ	ϕ, φ
Kappa	K	κ	Chi	X	χ
Lambda	Λ	λ	Psi	Ψ	ψ
Mu	M	μ	Omega	Ω	ω

Values of Some Numbers

π	= 3.1415927	$\sqrt{2}$	= 1.4142136	ln 2	= 0.6931472	$\log_{10} e$	= 0.4342945
e	= 2.7182818	$\sqrt{3}$	= 1.7320508	ln 10	= 2.3025851	1 rad	= 57.2957795°

Mathematical Signs and Symbols

$\propto$	is proportional to		$\leq$	is less than or equal to
$=$	is equal to		$\geq$	is greater than or equal to
$\approx$	is approximately equal to		Σ	sum of
$\neq$	is not equal to		$\bar{x}$	average value of x
$>$	is greater than		Δx	change in x
$\gg$	is much greater than		$\Delta x \to 0$	Δx approaches zero
$<$	is less than		$n!$	$n(n-1)(n-2)\ldots(1)$
$\ll$	is much less than			

Unit Conversions (Equivalents)

Length

1 in. = 2.54 cm
1 cm = 0.394 in.
1 ft = 30.5 cm
1 m = 39.37 in. = 3.28 ft
1 mi = 5280 ft = 1.61 km
1 km = 0.621 mi
1 nautical mile (U.S.) = 1.15 mi = 6076 ft = 1.852 km
1 fermi = 1 femtometer (fm) = 10^{-15} m
1 angstrom (Å) = 10^{-10} m
1 light-year (ly) = 9.46×10^{15} m
1 parsec = 3.26 ly = 3.09×10^{16} m

Volume

1 liter (L) = 1000 mL = 1000 cm^3 = 1.0×10^{-3} m^3 = 1.057 quart (U.S.) = 54.6 in.^{3}
1 gallon (U.S.) = 4 qt (U.S.) = 231 in.^{3} = 3.78 L = 0.83 gal (Imperial)
1 m^3 = 35.31 ft^3

Speed

1 mi/h = 1.47 ft/s = 1.609 km/h = 0.447 m/s
1 km/h = 0.278 m/s = 0.621 mi/h
1 ft/s = 0.305 m/s = 0.682 mi/h
1 m/s = 3.28 ft/s = 3.60 km/h
1 knot = 1.151 mi/h = 0.5144 m/s

Angle

1 radian (rad) = 57.30° = 57°18′
1° = 0.01745 rad
1 rev/min (rpm) = 0.1047 rad/s

Time

1 day = 8.64×10^4 s
1 year = 3.156×10^7 s

Mass

1 atomic mass unit (u) = 1.6605×10^{-27} kg
1 kg = 0.0685 slug
[1 kg has a weight of 2.20 lb where g = 9.81 m/s^2.]

Force

1 lb = 4.45 N
1 N = 10^5 dyne = 0.225 lb

Energy and Work

1 J = 10^7 ergs = 0.738 ft·lb
1 ft·lb = 1.36 J = 1.29×10^{-3} Btu = 3.24×10^{-4} kcal
1 kcal = 4.18×10^3 J = 3.97 Btu
1 eV = 1.602×10^{-19} J
1 kWh = 3.60×10^6 J = 860 kcal

Power

1 W = 1 J/s = 0.738 ft·lb/s = 3.42 Btu/h
1 hp = 550 ft·lb/s = 746 W

Pressure

1 atm = 1.013 bar = 1.013×10^5 N/m^2 = 14.7 lb/in.^{2} = 760 torr
1 lb/in.^{2} = 6.90×10^3 N/m^2
1 Pa = 1 N/m^2 = 1.45×10^{-4} lb/in.^{2}

SI Derived Units and Their Abbreviations

Quantity	Unit	Abbreviation	In Terms of Base Units[†]
Force	newton	N	kg·m/s^2
Energy and work	joule	J	kg·m^2/s^2
Power	watt	W	kg·m^2/s^3
Pressure	pascal	Pa	kg/(m·s^2)
Frequency	hertz	Hz	s^{-1}
Electric charge	coulomb	C	A·s
Electric potential	volt	V	kg·m^2/(A·s^3)
Electric resistance	ohm	Ω	kg·m^2/(A^2·s^3)
Capacitance	farad	F	A^2·s^4/(kg·m^2)
Magnetic field	tesla	T	kg/(A·s^2)
Magnetic flux	weber	Wb	kg·m^2/(A·s^2)
Inductance	henry	H	kg·m^2/(s^2·A^2)

[†]kg = kilogram (mass), m = meter (length), s = second (time), A = ampere (electric current).

Metric (SI) Multipliers

Prefix	Abbreviation	Value
exa	E	10^{18}
peta	P	10^{15}
tera	T	10^{12}
giga	G	10^9
mega	M	10^6
kilo	k	10^3
hecto	h	10^2
deka	da	10^1
deci	d	10^{-1}
centi	c	10^{-2}
milli	m	10^{-3}
micro	μ	10^{-6}
nano	n	10^{-9}
pico	p	10^{-12}
femto	f	10^{-15}
atto	a	10^{-18}

PHYSICS
for
SCIENTISTS & ENGINEERS

Part 3
PHYSICS
for
SCIENTISTS & ENGINEERS
Third Edition

DOUGLAS C. GIANCOLI

PRENTICE HALL
Upper Saddle River, New Jersey 07458

Editor-in-Chief: Paul F. Corey
Production Editor: Susan Fisher
Executive Editor: Alison Reeves
Development Editor: David Chelton
Director of Marketing: John Tweedale
Senior Marketing Manager: Erik Fahlgren
Assistant Vice President of Production and Manufacturing: David W. Riccardi
Executive Managing Editor: Kathleen Schiaparelli
Manufacturing Manager: Trudy Pisciotti
Art Manager: Gus Vibal
Director of Creative Services: Paul Belfanti
Advertising and Promotions Manager: Elise Schneider
Editor in Chief of Development: Ray Mullaney
Project Manager: Elizabeth Kell
Photo Research: Mary Teresa Giancoli
Photo Research Administrator: Melinda Reo
Copy Editor: Jocelyn Phillips
Editorial Assistant: Marilyn Coco
Cover photo: Onne van der Wal/Young America
Composition: Emilcomp srl / Preparé Inc.

 © 2000, 1989, 1984 by Douglas C. Giancoli
Published by Prentice Hall
Upper Saddle River, NJ 07458

All rights reserved. No part of this book may be
reproduced, in any form or by any means,
without permission in writing from the publisher.

Photo credits appear at the end of the book, and
constitute a continuation of the copyright page.

Printed in the United States of America

10 9 8 7 6 5 4 3 2 1

ISBN 0-13-029096-3

Prentice-Hall International (UK) Limited, *London*
Prentice-Hall of Australia Pty. Limited, *Sydney*
Prentice-Hall Canada Inc., *Toronto*
Prentice-Hall Hispanoamericana, S.A., *Mexico City*
Prentice-Hall of India Private Limited, *New Delhi*
Prentice-Hall of Japan, Inc., *Tokyo*
Prentice-Hall (*Singapore*) Pte. Ltd.
Editora Prentice-Hall do Brasil, Ltda., *Rio de Janeiro*

Contents

Preface		xvii
Supplements		xxviii
Notes to Students and Instructors on the Format		xxx
Use of Color		xxxi

PART 1

1 Introduction, Measurement, Estimating — 1

1-1	The Nature of Science	2
1-2	Models, Theories, and Laws	3
1-3	Measurement and Uncertainty; Significant Figures	4
1-4	Units, Standards, and the SI System	6
1-5	Converting Units	8
1-6	Order of Magnitude: Rapid Estimating	9
*1-7	Dimensions and Dimensional Analysis	12

SUMMARY 13 QUESTIONS 13
PROBLEMS 14 GENERAL PROBLEMS 15

2 Describing Motion: Kinematics in One Dimension — 16

2-1	Reference Frames and Displacement	17
2-2	Average Velocity	18
2-3	Instantaneous Velocity	20
2-4	Acceleration	23
2-5	Motion at Constant Acceleration	26
2-6	Solving Problems	28
2-7	Falling Objects	31
*2-8	Use of Calculus; Variable Acceleration	36

SUMMARY 38 QUESTIONS 38
PROBLEMS 39 GENERAL PROBLEMS 42

3 Kinematics in Two Dimensions; Vectors — 45

3-1	Vectors and Scalars	45
3-2	Addition of Vectors—Graphical Methods	46
3-3	Subtraction of Vectors, and Multiplication of a Vector by a Scalar	48
3-4	Adding Vectors by Components	48
3-5	Unit Vectors	52
3-6	Vector Kinematics	53
3-7	Projectile Motion	55
3-8	Solving Problems in Projectile Motion	58
3-9	Uniform Circular Motion	63
3-10	Relative Velocity	66

SUMMARY 68 QUESTIONS 69
PROBLEMS 70 GENERAL PROBLEMS 74

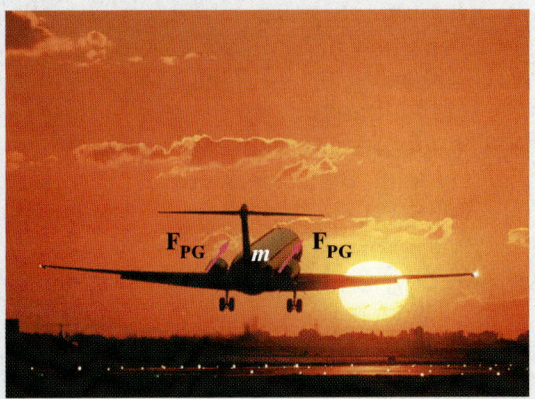

4 Dynamics: Newton's Laws of Motion — 77

4-1	Force	77
4-2	Newton's First Law of Motion	78
4-3	Mass	79
4-4	Newton's Second Law of Motion	80
4-5	Newton's Third Law of Motion	82
4-6	Weight—the Force of Gravity; and the Normal Force	85
4-7	Solving Problems with Newton's Laws: Free-Body Diagrams	88
4-8	Problem Solving—A General Approach	96

SUMMARY 97 QUESTIONS 97
PROBLEMS 98 GENERAL PROBLEMS 103

5 Further Applications of Newton's Laws — 106

- 5-1 Applications of Newton's Laws Involving Friction — 106
- 5-2 Dynamics of Uniform Circular Motion — 114
- 5-3 Highway Curves, Banked and Unbanked — 118
- *5-4 Nonuniform Circular Motion — 121
- *5-5 Velocity-Dependent Forces; Terminal Velocity — 122

SUMMARY 124 QUESTIONS 124
PROBLEMS 125 GENERAL PROBLEMS 129

6 Gravitation and Newton's Synthesis — 133

- 6-1 Newton's Law of Universal Gravitation — 133
- 6-2 Vector Form of Newton's Law of Universal Gravitation — 136
- 6-3 Gravity Near the Earth's Surface; Geophysical Applications — 137
- 6-4 Satellites and "Weightlessness" — 139
- 6-5 Kepler's Laws and Newton's Synthesis — 143
- 6-6 Gravitational Field — 146
- 6-7 Types of Forces in Nature — 147
- *6-8 Gravitational Versus Inertial Mass; the Principle of Equivalence — 148
- *6-9 Gravitation as Curvature of Space; Black Holes — 149

SUMMARY 150 QUESTIONS 150
PROBLEMS 151 GENERAL PROBLEMS 153

7 Work and Energy — 155

- 7-1 Work Done by a Constant Force — 156
- 7-2 Scalar Product of Two Vectors — 159
- 7-3 Work Done by a Varying Force — 161
- 7-4 Kinetic Energy and the Work–Energy Principle — 164
- *7-5 Kinetic Energy at Very High Speed — 169

SUMMARY 170 QUESTIONS 170
PROBLEMS 171 GENERAL PROBLEMS 174

8 Conservation of Energy — 176

- 8-1 Conservative and Nonconservative Forces — 177
- 8-2 Potential Energy — 178
- 8-3 Mechanical Energy and Its Conservation — 182
- 8-4 Problem Solving Using Conservation of Mechanical Energy — 184
- 8-5 The Law of Conservation of Energy — 189
- 8-6 Energy Conservation with Dissipative Forces: Solving Problems — 190
- 8-7 Gravitational Potential Energy and Escape Velocity — 192
- 8-8 Power — 195
- *8-9 Potential Energy Diagrams; Stable and Unstable Equilibrium — 197

SUMMARY 198 QUESTIONS 199
PROBLEMS 200 GENERAL PROBLEMS 204

9 LINEAR MOMENTUM AND COLLISIONS — 206

- 9-1 Momentum and Its Relation to Force — 206
- 9-2 Conservation of Momentum — 208
- 9-3 Collisions and Impulse — 211
- 9-4 Conservation of Energy and Momentum in Collisions — 214
- 9-5 Elastic Collisions in One Dimension — 214
- 9-6 Inelastic Collisions — 217
- 9-7 Collisions in Two or Three Dimensions — 219
- 9-8 Center of Mass (CM) — 221
- 9-9 Center of Mass and Translational Motion — 225
- *9-10 Systems of Variable Mass; Rocket Propulsion — 227

SUMMARY 230 QUESTIONS 230
PROBLEMS 231 GENERAL PROBLEMS 236

10 ROTATIONAL MOTION ABOUT A FIXED AXIS — 239

- 10-1 Angular Quantities — 240
- 10-2 Kinematic Equations for Uniformly Accelerated Rotational Motion — 243
- 10-3 Rolling Motion (without Slipping) — 244
- 10-4 Vector Nature of Angular Quantities — 246
- 10-5 Torque — 247
- 10-6 Rotational Dynamics; Torque and Rotational Inertia — 249
- 10-7 Solving Problems in Rotational Dynamics — 250
- 10-8 Determining Moments of Inertia — 254
- 10-9 Angular Momentum and Its Conservation — 256
- 10-10 Rotational Kinetic Energy — 260
- 10-11 Rotational Plus Translational Motion; Rolling — 262
- *10-12 Why Does a Rolling Sphere Slow Down? — 268

SUMMARY 268 QUESTIONS 269
PROBLEMS 270 GENERAL PROBLEMS 276

11 GENERAL ROTATION — 279

- 11-1 Vector Cross Product — 279
- 11-2 The Torque Vector — 280
- 11-3 Angular Momentum of a Particle — 281
- 11-4 Angular Momentum and Torque for a System of Particles; General Motion — 283
- 11-5 Angular Momentum and Torque for a Rigid Body — 285
- *11-6 Rotational Imbalance — 287
- 11-7 Conservation of Angular Momentum — 288
- *11-8 The Spinning Top — 290
- *11-9 Rotating Frames of Reference; Inertial Forces — 291
- *11-10 The Coriolis Effect — 292

SUMMARY 294 QUESTIONS 294
PROBLEMS 295 GENERAL PROBLEMS 298

12 STATIC EQUILIBRIUM; ELASTICITY AND FRACTURE — 300

- 12-1 Statics—The Study of Forces in Equilibrium — 300
- 12-2 The Conditions for Equilibrium — 301
- 12-3 Solving Statics Problems — 303
- 12-4 Stability and Balance — 308
- 12-5 Elasticity; Stress and Strain — 309
- 12-6 Fracture — 312
- *12-7 Trusses and Bridges — 315
- *12-8 Arches and Domes — 319

SUMMARY 321 QUESTIONS 321
PROBLEMS 322 GENERAL PROBLEMS 328

Contents vii

PART 2

13 FLUIDS 332

- 13-1 Density and Specific Gravity 332
- 13-2 Pressure in Fluids 333
- 13-3 Atmospheric Pressure and Gauge Pressure 337
- 13-4 Pascal's Principle 337
- 13-5 Measurement of Pressure; Gauges and the Barometer 338
- 13-6 Buoyancy and Archimedes' Principle 340
- 13-7 Fluids in Motion; Flow Rate and the Equation of Continuity 343
- 13-8 Bernoulli's Equation 345
- 13-9 Applications of Bernoulli's Principle: From Torricelli to Sailboats, Airfoils, and TIA 347
- *13-10 Viscosity 350
- *13-11 Flow in Tubes: Poiseuille's Equation 351
- *13-12 Surface Tension and Capillarity 351
- *13-13 Pumps 353

SUMMARY 354   QUESTIONS 354
PROBLEMS 356   GENERAL PROBLEMS 360

14 OSCILLATIONS 362

- 14-1 Oscillations of a Spring 363
- 14-2 Simple Harmonic Motion 364
- 14-3 Energy in the Simple Harmonic Oscillator 369
- 14-4 Simple Harmonic Motion Related to Uniform Circular Motion 371
- 14-5 The Simple Pendulum 371
- *14-6 Physical Pendulum and Torsion Pendulum 373
- 14-7 Damped Harmonic Motion 374
- 14-8 Forced Vibrations; Resonance 378

SUMMARY 380   QUESTIONS 381
PROBLEMS 381   GENERAL PROBLEMS 386

15 WAVE MOTION 388

- 15-1 Characteristics of Wave Motion 389
- 15-2 Wave Types 391
- 15-3 Energy Transported by Waves 395
- 15-4 Mathematical Representation of a Traveling Wave 396
- *15-5 The Wave Equation 399
- 15-6 The Principle of Superposition 401
- 15-7 Reflection and Transmission 402
- 15-8 Interference 404
- 15-9 Standing Waves; Resonance 405
- *15-10 Refraction 408
- *15-11 Diffraction 410

SUMMARY 410   QUESTIONS 411
PROBLEMS 412   GENERAL PROBLEMS 415

16 SOUND 417

- 16-1 Characteristics of Sound 417
- 16-2 Mathematical Representation of Longitudinal Waves 419
- 16-3 Intensity of Sound; Decibels 420
- 16-4 Sources of Sound: Vibrating Strings and Air Columns 424
- *16-5 Quality of Sound, and Noise 429
- 16-6 Interference of Sound Waves; Beats 429
- 16-7 Doppler Effect 432
- *16-8 Shock Waves and the Sonic Boom 435
- *16-9 Applications; Ultrasound and Ultrasound Imaging 437

SUMMARY 438   QUESTIONS 438
PROBLEMS 439   GENERAL PROBLEMS 443

17 TEMPERATURE, THERMAL EXPANSION, AND THE IDEAL GAS LAW 445

17–1	Atomic Theory of Matter	446
17–2	Temperature and Thermometers	447
17–3	Thermal Equilibrium and the Zeroth Law of Thermodynamics	449
17–4	Thermal Expansion	450
*17–5	Thermal Stresses	454
17–6	The Gas Laws and Absolute Temperature	454
17–7	The Ideal Gas Law	456
17–8	Problem Solving with the Ideal Gas Law	457
17–9	Ideal Gas Law in Terms of Molecules: Avogadro's Number	459
*17–10	Ideal Gas Temperature Scale—A Standard	460

SUMMARY 461 QUESTIONS 461
PROBLEMS 462 GENERAL PROBLEMS 464

18 KINETIC THEORY OF GASES 466

18–1	The Ideal Gas Law and the Molecular Interpretation of Temperature	466
18–2	Distribution of Molecular Speeds	470
18–3	Real Gases and Changes of Phase	473
*18–4	Vapor Pressure and Humidity	474
*18–5	Van der Waals Equation of State	477
*18–6	Mean Free Path	478
*18–7	Diffusion	479

SUMMARY 481 QUESTIONS 481
PROBLEMS 482 GENERAL PROBLEMS 484

19 HEAT AND THE FIRST LAW OF THERMODYNAMICS 485

19–1	Heat as Energy Transfer	485
19–2	Internal Energy	487
19–3	Specific Heat	488
19–4	Calorimetry—Solving Problems	489
19–5	Latent Heat	490
19–6	The First Law of Thermodynamics	493
19–7	Applying the First Law of Thermodynamics; Calculating the Work	495
19–8	Molar Specific Heats for Gases, and the Equipartition of Energy	498
19–9	Adiabatic Expansion of a Gas	502
19–10	Heat Transfer: Conduction, Convection, Radiation	503

SUMMARY 508 QUESTIONS 509
PROBLEMS 510 GENERAL PROBLEMS 514

20 SECOND LAW OF THERMODYNAMICS; HEAT ENGINES 516

20–1	The Second Law of Thermodynamics—Introduction	516
20–2	Heat Engines	517
20–3	Reversible and Irreversible Processes; The Carnot Engine	520
20–4	Refrigerators, Air Conditioners, and Heat Pumps	525
20–5	Entropy	528
20–6	Entropy and the Second Law of Thermodynamics	529
20–7	Order to Disorder	533
20–8	Energy Availability; Heat Death	534
*20–9	Statistical Interpretation of Entropy and the Second Law	535
*20–10	Thermodynamic Temperature Scale; Absolute Zero and the Third Law of Thermodynamics	537

SUMMARY 539 QUESTIONS 539
PROBLEMS 540 GENERAL PROBLEMS 543

PART 3

21 Electric Charge and Electric Field 545

21-1	Static Electricity; Electric Charge and Its Conservation	546
21-2	Electric Charge in the Atom	547
21-3	Insulators and Conductors	547
21-4	Induced Charge; the Electroscope	548
21-5	Coulomb's Law	549
21-6	The Electric Field	554
21-7	Electric Field Calculations for Continuous Charge Distributions	558
21-8	Field Lines	561
21-9	Electric Fields and Conductors	562
21-10	Motion of a Charged Particle in an Electric Field	564
21-11	Electric Dipoles	565

SUMMARY 567   QUESTIONS 568
PROBLEMS 569   GENERAL PROBLEMS 572

22 Gauss's Law 575

22-1	Electric Flux	576
22-2	Gauss's Law	578
22-3	Applications of Gauss's Law	580
*22-4	Experimental Basis of Gauss's and Coulomb's Law	586

SUMMARY 586   QUESTIONS 587
PROBLEMS 587   GENERAL PROBLEMS 590

23 Electric Potential 591

23-1	Electric Potential and Potential Difference	591
23-2	Relation Between Electric Potential and Electric Field	595
23-3	Electric Potential Due to Point Charges	597
23-4	Potential Due to Any Charge Distribution	599
23-5	Equipotential Surfaces	600
23-6	Electric Dipoles	601
23-7	$\mathbf{E}$ Determined from V	602
23-8	Electrostatic Potential Energy; the Electron Volt	603
*23-9	Cathode Ray Tube: TV and Computer Monitors, Oscilloscope	605

SUMMARY 607   QUESTIONS 607
PROBLEMS 608   GENERAL PROBLEMS 611

24 Capacitance, Dielectrics, Electric Energy Storage 613

24-1	Capacitors	613
24-2	Determination of Capacitance	614
24-3	Capacitors in Series and Parallel	617
24-4	Electric Energy Storage	620
24-5	Dielectrics	621
*24-6	Molecular Description of Dielectrics	624

SUMMARY 627   QUESTIONS 627
PROBLEMS 628   GENERAL PROBLEMS 632

25 Electric Currents and Resistance — 634

25-1	The Electric Battery	635
25-2	Electric Current	636
25-3	Ohm's Law: Resistance and Resistors	638
25-4	Resistivity	640
25-5	Electric Power	642
25-6	Power in Household Circuits	644
25-7	Alternating Current	645
25-8	Microscopic View of Electric Current: Current Density and Drift Velocity	647
*25-9	Superconductivity	650
25-10	Electric Hazards; Leakage Currents	651
	SUMMARY 653 QUESTIONS 653	
	PROBLEMS 654 GENERAL PROBLEMS 656	

26 DC Circuits — 658

26-1	EMF and Terminal Voltage	659
26-2	Resistors in Series and in Parallel	660
26-3	Kirchhoff's Rules	664
26-4	Circuits Containing Resistor and Capacitor (*RC* Circuits)	669
*26-5	DC Ammeters and Voltmeters	674
*26-6	Transducers and the Thermocouple	676
	SUMMARY 678 QUESTIONS 678	
	PROBLEMS 679 GENERAL PROBLEMS 683	

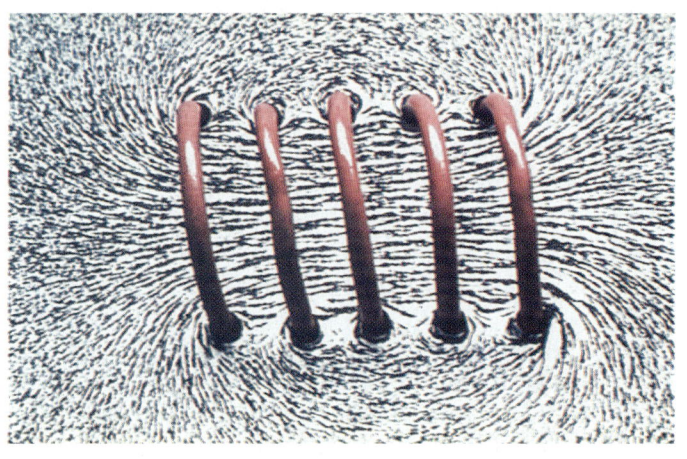

27 Magnetism — 686

27-1	Magnets and Magnetic Fields	686
27-2	Electric Currents Produce Magnetism	689
27-3	Force on an Electric Current in a Magnetic Field; Definition of **B**	689
27-4	Force on an Electric Charge Moving in a Magnetic Field	692
27-5	Torque on a Current Loop; Magnetic Dipole Moment	695
*27-6	Applications: Galvanometers, Motors, Loudspeakers	697
27-7	Discovery and Properties of the Electron	699
*27-8	The Hall Effect	701
*27-9	Mass Spectrometer	702
	SUMMARY 702 QUESTIONS 703	
	PROBLEMS 704 GENERAL PROBLEMS 707	

28 Sources of Magnetic Field — 709

28-1	Magnetic Field Due to a Straight Wire	710
28-2	Force Between Two Parallel Wires	710
28-3	Operational Definitions of the Ampere and the Coulomb	712
28-4	Ampère's Law	712
28-5	Magnetic Field of a Solenoid and a Toroid	716
28-6	Biot-Savart Law	719
*28-7	Magnetic Materials—Ferromagnetism	722
*28-8	Electromagnets and Solenoids	723
*28-9	Magnetic Fields in Magnetic Materials; Hysteresis	724
*28-10	Paramagnetism and Diamagnetism	725
	SUMMARY 727 QUESTIONS 727	
	PROBLEMS 728 GENERAL PROBLEMS 732	

29 Electromagnetic Induction and Faraday's Law — 734

29-1	Induced EMF	734
29-2	Faraday's Law of Induction; Lenz's Law	736
29-3	EMF Induced in a Moving Conductor	739
29-4	Electric Generators	740
*29-5	Counter EMF and Torque; Eddy Currents	742
29-6	Transformers and Transmission of Power	744
29-7	A Changing Magnetic Flux Produces an Electric Field	747
*29-8	Applications of Induction: Sound Systems, Computer Memory, the Seismograph	749
	SUMMARY 750 QUESTIONS 750	
	PROBLEMS 751 GENERAL PROBLEMS 754	

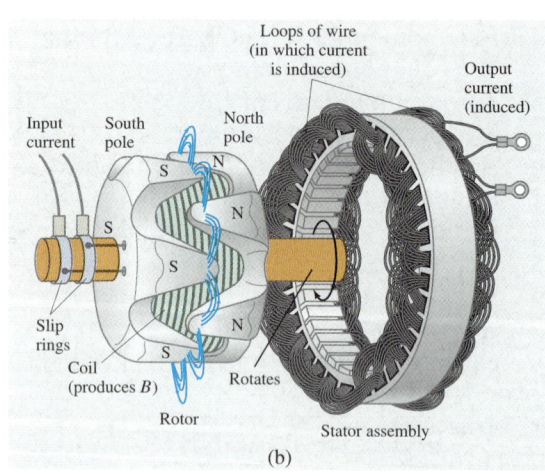

30 Inductance; and Electromagnetic Oscillations — 756

30-1	Mutual Inductance	756
30-2	Self-Inductance	758
30-3	Energy Stored in a Magnetic Field	760
30-4	LR Circuits	762
30-5	LC Circuits and Electromagnetic Oscillations	764
30-6	LC Oscillations with Resistance (LRC Circuit)	766
	SUMMARY 767 QUESTIONS 768	
	PROBLEMS 768 GENERAL PROBLEMS 770	

31 AC Circuits — 772

31-1	Introduction: AC Circuits	772
31-2	AC Circuit Containing only Resistance R	773
31-3	AC Circuit Containing only Inductance L	773
31-4	AC Circuit Containing only Capacitance C	774
31-5	LRC Series AC Circuit	776
31-6	Resonance in AC Circuits	780
*31-7	Impedance Matching	781
*31-8	Three-Phase AC	782
	SUMMARY 783 QUESTIONS 783	
	PROBLEMS 784 GENERAL PROBLEMS 785	

32 Maxwell's Equations and Electromagnetic Waves — 787

32-1	Changing Electric Fields Produce Magnetic Fields; Ampère's Law and Displacement Current	788
32-2	Gauss's Law for Magnetism	791
32-3	Maxwell's Equations	792
32-4	Production of Electromagnetic Waves	792
32-5	Electromagnetic Waves, and Their Speed, from Maxwell's Equations	794
32-6	Light as an Electromagnetic Wave and the Electromagnetic Spectrum	798
32-7	Energy in EM Waves; the Poynting Vector	800
*32-8	Radiation Pressure	802
*32-9	Radio and Television	803
	SUMMARY 806 QUESTIONS 807	
	PROBLEMS 807 GENERAL PROBLEMS 809	

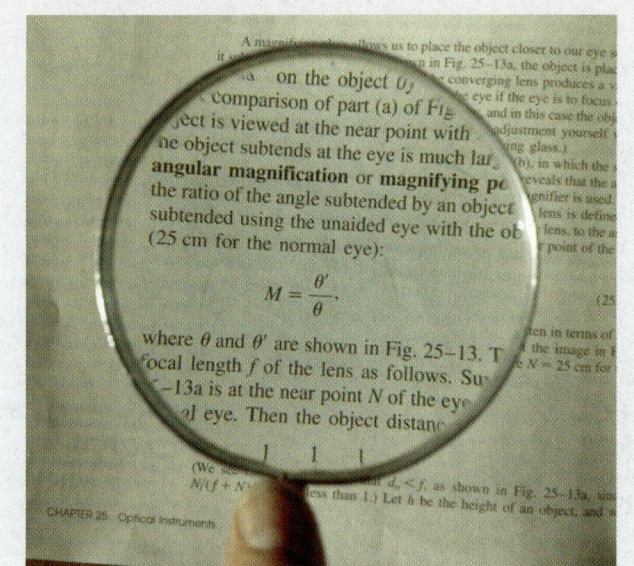

PART 4

33 Light: Reflection and Refraction — 810

- 33-1 The Ray Model of Light — 811
- 33-2 The Speed of Light and Index of Refraction — 811
- 33-3 Reflection; Image Formation by a Plane Mirror — 812
- 33-4 Formation of Images by Spherical Mirrors — 816
- 33-5 Refraction: Snell's Law — 822
- 33-6 Visible Spectrum and Dispersion — 824
- 33-7 Total Internal Reflection; Fiber Optics — 826
- *33-8 Refraction at a Spherical Surface — 828
  - SUMMARY 830 QUESTIONS 831
  - PROBLEMS 832 GENERAL PROBLEMS 835

34 Lenses and Optical Instruments — 836

- 34-1 Thin Lenses; Ray Tracing — 837
- 34-2 The Lens Equation — 840
- 34-3 Combinations of Lenses — 843
- 34-4 Lensmaker's Equation — 845
- *34-5 Cameras — 848
- 34-6 The Human Eye; Corrective Lenses — 850
- 34-7 Magnifying Glass — 853
- *34-8 Telescopes — 854
- *34-9 Compound Microscope — 856
- *34-10 Aberrations of Lenses and Mirrors — 858
  - SUMMARY 860 QUESTIONS 860
  - PROBLEMS 861 GENERAL PROBLEMS 864

35 Wave Nature of Light; Interference — 866

- 35-1 Huygens' Principle and Diffraction — 867
- 35-2 Huygens' Principle and the Law of Refraction — 867
- 35-3 Interference—Young's Double-Slit Experiment — 870
- 35-4 Coherence — 873
- 35-5 Intensity in the Double-Slit Interference Pattern — 874
- 35-6 Interference in Thin Films — 877
- *35-7 Michelson Interferometer — 881
- *35-8 Luminous Intensity — 882
  - SUMMARY 883 QUESTIONS 883
  - PROBLEMS 884 GENERAL PROBLEMS 885

36 Diffraction and Polarization — 887

- 36-1 Diffraction by a Single Slit — 888
- 36-2 Intensity in Single-Slit Diffraction Pattern — 890
- *36-3 Diffraction in the Double-Slit Experiment — 893
- 36-4 Limits of Resolution; Circular Apertures — 896
- 36-5 Resolution of Telescopes and Microscopes; the λ Limit — 898
- *36-6 Resolution of the Human Eye and Useful Magnification — 899
- 36-7 Diffraction Grating — 900
- *36-8 The Spectrometer and Spectroscopy — 901
- *36-9 Peak Widths and Resolving Power for a Diffraction Grating — 903
- *36-10 X-Rays and X-Ray Diffraction — 905
- 36-11 Polarization — 907
- *36-12 Scattering of Light by the Atmosphere — 911
  - SUMMARY 911 QUESTIONS 912
  - PROBLEMS 913 GENERAL PROBLEMS 915

37 SPECIAL THEORY OF RELATIVITY 916

- 37-1 Galilean–Newtonian Relativity 917
- *37-2 The Michelson–Morley Experiment 919
- 37-3 Postulates of the Special Theory of Relativity 922
- 37-4 Simultaneity 924
- 37-5 Time Dilation and the Twin Paradox 926
- 37-6 Length Contraction 930
- 37-7 Four-Dimensional Space–Time 932
- 37-8 Galilean and Lorentz Transformations 932
- 37-9 Relativistic Momentum and Mass 936
- 37-10 The Ultimate Speed 938
- 37-11 Energy and Mass; $E = mc^2$ 938
- 37-12 Doppler Shift for Light 942
- 37-13 The Impact of Special Relativity 943

SUMMARY 944 QUESTIONS 945
PROBLEMS 945 GENERAL PROBLEMS 947

PART 5

38 EARLY QUANTUM THEORY AND MODELS OF THE ATOM 949

- 38-1 Planck's Quantum Hypothesis 949
- 38-2 Photon Theory of Light and the Photoelectric Effect 952
- 38-3 Photons and the Compton Effect 955
- 38-4 Photon Interactions; Pair Production 957
- 38-5 Wave-Particle Duality; The Principle of Complementarity 958
- 38-6 Wave Nature of Matter 959
- *38-7 Electron Microscopes 961
- 38-8 Early Models of the Atom 962
- 38-9 Atomic Spectra: Key to the Structure of the Atom 963
- 38-10 The Bohr Model 965
- 38-11 De Broglie's Hypothesis Applied to Atoms 971

SUMMARY 972 QUESTIONS 973
PROBLEMS 974 GENERAL PROBLEMS 976

39 QUANTUM MECHANICS 977

- 39-1 Quantum Mechanics—A New Theory 978
- 39-2 The Wave Function and Its Interpretation; the Double-Slit Experiment 979
- 39-3 The Heisenberg Uncertainty Principle 981
- 39-4 Philosophic Implications; Probability Versus Determinism 984
- 39-5 The Schrödinger Equation in One Dimension—Time-Independent Form 985
- *39-6 Time-Dependent Schrödinger Equation 988
- 39-7 Free Particles; Plane Waves and Wave Packets 989
- 39-8 Particle in an Infinitely Deep Square Well Potential (a Rigid Box) 990
- *39-9 Finite Potential Well 994
- 39-10 Tunneling through a Barrier 996

SUMMARY 999 QUESTIONS 1000
PROBLEMS 1000 GENERAL PROBLEMS 1002

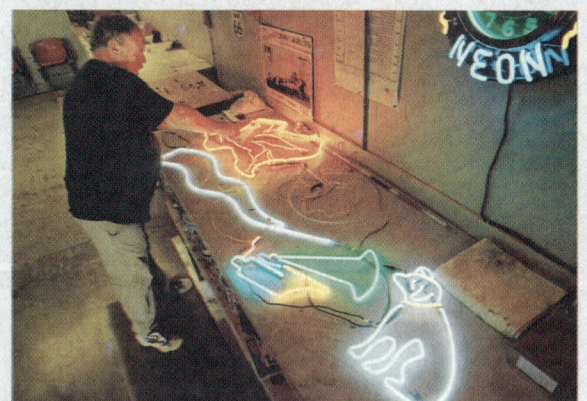

40 QUANTUM MECHANICS OF ATOMS 1003

- 40-1 Quantum-Mechanical View of Atoms 1003
- 40-2 Hydrogen Atom: Schrödinger Equation and Quantum Numbers 1004
- 40-3 Hydrogen Atom Wave Functions 1007
- 40-4 Complex Atoms; the Exclusion Principle 1010
- 40-5 The Periodic Table of Elements 1012
- 40-6 X-Ray Spectra and Atomic Number 1013
- *40-7 Magnetic Dipole Moments; Total Angular Momentum 1015
- *40-8 Fluorescence and Phosphorescence 1019
- *40-9 Lasers 1019
- *40-10 Holography 1023

SUMMARY 1025 QUESTIONS 1025
PROBLEMS 1026 GENERAL PROBLEMS 1028

41 Molecules and Solids — 1030

41-1	Bonding in Molecules	1030
41-2	Potential-Energy Diagrams for Molecules	1033
41-3	Weak (van der Waals) Bonds	1036
41-4	Molecular Spectra	1037
41-5	Bonding in Solids	1044
41-6	Free-Electron Theory of Metals	1045
41-7	Band Theory of Solids	1048
41-8	Semiconductors and Doping	1051
*41-9	Semiconductor Diodes	1052
*41-10	Transistors and Integrated Circuits	1054

SUMMARY 1055   QUESTIONS 1056
PROBLEMS 1057   GENERAL PROBLEMS 1059

42 Nuclear Physics and Radioactivity — 1061

42-1	Structure and Properties of the Nucleus	1061
42-2	Binding Energy and Nuclear Forces	1064
42-3	Radioactivity	1067
42-4	Alpha Decay	1068
42-5	Beta Decay	1070
42-6	Gamma Decay	1072
42-7	Conservation of Nucleon Number and Other Conservation Laws	1073
42-8	Half-Life and Rate of Decay	1073
42-9	Decay Series	1077
42-10	Radioactive Dating	1078
42-11	Detection of Radiation	1080

SUMMARY 1081   QUESTIONS 1081
PROBLEMS 1082   GENERAL PROBLEMS 1083

43 Nuclear Energy: Effects and Uses of Radiation — 1085

43-1	Nuclear Reactions and the Transmutation of Elements	1085
43-2	Cross Section	1088
43-3	Nuclear Fission; Nuclear Reactors	1090
43-4	Fusion	1095
43-5	Passage of Radiation through Matter; Radiation Damage	1100
43-6	Measurement of Radiation—Dosimetry	1101
*43-7	Radiation Therapy	1104
*43-8	Tracers	1104
*43-9	Imaging by Tomography: CAT Scans, and Emission Tomography	1105
*43-10	Nuclear Magnetic Resonance (NMR) and Magnetic Resonance Imaging (MRI)	1107

SUMMARY 1109   QUESTIONS 1110
PROBLEMS 1110   GENERAL PROBLEMS 1113

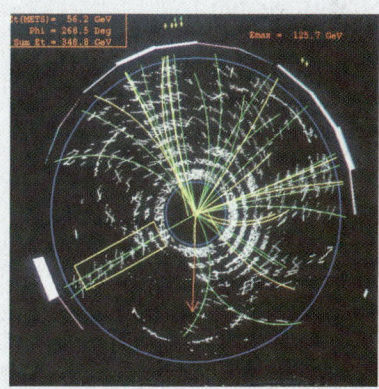

44 Elementary Particles — 1114

44-1	High-Energy Particles	1115
44-2	Particle Accelerators and Detectors	1115
44-3	Beginnings of Elementary Particle Physics—Particle Exchange	1121
44-4	Particles and Antiparticles	1124
44-5	Particle Interactions and Conservation Laws	1124
44-6	Particle Classification	1126
44-7	Particle Stability and Resonances	1127
44-8	Strange Particles	1129
44-9	Quarks	1130
44-10	The "Standard Model": Quantum Chromodynamics (QCD) and the Electroweak Theory	1132
44-11	Grand Unified Theories	1134

SUMMARY 1136   QUESTIONS 1137
PROBLEMS 1137   GENERAL PROBLEMS 1138

45 ASTROPHYSICS AND AND COSMOLOGY   1140

45-1	Stars and Galaxies	1141
45-2	Stellar Evolution; the Birth and Death of Stars	1145
45-3	General Relativity: Gravity and the Curvature of Space	1151
45-4	The Expanding Universe	1156
45-5	The Big Bang and the Cosmic Microwave Background	1159
45-6	The Standard Cosmological Model: The Early History of the Universe	1161
45-7	The Future of the Universe?	1165

SUMMARY 1168   QUESTIONS 1169
PROBLEMS 1170   GENERAL PROBLEMS 1171

APPENDICES

A	**MATHEMATICAL FORMULAS**	A–1
A-1	Quadratic Formula	A–1
A-2	Binomial Expansion	A–1
A-3	Other Expansions	A–1
A-4	Areas and Volumes	A–1
A-5	Plane Geometry	A–2
A-6	Trigonometric Functions and Identities	A–2
A-7	Logarithms	A–3
B	**DERIVATIVES AND INTEGRALS**	A–4
C	**GRAVITATIONAL FORCE DUE TO A SPHERICAL MASS DISTRIBUTION**	A–6
D	**SELECTED ISOTOPES**	A–9

ANSWERS TO ODD-NUMBERED PROBLEMS	A–14
INDEX	A–33
PHOTO CREDITS	A–57

PREFACE

A Brand New Third Edition

It has been more than ten years since the second edition of this calculus-based introductory physics textbook was published. A lot has changed since then, not only in physics itself, but also in how physics is presented. Research in how students learn has provided textbook authors new opportunities to help students learn physics and learn it well.

This third edition comes in two versions. The standard version covers all of classical physics plus a chapter on special relativity and one on the early quantum theory. The extended version, with modern physics, contains a total of nine detailed chapters on modern physics, ending with astrophysics and cosmology. This book retains the original approach: in-depth physics, concrete and nondogmatic, readable.

This new third edition has many improvements in the physics and its applications. Before discussing those changes in detail, here is a list of some of the overall changes that will catch the eye immediately.

Full color throughout is not just cosmetic, although fine color photographs do help to attract the student readers. More important, full color diagrams allow the physics to be displayed with much greater clarity. We have not stopped at a 4-color process; this book has actually been printed in 5 pure colors (5 passes through the presses) to provide better variety and definition for illustrating vectors and other physics concepts such as rays and fields. I want to emphasize that color is used pedagogically to bring out the physics. For example, different types of vectors are given different colors—see the chart on page xxxi.

Many more diagrams, almost double the number in the previous edition, have all been done or redone carefully using full color; there are many more graphs and many more photographs throughout. See for example in optics where new photographs show lenses and the images they make.

Marginal notes have been added as an aid to students to (i) point out what is truly important, (ii) serve as a sort of outline, and (iii) help students find details about something referred to later that they may not remember so well. Besides such "normal" marginal notes, there are also marginal notes that point out brief *problem solving* hints, and others that point out interesting *applications*.

The great laws of physics are emphasized by giving them a marginal note all in capital letters and enclosed in a rectangle. The most important equations, especially those expressing the great laws, are further emphasized by a tan-colored screen behind them.

Chapter opening photographs have been chosen to illustrate aspects of each chapter. Each was chosen with an eye to writing a caption which could serve as a kind of summary of what is in that chapter, and sometimes offer a challenge. Some chapter-opening photos have vectors or other analysis superimposed on them.

Page layout: complete derivations. Serious attention has been paid to how each page was formatted, especially for page turns. Great effort has been made to keep important derivations and arguments on facing pages. Students then don't have to turn back to check. More important, readers repeatedly see before them, on two facing pages, an important slice of physics.

Two new kinds of Examples: Conceptual Examples and **Estimates.**

New Physics

The whole idea of a new edition is to improve, to bring in new material, and to delete material that is verbose and only makes the book longer or is perhaps too advanced and not so useful. Here is a brief summary of a few of the changes involving the physics iself. These lists are selections, not complete lists.

New discoveries:
- planets revolving around distant stars
- Hubble Space Telescope
- updates in particle physics and cosmology, such as inflation and the age of the universe

New physics topics added:
- new treatment of how to make estimates (Chapter 1), including new Estimating Examples throughout (in Chapter 1, estimating the volume of a lake, and the radius of the Earth)
- symmetry used much more, including for solving problems
- new Tables illustrating the great range of lengths, time intervals, masses, voltages
- gravitation as curvature of space, and black holes (Chapter 6)
- engine efficiency (Chapter 8 as well as Chapter 20)
- rolling with and without slipping, and other useful details of rotational motion (Chapter 10)
- forces in structures including trusses, bridges, arches, and domes (Chapter 12)
- square wave (Chapter 15)
- using the Maxwell distribution (Chapter 18)
- Otto cycle (Chapter 20)
- statistical calculation of entropy change in free expansion (Chapter 20)
- effects of dielectrics on capacitor connected and not (Chapter 24)
- grounding to avoid electric hazards (Chapter 25)
- three phase ac (Chapter 31)
- equal energy in **E** and **B** of EM wave (Chapter 32)
- radiation pressure, EM wave (Chapter 32)
- photos of lenses and mirrors with their images (Chapter 33)
- detailed outlines for ray tracing with mirrors and lenses (Chapters 33, 34)
- lens combinations (Chapter 34)
- new radiation standards (Chapter 43)
- Higgs boson, supersymmetry (Chapter 44)

Modern physics. A number of modern physics topics are discussed in the framework of classical physics. Here are some highlights:

- gravitation as curvature of space, and black holes (Chapter 6)
- planets revolving around distant stars (Chapter 6)
- kinetic energy at relativistic speeds (Chapter 7)
- nuclear collisions (Chapter 9)
- star collapse (Chapter 10)
- galaxy red shift, Doppler (Chapter 16)
- atoms, theory of (Chapters 17, 18, 21)
- atomic theory of thermal expansion (Chapter 17)
- mass of hydrogen atom (Chapter 17)
- atoms and molecules in gases (Chapters 17, 18)
- molecular speeds (Chapter 18)
- equipartition of energy; molar specific heats (Chapter 19)
- star size (Chapter 19)
- molecular dipoles (Chapters 21, 23)
- cathode ray tube (Chapters 23, 27)
- electrons in a wire (Chapter 25)
- superconductivity (Chapter 25)
- discovery and properties of the electron, e/m, oil drop experiment (Chapter 27)
- Hall effect (Chapter 27)

- magnetic moment of electrons (Chapter 27)
- mass spectrometer (Chapter 27)
- velocity selector (Chapter 27)
- electron spin in magnetic materials (Chapter 28)
- light and EM wave emission (Chapter 32)
- spectroscopy (Chapter 36)

Many other examples of modern physics are found as Problems, even in early chapters. Chapters 37 and 38 contain the modern physics topics of Special Relativity, and an introduction to Quantum Theory and Models of the Atom. The longer version of this text, "with Modern Physics," contains an additional seven chapters (for a total of nine) which present a detailed and extremely up-to-date treatment of modern physics: Quantum Mechanics of Atoms (Chapters 38 to 40); Molecules and Condensed Matter (Chapter 41); Nuclear Physics (Chapter 42 and 43); Elementary Particles (Chapter 44); and finally Astrophysics, General Relativity, and Cosmology (Chapter 45).

Revised physics and reorganizations. First of all, a major effort has been made to not throw everything at the students in the first few chapters. The basics have to be learned first; many aspects can come later, when the students are more prepared. Secondly, a great part of this book has been rewritten to make it clearer and more understandable to students. Clearer does not always mean simpler or easier. Sometimes making it "easier" actually makes it harder to understand. Often a little more detail, without being verbose, can make an explanation clearer. Here are a few of the changes, big and small:

- new graphs and diagrams to clarify velocity and acceleration; deceleration carefully treated.
- unit conversion now a new Section in Chapter 1, instead of interrupting kinematics.
- circular motion: Chapter 3 now gives only the basics, with more complicated treatment coming later: non-uniform circular motion in Chapter 5, angular variables in Chapter 10.
- Newton's second law now written throughout as $m\mathbf{a} = \Sigma \mathbf{F}$, to emphasize inclusion of all forces acting on a body.
- Newton's third law follows the second directly, with inertial reference frames placed earlier. New careful discussions to head off confusion when using Newton's third law.
- careful rewriting of chapters on Work and Energy, especially potential energy, conservative and nonconservative forces, and the conservation of energy.
- renewed emphasis that $\Sigma \tau = I\alpha$ is not always valid: only for an axis fixed in an inertial frame or if axis is through the CM (Chapters 10 and 11).
- rolling motion introduced early in Chapter 10, with more details later, including rolling with and without slipping.
- rotating frames of reference, and Coriolis, moved later, to Chapter 11, shortened, optional, but still including why an object does not fall straight down on Earth.
- fluids reduced to a single chapter (13); some topics and details dropped or greatly shortened.
- clearer details on how an object floats (Chapter 13).
- distinction between wave interference in space, and in time (beats) (Chapter 16).
- thermodynamics reduced to four chapters; the old chapters on Heat and on the First Law of Thermodynamics have been combined into one (19), with some topics shortened and a more rational sequence of topics achieved.
- heat transfer now follows the first law of thermodynamics (Chapter 19).
- electric potential carefully rewritten for accuracy (Chapter 23).
- CRT, computer monitors, TV, treated earlier (Chapter 23).
- use of Q_{encl} and I_{encl} for Gauss's and Ampère's laws, with subscripts meaning "enclosed".
- Ohm's law and definition of resistance carefully redone (Chapter 25).
- sources of magnetic field, Chapter 28, reorganized for ease of understanding, with some new material, and deletion of the advanced topic on magnetization vector.
- circuits with $L, C,$ and/or R now introduced via Kirchhoff's loop rule, and clarified in other ways too (Chapters 30, 31).
- streamlined Maxwell's equations, with displacement current downplayed (Chapter 32).
- optics reduced to four chapters; polarization is now placed in the same chapter as diffraction.

New Pedagogy

All of the above mentioned revisions, rewritings, and reorganizations are intended to help students learn physics better. They were done in response to contemporary research in how students learn, as well as to kind and generous input from professors who have read, reviewed, or used the previous editions. This new edition also contains some new elements, especially an increased emphasis on conceptual development:

Conceptual Examples, typically 1 or 2 per chapter, sometimes more, are each a sort of brief Socratic question and answer. It is intended that students will be stimulated by the question to think, or reflect, and come up with a response—before reading the Response given. Here are a few:
- using symmetry (Chapters 1, 44, and elsewhere)
- ball moving upward: misconceptions (Chapter 2)
- reference frames and projectile motion: where does the apple land? (Chapter 3)
- what exerts the force that makes a car move? (Chapter 4)
- Newton's third law clarification: pulling a sled (Chapter 4)
- free-body diagram for a hockey puck (Chapter 4)
- advantage of a pulley (Chapter 4), and of a lever (Chapter 12)
- to push or to pull a sled (Chapter 5)
- which object rolls down a hill faster? (Chapter 10)
- moving the axis of a spinning wheel (Chapter 11)
- tragic collapse (Chapter 12)
- finger at top of a full straw (Chapter 13)
- suction cups on a spacecraft (Chapter 13)
- doubling amplitude of SHM (Chapter 14)
- do holes expand thermally? (Chapter 17)
- simple adiabatic process: stretching a rubber band (Chapter 19)
- charge inside a conductor's cavity (Chapter 22)
- how stretching a wire changes its resistance (Chapter 25)
- series or parallel (Chapter 26)
- bulb brightness (Chapter 26)
- spiral path in magnetic field (Ch. 27)
- practice with Lenz's law (Chapter 29)
- motor overload (Chapter 29)
- emf direction in inductor (Chapter 30)
- photo with reflection—is it upside down? (Chapter 33)
- reversible light rays (Chapter 33)
- how tall must a full-length mirror be? (Chapter 33)
- diffraction spreading (Chapter 36)

Estimating Examples, roughly 10% of all Examples, also a new feature of this edition, are intended to develop the skills for making order-of-magnitude estimates, even when the data are scarce, and even when you might never have guessed that any result was possible at all. See, for example, Section 1–6, Examples 1–5 to 1–8.

Problem Solving, with New and Improved Approaches

Learning how to approach and solve problems is a basic part of any physics course. It is a highly useful skill in itself, but is also important because the process helps bring understanding of the physics. Problem solving in this new edition has a significantly increased emphasis, including some new features.

Problem-solving boxes, about 20 of them, are new to this edition. They are more concentrated in the early chapters, but are found throughout the book. They each outline a step-by-step approach to solving problems in general, and/or specifically for the material being covered. The best students may find these separate "boxes" unnecessary (they can skip them), but many students will find it helpful to be reminded of the general approach and of steps they can take to get started; and, I think, they help to build confidence. The general problem solving box in Section 4–8 is placed there, after students have had some experience wrestling with problems, and so may be strongly motivated to read it with close attention. Section 4–8 can, of course, be covered earlier if desired.

Problem-solving Sections occur in many chapters, and are intended to provide extra drill in areas where solving problems is especially important or detailed.

Examples. This new edition has many more worked-out Examples, and they all now have titles for interest and for easy reference. There are even two new categories of Example: Conceptual, and Estimates, as described above. Regular Examples serve as "practice problems". Many new ones have been added, some of the old ones have been dropped, and many have been reworked to provide greater clarity and detail: more steps are spelled out, more of "why we do it this way", and more discussion of the reasoning and approach. In sum, the idea is "to think aloud with the students", leading them to develop insight. The total number of worked-out Examples is about 30% greater than in the previous edition, for an average of 12 to 15 per chapter. There is a significantly higher concentration of Examples in the early chapters, where drill is especially important for developing skills and a variety of approaches. The level of the worked-out Examples for most topics increases gradually, with the more complicated ones being on a par with the most difficult Problems at the end of each chapter, so that students can see how to approach complex problems. Many of the new Examples, and improvements to old ones, provide relevant applications to engineering, other related fields, and to everyday life.

Problems at the end of each chapter have been greatly increased in quality and quantity. There are over 30% more Problems than in the second edition. Many of the old ones have been replaced, or rewritten to make them clearer, and/or have had their numerical values changed. Each chapter contains a large group of Problems arranged by Section and graded according to difficulty: level I Problems are simple, designed to give students confidence; level II are "normal" Problems, providing more of a challenge and often the combination of two different concepts; level III are the most complex, typically combining different issues, and will challenge even superior students. The arrangement by Section number means only that those Problems depend on material up to and including that Section: earlier material may also be relied upon. The ranking of Problems by difficulty (I, II, III) is intended only as a guide.

General Problems. About 70% of Problems are ranked by level of difficulty (I, II, III) and arranged by Section. New to this edition are General Problems that are unranked and grouped together at the end of each chapter, and account for about 30% of all problems. The average total number of Problems per chapter is about 90. Answers to odd-numbered Problems are given at the back of the book.

Complete Physics Coverage, with Options

This book is intended to give students the opportunity to obtain a thorough background in all areas of basic physics. There is great flexibility in choice of topics so that instructors can choose which topics they cover and which they omit. Sections marked with an asterisk can be considered optional, as discussed more fully on p. xxv. Here I want to emphasize that topics not covered in class can still be read by serious students for their own enrichment, either immediately or later. Here is a partial list of physics topics, not the standard ones, but topics that might not usually be covered, and that represent how thorough this book is in its coverage of basic physics. Section numbers are given in parentheses.

- use of calculus; variable acceleration (2–8)
- nonuniform circular motion (5–4)
- velocity-dependent forces (5–5)
- gravitational versus inertial mass; principle of equivalence (6–8)
- gravitation as curvature of space; black holes (6–9)
- kinetic energy at very high speed (7–5)
- potential energy diagrams (8–9)
- systems of variable mass (9–10)
- rotational plus translational motion (10–11)
- using $\Sigma \tau_{CM} = I_{CM} \alpha_{CM}$ (10–11)
- derivation of $K = K_{CM} + K_{rot}$ (10–11)
- why does a rolling sphere slow down? (10–12)
- angular momentum and torque for a system (11–4)
- derivation of $d\mathbf{L}_{CM}/dt = \Sigma \boldsymbol{\tau}_{CM}$ (11–4)
- rotational imbalance (11–6)
- the spinning top (11–8)
- rotating reference frames; inertial forces (11–9)
- coriolis effect (11–10)
- trusses (12–7)
- flow in tubes: Poiseuille's equation (13–11)
- surface tension and capillarity (13–12)
- physical pendulum; torsion pendulum (14–6)
- damped harmonic motion: finding the solution (14–7)
- forced vibrations; equation of motion and its solution; Q-value (14–8)
- the wave equation (15–5)
- mathematical representation of waves; pressure wave derivation (16–2)
- intensity of sound related to amplitude (16–3)
- interference in space and in time (16–6)
- atomic theory of expansion (17–4)
- thermal stresses (17–5)
- ideal gas temperature scale (17–10)
- calculations using the Maxwell distribution of molecular speeds (18–2)
- real gases (18–3)
- vapor pressure and humidity (18–4)
- van der Waals equation of state (18–5)
- mean free path (18–6)
- diffusion (18–7)
- equipartition of energy (19–8)
- energy availability; heat death (20–8)
- statistical interpretation of entropy and the second law (20–9)
- thermodynamic temperature scale; absolute zero and the third law (20–10)
- electric dipoles (21–11, 23–6)
- experimental basis of Gauss's and Coulomb's laws (22–4)
- general relation between electric potential and electric field (23–2, 23–8)
- electric fields in dielectrics (24–5)
- molecular description of dielectrics (24–6)
- current density and drift velocity (25–8)
- superconductivity (25–9)
- RC circuits (26–4)
- use of voltmeters and ammeters; effects of meter resistance (26–5)
- transducers (26–6)
- magnetic dipole moment (27–5)
- Hall effect (27–8)
- operational definition of the ampere and coulomb (28–3)
- magnetic materials—ferromagnetism (28–7)
- electromagnets and solenoids (28–8)
- hysteresis (28–9)
- paramagnetism and diamagnetism (28–10)
- counter emf and torque; eddy currents (29–5)
- Faraday's law—general form (29–7)
- force due to changing $\mathbf{B}$ is nonconservative (29–7)
- LC circuits and EM oscillations (30–5)
- AC resonance; oscillators (31–6)
- impedance matching (31–7)
- three phase AC (31–8)
- changing electric fields produce magnetic fields (32–1)
- speed of light from Maxwell's equations (32–5)
- radiation pressure (32–8)
- fiber optics (33–7)
- lens combinations (34–3)
- aberrations of lenses and mirrors (34–10)
- coherence (35–4)
- intensity in double-slit pattern (35–5)
- luminous intensity (35–8)
- intensity for single-slit (36–2)
- diffraction for double-slit (36–3)
- limits of resolution, the λ limit (36–4, 36–5)
- resolution of the human eye and useful magnification (36–6)
- spectroscopy (36–8)
- peak widths and resolving power for a diffraction grating (36–9)
- x-rays and x-ray diffraction (36–10)
- scattering of light by the atmosphere (36–12)
- time–dependent Schrödinger equation (39–6)
- wave packets (39–7)
- tunneling through a barrier (39–9)
- free-electron theory of metals (41–6)
- semiconductor electronics (41–9)
- standard model, symmetry, QCD, GUT (44–9, 44–10)
- astrophysics, cosmology (Ch. 45)

New Applications

Relevant applications to everyday life, to engineering, and to other fields such as geology and medicine, provide students with motivation and offer the instructor the opportunity to show the relevance of physics. Applications are a good response to students who ask "Why study physics?" Many new applications have been added in this edition. Here are some highlights:

- airbags (Chapter 2)
- elevator and counterweight (Chapter 4)
- antilock brakes and skidding (Chapter 5)
- geosynchronous satellites (Chapter 6)
- hard drive and bit speed (Chapter 10)
- star collapse (Chapter 10)
- forces within trusses, bridges, arches, domes (Chapter 12)
- the Titanic (Chapter 12)
- Bernoulli's principle: wings, sailboats, TIA, plumbing traps and bypasses (Chapter 13)
- pumps (Chapter 13)
- car springs, shock absorbers, building dampers for earthquakes (Chapter 14)
- loudspeakers (Chapters 14, 16, 27)
- autofocusing cameras (Chapter 16)
- sonar (Chapter 16)
- ultrasound imaging (Chapter 16)
- thermal stresses (Chapter 17)
- R-values, thermal insulation (Ch. 19)
- engines (Chapter 20)
- heat pumps, refrigerators, AC; coefficient of performance (Chapter 20)
- thermal pollution (Chapter 20)
- electric shielding (Chapters 21, 28)
- photocopier (Chapter 21)
- superconducting cables (Chapter 25)
- jump starting a car (Chapter 26)
- aurora borealis (Chapter 27)
- solenoids and electromagnetics (Ch. 28)
- computer memory and digital information (Chapter 29)
- seismograph (Chapter 29)
- tape recording (Chapter 29)
- loudspeaker cross-over network (Ch. 31)
- antennas, for **E** or **B** (Chapter 32)
- TV and radio; AM and FM (Chapter 32)
- eye and corrective lenses (Chapter 34)
- mirages (Chapter 35)
- liquid crystal displays (Chapter 36)
- CAT scans, PET, MRI (Chapter 43)

Some old favorites retained (and improved):

- pressure gauges (Chapter 13)
- musical instruments (Chapter 16)
- humidity (Chapter 18)
- CRT, TV, computer monitors (Ch. 23, 27)
- electric hazards (Chapter 25)
- power in household circuits (Chapter 25)
- ammeters and voltmeters (Chapter 26)
- microphones (Chapters 26, 29)
- transducers (Chapter 26, and elsewhere)
- electric motors (Chapter 27)
- car alternator (Chapter 29)
- electric power transmission (Chapter 29)
- capacitors as filters (Chapter 31)
- impedance matching (Chapter 31)
- fiber optics (Chapter 33)
- cameras, telescopes, microscopes, other optical instruments (Chapter 34)
- lens coatings (Chapter 35)
- spectroscopy (Chapter 36)
- electron microscopes (Chapter 38)
- lasers, holography, CD players (Ch. 40)
- semiconductor electronics (Chapter 41)
- radioactivity (Chapters 42 and 43)

Deletions

Something had to go, or the book would have been too long. Lots of subjects were shortened—the detail simply isn't necessary at this level. Some topics were dropped entirely: polar coordinates; center-of-momentum reference frame; Reynolds number (now a Problem); object moving in a fluid and sedimentation; derivation of Poiseuille's equation; Stoke's equation; waveguide and transmission line analysis; electric polarization and electric displacement vectors; potentiometer (now a Problem); negative pressure; combinations of two harmonic motions; adiabatic character of sound waves; central forces.

Many topics have been shortened, often a lot, such as: velocity-dependent forces; variable acceleration; instantaneous axis; surface tension and capillarity; optics topics such as some aspects of light polarizarion. Many of the brief historical and philosophical issues have been shortened as well.

General Approach

This book offers an in-depth presentation of physics, and retains the basic approach of the earlier editions. Rather than using the common, dry, dogmatic approach of treating topics formally and abstractly first, and only later relating the material to the students' own experience, my approach is to recognize that physics is a description of reality and thus to start each topic with concrete observations and experiences that students can directly relate to. Then we move on to the generalizations and more formal treatment of the topic. Not only does this make the material more interesting and easier to understand, but it is closer to the way physics is actually practiced.

This new edition, even more than previous editions, aims to explain the physics in a readable and interesting manner that is accessible and clear. It aims to teach students by anticipating their needs and difficulties, but without oversimplifying. Physics is all about us. Indeed, it is the goal of this book to help students "see the world through eyes that know physics."

As mentioned above, this book includes of a wide range of Examples and applications from technology, engineering, architecture, earth sciences, the environment, biology, medicine, and daily life. Some applications serve only as examples of physical principles. Others are treated in depth. But applications do not dominate the text—this is, after all, a physics book. They have been carefully chosen and integrated into the text so as not to interfere with the development of the physics but rather to illuminate it. You won't find essay sidebars here. The applications are integrated right into the physics. To make it easy to spot the applications, a new *Physics Applied* marginal note is placed in the margin (except where diagrams in the margin prevent it).

It is assumed that students have started calculus or are taking it concurrently. Calculus is treated gently at first, usually in an optional Section so as not to burden students taking calculus concurrently. For example, using the integral in kinematics, Chapter 2, is an optional Section. But in Chapter 7, on work, the integral is discussed fully for all readers.

Throughout the text, *Système International* (SI) units are used. Other metric and British units are defined for informational purposes. Careful attention is paid to significant figures. When a certain value is given as, say, 3, with its units, it is meant to be 3, not assumed to be 3.0 or 3.00. When we mean 3.00 we write 3.00. It is important for students to be aware of the uncertainty in any measured value, and not to overestimate the precision of a numerical result.

Rather than start this physics book with a chapter on mathematics, I have instead incorporated many mathematical tools, such as vector addition and multiplication, directly in the text where first needed. In addition, the Appendices contain a review of many mathematical topics such as trigonometric identities, integrals, and the binomial (and other) expansions. One advanced topic is also given an Appendix: integrating to get the gravitational force due to a spherical mass distribution.

It is necessary, I feel, to pay careful attention to detail, especially when deriving an important result. I have aimed at including all steps in a derivation, and have tried to make clear which equations are general, and which are not, by explicitly stating the limitations of important equations in brackets next to the equation, such as

$$x = x_0 + v_0 t + \tfrac{1}{2} a t^2. \qquad \text{[constant acceleration]}$$

The more detailed introduction to Newton's laws and their use is of crucial pedagogic importance. The many new worked-out Examples include initially fairly simple ones that provide careful step-by-step analysis of how to proceed in solving dynamics problems. Each succeeding Example adds a new element or a new twist that introduces greater complexity. It is hoped that this strategy will enable even less-well-prepared students to acquire the tools for using Newton's laws correctly. If students don't surmount this crucial hurdle, the rest of physics may remain forever beyond their grasp.

Rotational motion is difficult for most students. As an example of attention to detail (although this is not really a "detail"), I have carefully distinguished the position vector (**r**) of a point and the perpendicular distance of that point from an axis, which is called R in this book (see Fig. 10–2). This distinction, which enters particularly in connection with torque, moment of inertia, and angular momentum, is often not made clear—it is a disservice to students to use **r** or r for both without distinguishing. Also, I have made clear that it is not always true that $\Sigma\tau = I\alpha$. It depends on the axis chosen (valid if axis is fixed in an inertial reference frame, or through the CM). To not tell this to students can get them into serious trouble. (See pp. 250, 283, 284.) I have treated rotational motion by starting with the simple instance of rotation about an axis (Chapter 10), including the concepts of angular momentum and rotational kinetic energy. Only in Chapter 11 is the more general case of rotation about a point dealt with, and this slightly more advanced material can be omitted if desired (except for Sections 11–1 and 11–2 on the vector product and the torque vector). The end of Chapter 10 has an optional subsection containing three slightly more advanced Examples, using $\Sigma\tau_{CM} = I_{CM}\alpha_{CM}$: car braking distribution, a falling yo-yo, and a sphere rolling with and without slipping.

Among other special treatments is Chapter 28, Sources of Magnetic Field: here, in one chapter, are discussed the magnetic field due to currents (including Ampère's law and the law of Biot-Savart) as well as magnetic materials, ferromagnetism, paramagnetism, and diamagnetism. This presentation is clearer, briefer, and more of a whole, and all the content is there.

Organization

The general outline of this new edition retains a traditional order of topics: mechanics (Chapters 1 to 12); fluids, vibrations, waves, and sound (Chapter 13 to 16); kinetic theory and thermodynamics (Chapters 17 to 20). In the two-volume version of this text, volume I ends here, after Chapter 20. The text continues with electricity and magnetism (Chapters 21 to 32), light (Chapters 33 to 36), and modern physics (Chapters 37 and 38 in the short version, Chapters 37 to 45 in the extended version "with Modern Physics"). Nearly all topics customarily taught in introductory physics courses are included. A number of topics from modern physics are included with the classical physics chapters as discussed earlier.

The tradition of beginning with mechanics is sensible, I believe, because it was developed first, historically, and because so much else in physics depends on it. Within mechanics, there are various ways to order topics, and this book allows for considerable flexibility. I prefer, for example, to cover statics after dynamics, partly because many students have trouble working with forces without motion. Besides, statics is a special case of dynamics—we study statics so that we can prevent structures from becoming dynamic (falling down)—and that sense of being at the limit of dynamics is intuitively helpful. Nonetheless statics (Chapter 12) can be covered earlier, if desired, before dynamics, after a brief introduction to vector addition. Another option is light, which I have placed after electricity and magnetism and EM waves. But light could be treated immediately after the chapters on waves (Chapters 15 and 16). Special relativity is Chapter 37, but could instead be treated along with mechanics—say, after Chapter 9.

Not every chapter need be given equal weight. Whereas Chapter 4 might require $1\frac{1}{2}$ to 2 weeks of coverage, Chapter 16 or 22 may need only $\frac{1}{2}$ week.

Some instructors may find that this book contains more material than can be covered completely in their courses. But the text offers great flexibility in choice of topics. Sections marked with a star (asterisk) are considered optional. These Sections contain slightly more advanced physics material, or material not usually covered in typical courses, and/or interesting applications. They contain no material needed in later chapters (except perhaps in later optional Sections). This does not imply that all nonstarred Sections must be covered: there still remains considerable flexibility in the choice of material. For a brief course, all optional material could be dropped as well as major parts of Chapters 11, 13, 16, 26, 30, 31, and 36 as well as selected parts of Chapters 9, 12, 19, 20, 32, 34, and the modern physics chapters. Topics not covered in class can be a valuable resource for later study; indeed, this text can serve as a useful reference for students for years because of its wide range of coverage.

Thanks

Some 60 physics professors provided input or direct feedback on every aspect of this textbook. The reviewers and contributors to this third edition are listed below. I owe each a debt of gratitude.

Ralph Alexander, University of Missouri at Rolla
Zaven Altounian, McGill University
Charles R. Bacon, Ferris State University
Bruce Birkett, University of California, Berkeley
Art Braundmeier, Southern Illinois University at Edwardsville
Wayne Carr, Stevens Institute of Technology
Edward Chang, University of Massachusetts, Amherst
Charles Chiu, University of Texas at Austin
Lucien Crimaldi, University of Mississippi
Robert Creel, University of Akron
Alexandra Cowley, Community College of Philadelphia
Timir Datta, University of South Carolina
Gary DeLeo, Lehigh University
John Dinardo, Drexel University
Paul Draper, University of Texas, Arlington
Alex Dzierba, Indiana University
William Fickinger, Case Western University
Jerome Finkelstein, San Jose State University
Donald Foster, Wichita State University
Gregory E. Frances, Montana State University
Lothar Frommhold, University of Texas at Austin
Thomas Furtak, Colorado School of Mines
Edward Gibson, California State University, Sacramento
Christopher Gould, University of Southern California
John Gruber, San Jose State University
Martin den Boer, Hunter College
Greg Hassold, General Motors Institute
Joseph Hemsky, Wright State University
Laurent Hodges, Iowa State University
Mark Holtz, Texas Tech University
James P. Jacobs, University of Montana
James Kettler, Ohio University Eastern Campus

Jean Krisch, University of Michigan
Mark Lindsay, University of Louisville
Eugene Livingston, University of Notre Dame
Bryan Long, Columbia State Community College
Daniel Mavlow, Princeton University
Pete Markowitz, Florida International University
John McCullen, University of Arizona, Tucson
Peter Nemeth, New York University
Hon-Kie Ng, Florida State University
Eugene Patroni, Georgia Institute of Technology
Robert Pelcovits, Brown University
William Pollard, Valdosta State University
Joseph Priest, Miami University
Carl Rotter, West Virginia University
Lawrence Rees, Brigham Young University
Peter Riley, University of Texas at Austin
Roy Rubins, University of Texas at Arlington
Mark Semon, Bates College
Robert Simpson, University of New Hampshire
Mano Singham, Case Western University
Harold Slusher, University of Texas at El Paso
Don Sparks, Los Angeles Pierce Community College
Michael Strauss, University of Oklahoma
Joseph Strecker, Wichita State University
William Sturrus, Youngstown State University
Arthur Swift, University of Massachusetts, Amherst
Leo Takahasi, The Pennsylvania State University
Edward Thomas, Georgia Institute of Technology
Som Tyagi, Drexel University
John Wahr, University of Colorado
Robert Webb, Texas A & M University
James Whitmore, The Pennsylvania State University
W. Steve Quon, Ventura College

I owe special thanks to Irv Miller, not only for many helpful physics discussions, but for having worked out all the Problems and managed the team that also worked out the Problems, each checking the other, and finally for producing the Solutions Manual and all the answers to the odd-numbered Problems at the end of this book. He was ably assisted by Zaven Altounian and Anand Batra.

I am particularly grateful to Robert Pelcovits and Peter Riley, as well as to Paul Draper and James Jacobs, who inspired many of the new Examples, Conceptual Examples, and Problems.

Crucial for rooting out errors, as well as providing excellent suggestions, were the perspicacious Edward Gibson and Michael Strauss, both of whom carefully checked all aspects of the physics in page proof.

Special thanks to Bruce Birkett for input of every kind, from illuminating discussions on pedagogy to a careful checking of details in many sections of this book. I wish also to thank Professors Howard Shugart, Joe Cerny, Roger Falcone and Buford Price for helpful discussions, and for hospitality at the University of California, Berkeley. Many thanks also to Prof. Tito Arecchi at the Istituto Nazionale di Ottica, Florence, Italy, and to the staff of the Institute and Museum for the History of Science, Florence, for their hospitality.

Finally, I wish to thank the superb editorial and production work provided by all those with whom I worked directly at Prentice Hall: Susan Fisher, Marilyn Coco, David Chelton, Kathleen Schiaparelli, Trudy Pisciotti, Gus Vibal, Mary Teresa Giancoli, and Jocelyn Phillips.

The biggest thanks of all goes to Paul Corey, whose constant encouragement and astute ability to get things done, provided the single strongest catalyst.

The final responsibility for all errors lies with me, of course. I welcome comments and corrections.

D.C.G.

AVAILABLE SUPPLEMENTS

For the Student

Student Study Guide and Solutions Manual
Douglas Brandt, Eastern Illinois University. (0-13-021475-2)
Contains chapter objectives, summaries with additional examples, self-study quizzes, key mathematical equations, and complete worked-out solutions to alternate odd problems in the text.

Doing Physics with Spreadsheets: A Workbook
Gordon Aubrecht, T. Kenneth Bolland, and Michael Ziegler, all of The Ohio State University.
(0-13-021474-4)
Designed to introduce students to the use of spreadsheets for solving simple and complex physics problems. Students are either provided with spreadsheets or must construct their own, then use the model to most closely approximate natural behavior. The amount of spreadsheet construction and the complexity of the spreadsheet increases as the student gains experience.

Science on the Internet: A Student's Guide, 1999
Andrew Stull and Carl Adler (0-13-021308-X)
The perfect tool to help students take advantage of the *Physics for Scientists and Engineers, Third Edition* Web page. This useful resource gives clear steps to access Prentice Hall's regularly updated physics resources, along with an overview of general World Wide Web navigation strategies. Available FREE for students when packaged with the text.

Prentice Hall/*New York Times* Themes of the Times — Physics
This unique newspaper supplement brings together a collection of the latest physics-related articles from the pages of *The New York Times*. Updated twice per year and available FREE to students when packaged with the text.

For the Instructor

Instructor's Solutions Manual
Irvin A. Miller, Drexel University.
Print version (0-13-021381-0); Electronic (CD-ROM) version (0-13-021481-7)
Contains detailed worked solutions to every problem in the text. Electronic versions are available in CD-ROM (dual platform for both Windows and Macintosh systems) for instructors with Microsoft Word or Word-compatible software.

Test Item File
Robert Pelcovits, Brown University; David Curott, University of North Alabama; and Edward Oberhofer, University of North Carolina at Charlotte (0-13-021482-5)
Contains over 2200 multiple choice questions, about 25% conceptual in nature. All are referenced to the corresponding Section in the text and ranked by difficulty.

Prentice Hall Custom Test Windows (0-13-021477-9); Macintosh (0-13-021476-0)
Based on the powerful testing technology developed by Engineering Software Associates, Inc. (ESA), Prentice Hall Custom Test includes all questions from the Test Item File and allows instructors to create and tailor exams to their own needs. With the Online Testing Program, exams can also be administered on line and data can then be automatically transferred for evaluation. A comprehensive desk reference guide is included along with online assistance.

Transparency Pack (0-13-021470-1)
Includes approximately 400 full color transparencies of images from the text.

Media Supplements

Physics for Scientists and Engineers Web Site www.prenhall.com/giancoli
A FREE innovative online resource that provides students with a wealth of activities and exercises for each text chapter. Features on the site include:
- Practice Questions, Destinations (links to related sites), NetSearch keywords and algorithmically generated numeric Practice Problems by Carl Adler of East Carolina University.
- Physlet Problems (Java-applet simulations) by Wolfgang Christian of Davidson College.
- Warmups and Puzzles essay questions and Applications from Gregor Novak and Andrew Gavrin at Indiana University-Purdue University, Indianapolis.
- Ranking Task Exercises edited by Tom O'Kuma of Lee College, Curtis Hieggelke of Joliet Junior College and David Maloney of Indiana University-Purdue University, Fort Wayne.

Using Prentice Hall CW '99 technology, the website grades and scores all objective questions, and results can be automatically e-mailed directly to the instructors if so desired. Instructors can also create customized syllabi online and link directly to activities on the Giancoli website.

Presentation Manager CD-ROM
Dual Platform (Windows/Macintosh; 0-13-214479-5)
This CD-ROM enables instructors to build custom sequences of Giancoli text images and Prentice Hall digital media for playback in lecture presentations. The CD-ROM contains all text illustrations, digitized segments from the Prentice Hall *Physics You Can See* videotape as well as additional lab and demonstration videos and animations from the Prentice Hall *Interactive Journey Through Physics* CD-ROM. Easy to navigate with Prentice Hall Presentation Manager software, instructors can preview, sequence, and play back images, as well as perform keyword searches, add lecture notes, and incorporate their own digital resources.

Physics You Can See *Video*
(0-205-12393-7)
Contains eleven two- to five-minute demonstrations of classical physics experiments. It includes segments such as "Coin and Feather" (acceleration due to gravity), "Monkey and Gun" (projectile motion), "Swivel Hips" (force pairs), and "Collapse a Can" (atmospheric pressure).

CAPA: A Computer-Assisted Personalized Approach to Assignments, Quizzes, and Exams
CAPA is an on-line homework system developed at Michigan State University that instructors can use to deliver problem sets with randomized variables for each student. The system gives students immediate feedback on their answers to problems, and records their participation and performance. Prentice Hall has arranged to have half of the even-numbered problems of Giancoli, *Physics for Scientists and Engineers, Third Edition*, coded for use with the CAPA system. For additional information about the CAPA system, please visit the web site at http://www.pa.msu.edu/educ/CAPA/.

WebAssign
WebAssign is a web-based homework delivery, collection, grading, and recording service developed and hosted by North Carolina State University. Prentice Hall will arrange for end-of-chapter problems from Giancoli, *Physics for Scientists and Engineers, Third Edition* to be coded for use with the *WebAssign* system for instructors who wish to take advantage of this service. For more information on the *WebAssign* system and its features, please visit http://webassign.net/info or e-mail webassign@ncsu.edu.

NOTES TO STUDENTS AND INSTRUCTORS ON THE FORMAT

1. Sections marked with a star (*) are considered optional. They can be omitted without interrupting the main flow of topics. No later material depends on them except possibly later starred sections. They may be fun to read though.
2. The customary conventions are used: symbols for quantities (such as m for mass) are italicized, whereas units (such as m for meter) are not italicized. Boldface (**F**) is used for vectors.
3. Few equations are valid in all situations. Where practical, the limitations of important equations are stated in square brackets next to the equation. The equations that represent the great laws of physics are displayed with a tan background, as are a few other equations that are so useful that they are indispensable.
4. The number of significant figures (see Section 1–3) should not be assumed to be greater than given: if a number is stated as (say) 6, with its units, it is meant to be 6 and not 6.0 or 6.00.
5. At the end of each chapter is a set of Questions that students should attempt to answer (to themselves at least). These are followed by Problems which are ranked as level I, II, or III, according to estimated difficulty, with level I Problems being easiest. These Problems are arranged by Section, but Problems for a given Section may depend on earlier material as well. There follows a group of General Problems, which are not arranged by Section nor ranked as to difficulty. Questions and Problems that relate to optional Sections are starred.
6. Being able to solve problems is a crucial part of learning physics, and provides a powerful means for understanding the concepts and principles. This book contains many aids to problem solving: (a) worked-out Examples and their solutions in the text, which are set off with a vertical blue line in the margin, and should be studied as an integral part of the text; (b) special "Problem-solving boxes" placed throughout the text to suggest ways to approach problem solving for a particular topic—but don't get the idea that every topic has its own "techniques," because the basics remain the same; (c) special problem-solving Sections (marked in blue in the Table of Contents); (d) "Problem solving" marginal notes (see point 8 below) which refer to hints for solving problems within the text; (e) some of the worked-out Examples are Estimation Examples, which show how rough or approximate results can be obtained even if the given data are sparse (see Section 1–6); and finally (f) the Problems themselves at the end of each chapter (point 5 above).
7. Conceptual Examples look like ordinary Examples but are conceptual rather than numerical. Each proposes a question or two, which hopefully starts you to think and come up with a response. Give yourself a little time to come up with your own response before reading the Response given.
8. Marginal notes: brief notes in the margin of almost every page are printed in blue and are of four types: (a) ordinary notes (the majority) that serve as a sort of outline of the text and can help you later locate important concepts and equations; (b) notes that refer to the great laws and principles of physics, and these are in capital letters and in a box for emphasis; (c) notes that refer to a problem-solving hint or technique treated in the text, and these say "Problem Solving"; (d) notes that refer to an application of physics, in the text or an Example, and these say "Physics Applied."
9. This book is printed in full color. But not simply to make it more attractive. The color is used above all in the Figures, to give them greater clarity for our analysis, and to provide easier learning of the physical principles involved. The Table on the next page is a summary of how color is used in the Figures, and shows which colors are used for the different kinds of vectors, for field lines, and for other symbols and objects. These colors are used consistently throughout the book.
10. Appendices include useful mathematical formulas (such as derivatives and integrals, trigonometric identities, areas and volumes, expansions), and a table of isotopes with atomic masses and other data. Tables of useful data are located inside the front and back covers.

USE OF COLOR

Vectors

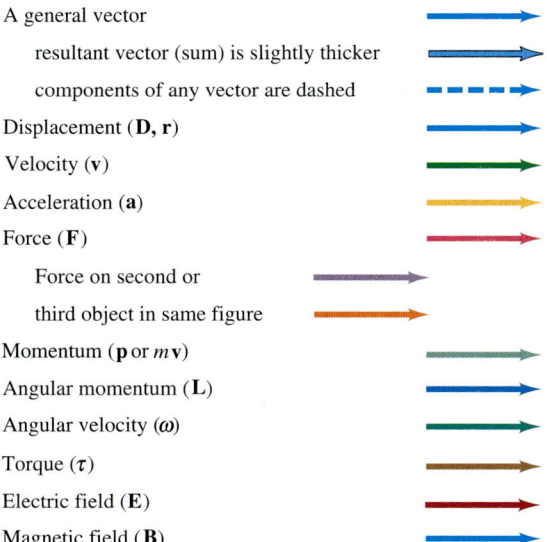

A general vector	
resultant vector (sum) is slightly thicker	
components of any vector are dashed	
Displacement (**D**, **r**)	
Velocity (**v**)	
Acceleration (**a**)	
Force (**F**)	
Force on second or third object in same figure	
Momentum (**p** or m**v**)	
Angular momentum (**L**)	
Angular velocity (ω)	
Torque (τ)	
Electric field (**E**)	
Magnetic field (**B**)	

Electricity and magnetism

- Electric field lines
- Equipotential lines
- Magnetic field lines
- Electric charge (+) or +
- Electric charge (−) or −

Electric circuit symbols

- Wire
- Resistor
- Capacitor
- Inductor
- Battery

Optics

- Light rays
- Object
- Real image (dashed)
- Virtual image (dashed and paler)

Other

- Energy level (atom, etc.)
- Measurement lines |←1.0 m→|
- Path of a moving object
- Direction of motion or current

Preface **xxxi**

This comb has been passed through hair, or rubbed by a cloth or paper towel, which gave it a static electric charge. The electrical charge on the comb induces a polarization (separation of charge) in all those scraps of paper, and thus attracts them.

Our introduction to electricity in this chapter covers conductors and insulators, and Coulomb's law which relates the force between two point charges as a function of their distance apart. We introduce the concept of electric field, which is easy to understand and powerful to use.

CHAPTER 21

Electric Charge and Electric Field

The word "electricity" may evoke an image of complex modern technology: computers, lights, motors, electric power. But the electric force would seem to play an even deeper role in our lives. According to atomic theory, electric forces between atoms and molecules hold them together to form liquids and solids, and electric forces are also involved in the metabolic processes that occur within our bodies. Many of the forces we have dealt with so far, such as elastic forces, the normal force, and other contact forces (pushes and pulls) are now considered to result from electric forces acting at the atomic level. Gravity, on the other hand, is a separate force.[†]

The earliest studies on electricity date back to the ancients, but it has been only in the past two centuries that electricity was studied in detail. We will discuss the development of ideas about electricity, including practical devices, as well as the relation to magnetism, in the next twelve chapters.

[†] As we discussed in Section 6–7, physicists in this century came to recognize four different fundamental forces in nature: (1) gravitational force, (2) electromagnetic force (we will see later that electric and magnetic forces are intimately related), (3) strong nuclear force, and (4) weak nuclear force. The last two forces operate at the level of the nucleus of an atom. A recent theory has combined the electromagnetic and weak nuclear forces so they are now considered to have a common origin known as the electroweak force.

FIGURE 21–1 Rub a plastic ruler (a) and bring it close (b) to some tiny pieces of paper.

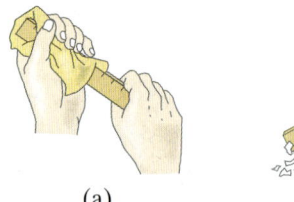
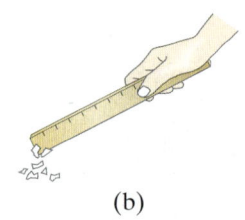

(a)   (b)

21–1 Static Electricity; Electric Charge and Its Conservation

The word *electricity* comes from the Greek word *elektron*, which means "amber." Amber is petrified tree resin, and the ancients knew that if you rub an amber rod with a piece of cloth, the amber attracts small pieces of leaves or dust. A piece of hard rubber, a glass rod, or a plastic ruler rubbed with a cloth will also display this "amber effect," or **static electricity** as we call it today. You can readily pick up small pieces of paper with a plastic comb or ruler that you've just vigorously rubbed with even a paper towel. See the photo on the previous page and Fig. 21–1. You have probably experienced static electricity when combing your hair or when taking a synthetic blouse or shirt from a clothes dryer. And you may have felt a shock when you touched a metal doorknob after sliding across a car seat or walking across a nylon carpet. In each case, an object becomes "charged" due to a rubbing process and is said to possess a net **electric charge**.

Is all electric charge the same, or is it possible that there is more than one type? In fact, there are *two* types of electric charge, as the following simple experiments show. A plastic ruler is suspended by a thread and rubbed vigorously with a cloth to charge it. When a second ruler, which has also been charged in the same way, is brought close to the first, it is found that the one ruler *repels* the other. This is shown in Fig. 21–2a. Similarly, if a rubbed glass rod is brought close to a second charged glass rod, again a repulsive force is seen to act, Fig. 21–2b. However, if the charged glass rod is brought close to the charged plastic ruler, it is found that they *attract* each other, Fig. 21–2c. The charge on the glass must therefore be different from that on the plastic. Indeed, it is found experimentally that all charged objects fall into one of two categories. Either they are attracted to the plastic and repelled by the glass, just as glass is; or they are repelled by the plastic and attracted to the glass, just as the plastic ruler is. Thus there seem to be two, and only two, types of electric charge. Each type of charge repels the same type but attracts the opposite type. That is: **unlike charges attract; like charges repel**.

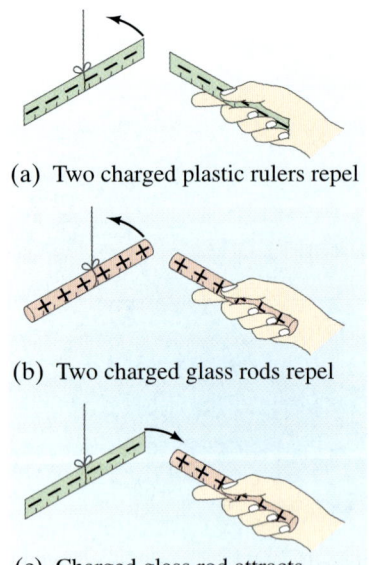

FIGURE 21–2 Unlike charges attract, whereas like charges repel one another.

(a) Two charged plastic rulers repel

(b) Two charged glass rods repel

(c) Charged glass rod attracts charged plastic ruler

Likes repel; unlikes attract

The two types of electric charge were referred to as *positive* and *negative* by the American statesman, philosopher, and scientist Benjamin Franklin (1706–1790). The choice of which name went with which type of charge was of course arbitrary. Franklin's choice set the charge on the rubbed glass rod to be positive charge, so the charge on a rubbed plastic ruler (or amber) is called negative charge. We still follow this convention today.

Franklin argued that whenever a certain amount of charge is produced on one body in a process, an equal amount of the opposite type of charge is produced on another body. The positive and negative are to be treated *algebraically*, so that during any process, the net change in the amount of charge produced is zero. For example, when a plastic ruler is rubbed with a paper towel, the plastic acquires a negative charge and the towel an equal amount of positive charge. The charges are separated, but the sum of the two is zero. This is an example of a law that is now well established: the **law of conservation of electric charge**, which states that

the net amount of electric charge produced in any process is zero.

LAW OF CONSERVATION OF ELECTRIC CHARGE

If one object or one region of space acquires a positive charge, then an equal amount of negative charge will be found in neighboring areas or objects. No violations have ever been found, and this conservation law is as firmly established as those for energy and momentum.

21-2 Electric Charge in the Atom

Only within the past century has it become clear that an understanding of electricity begins inside the atom itself. In later chapters we will discuss atomic structure and the ideas that led to our present view of the atom in more detail. But it will help our understanding of electricity if we discuss it briefly now.

A simplified model of an atom shows it as having a tiny but heavy, positively charged nucleus surrounded by one or more negatively charged electrons (Fig. 21–3). The nucleus contains protons, which are positively charged, and neutrons, which have no net electric charge. All protons and all electrons have exactly the same magnitude of electric charge, but their signs are opposite. Hence, neutral atoms, having no net charge, contain equal numbers of protons and electrons. Sometimes, as we shall see, an atom may lose one or more of its electrons, or may gain extra electrons. In this case the atom will have a net positive or negative charge, and is called an **ion**.

In solid materials the nuclei tend to remain close to fixed positions, whereas some of the electrons move quite freely. The charging of a solid object by rubbing can be explained by the transfer of electrons from one material to the other. When a plastic ruler becomes negatively charged by rubbing with a paper towel, the transfer of electrons from the towel to the plastic leaves the towel with a positive charge equal in magnitude to the negative charge acquired by the plastic. In liquids and gases, nuclei or ions can move as well as electrons.

Normally when objects are charged by rubbing, they hold their charge only for a limited time and eventually return to the neutral state. Where does the charge go? In some cases it is neutralized by charged ions in the air (formed, for example, by collisions with charged particles known as cosmic rays that reach the Earth from space). Often more importantly, the charge can "leak off" onto water molecules in the air. This is because water molecules are **polar**—that is, even though they are neutral, their charge is not distributed uniformly, Fig. 21–4. Thus the extra electrons on, say, a charged plastic ruler can "leak off" into the air because they are attracted to the positive end of water molecules. A positively charged object, on the other hand, can be neutralized by transfer of loosely held electrons from water molecules in the air. On dry days, static electricity is much more noticeable since the air contains fewer water molecules to allow leakage. On humid or rainy days, it is difficult to make any object hold a net charge for long.

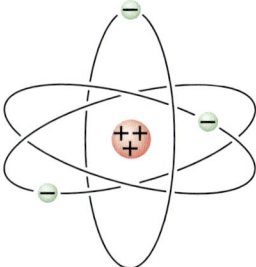

FIGURE 21–3 Simple model of the atom.

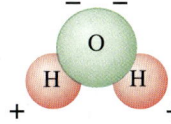

FIGURE 21–4 Diagram of a water molecule. Because it has opposite charges on different ends, it is called a "polar" molecule.

21-3 Insulators and Conductors

Suppose we have two metal spheres, one highly charged and the other electrically neutral (Fig. 21–5a). If we now place a metal object, such as a nail, so that it touches both the spheres (Fig. 21–5b), it is found that the previously uncharged sphere quickly becomes charged. If, instead, we connect the two spheres together by a wooden rod or a piece of rubber (Fig. 21–5c), the uncharged ball does not become noticeably charged. Materials like the iron nail are said to be **conductors** of electricity, whereas wood and rubber are **nonconductors** or **insulators**.

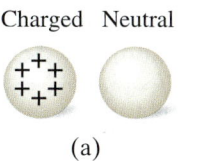
(a)

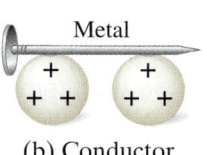
(b) Conductor

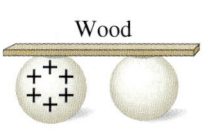
(c) Insulator

FIGURE 21–5 (a) A charged metal sphere and a neutral metal sphere. (b) The two spheres connected by a metal nail, which conducts charge from one sphere to the other. (c) The two spheres connected by an insulator (wood); almost no charge is conducted.

Metals are generally good conductors whereas most other materials are insulators (although even insulators conduct electricity very slightly). It is interesting that nearly all natural materials fall into one or the other of these two quite distinct categories. There are some materials, however (notably silicon, germanium, and carbon), that fall into an intermediate (but distinct) category known as **semiconductors**.

From the atomic point of view, the electrons in an insulating material are bound very tightly to the nuclei. In a good conductor, on the other hand, some of the electrons are bound very loosely and can move about freely within the material (although they cannot *leave* the object easily) and are often referred to as *free electrons* or *conduction electrons*. When a positively charged object is brought close to or touches a conductor, the free electrons in the conductor are attracted by this positive charge and move quickly toward it. On the other hand, the free electrons move swiftly away from a negative charge that is brought close. In a semiconductor, there are very few free electrons, and in an insulator, almost none.

Metals are good conductors

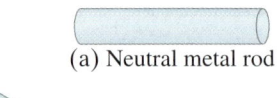
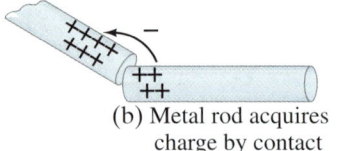

FIGURE 21–6 (a) Neutral metal rod acquires a charge (b) when placed in contact with a charged metal object.

FIGURE 21–7 Charging by induction.

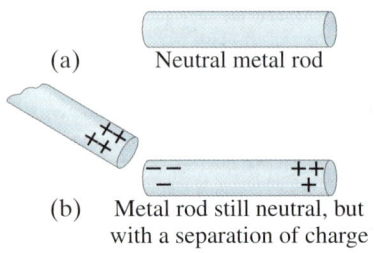

FIGURE 21–8 Inducing a charge on an object connected to ground.

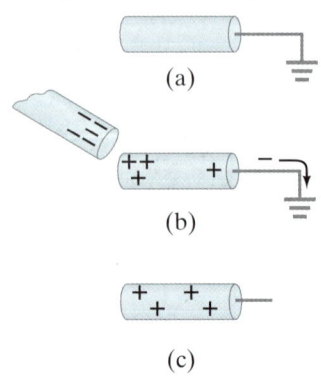

21–4 Induced Charge; the Electroscope

Suppose a positively charged metal object is brought close to an uncharged metal object. If the two touch, the free electrons in the neutral one are attracted to the positively charged object and some will pass over to it, Fig. 21–6. Since the second object is now missing some of its negative electrons, it will have a net positive charge. This process is called "charging by conduction," or "by contact," and the two objects end up with the same sign of charge.

Now suppose a positively charged object is brought close to a neutral metal rod, but does not touch it. Although the electrons of the metal rod do not leave the rod, they still move within the metal toward the charged object, which leaves a positive charge at the opposite end, Fig. 21–7. A charge is said to have been *induced* at the two ends of the metal rod. Of course no net charge has been created in the rod; charges have merely been *separated*. The net charge on the metal rod is still zero. However, if the metal were broken into two pieces, we could have two charged objects, one charged positively and one charged negatively.

Another way to induce a net charge on a metal object is to connect it with a conducting wire to the ground (or a conducting pipe leading into the ground) as shown in Fig. 21–8a (⏚ means "ground"). The object is then said to be "grounded" or "earthed." Now the Earth, since it is so large and can conduct, can easily accept or give up electrons; hence it acts like a reservoir for charge. If a charged object— say negative this time—is brought up close to the metal, free electrons in the metal are repelled and many of them move down the wire into the Earth, Fig. 21–8b. This leaves the metal positively charged. If the wire is now cut, the metal will have a positive induced charge on it (Fig. 21–8c). If the wire were cut after the negative object is moved away, the electrons would all have moved back into the metal and it would be neutral.

An **electroscope** is a device that can be used for detecting charge. As shown in Fig. 21–9, inside of a case are two movable metal leaves, often made of gold. (Sometimes only one leaf is movable.) The leaves are connected by a conductor to a metal knob on the outside of the case, but are insulated from the case itself. If a positively charged object is brought close to the knob, a separation of charge is induced, as electrons are attracted up into the knob, leaving the leaves positively charged, Fig. 21–10a. The two leaves repel each other as shown. If, instead, the

548 CHAPTER 21 Electric Charge and Electric Field

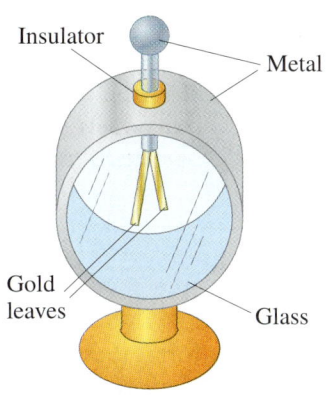

FIGURE 21–9 Electroscope.

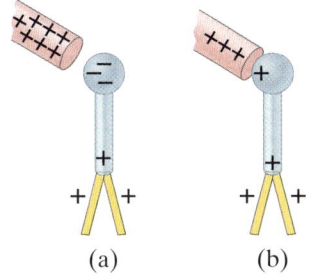

FIGURE 21–10 Electroscope charged (a) by induction, (b) by conduction.

knob is charged by conduction, the whole apparatus acquires a net charge as shown in Fig. 21–10b. In either case, the greater the amount of charge, the greater the separation of the leaves.

Note, however that you cannot tell the sign of the charge in this way, since a negative charge will cause the leaves to separate just as much as an equal magnitude positive charge—in either case the two leaves repel each other. An electroscope can, however, be used to determine the sign of the charge if it is first charged by conduction, say negatively, as in Fig. 21–11a. Now if a negative object is brought close, as in Fig. 21–11b, more electrons are induced to move down into the leaves and they separate further. On the other hand, if a positive charge is brought close, the electrons are induced to flow upward, leaving the leaves less negative and their separation is reduced, Fig. 21–11c.

The electroscope was much used in the early studies of electricity. The same principle, aided by some electronics, is used in much more sensitive modern **electrometers**.

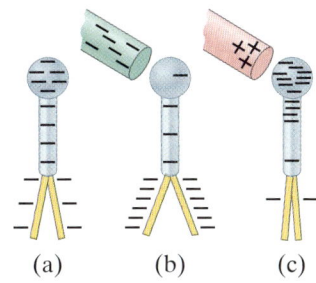

FIGURE 21–11 A previously charged electroscope can be used to determine the sign of a given charge.

21–5 Coulomb's Law

We have seen that an electric charge exerts a force on other electric charges. What factors affect the magnitude of this force? To answer this, the French physicist Charles Coulomb (1736–1806) investigated electric forces in the 1780s using a torsion balance (Fig. 21–12) much like that used by Cavendish for his studies of the gravitational force (Chapter 6).

Although precise instruments for the measurement of electric charge were not available in Coulomb's time, he was able to prepare small spheres with different magnitudes of charge in which the *ratio* of the charges was known. He reasoned that if a charged conducting sphere is placed in contact with an identical uncharged sphere, the charge on the first would be shared equally by the two of them because of symmetry. He thus had a way to produce charges equal to $\frac{1}{2}, \frac{1}{4}$, and so on, of the original charge. Although he had some difficulty with induced charges, Coulomb was able to argue that the force one tiny charged object exerted on a second tiny charged object is directly proportional to the charge on each of them. That is, if the charge on either one of the objects was doubled, the force was doubled; and if the charge on both of the objects was doubled, the force increased to four times the original value. This was the case when the distance between the two charges remained the same. If the distance between them was allowed to increase, he found that the force decreased with the *square of the distance* between

FIGURE 21–12 Principle of Coulomb's apparatus. It is similar to Cavendish's, which was used for the gravitational force. When a charged sphere is placed close to the one on the suspended bar, the bar rotates slightly. The suspending fiber resists the twisting motion and the angle of twist is proportional to the force applied. By the use of this apparatus, Coulomb investigated how the electric force varies as a function of the magnitude of the charges and of the distance between them.

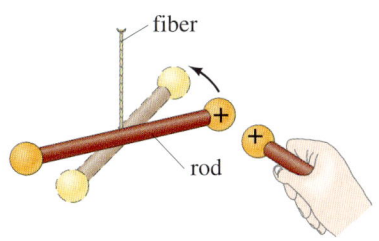

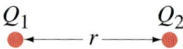

FIGURE 21–13 Coulomb's law, Eq. 21–1, gives the force between two point charges, Q_1 and Q_2, a distance r apart.

COULOMB'S LAW

them. That is, if the distance was doubled, the force fell to one-fourth of its original value. Thus, Coulomb concluded, the force one tiny charged object exerts on a second one is proportional to the product of the magnitude of the charge on one, Q_1, times the magnitude of the charge on the other, Q_2, and inversely proportional to the square of the distance r between them (Fig. 21–13). As an equation, we can write **Coulomb's law** as

$$F = k\frac{Q_1 Q_2}{r^2}, \tag{21-1}$$

where k is a proportionality constant.

Equation 21–1 gives the *magnitude* of the electric force that either object exerts on the other. The *direction* of the electric force *is always along the line joining the two objects*. If the two charges have the same sign, the force on either object is directed away from the other. If the two charges have opposite signs, the force on one is directed toward the other, Fig. 21–14. Notice that the force one charge exerts on the second is equal but opposite to that exerted by the second on the first, in accord with Newton's third law.

The validity of Coulomb's law today rests on precision measurements that are much more sophisticated than Coulomb's original experiment. The exponent, 2, in Coulomb's law has been shown to be accurate to 1 part in 10^{16} [that is, $2 \pm (1 \times 10^{-16})$].

Since we are dealing here with a new quantity (electric charge), we could choose its unit so that the proportionality constant k in Eq. 21–1 would be one. Indeed, such a system of units was once common.† However, the most widely used unit now is the **coulomb** (C), which is the SI unit. The precise definition of the coulomb today is in terms of electric current and magnetic field, and will be discussed later (Section 28–3). In SI units, k has the value

$$k = 8.988 \times 10^9 \, \text{N} \cdot \text{m}^2/\text{C}^2 \approx 9.0 \times 10^9 \, \text{N} \cdot \text{m}^2/\text{C}^2.$$

Unit for charge: the coulomb

FIGURE 21–14 Direction of the force depends on whether the charges have the same sign, (a) and (b), or opposite signs (c).

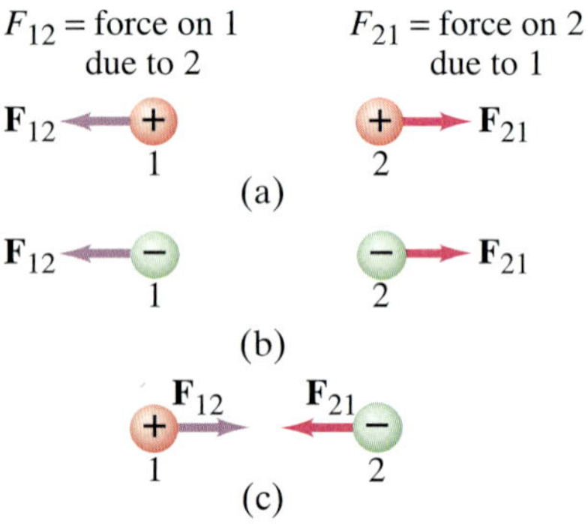

F_{12} = force on 1 due to 2 F_{21} = force on 2 due to 1

$\mathbf{F}_{12}$ ←⊕ ⊕→ $\mathbf{F}_{21}$
       1   2
        (a)

$\mathbf{F}_{12}$ ←⊖ ⊖→ $\mathbf{F}_{21}$
       1   2
        (b)

  ⊕→$\mathbf{F}_{12}$ $\mathbf{F}_{21}$←⊖
   1   2
        (c)

Charge on electron (the elementary charge)

Thus, 1 C is that amount of charge which, if placed on each of two point objects that are 1.0 m apart, will result in each object exerting a force of $(9.0 \times 10^9 \, \text{N} \cdot \text{m}^2/\text{C}^2)(1.0 \, \text{C})(1.0 \, \text{C})/(1.0 \, \text{m})^2 = 9.0 \times 10^9 \, \text{N}$ on the other. This would be an enormous force, equal to the weight of almost a million tons. We don't normally encounter charges as large as a coulomb.

Charges produced by rubbing ordinary objects (such as a comb or plastic ruler) are typically around a microcoulomb ($1 \, \mu\text{C} = 10^{-6} \, \text{C}$) or less. Objects that carry a positive charge have a deficit of electrons, whereas negatively charged objects have an excess of electrons. The magnitude of the charge on one electron has been determined to be about $1.602 \times 10^{-19} \, \text{C}$, and its sign is negative. This is the smallest charge found in nature,‡ and because of its fundamental nature, it is given the symbol e and is often referred to as the *elementary charge*:

$$e = 1.602 \times 10^{-19} \, \text{C}.$$

Note that e is defined as a positive number, so the charge on the electron is $-e$. (The charge on a proton, on the other hand, is $+e$). Since an object cannot gain or lose a fraction of an electron, the net charge on any object must be an integral multiple of this charge. Electric charge is thus said to be **quantized** (existing only in discrete amounts: $1e, 2e, 3e$, etc.). Because e is so small, however, we normally don't notice this discreteness in macroscopic charges ($1 \, \mu\text{C}$ requires about 10^{13} electrons), which thus seem continuous.

Electric charge is quantized

---

†This is a cgs system of units, and the unit of electric charge is called the *electrostatic unit* (esu) or the statcoulomb. One esu is defined as that charge, on each of two point objects 1 cm apart, that gives rise to a force of 1 dyne.

‡According to the standard model of elementary particle physics, subnuclear particles called quarks have a smaller charge than that on the electron, equal to $\frac{1}{3}e$ or $\frac{2}{3}e$. Quarks have not been detected directly as isolated objects, and theory indicates that free quarks may not be detectable.

Note the similarity of Coulomb's law to the law of universal gravitation, Eq. 6–1. Both are inverse square laws ($F \propto 1/r^2$). Both also have a proportionality to a product of a property of each body—mass for gravity, electric charge for electricity. A major difference between the two laws is that gravity is always an attractive force, whereas the electric force can be either attractive or repulsive.

The constant k in Eq. 21–1 is often written in terms of another constant, ϵ_0, called the **permittivity of free space**. It is related to k by $k = 1/4\pi\epsilon_0$. Coulomb's law can then be written

$$F = \frac{1}{4\pi\epsilon_0}\frac{Q_1 Q_2}{r^2}, \qquad (21\text{-}2)$$

COULOMB'S LAW
(in terms of ϵ_0)

where

$$\epsilon_0 = \frac{1}{4\pi k} = 8.85 \times 10^{-12} \, C^2/N \cdot m^2.$$

Equation 21–2 looks more complicated than Eq. 21–1, but other fundamental equations we haven't seen yet are simpler in terms of ϵ_0 rather than k. It doesn't matter which form we use, of course, since Eqs. 21–1 and 21–2 are equivalent.

It should be recognized that Eqs. 21–1 and 21–2 apply to objects whose size is much smaller than the distance between them. Ideally, it is precise for **point charges** (spatial size negligible compared to other distances). For finite-sized objects, it is not always clear what value to use for r, particularly since the charge may not be distributed uniformly on the objects. If the two objects are spheres and the charge is known to be distributed uniformly on each, then r is the distance between their centers.

Coulomb's law describes the force between two charges when they are at rest. Additional forces come into play when charges are in motion, and these will be discussed in later chapters. In this chapter we discuss only charges at rest, the study of which is called **electrostatics**.

When calculating with Coulomb's law, we can usually ignore the signs of the charges and determine direction based on whether the force is attractive or repulsive.

EXAMPLE 21–1 **Electric force on electron by proton.** Determine the magnitude of the electric force on the electron of a hydrogen atom exerted by the single proton ($Q_2 = +e$) that is its nucleus. Assume the electron "orbits" the proton at its average distance of $r = 0.53 \times 10^{-10}$ m, Fig. 21–15.

SOLUTION We use Coulomb's law, $F = k Q_1 Q_2/r^2$ (Eq. 21–1), with $r = 0.53 \times 10^{-10}$ m, and $Q_1 = Q_2 = 1.6 \times 10^{-19}$ C (ignoring the signs of the charges):

$$F = \frac{(9.0 \times 10^9 \, N \cdot m^2/C^2)(1.6 \times 10^{-19} \, C)(1.6 \times 10^{-19} \, C)}{(0.53 \times 10^{-10} \, m)^2}$$

$$= 8.2 \times 10^{-8} \, N.$$

The direction of the force on the electron is toward the proton, since the charges have opposite signs and the force is attractive.

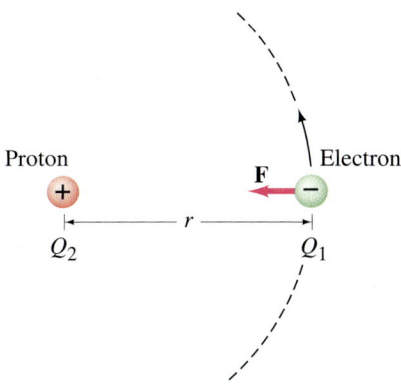

FIGURE 21–15 Example 21–1.

CONCEPTUAL EXAMPLE 21–2 **Which charge exerts the greater force?** Two positive point charges, $Q_1 = 50 \, \mu C$ and $Q_2 = 1 \, \mu C$, are separated by a distance l, Fig. 21–16. Which is larger in magnitude, the force that Q_1 exerts on Q_2, or the force that Q_2 exerts on Q_1?

RESPONSE From Coulomb's law, the force on Q_1 exerted by Q_2 is:

$$F_{12} = k\frac{Q_1 Q_2}{l^2}.$$

The force on Q_2 exerted by Q_1 is the same except that Q_1 and Q_2 are reversed. The equation is symmetric with respect to the two charges, so $F_{21} = F_{12}$. Newton's third law also tells us that these two forces must have equal magnitude.

FIGURE 21–16 Example 21–2.

$Q_1 = 50 \mu C$ $\qquad\qquad Q_2 = 1\mu C$

Superposition principle: electric forces add as vectors

It is very important to keep in mind that Eq. 21–1 (or 21–2) gives the force on a charge due to only *one* other charge. If several (or many) charges are present, the *net force on any one of them will be the vector sum of the forces due to each of the others*. This **principle of superposition** is based on experiment, and tells us that electric force vectors add like any other vector. For continuous distributions of charge, the sum becomes an integral.

When dealing with several charges, it is helpful to use double subscripts on each of the forces involved. The first subscript refers to the particle *on* which the force acts; the second refers to the particle that exerts the force. For example, if we have three charges, $\mathbf{F}_{31}$ means the force exerted *on* particle 3 *by* particle 1.

As in all problem solving, it is very important to draw a diagram, in particular a free-body diagram (Chapter 4) for each body, showing all the forces acting on that body. In applying Coulomb's law, we can deal with charge magnitudes only (leaving out minus signs) to get the magnitude of each force. Then determine the direction of the force physically (along the line joining the two particles: like charges repel, unlike attract), and show the force on the diagram. Finally, add all the forces on one object together as vectors.

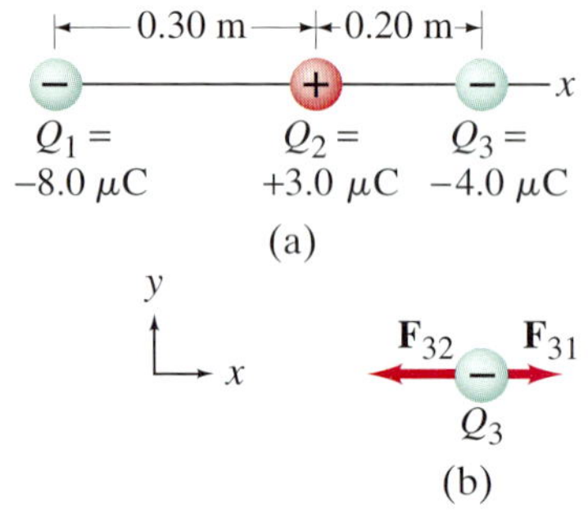

FIGURE 21–17 Diagram for Example 21–3.

EXAMPLE 21–3 Three charges in a line. Three charged particles are arranged in a line, as shown in Fig. 21–17a. Calculate the net electrostatic force on particle 3 (the $-4.0\,\mu\mathrm{C}$ on the right) due to the other two charges.

SOLUTION The net force on particle 3 will be the vector sum of the force $\mathbf{F}_{31}$ exerted by particle 1 and the force $\mathbf{F}_{32}$ exerted by particle 2: $\mathbf{F} = \mathbf{F}_{31} + \mathbf{F}_{32}$. The magnitudes of these two forces are

$$F_{31} = \frac{(9.0 \times 10^9\,\mathrm{N\cdot m^2/C^2})(4.0 \times 10^{-6}\,\mathrm{C})(8.0 \times 10^{-6}\,\mathrm{C})}{(0.50\,\mathrm{m})^2} = 1.2\,\mathrm{N}$$

$$F_{32} = \frac{(9.0 \times 10^9\,\mathrm{N\cdot m^2/C^2})(4.0 \times 10^{-6}\,\mathrm{C})(3.0 \times 10^{-6}\,\mathrm{C})}{(0.20\,\mathrm{m})^2} = 2.7\,\mathrm{N}.$$

Since we were calculating the magnitudes of the forces, we omitted the signs of the charges; but we must be aware of them to get the direction of each force. Let the line joining the particles be the *x* axis, and we take it positive to the right. Then, because $\mathbf{F}_{31}$ is repulsive and $\mathbf{F}_{32}$ is attractive, the directions of the forces are as shown in Fig. 21–17b: F_{31} points in the positive *x* direction and F_{32} points in the negative *x* direction. The net force on particle 3 is then

$$F = -F_{32} + F_{31} = -2.7\,\mathrm{N} + 1.2\,\mathrm{N} = -1.5\,\mathrm{N}.$$

The magnitude of the net force is 1.5 N, and it points to the left.

Notice in this Example that the charge in the middle (Q_2) in no way blocks the effect of the other charge (Q_1).

EXAMPLE 21–4 Electric force using vector components. Calculate the net electrostatic force on charge Q_3 shown in Fig. 21–18a due to the charges Q_1 and Q_2.

SOLUTION The forces $\mathbf{F}_{31}$ and $\mathbf{F}_{32}$ have the directions shown in the diagram since Q_1 exerts an attractive force and Q_2 a repulsive force. The magnitudes of $\mathbf{F}_{31}$ and $\mathbf{F}_{32}$ are (ignoring signs since we know the directions)

$$F_{31} = \frac{(9.0 \times 10^9\,\mathrm{N\cdot m^2/C^2})(6.5 \times 10^{-5}\,\mathrm{C})(8.6 \times 10^{-5}\,\mathrm{C})}{(0.60\,\mathrm{m})^2} = 140\,\mathrm{N},$$

$$F_{32} = \frac{(9.0 \times 10^9\,\mathrm{N\cdot m^2/C^2})(6.5 \times 10^{-5}\,\mathrm{C})(5.0 \times 10^{-5}\,\mathrm{C})}{(0.30\,\mathrm{m})^2} = 330\,\mathrm{N}.$$

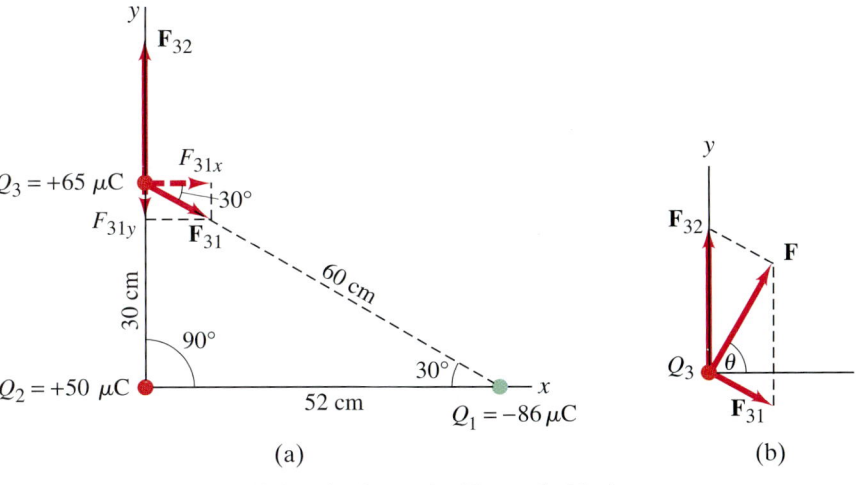

FIGURE 21–18 Determining the forces for Example 21–4.

We resolve $\mathbf{F}_{31}$ into its components along the x and y axes, as shown in Fig. 21–18a:

$$F_{31x} = F_{31} \cos 30° = 120 \text{ N},$$

$$F_{31y} = -F_{31} \sin 30° = -70 \text{ N}.$$

The force $\mathbf{F}_{32}$ has only a y component. So the net force $\mathbf{F}$ on Q_3 has components

$$F_x = F_{31x} = 120 \text{ N}$$

$$F_y = F_{32} + F_{31y} = 330 \text{ N} - 70 \text{ N} = 260 \text{ N}.$$

Thus the magnitude of the net force is

$$F = \sqrt{F_x^2 + F_y^2} = \sqrt{(120 \text{ N})^2 + (260 \text{ N})^2} = 290 \text{ N};$$

and it acts at an angle θ (see Fig. 21–18b) given by $\tan \theta = F_y/F_x = 260 \text{ N}/120 \text{ N} = 2.2$, so $\theta = 65°$.

*Vector Form of Coulomb's Law

Coulomb's law can be written in vector form (as we did for Newton's law of universal gravitation in Chapter 6, Section 6–2) as

$$\mathbf{F}_{12} = k \frac{Q_1 Q_2}{r_{21}^2} \hat{\mathbf{r}}_{21},$$

where $\mathbf{F}_{12}$ is the vector force on charge Q_1 due to Q_2 and $\hat{\mathbf{r}}_{21}$ is the unit vector pointing from Q_2 toward Q_1. That is $\mathbf{r}_{21}$ points from the "source" charge (Q_2) toward the charge on which we want to know the force (Q_1). See Fig. 21–19. The charges Q_1 and Q_2 can be either positive or negative, and this will affect the direction of the electric force. If Q_1 and Q_2 have the same sign, the product $Q_1 Q_2 > 0$ and the force on Q_1 points away from Q_2—that is, it is repulsive. If Q_1 and Q_2 have opposite signs, $Q_1 Q_2 < 0$ and $\mathbf{F}_{12}$ points toward Q_2—that is, it is attractive.

(F_{12} is in same or opposite direction as $\hat{r}_{21}$)

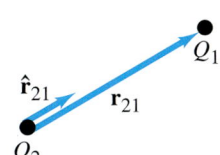

FIGURE 21–19 Determining the force on Q_1 due to Q_2, showing the direction of the unit vector $\hat{\mathbf{r}}_{21}$.

21–6 The Electric Field

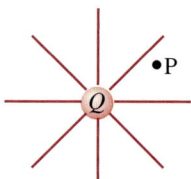

FIGURE 21–20 An electric field surrounds every charge. P is an arbitrary point.

FIGURE 21–21 Force exerted by charge $+Q$ on a small test charge, q, placed at points a, b, and c.

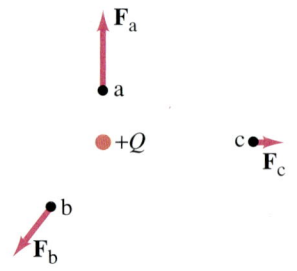

Both the gravitational force and the electrical force act over a distance: there is a force even when the two objects are not touching. The idea of a force *acting at a distance* was a difficult one for early thinkers. Newton himself felt uneasy with this idea when he published his law of universal gravitation. A helpful way to look at the situation uses the idea of the **field**, developed by the British scientist Michael Faraday (1791–1867). In the electrical case, according to Faraday, an *electric field* extends outward from every charge and permeates all of space (Fig. 21–20). When a second charge is placed near the first charge, it feels a force because of the electric field that is there (say, at point P in Fig. 21–20). The electric field at the location of the second charge is considered to interact directly with this charge to produce the force.

We can in principle investigate the electric field surrounding a charge or group of charges by measuring the force on a small positive **test charge**. By a test charge we mean a charge so small that the force it exerts does not significantly alter the distribution of the charges that create the field being measured. The force on a tiny positive test charge q placed at various locations in the vicinity of a single positive charge Q would be as shown in Fig. 21–21. The force at b is less than at a because the distance is greater (Coulomb's law); and the force at c is smaller still. In each case, the force is directed radially outward from Q. The electric field is defined in terms of the force on such a positive test charge. In particular, the **electric field, E**, at any point in space is defined as the force **F** exerted on a tiny positive test charge at that point divided by the magnitude of the test charge q:

Definition of electric field

$$\mathbf{E} = \frac{\mathbf{F}}{q}. \quad (21\text{–}3)$$

Ideally, **E** is defined as the limit of $\mathbf{F}/q$ as q is taken smaller and smaller, approaching zero. From this definition (Eq. 21–3), we see that the electric field at any point in space is a vector whose direction is the direction of the force on a positive test charge at that point, and whose magnitude is the *force per unit charge*. Thus **E** is measured in units of newtons per coulomb (N/C).

The reason for defining **E** as $\mathbf{F}/q$ (with $q \to 0$) is so that **E** does not depend on the magnitude of the test charge q. This means that **E** describes only the effect of the charges creating the electric field at that point.

The electric field at any point in space can be measured, based on the definition, Eq. 21–3. For simple situations involving one or several point charges, we can calculate what **E** will be. For example, the electric field at a distance r from a single point charge Q would have magnitude

$$E = \frac{F}{q} = \frac{kqQ/r^2}{q}$$

$$= k\frac{Q}{r^2}; \qquad \text{[single point charge]} \quad (21\text{–}4\text{a})$$

or, in terms of ϵ_0 as in Eq. 21–2 ($k = 1/4\pi\epsilon_0$):

Electric field due to one point charge

$$E = \frac{1}{4\pi\epsilon_0}\frac{Q}{r^2}. \qquad \text{[single point charge]} \quad (21\text{–}4\text{b})$$

Notice that E is independent of q—that is, it depends only on the charge Q which produces the field, and not on the value of the test charge q. Equations 21–4 could be referred to as the electric field form of Coulomb's law.

EXAMPLE 21–5 **Electrostatic copier.** An electrostatic copier works by selectively arranging positive charges (in a pattern to be copied) on the surface of a nonconducting drum, then gently sprinkling negatively charged dry toner (ink) particles onto the drum. The toner particles temporarily stick to the pattern on the drum and are later transferred to paper and "melted" to produce the copy. Suppose each toner particle has a mass of 9.0×10^{-16} kg and carries an average of 20 extra electrons to provide an electric charge. Assuming that the electric force on a toner particle must exceed twice its weight in order to ensure sufficient attraction, compute the required electric field strength near the surface of the drum. See Fig. 21–22.

SOLUTION The minimum value of electric field satisfies the relation

$$qE = 2mg$$

where $q = 20e$. Hence

$$E = \frac{2mg}{q} = \frac{2(9.0 \times 10^{-16} \text{ kg})(9.8 \text{ m/s}^2)}{20(1.6 \times 10^{-19} \text{ C})}$$
$$= 5.5 \times 10^3 \text{ N/C}.$$

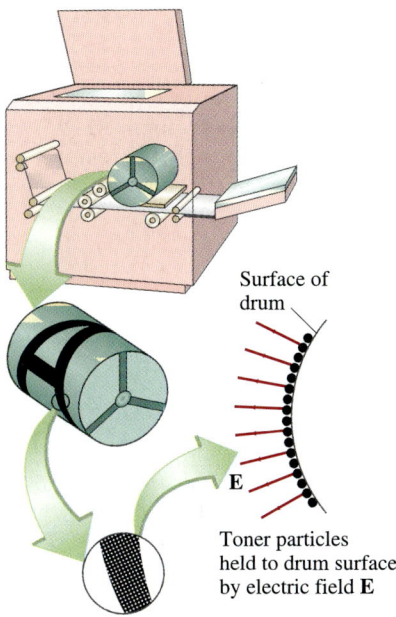

FIGURE 21–22 Example 21–5.

EXAMPLE 21–6 **Electric field of a single point charge.** Calculate the magnitude and direction of the electric field at a point P which is 30 cm to the right of a point charge $Q = -3.0 \times 10^{-6}$ C.

SOLUTION The magnitude of the electric field due to a single point charge is given by Eq. 21–4:

$$E = k\frac{Q}{r^2} = \frac{(9.0 \times 10^9 \text{ N} \cdot \text{m}^2/\text{C}^2)(3.0 \times 10^{-6} \text{ C})}{(0.30 \text{ m})^2} = 3.0 \times 10^5 \text{ N/C}.$$

The direction of the electric field is *toward* the charge Q as shown in Fig. 21–23a since we defined the direction as that of the force on a positive test charge. If Q had been positive, the electric field would have pointed away, as in Fig. 21–23b.

FIGURE 21–23 Example 21–6. Electric field at point P (a) due to a negative charge Q, and (b) due to a positive charge Q.

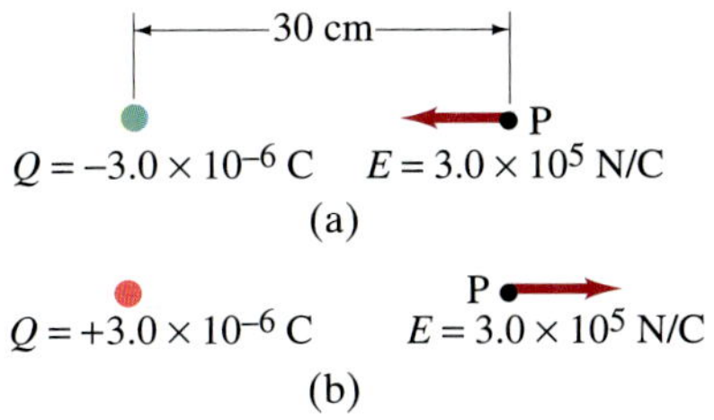

|←—30 cm—→|

$Q = -3.0 \times 10^{-6}$ C $E = 3.0 \times 10^5$ N/C
(a)

$Q = +3.0 \times 10^{-6}$ C $E = 3.0 \times 10^5$ N/C
(b)

This Example illustrates a general result: The electric field due to a positive charge points away from the charge, whereas **E** due to a negative charge points toward that charge.

If the field is due to more than one charge, the individual fields (call them $\mathbf{E}_1$, $\mathbf{E}_2$, etc.) due to each charge are added vectorially to get the total field at any point:

$$\mathbf{E} = \mathbf{E}_1 + \mathbf{E}_2 + \cdots. \tag{21–5}$$

The validity of this **superposition principle** for electric fields is fully confirmed by experiment.

If we are given the electric field **E** at a given point in space, then we can calculate the force **F** on any charge q (even if not small) placed at that point by writing (see Eq. 21–3):

$$\mathbf{F} = q\mathbf{E}.$$

If q is positive, **F** and **E** will point in the same direction. If q is negative, **F** and **E** point in opposite directions. See Fig. 21–24.

FIGURE 21–24 (a) Electric field at a given point in space. (b) Force on a positive charge. (c) Force on a negative charge.

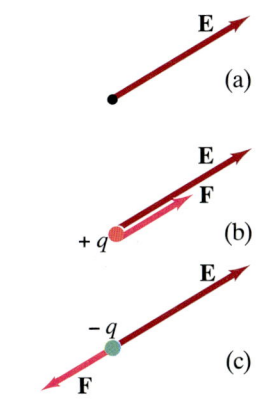

SECTION 21–6 The Electric Field 555

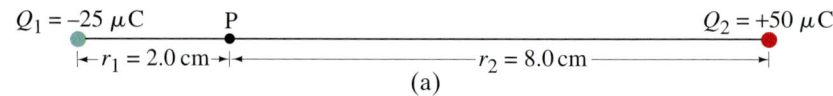
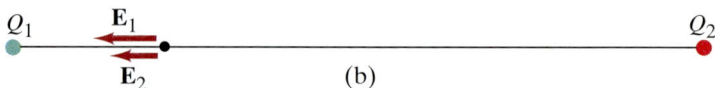

FIGURE 21–25 Example 21–7. In (b), we don't know the relative lengths of $\mathbf{E}_1$ and $\mathbf{E}_2$ until we do the calculation.

EXAMPLE 21–7 E in between two point charges. Two point charges are separated by a distance of 10.0 cm. One has a charge of $-25\,\mu\text{C}$ and the other $+50\,\mu\text{C}$. (a) What is the direction and magnitude of the electric field at a point P in between them, that is 2.0 cm from the negative charge (Fig. 21–25a)? (b) If an electron is placed at rest at P, what will its acceleration (direction and magnitude) be initially?

SOLUTION (a) The field will be a combination of two fields both pointing to the left: the field due to the negative charge Q_1 points toward Q_1, and the field due to the positive charge Q_2 points away from Q_2, again to the left, Fig. 21–25b. Thus, we can add the magnitudes of the two fields together algebraically:

$$E = k\frac{Q_1}{r_1^2} + k\frac{Q_2}{r_2^2} = k\left(\frac{Q_1}{r_1^2} + \frac{Q_2}{r_2^2}\right) = k\frac{Q_1}{r_1^2}\left[1 + \frac{(Q_2/Q_1)}{(r_2^2/r_1^2)}\right]$$

where in the last step we factored out (Q_1/r_1^2). We substitute in $r_1 = 2.0\,\text{cm} = 2.0 \times 10^{-2}\,\text{m}$ and $r_2 = 8.0 \times 10^{-2}\,\text{m}$:

$$E = (9.0 \times 10^9\,\text{N}\cdot\text{m}^2/\text{C}^2)\frac{(25 \times 10^{-6}\,\text{C})}{(2.0 \times 10^{-2}\,\text{m})^2}\left[1 + \frac{(50/25)}{(8.0/2.0)^2}\right]$$

$$= 5.6 \times 10^8\left[1 + \tfrac{1}{8}\right]\text{N/C} = 6.3 \times 10^8\,\text{N/C}.$$

Notice how factoring out Q_1/r_1^2 on the first line allowed us to see the relative strengths of the two contributing fields—namely that Q_2's field is only $\tfrac{1}{8}$ of Q_1's (or $\tfrac{1}{9}$ of the total).

(b) The electron will feel a force to the *right* since it is negatively charged and the acceleration will therefore be to the right. From the definition of electric field, Eq. 21–3, the force on any charge q (even if not small) placed in an electric field E is given by $F = qE$. Hence the magnitude of the acceleration is

$$a = \frac{F}{m} = \frac{qE}{m} = \frac{(1.60 \times 10^{-19}\,\text{C})(6.3 \times 10^8\,\text{N/C})}{9.1 \times 10^{-31}\,\text{kg}} = 1.1 \times 10^{20}\,\text{m/s}^2.$$

EXAMPLE 21–8 E above two point charges. Calculate the total electric field (a) at point A and (b) at point B in Fig. 21–26 due to both charges, Q_1 and Q_2.

SOLUTION (a) The calculation is much like that of Example 21–4, but now we are dealing with electric fields. The electric field at A is the vector sum of the fields $\mathbf{E}_{A1}$ due to Q_1, and $\mathbf{E}_{A2}$ due to Q_2; for each point charge $E = kQ/r^2$, so

$$E_{A1} = \frac{(9.0 \times 10^9\,\text{N}\cdot\text{m}^2/\text{C}^2)(50 \times 10^{-6}\,\text{C})}{(0.60\,\text{m})^2} = 1.25 \times 10^6\,\text{N/C},$$

$$E_{A2} = \frac{(9.0 \times 10^9\,\text{N}\cdot\text{m}^2/\text{C}^2)(50 \times 10^{-6}\,\text{C})}{(0.30\,\text{m})^2} = 5.0 \times 10^6\,\text{N/C}.$$

The directions are as shown, so the total electric field at A, $\mathbf{E}_A$, has components

$$E_{Ax} = E_{A1}\cos 30° = 1.1 \times 10^6\,\text{N/C},$$
$$E_{Ay} = E_{A2} - E_{A1}\sin 30° = 4.4 \times 10^6\,\text{N/C}.$$

> **PROBLEM SOLVING**
>
> *Ignore signs of charges and determine direction physically, showing directions on diagram*

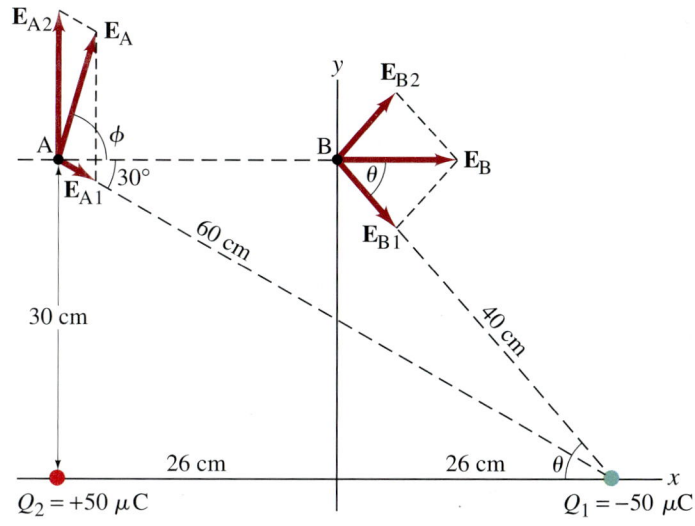

FIGURE 21–26 Calculation of the electric field at points A and B for Example 21–8.

Thus the magnitude of $\mathbf{E}_A$ is

$$E_A = \sqrt{(1.1)^2 + (4.4)^2} \times 10^6 \text{ N/C} = 4.5 \times 10^6 \text{ N/C},$$

and its direction is ϕ given by $\tan\phi = E_{Ay}/E_{Ax} = 4.4/1.1 = 4.0$, so $\phi = 76°$.
(b) Because B is equidistant (40 cm by the Pythagorean theorem) from the two equal charges, the magnitudes of E_{B1} and E_{B2} are the same; that is,

$$E_{B1} = E_{B2} = \frac{kQ}{r^2} = \frac{(9.0 \times 10^9 \text{ N}\cdot\text{m}^2/\text{C}^2)(50 \times 10^{-6} \text{ C})}{(0.40 \text{ m})^2}$$

$$= 2.8 \times 10^6 \text{ N/C}.$$

Also, because of the symmetry, the y components are equal and opposite. Hence the total field E_B is horizontal and equals $E_{B1}\cos\theta + E_{B2}\cos\theta = 2E_{B1}\cos\theta$; from the diagram, $\cos\theta = 26 \text{ cm}/40 \text{ cm} = 0.65$. Then

$$E_B = 2E_{B1}\cos\theta = 2(2.8 \times 10^6 \text{ N/C})(0.65) = 3.6 \times 10^6 \text{ N/C},$$

and the direction of $\mathbf{E}_B$ is along the $+x$ direction.

> **PROBLEM SOLVING**
> *Use symmetry to save work, when possible*

PROBLEM SOLVING   Electrostatics

Solving electrostatics problems follows, to a large extent, the general problem-solving procedure discussed in Section 4–8. In particular,

1. Draw a careful diagram—namely, a free-body diagram for each object, showing all the forces acting on that object, or the electric field at a point due to all sources.
2. Apply Coulomb's law to get the magnitude of the force that each contributing charge exerts on a charged object, or the electric field at a point. Deal only with magnitudes of charges (leaving out minus signs), and obtain the magnitude of each force or electric field. Then determine the direction of each force or electric field physically (like charges repel each other, unlike charges attract). Show and label each vector force or field on your diagram. Then add vectorially all the forces on an object, or the contributing fields at a point, to get the resultant.
3. Use symmetry (say, in the geometry) whenever possible.

*21–7 Electric Field Calculations for Continuous Charge Distributions

In many cases we can treat[†] charge as being distributed continuously. We can divide up a charge distribution into infinitesimal charges dQ, each of which will act as a tiny point charge. The contribution to the electric field at a distance r from each dQ is

$$dE = \frac{1}{4\pi\epsilon_0}\frac{dQ}{r^2}. \qquad (21\text{–}6a)$$

Then the electric field, **E**, at any point is obtained by summing over all the infinitesimal contributions, which is the integral

$$\mathbf{E} = \int d\mathbf{E}. \qquad (21\text{–}6b)$$

Note that $d\mathbf{E}$ is a vector (Eq. 21–6a gives its magnitude). [In situations where Eq. 21–6b is difficult to evaluate, other techniques (discussed in the next two chapters) can often be used instead to determine **E**. Numerical integration can also be used in many cases.]

EXAMPLE 21–9 **A ring of charge.** A thin, ring-shaped object of radius a holds a total charge Q distributed uniformly around it. Determine the electric field at a point P on its axis, a distance x from the center. See Fig. 21–27. Let λ be the charge per unit length (C/m).

SOLUTION The electric field, $d\mathbf{E}$, due to a particular segment of the ring of length dl has magnitude

$$dE = \frac{1}{4\pi\epsilon_0}\frac{dQ}{r^2}.$$

The whole ring has length (circumference) of $2\pi a$, so the charge on a length dl is

$$dQ = Q\left(\frac{dl}{2\pi a}\right) = \lambda\, dl$$

where $\lambda = Q/2\pi a$ is the charge per unit length. Now we write dE as

$$dE = \frac{1}{4\pi\epsilon_0}\frac{\lambda\, dl}{r^2}.$$

The vector $d\mathbf{E}$ has components dE_x along the x axis and $dE_\perp$ perpendicular to the x axis (Fig. 21–27). We are going to sum (integrate) around the entire ring. We note that an equal-length segment diametrically opposite the dl shown will produce a $d\mathbf{E}$ whose component perpendicular to the x axis will just cancel the $dE_\perp$ shown. This is true for all segments of the ring, so by symmetry **E** will be

→ **PROBLEM SOLVING**
Use symmetry when possible

[†]Because we believe there is a minimum charge (e), the treatment here is only for convenience; it is nonetheless useful and accurate since e is usually very much smaller than macroscopic charges.

FIGURE 21–27 Example 21–9.

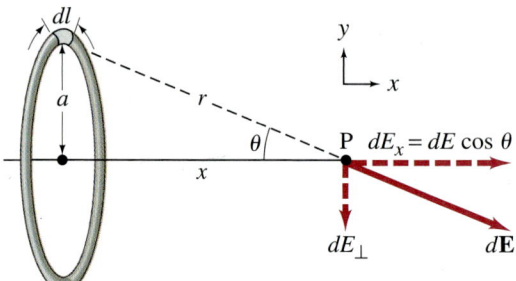

directed along the x axis and we need only sum the x components, dE_x. The total field is then

$$E = E_x = \int dE_x = \int dE \cos\theta = \frac{1}{4\pi\epsilon_0} \lambda \int \frac{dl}{r^2} \cos\theta.$$

Since $\cos\theta = x/r$, where $r = (x^2 + a^2)^{\frac{1}{2}}$, we have

$$E = \frac{\lambda}{(4\pi\epsilon_0)} \frac{x}{(x^2 + a^2)^{\frac{3}{2}}} \int_0^{2\pi a} dl = \frac{1}{4\pi\epsilon_0} \frac{\lambda x(2\pi a)}{(x^2 + a^2)^{\frac{3}{2}}}$$

$$= \frac{1}{4\pi\epsilon_0} \frac{Qx}{(x^2 + a^2)^{\frac{3}{2}}}.$$

At great distances, $x \gg a$, this reduces to $E = Q/4\pi\epsilon_0 x^2$. We would expect this result since at great distances the ring would appear to be a point charge ($1/r^2$ dependence).

➡ **PROBLEM SOLVING**
Check result by noting that at a great distance the ring looks like a point charge

Note in this Example three important problem solving techniques or "tricks" that can be used elsewhere: (1) The use of symmetry to reduce the complexity of the problem; (2) expressing the charge dQ in terms of a charge density (here linear, $\lambda = Q/2\pi a$); and (3) the checking of the answer at the limit of large r, which serves as an indication (but not proof) of the correctness of the answer—if the result did not check at large r, the result no doubt would be wrong entirely.

➡ **PROBLEM SOLVING**
Hints

EXAMPLE 21–10 **Long line of charge.** Determine the magnitude of the electric field at any point P a distance x from a very long line (a wire, say) of uniformly distributed charge, Fig. 21–28. Assume x is much smaller than the length of the wire, and let λ be the charge per unit length (C/m).

SOLUTION We set up a coordinate system so the wire is on the y axis with origin 0 as shown. A segment of wire dy has charge $dQ = \lambda\, dy$. The field $d\mathbf{E}$ at P due to such a length of wire at y has magnitude

$$dE = \frac{1}{4\pi\epsilon_0} \frac{dQ}{r^2} = \frac{1}{4\pi\epsilon_0} \frac{\lambda\, dy}{(x^2 + y^2)},$$

where $r = (x^2 + y^2)^{\frac{1}{2}}$ as shown in Fig. 21–28. The vector $d\mathbf{E}$ has components dE_x and dE_y as shown where $dE_x = dE \cos\theta$ and $dE_y = dE \sin\theta$. If the wire is extremely long in both directions (so distant contributions have little effect compared to nearby ones), or if 0 is at the midpoint of the wire (even if the wire is short), the y component of $\mathbf{E}$ will be zero since there will be equal contributions to $E_y = \int dE_y$ from above and below point 0, so

$$E_y = \int dE \sin\theta = 0.$$

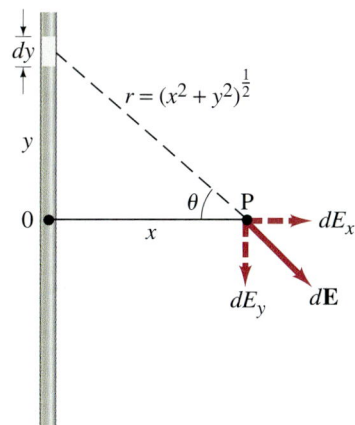

FIGURE 21–28 Example 21–10.

Then we have

$$E = E_x = \int dE \cos\theta = \frac{\lambda}{4\pi\epsilon_0} \int \frac{\cos\theta\, dy}{x^2 + y^2}.$$

The integration here is over y, along the wire, with x treated as constant. We must now write θ as a function of y, or y as a function of θ. We do the latter: since $y = x \tan\theta$, $dy = x\, d\theta/\cos^2\theta$ and $(x^2 + y^2) = x^2/\cos^2\theta$. Then

$$E = \frac{\lambda}{4\pi\epsilon_0} \frac{1}{x} \int_{-\pi/2}^{\pi/2} \cos\theta\, d\theta = \frac{\lambda}{4\pi\epsilon_0 x} (\sin\theta)\bigg|_{-\pi/2}^{\pi/2} = \frac{1}{2\pi\epsilon_0} \frac{\lambda}{x},$$

where we have assumed the wire is extremely long in both directions ($y \to \pm\infty$) which corresponds to the limits $\theta = \pm\pi/2$. Thus the field of a long straight line of charge decreases inversely as the first power of the distance from the wire. This result, obtained for an infinite wire, is a good approximation for a wire of finite length as long as x is small compared to the distance of P from the ends of the wire.

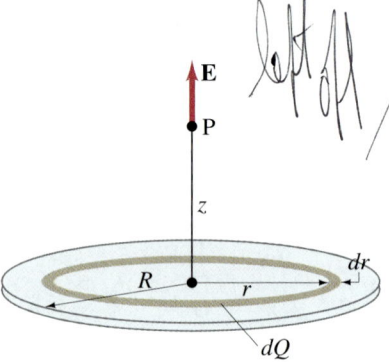

FIGURE 21–29 Example 21–11: a uniformly charged flat disk of radius R.

EXAMPLE 21–11 Uniformly charged disk. Charge is distributed uniformly over a thin circular disk of radius R. The charge per unit area (C/m^2) is σ. Calculate the electric field at a point P on the axis of the disk, a distance z above its center, Fig. 21–29.

SOLUTION We can think of the disk as a set of concentric rings. We can then apply the result of Example 21–9 to each of these rings, and then sum over all the rings. For the ring of radius r shown in Fig. 21–29, the electric field has magnitude

$$dE = \frac{1}{4\pi\epsilon_0} \frac{z\, dQ}{(z^2 + r^2)^{\frac{3}{2}}}$$

where we have used the result of Example 21–9 and have written dE (instead of E) for this thin ring of total charge dQ. The ring has area $(dr)(2\pi r)$ and we insert the charge per unit area $\sigma = dQ/(2\pi r\, dr)$:

$$dE = \frac{1}{4\pi\epsilon_0} \frac{z\sigma 2\pi r\, dr}{(z^2 + r^2)^{\frac{3}{2}}} = \frac{z\sigma r\, dr}{2\epsilon_0 (z^2 + r^2)^{\frac{3}{2}}}.$$

Now we sum over all the rings, starting at $r = 0$ out to the largest with $r = R$:

$$E = \frac{z\sigma}{2\epsilon_0} \int_0^R \frac{r\, dr}{(z^2 + r^2)^{\frac{3}{2}}} = \frac{z\sigma}{2\epsilon_0} \left[-\frac{1}{(z^2 + r^2)^{\frac{1}{2}}} \right]_0^R$$

$$= \frac{\sigma}{2\epsilon_0} \left[1 - \frac{z}{(z^2 + R^2)^{\frac{1}{2}}} \right].$$

This gives the magnitude of **E** at any point z along the axis of the disk. The direction of each $d\mathbf{E}$ due to each ring is along the z axis (as in Example 21–9), and therefore the direction of **E** is along z. If Q (and σ) are positive, **E** points away from the disk; if Q (and σ) are negative, **E** points toward the disk.

If the radius of the disk in Example 21–11 is much greater than the distance of our point P from the disk (i.e., $z \ll R$) then we can obtain a very useful result: the second term in the solution above becomes very small, so

Electric field near a thin large surface

$$E = \frac{\sigma}{2\epsilon_0}. \qquad \text{[infinite plane]} \quad (21\text{–}7)$$

This result is valid for any point above (or below) an infinite plane of any shape holding a uniform charge density σ. It is also valid for points close to a finite plane, as long as the point is close to the plane compared to the distance to the plane's edge. Thus the field near a large uniformly charged plane is uniform, and directed outward if the plane is positively charged.

It is interesting to compare here the distance dependence of the electric field due to a point charge $(E \sim 1/r^2)$, due to a very long uniform line of charge $(E \sim 1/r)$, and due to a very large uniform plane of charge (E does not depend on r).

EXAMPLE 21–12 Two parallel plates. Determine the electric field between two large parallel plates or sheets, which are very thin and are separated by a distance d which is small compared to their height and width. One plate carries a uniform surface charge density σ and the other carries a uniform surface charge density $-\sigma$, as shown in Fig. 21–30.

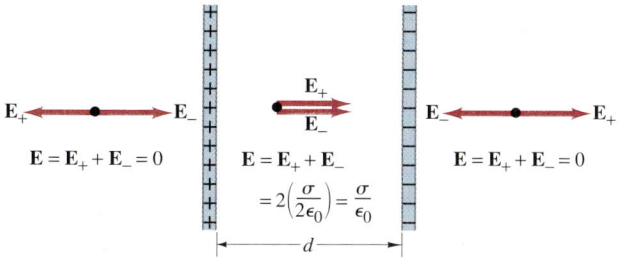

FIGURE 21–30 Example 21–12.

SOLUTION From Eq. 21–7, each plate sets up an electric field of magnitude $E = \pm \sigma/2\epsilon_0$. The field due to the positive plate points away from that plate whereas the field due to the negative plate points toward that plate. Hence, in the region between the plates, the fields add together as shown:

$$E = E_+ + E_- = \frac{\sigma}{2\epsilon_0} + \frac{\sigma}{2\epsilon_0} = \frac{\sigma}{\epsilon_0}.$$

The field is uniform, since the plates are very large compared to their separation, so this result is valid for any point, whether near one or the other of the plates, or midway between them as long as the point is far from the ends. Outside the plates, the fields cancel,

$$E = E_+ + E_- = \frac{\sigma}{2\epsilon_0} - \frac{\sigma}{2\epsilon_0} = 0,$$

as shown in the diagram. These results are valid ideally for infinitely large plates; they are a good approximation for finite plates if the separation is much less than the dimensions of the plate and for points not too close to the edge. These useful and extraordinary results illustrate the principle of superposition and its great power.

Electric field between two large oppositely charged parallel plates

21–8 Field Lines

Since the electric field is a vector, it is sometimes referred to as a *vector field*. We could indicate the electric field with arrows at various points in a given situation, such as at a, b, and c in Fig. 21–31. The directions of $\mathbf{E}_a$, $\mathbf{E}_b$, and $\mathbf{E}_c$ are the same as that of the forces shown earlier in Fig. 21–21, but the lengths (magnitudes) are different since we divide by q. However, the relative lengths of $\mathbf{E}_a$, $\mathbf{E}_b$, and $\mathbf{E}_c$ are the same as for the forces since we divide by the same q each time. To indicate the electric field in such a way at *many* points, however, would result in many arrows, which might appear complicated or confusing. To avoid this, we use another technique, that of field lines.

In order to visualize the electric field, we draw a series of lines to indicate the direction of the electric field at various points in space. These **electric field lines** (sometimes called *lines of force*) are drawn so that they indicate the direction of the force due to the given field on a positive test charge. The lines of force due to a single isolated positive charge are shown in Fig. 21–32a and for a single isolated negative charge in Fig. 21–32b. In part (a) the lines point radially outward from the charge, and in part (b) they point radially inward toward the charge because that is the direction the force would be on a positive test charge in each case (as in Fig. 21–23). Only a few representative lines are shown. One could just as well draw lines in between those shown since the electric field exists there as well. We can always draw the lines so that the *number of lines starting on a positive charge, or ending on a negative charge, is proportional to the magnitude of the charge*. Notice that near the charge, where the electric field is greatest, the lines are closer together. This is a general property of electric field lines: *the closer the lines are together, the stronger the electric field in that region*. In fact the lines can always be drawn so that the number of lines crossing unit area perpendicular to **E** is proportional to the magnitude of the electric field.

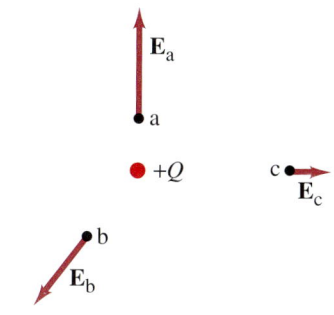

FIGURE 21–31 Electric field vector shown at three points, due to a single point charge Q. (Compare to Fig. 21–21.)

FIGURE 21–32 Electric field lines (a) near a single positive point charge, (b) near a single negative point charge.

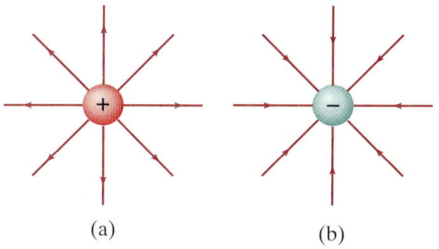

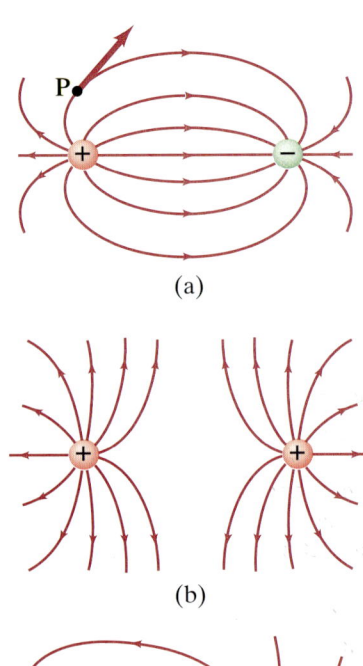

(a)

(b)

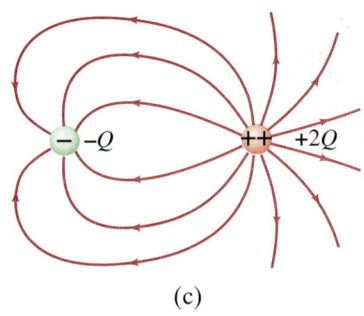

(c)

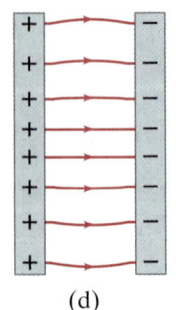

(d)

FIGURE 21–33 Electric field lines for four arrangements of charges.

FIGURE 21–34 The Earth's gravitational field.

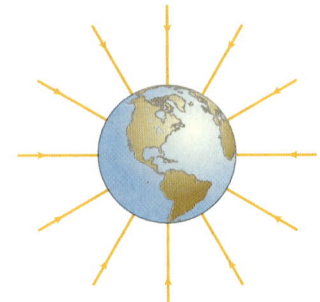

Figure 21–33a shows the electric field lines due to two charges of opposite sign. The electric field lines are curved in this case and they are directed from the positive charge to the negative charge. The direction of the field at any point is directed tangentially as shown by the arrow at point P. To satisfy yourself that this is the correct pattern for the electric field lines, you can make a few calculations such as those done in Example 21–8 for just this case (see Fig. 21–26). Figures 21–33b and c show the electric field lines (b) for two equal positive charges, and (c) for unequal charges, $+2Q$ and $-Q$; note that twice as many lines leave $+2Q$ as there are lines entering $-Q$ (number of lines is proportional to magnitude of Q). Finally, in (d), we see the field between two oppositely charged parallel plates. Notice that the electric field lines between the two plates start out perpendicular to the surface of the metal plates (we'll see why this is always true in the next Section) and go directly from one plate to the other, as we expect because a positive test charge placed between the plates would feel a strong repulsion from the positive plate and a strong attraction to the negative plate. The field lines between the plates are parallel and equally spaced in the central region far from the edges (as in Example 21–12), but fringe outward near the edges. Thus, in the central region, the electric field has the same magnitude at all points and we can write (see Example 21–12)

$$E = \text{constant} = \frac{\sigma}{\epsilon_0}. \quad \text{[between two closely spaced parallel plates]} \quad (21\text{–}8)$$

The fringing of the field near the edges can often be ignored, particularly if the separation of the plates is small compared to their size.

We summarize the properties of field lines as follows:

1. The field lines indicate the direction of the electric field; the field points in the direction tangent to the field line at any point.
2. The lines are drawn so that the magnitude of the electric field, E, is proportional to the number of lines crossing unit area perpendicular to the lines. The closer the lines, the stronger the field.
3. Electric field lines start on positive charges and end on negative charges; and the number starting or ending is proportional to the magnitude of the charge.

Also note that field lines never cross. Why not? Because it would not make sense for the electric field to have two values at the same point.

The field concept can also be applied to the gravitational force as mentioned in Chapter 6. Thus we can say that a **gravitational field** exists for every object that has mass. One object attracts another by means of the gravitational field. The Earth, for example, can be said to possess a gravitational field (Fig. 21–34) which is responsible for the gravitational force on objects. The *gravitational field* is defined as the *force per unit mass*. The magnitude of the Earth's gravitational field at any point is then (GM_E/r^2), where M_E is the mass of the Earth, r is the distance of the point from the Earth's center, and G is the gravitational constant (Chapter 6). At the Earth's surface, r is simply the radius of the Earth and the gravitational field is simply equal to g, the acceleration due to gravity (since $F/m = mg/m = g$). Beyond the Earth, the gravitational field can be calculated at any point as a sum of terms due to Earth, Sun, Moon, and other bodies that contribute significantly.

21–9 Electric Fields and Conductors

We now discuss some properties of conductors. First, *the electric field inside a conductor is zero in the static situation*—that is, when the charges are at rest. If there were an electric field within a conductor, there would be a force on its free electrons. The electrons would move until they reached positions where the electric field, and therefore the electric force on them, did become zero.

This reasoning has some interesting consequences. For one, *any net charge on a conductor distributes itself on the surface*. For a negatively charged conductor, you can imagine that the negative charges repel one another and race to the surface to get as far from one another as possible. Another consequence is the following. Suppose that a positive charge Q is surrounded by an isolated uncharged metal conductor whose shape is a spherical shell, Fig. 21–35. Because there can be no field within the metal, the lines leaving the positive charge must end on negative charges on the inner surface of the metal. Thus an equal amount of negative charge, −Q, is induced on the inner surface of the spherical shell. Then, since the shell is neutral, a positive charge of the same magnitude, +Q, must exist on the outer surface of the shell. Thus, although no field exists in the metal itself, an electric field exists outside of it, as shown in Fig. 21–35, as if the metal were not even there.

A related property of static electric fields and conductors is that *the electric field is always perpendicular to the surface outside of a conductor*. If there were a component of **E** parallel to the surface (Fig. 21–36), electrons at the surface would move along the surface in response to this force until they reached positions where no net force was exerted on them parallel to the surface—that is, until the electric field was perpendicular to the surface.

These properties pertain only to conductors. Inside a nonconductor, which does not have free electrons, an electric field can exist as we will see later, in Chapter 24. And the electric field outside a nonconductor does not necessarily make an angle of 90° to the surface.

CONCEPTUAL EXAMPLE 21–13 **Shielding, and safety in a storm.** A hollow metal box is placed between two parallel charged plates as shown in Fig. 21–37a. What is the field like inside the box?

RESPONSE If our metal box were solid, and not hollow, the electrons in the box, even if it were neutral overall, would redistribute themselves along the surface so that the field lines would not penetrate the conducting metal of the box. For a hollow box, the external field is not changed since the electrons in the metal can move just as freely as before to the surface. Hence we conclude that the field inside the hollow metal box is zero. So the field lines are something like those shown in Fig. 21–37b. A conducting box used in this way is an effective device for shielding delicate instruments and electronic circuits from unwanted external electric fields. We also can see that a relatively safe place to be during a lightning storm is inside a car, surrounded by metal. See also Fig. 21–38, where we see that a person inside a porous "cage" is protected from a strong electric discharge.

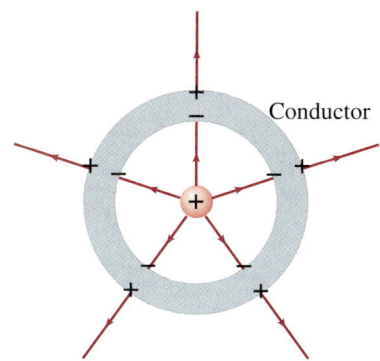

FIGURE 21–35 A charge placed inside a spherical shell. Charges are induced on the conductor surfaces. The electric field exists even beyond the shell but not within the conductor itself.

FIGURE 21–36 If the electric field **E** at the surface of a conductor had a component parallel to the surface, $E_\parallel$, the latter would accelerate electrons into motion. In the static case (no charges are in motion), $E_\parallel$ must be zero, and so the electric field must be perpendicular to the conductor's surface: $\mathbf{E} = \mathbf{E}_\perp$.

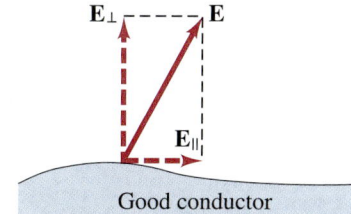

FIGURE 21–37 Example 21–13.

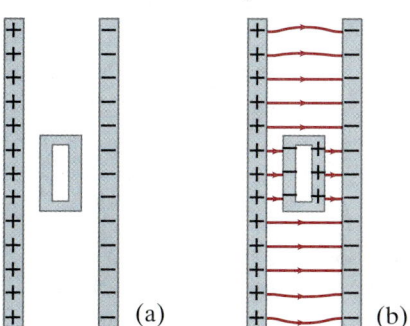

FIGURE 21–38 A strong electric field exists in the vicinity of this "Faraday cage," so strong that electrons are pulled off atoms of the air, and charge flows to (or from) the metal cage. Yet the person inside the cage is not affected.

21–10 Motion of a Charged Particle in an Electric Field

If an object having an electric charge q is at a point in space where the electric field is **E**, the force on the object is given by

$$\mathbf{F} = q\mathbf{E}$$

(see Eq. 21–3). In the past few Sections we have seen how to determine **E** for some particular situations. Now let us suppose we know **E** and we want to find the force on a charged object and its subsequent motion.

EXAMPLE 21–14 Electron accelerated by electric field. An electron (mass $m = 9.1 \times 10^{-31}$ kg) is accelerated in the uniform field **E** ($E = 2.0 \times 10^4$ N/C) between two parallel charged plates. The separation of the plates is 1.5 cm. The electron is accelerated from rest near the negative plate and passes through a tiny hole in the positive plate, Fig. 21–39. (a) With what speed does it leave the hole? (b) Show that the gravitational force can be ignored. Assume the hole is so small that it does not affect the uniform field between the plates.

SOLUTION (a) The magnitude of the force on the electron is

$$F = qE$$

and is directed to the right. The magnitude of the electron's acceleration is

$$a = \frac{F}{m} = \frac{qE}{m}.$$

Between the plates **E** is uniform so the electron undergoes uniformly accelerated motion with acceleration

$$a = \frac{(1.6 \times 10^{-19}\,\text{C})(2.0 \times 10^4\,\text{N/C})}{(9.1 \times 10^{-31}\,\text{kg})} = 3.5 \times 10^{15}\,\text{m/s}^2.$$

It travels a distance $x = 1.5 \times 10^{-2}$ m before reaching the hole, and since its initial speed was zero, we can use the kinematic equation, $v^2 = v_0^2 + 2ax$ (Eq. 2–12c), with $v_0 = 0$:

$$v = \sqrt{2ax} = \sqrt{2(3.5 \times 10^{15}\,\text{m/s}^2)(1.5 \times 10^{-2}\,\text{m})} = 1.0 \times 10^7\,\text{m/s}.$$

There is no electric field outside the plates, so after passing through the hole, the electron moves with this speed, which is now constant.
(b) The magnitude of the electric force on the electron is

$$qE = (1.6 \times 10^{-19}\,\text{C})(2.0 \times 10^4\,\text{N/C}) = 3.2 \times 10^{-15}\,\text{N}.$$

The gravitational force is

$$mg = (9.1 \times 10^{-31}\,\text{kg})(9.8\,\text{m/s}^2) = 8.9 \times 10^{-30}\,\text{N},$$

which is 10^{14} times smaller! Note that the electric field due to the electron does not enter the problem (since a particle cannot exert a force on itself).

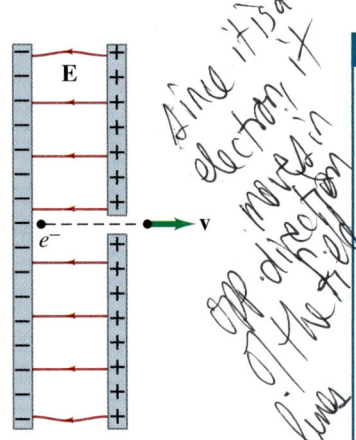

FIGURE 21–39 Example 21–14.

EXAMPLE 21–15 Electron moving perpendicular to E. Suppose an electron (say from the Example 21–14) traveling with speed $v_0 = 1.0 \times 10^7 \text{ m/s}$ enters a uniform electric field **E** at right angles to $\mathbf{v}_0$ as shown in Fig. 21–40. Describe its motion by giving the equation of its path while in the electric field. Ignore gravity.

SOLUTION When the electron enters the electric field (at $x = y = 0$) it has velocity $\mathbf{v}_0 = v_0\mathbf{i}$ in the x direction. The electric field **E**, pointing vertically upward, imparts a uniform vertical acceleration to the electron of

$$a_y = \frac{F}{m} = \frac{qE}{m} = -\frac{eE}{m},$$

where we set $q = -e$ for the electron. Its vertical position is given by

$$y = \frac{1}{2}a_y t^2 = -\frac{eE}{2m}t^2$$

since the motion is at constant acceleration. The horizontal position is given by

$$x = v_0 t$$

since $a_x = 0$. We eliminate t between these two equations and obtain

$$y = -\frac{eE}{2mv_0^2}x^2,$$

which is the equation of a parabola (just as in projectile motion, Section 3–7).

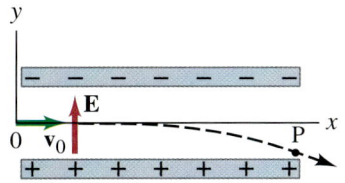

FIGURE 21–40 Example 21–15.

21–11 Electric Dipoles

The combination of two equal charges of opposite sign, $+Q$ and $-Q$, separated by a distance l, is referred to as an **electric dipole**. The quantity Ql is called the **dipole moment** and is represented[†] by the symbol p. The dipole moment can be considered to be a vector **p**, of magnitude Ql, that points from the negative to the positive charge as shown in Fig. 21–41. Many molecules, such as the diatomic molecule CO, have a dipole moment (C has a small positive charge and O a small negative charge of equal magnitude), and are referred to as **polar molecules**. Even though the molecule as a whole is neutral, there is a separation of charge that results from an uneven sharing of electrons by the two atoms.[‡] (Symmetric diatomic molecules, like O_2, have no dipole moment.) The water molecule, with its uneven sharing of electrons (O is negative, the two H are positive), also has a dipole moment—see Figs. 21–4 and 21–42.

Dipole moment

FIGURE 21–41 A dipole consists of equal but opposite charges, $+Q$ and $-Q$, separated by a distance l. The dipole moment $\mathbf{p} = Ql$ and points from the negative to the positive charge.

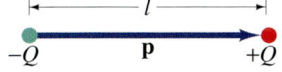

[†]Be careful not to confuse this p for dipole moment with p for momentum.
[‡]The value of the separated charges may be a fraction of e (say $\pm 0.2e$ or $\pm 0.4e$) but note that such charges don't violate what we said about e being the smallest charge. These charges less than e cannot be isolated and merely represent how much time electrons spend around one atom or the other.

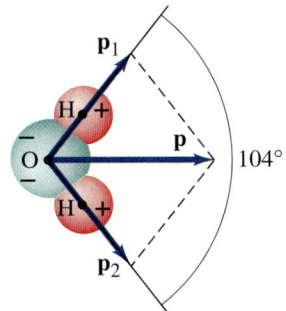

FIGURE 21–42 In the water molecule (H_2O), the electrons spend more time around the oxygen atom than around the two hydrogen atoms. The net dipole moment **p** can be considered as the vector sum of two dipole moments $\mathbf{p}_1$ and $\mathbf{p}_2$ that point from the O to each H as shown: $\mathbf{p} = \mathbf{p}_1 + \mathbf{p}_2$.

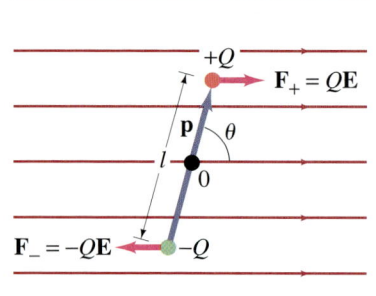

FIGURE 21–43 An electric dipole in a uniform electric field.

Dipole in an External Field

First let us consider a dipole, of dipole moment $p = Ql$, that is placed in a uniform electric field **E**, as shown in Fig. 21–43. If the field is uniform, the force QE on the positive charge and the force $-QE$ on the negative charge result in no net force on the dipole. There will, however, be a *torque* on the dipole which has magnitude (calculated about the center, 0, of the dipole)

$$\tau = QE\frac{l}{2}\sin\theta + QE\frac{l}{2}\sin\theta = pE\sin\theta. \quad (21\text{–}9a)$$

This can be written in vector notation as

$$\boldsymbol{\tau} = \mathbf{p} \times \mathbf{E}. \quad (21\text{–}9b)$$

The effect of the torque is to try to turn the dipole so **p** is parallel to **E**. The work, W, done on the dipole by the electric field to change the angle θ from θ_1 to θ_2, is given by (see Eq. 10–22)

$$W = \int_{\theta_1}^{\theta_2} \tau\, d\theta = pE\int_{\theta_1}^{\theta_2}\sin\theta\, d\theta = -pE\cos\theta\Big|_{\theta_1}^{\theta_2} = pE(\cos\theta_1 - \cos\theta_2).$$

Work done by the field decreases the potential energy, U, of the dipole in this field. If we choose $U = 0$ when **p** is perpendicular to **E** ($\theta = 90°$), then

$$U = -W = -pE\cos\theta = -\mathbf{p}\cdot\mathbf{E}. \quad (21\text{–}10)$$

Potential energy of dipole in electric field

If the electric field is *not* uniform, the force on the $+Q$ of the dipole may not have the same magnitude as the force on the $-Q$, so there may be a net force, as well as a torque.

Electric Field Produced by a Dipole

We have just seen how an external electric field affects an electric dipole. Now let us suppose that there is no external field, and we want to determine the electric field produced *by* the dipole. For brevity, we restrict ourselves to points that are on the perpendicular bisector of the dipole, such as point P in Fig. 21–44 which is a distance r above the midpoint of the dipole. Note that r in Fig. 21–44 is not the distance from either charge to P; the latter distance is $(r^2 + l^2/4)^{\frac{1}{2}}$ and this is what must be used in Eq. 21–4. The total field at P is

$$\mathbf{E} = \mathbf{E}_+ + \mathbf{E}_-,$$

where $\mathbf{E}_+$ and $\mathbf{E}_-$ are the fields due to the $+$ and $-$ charges respectively. The magnitudes E_+ and E_- are equal:

$$E_+ = E_- = \frac{1}{4\pi\epsilon_0}\frac{Q}{r^2 + l^2/4}.$$

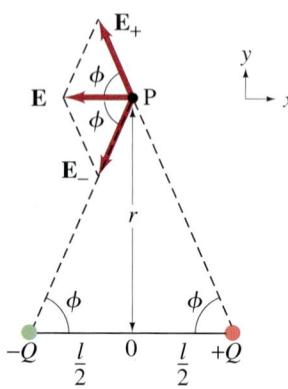

FIGURE 21–44 Electric field due to an electric dipole.

Their y components cancel at point P, so the magnitude of the total field **E** is

$$E = 2E_+\cos\phi = \frac{1}{2\pi\epsilon_0}\left(\frac{Q}{r^2 + l^2/4}\right)\frac{l}{2(r^2 + l^2/4)^{\frac{1}{2}}}$$

or

$$E = \frac{1}{4\pi\epsilon_0}\frac{p}{(r^2 + l^2/4)^{\frac{3}{2}}}. \quad \left[\begin{array}{c}\text{on perpendicular bisector}\\ \text{of dipole}\end{array}\right] \quad (21\text{–}11)$$

Far from the dipole, $r \gg l$, this reduces to

$$E = \frac{1}{4\pi\epsilon_0}\frac{p}{r^3}. \quad \left[\begin{array}{c}\text{on perpendicular bisector}\\ \text{of dipole};\ r \gg l\end{array}\right] \quad (21\text{–}12)$$

So the field decreases more rapidly for a dipole than for a single point charge ($1/r^3$ versus $1/r^2$), which we expect since at large distances the two opposite charges appear so close together as to neutralize each other. This $1/r^3$ dependence also applies for points not on the perpendicular bisector (see the Problems).

EXAMPLE 21–16 Dipole in a field. The dipole moment of a water molecule is 6.1×10^{-30} C·m. A water molecule is placed in a uniform electric field with magnitude 2.0×10^5 N/C. (*a*) What is the magnitude of the maximum torque that the field can exert on the molecule? (*b*) What is the potential energy when the torque is at its maximum? (*c*) In what position will the potential energy take on its greatest value? Why is this different than the position where the torque is maximized?

SOLUTION (*a*) From Eq. 21–9 we see that τ is maximized when θ is 90°. Then $\tau = pE = (6.1 \times 10^{-30}\text{ C·m})(2.0 \times 10^5\text{ N/C}) = 1.2 \times 10^{-24}$ N·m.
(*b*) The potential energy is given by Eq. 21–10. For $\theta = 90°$ the potential energy is zero. Note that the potential energy is negative for smaller values of θ, so U is not a minimum for $\theta = 90°$.
(*c*) The potential energy will be a maximum when $\cos\theta = -1$, so $\theta = 180°$, meaning **E** and **p** are antiparallel. The potential energy is maximized when the dipole is oriented so that it has to rotate through the largest angle, 180°, to reach the equilibrium position at $\theta = 0°$. The torque on the other hand is maximized when the electric forces are perpendicular to **p**.

Summary

There are two kinds of **electric charge**, positive and negative. These designations are to be taken algebraically—that is, any charge is plus or minus so many coulombs (C), in SI units.

Electric charge is **conserved**: if a certain amount of one type of charge is produced in a process, an equal amount of the opposite type is also produced; thus the *net* charge produced is zero.

We can better understand electricity by appealing to atomic theory, which theorizes an atom as consisting of a positively charged nucleus surrounded by negatively charged electrons. Each electron has a charge $-e = -1.6 \times 10^{-19}$ C.

Conductors are those materials in which many electrons are relatively free to move, whereas electric **insulators** are those in which very few electrons are free to move.

An object is negatively charged when it has an excess of electrons, and positively charged when it has less than its normal amount of electrons. The charge on any object is thus a whole number times $+e$ or $-e$. That is, charge is **quantized**.

An object can become charged by rubbing (in which electrons are transferred from one material to another), by conduction (which is transfer of charge from one charged object to another by touching), or by induction (the separation of charge within an object because of the close approach of another charged object but without touching).

Electric charges exert a force on each other. If two charges are of opposite types, one positive and one negative, they each exert an attractive force on the other. If the two charges are the same type, each repels the other.

The magnitude of the force one point charge exerts on another is proportional to the product of their charges, and inversely proportional to the square of the distance between them:

$$F = k\frac{Q_1 Q_2}{r^2};$$

this is **Coulomb's law**. In SI units, k is often written as $1/4\pi\epsilon_0$.

We think of an **electric field** as existing in space around any charge or group of charges. The force on another charged object is then said to be due to the electric field present at its location.

The *electric field*, **E**, at any point in space due to one or more charges, is defined as the force per unit charge that would act on a test charge q placed at that point:

$$\mathbf{E} = \frac{\mathbf{F}}{q}.$$

Electric fields are represented by **electric field lines** that start on positive charges and end on negative charges. Their direction indicates the direction the force would be on a tiny positive test charge placed at a point. The lines can be drawn so that the number per unit area is proportional to the magnitude of E.

The static electric field (that is, no charges moving) inside a conductor is zero, and the electric field lines just outside a charged conductor are perpendicular to its surface.

An **electric dipole** is a combination of two equal but opposite charges, $+Q$ and $-Q$, separated by a distance l. The **dipole moment** is $p = Ql$. A dipole placed in a uniform electric field feels no net force but does feel a net torque (unless **p** is parallel to **E**). The electric field produced by a dipole decreases as the third power of the distance r from the dipole ($E \propto 1/r^3$) for r large compared to l.

Questions

1. If you charge a pocket comb by rubbing it with a silk scarf, how can you determine if the comb is positively or negatively charged?

2. Why does a shirt or blouse taken from a clothes dryer sometimes cling to your body?

3. Explain why fog or rain droplets tend to form around ions or electrons in the air.

4. A positively charged rod is brought close to a neutral piece of paper, which it attracts. Draw a diagram showing the separation of charge and explain why attraction occurs.

5. Why does a plastic ruler that has been rubbed with a cloth have the ability to pick up small pieces of paper? Why is this difficult to do on a humid day?

6. Contrast the *net charge* on a conductor to the "free charges" in the conductor.

7. Figures 21–7 and 21–8 show how a charged rod placed near an uncharged metal object can attract (or repel) electrons. There are a great many electrons in the metal, yet only some of them move as shown. Why not all of them?

8. When an electroscope is charged, the two leaves repel each other and remain at an angle. What balances the electric force of repulsion so that the leaves don't separate further?

9. The form of Coulomb's law is very similar to that for Newton's law of universal gravitation. What are the differences between these two laws? Compare also gravitational mass and electric charge.

10. We are not normally aware of the gravitational or electric force between two ordinary objects. What is the reason in each case? Give an example where we are aware of each one and why.

11. Is the electric force a conservative force? Why or why not? (See Chapter 8.)

12. What experimental observations mentioned in the text rule out the possibility that the numerator in Coulomb's law contains the sum $(Q_1 + Q_2)$ rather than the product $Q_1 Q_2$?

13. When a charged ruler attracts small pieces of paper, sometimes a piece jumps quickly away after touching the ruler. Explain.

14. Explain why we use *small* test charges when measuring electric fields.

15. When determining an electric field, must we use a *positive* test charge, or would a negative one do as well? Explain.

16. Draw the electric field lines surrounding two negative electric charges a distance l apart.

17. Assume that the two opposite charges in Fig. 21–33a are 12.0 cm apart. Consider the magnitude of the electric field 2.5 cm from the positive charge. On which side of this charge—top, bottom, left, or right—is the electric field the strongest? The weakest?

18. Consider the electric field at the three points indicated by the letters A, B, and C in Fig. 21–45. First draw an arrow at each point indicating the direction of the net force that a positive test charge would experience if placed at that point, then list the letters in order of *decreasing* field strength (strongest first).

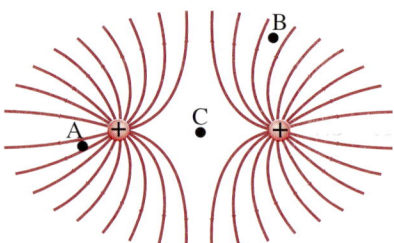

FIGURE 21–45 Question 18.

19. Why can electric field lines never cross?

20. Show, using the three rules for field lines given in Section 21–8, that the electric field lines starting or ending on a single point charge must be symmetrically spaced around the charge.

21. Given two point charges, Q and $2Q$, a distance l apart, is there a point along the straight line that passes through them where $E = 0$ when their signs are (a) opposite, (b) the same? If yes, state roughly where this point will be.

22. Suppose the ring of Fig. 21–27 has a uniformly distributed negative charge Q. What is the magnitude and direction of **E** at point P?

23. Consider a small positive test charge located on an electric field line at some point, such as point P in Fig. 21–33a. Is the direction of the velocity and/or acceleration of the test charge along this line? Discuss.

24. We wish to determine the electric field at a point near a positively charged metal sphere (a good conductor). We do so by bringing a small test charge, q_0, to this point and measure the force F_0 on it. Will F_0/q_0 be greater than, less than, or equal to, the electric field **E** as it was at that point before the test charge was present?

25. In what ways does the electron motion in Example 21–15 resemble projectile motion (Section 3–7)? In which ways not?

26. Describe the motion of the dipole shown in Fig. 21–43 if it is released from rest at the position shown.

27. Explain why there can be a net force on an electric dipole placed in a nonuniform electric field.

Problems

Section 21–5

1. (I) Calculate the magnitude of the force between two 2.50-C point charges 3.0 m apart.

2. (I) How many electrons make up a charge of $-30.0\,\mu\text{C}$?

3. (I) What is the magnitude of the electric force of attraction between an iron nucleus ($q = +26e$) and its innermost electron if the distance between them is 1.5×10^{-12} m?

4. (I) What is the repulsive electrical force between two protons in a nucleus that are 5.0×10^{-15} m apart from each other?

5. (I) What is the magnitude of the force a $+25$-μC charge exerts on a $+3.0$-mC charge 35 cm away? ($1\,\mu\text{C} = 10^{-6}$ C, 1 mC = 10^{-3} C.)

6. (II) Two charged smoke particles exert a force of 4.2×10^{-2} N on each other. What will be the force if they are moved so they are only one eighth as far apart?

7. (II) Two charged balls are 15.0 cm apart. They are moved, and the force on each of them is found to have been tripled. How far apart are they now?

8. (II) A person scuffing her feet on a wool rug on a dry day accumulates a net charge of $-40\,\mu\text{C}$. How many excess electrons does this person get, and by how much does her mass increase?

9. (II) What is the total charge of all the electrons in 1.0 kg of H_2O?

10. (II) Particles of charge $+70$, $+48$, and $-80\,\mu\text{C}$ are placed in a line (Fig. 21–46). The center one is 0.35 m from each of the others. Calculate the net force on each charge due to the other two.

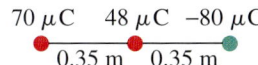

FIGURE 21–46 Problem 10.

11. (II) Three positive particles of charges 11.0 μC are located at the corners of an equilateral triangle of side 15.0 cm (Fig. 21–47). Calculate the magnitude and direction of the net force on each particle.

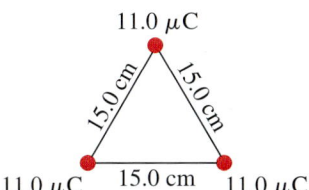

FIGURE 21–47 Problem 11.

12. (II) A charge of 6.00 mC is placed at each corner of a square 0.100 m on a side. Determine the magnitude and direction of the force on each charge.

13. (II) Repeat Problem 12 for the case when two of the positive charges, on opposite corners, are replaced by negative charges of the same magnitude (Fig. 21–48).

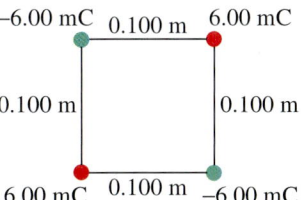

FIGURE 21–48 Problem 13.

14. (II) At each corner of a square of side l there are point charges of magnitude Q, $2Q$, $3Q$, and $4Q$ (Fig. 21–49). Determine the force on each charge due to the other three charges.

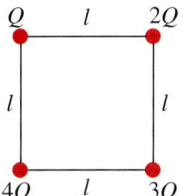

FIGURE 21–49 Problem 14.

15. (II) Repeat Problem 14 assuming the $3Q$ charge is replaced by a $-3Q$ charge (i.e., its sign is changed).

16. (II) Compare the electric force holding the electron in orbit ($r = 0.53 \times 10^{-10}$ m) around the proton nucleus of the hydrogen atom, with the gravitational force between the same electron and proton. What is the ratio of these two forces?

17. (II) Two positive point charges are a fixed distance apart. The sum of their charges is Q_T. What charge must each have in order to (a) maximize the electric force between them, and (b) minimize it?

18. (II) Two point charges have a total charge of 560 μC. When placed 1.10 m apart, the force each exerts on the other is 22.8 N and is repulsive. What is the charge on each?

19. (II) Two charges, $-Q_0$ and $-3Q_0$, are a distance l apart. These two charges are free to move but do not because there is a third charge nearby. What must be the charge and placement of the third charge for the first two to be in equilibrium?

20. (III) A $+7.7\,\mu\text{C}$ and a $-3.5\,\mu\text{C}$ charge are placed 18.5 cm apart. Where can a third charge be placed so that it experiences no net force?

21. (III) Two small nonconducting spheres have a total charge of 90.0 μC. When placed 1.16 m apart, the force each exerts on the other is 12.0 N and is repulsive. What is the charge on each? What if the force were attractive?

22. (III) Two small charged spheres hang from cords of equal length l as shown in Fig. 21–50 and make small angles θ_1 and θ_2 with the vertical. (a) If $Q_1 = Q$, $Q_2 = 2Q$, and $m_1 = m_2 = m$, determine the ratio θ_1/θ_2. (b) If $Q_1 = Q$, $Q_2 = 2Q$, $m_1 = m$, and $m_2 = 2m$, determine the ratio θ_1/θ_2. (c) Estimate the distance between the spheres for each case.

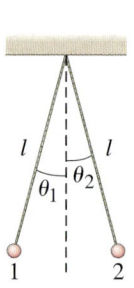

FIGURE 21–50 Problem 22.

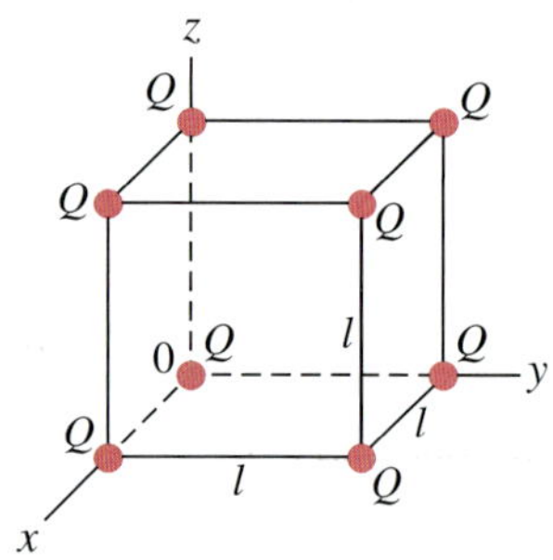

FIGURE 21–51 Problem 23.

23. (III) On each corner of a cube of side l there is a point charge Q. What is the force on the charge at the origin 0 due to the others? Give answer in vector notation for the charge at the origin in Fig. 21–51.

Sections 21–6 to 21–8

24. (I) What is the magnitude of the acceleration experienced by an electron in an electric field of 600 N/C? How does the direction of the acceleration depend on the direction of the field at that point?

25. (I) What is the magnitude and direction of the electric force on an electron in a uniform electric field of strength 1360 N/C that points due east?

26. (I) A proton is released in a uniform electric field, and it experiences an electric force of 2.75×10^{-14} N toward the south. What are the magnitude and direction of the electric field?

27. (I) What is the magnitude and direction of the electric field 20.0 cm directly above an isolated 33.0×10^{-6}-C charge?

28. (I) What is the magnitude and direction of the electric field at a point midway between a -8.0-μC and a $+7.0$-μC charge 8.0 cm apart? Assume no other charges are nearby.

29. (I) The electric force on a $+4.20$-μC charge is $\mathbf{F} = 5.85 \times 10^{-4}$ N$\mathbf{j}$. What is the electric field at the position of the charge?

30. (I) What is the electric field at a point when the force on a 1.25-μC charge placed at that point is $\mathbf{F} = (3.0\mathbf{i} - 5.0\mathbf{j}) \times 10^{-3}$ N?

31. (II) Electric field lines can always be drawn so that their number per unit area perpendicular to the lines is proportional to the electric field magnitude E. However, if Coulomb's law were not valid—that is, if the field due to a single point did not fall off precisely as $1/r^2$ (so the exponent on r was not precisely 2)—this property of field lines would not be true. Show why. [Hint: Take the example of a single point charge.]

32. (II) Draw, approximately, the electric field lines about two point charges, $+Q$ and $-3Q$, which are a distance l apart.

33. (II) Draw, approximately, the electric field lines emanating from a uniformly charged straight wire whose length l is not great. The spacing between lines near the wire should be somewhat less than l. [Hint: Also consider points very far from the wire.]

34. (II) What is the electric field strength at a point in space where a proton ($m = 1.67 \times 10^{-27}$ kg) experiences an acceleration of 1 million "g's"?

35. (II) An electron is released from rest in a uniform electric field and accelerates to the north at a rate of 145 m/s². What is the magnitude and direction of the electric field?

36. (II) The electric field midway between two equal but opposite point charges is 845 N/C, and the distance between the charges is 16.0 cm. What is the magnitude of the charge on each?

37. (II) Use Coulomb's law to determine the magnitude and direction of the electric field at points A and B in Fig. 21–52 due to the two positive charges ($Q = 7.0\,\mu$C) shown. Is your result consistent with Fig. 21–33b?

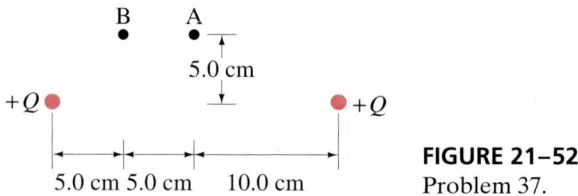

FIGURE 21–52 Problem 37.

38. (II) Calculate the electric field at the center of a square 52.5 cm on a side if one corner is occupied by a $+45.0$-μC charge and the other three are occupied by -27.0-μC charges.

39. (II) Calculate the electric field at one corner of a square 1.00 m on a side if the other three corners are occupied by 3.25×10^{-6}-C charges.

40. (II) (a) Determine the electric field $\mathbf{E}$ at the origin 0 in Fig. 21–53 due to the two charges at A and B. (b) Repeat, but let the charge at B be reversed in sign.

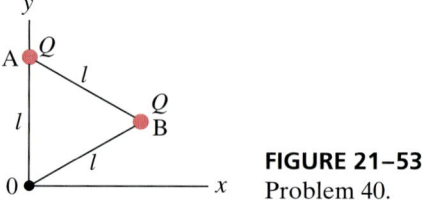

FIGURE 21–53 Problem 40.

41. (II) You are given two unknown point charges, Q_1 and Q_2. At a point on the line joining them, one-third of the way from Q_1 to Q_2, the electric field is zero (Fig. 21–54). What is the ratio Q_1/Q_2?

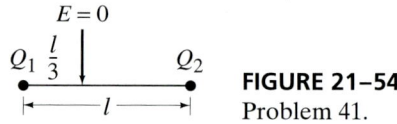

FIGURE 21–54 Problem 41.

42. (II) Two parallel circular rings of radius R have their centers on the x axis separated by a distance l as shown in Fig. 21–55. If each ring carries a uniformly distributed charge Q, find the electric field, $\mathbf{E}(x)$, at points along the x axis.

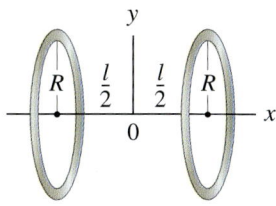

FIGURE 21–55
Problem 42.

43. (II) (a) Two equal charges Q are positioned at points $(x = l, y = 0)$ and $(x = -l, y = 0)$. Determine the electric field as a function of y for points along the y axis. (b) Show that the field is a maximum at $y = \pm l/\sqrt{2}$.

44. (II) At what position, $x = x_M$, is the magnitude of the electric field along the axis of the ring of Example 21–9 a maximum?

45. (II) Suppose the charge Q on the ring of Fig. 21–27 was all distributed uniformly on only the upper half of the ring, and no charge was on the lower half. Determine the electric field $\mathbf{E}$ at P. (Take y vertically upward.)

46. (II) Estimate the electric field at a point 2.8 cm perpendicular to the midpoint of a 2.0-m-long thin wire carrying a total charge of 4.75 μC.

47. (II) The uniformly charged straight wire in Fig. 21–28 has the length L, where point 0 is at the midpoint. Show that the field at point P, a perpendicular distance x from 0, is given by

$$E = \frac{\lambda}{2\pi\epsilon_0} \frac{L}{x(L^2 + 4x^2)^{\frac{1}{2}}},$$

where λ is the charge per unit length.

48. (III) A thin rod of length l carries a total charge Q distributed uniformly along its length. See Fig. 21–56. Determine the electric field along the axis of the rod starting at one end—that is, find $E(x)$ for $x \geq 0$ in Fig. 21–56.

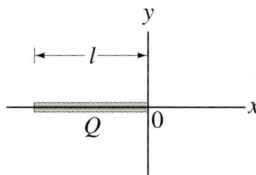

FIGURE 21–56
Problem 48.

49. (III) A thin rod bent into the shape of an arc of a circle of radius R carries a uniform charge per unit length λ. The arc subtends a total angle $2\theta_0$, symmetric about the x axis, as shown in Fig. 21–57. Determine the electric field $\mathbf{E}$ at the origin 0.

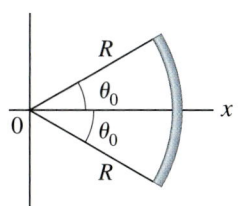

FIGURE 21–57
Problems 49 and 50.

50. (III) (a) Suppose the circular arc shown in Fig. 21–57, carries a charge per unit length λ that varies with θ as $\lambda = \lambda_0 \cos\theta$, where θ is measured from the x axis. Determine the electric field $\mathbf{E}$ at the origin 0. (b) Repeat, assuming $\lambda = \lambda_0 \sin\theta$.

51. (III) Suppose the uniformly charged wire of Fig. 21–28 starts at point 0 and rises vertically to a length L. (a) Determine the components of the electric field E_x and E_y at point P, a distance x from 0. (That is, calculate $\mathbf{E}$ near one end of a long wire, in the plane perpendicular to the wire.) (b) If the wire extends from $y = 0$ to $y = \infty$, so that $L = \infty$, show that $\mathbf{E}$ makes a 45° angle to the horizontal for any x.

52. (III) Suppose in Example 21–10 that $x = 0.250$ m, $Q = 3.15$ μC, and that the uniformly charged wire is only 6.0 m long and extends along the y axis from $y = -4.0$ m to $y = +2.0$ m. (a) Calculate E_x and E_y at point P. (b) Determine what the error would be if you simply used the result of Example 21–10, $E = \lambda/2\pi\epsilon_0 x$. Express this error as $(E_x - E)/E$ and E_y/E.

53. (III) *Uniform plane of charge.* Charge is distributed uniformly over a large square plane of side L, as shown in Fig. 21–58. The charge per unit area (C/m^2) is σ. Calculate the electric field at a point P a distance z above the center of the plane, in the limit $L \to \infty$. [*Hint*: Divide the plane into long narrow strips of width dy, and use the result of Example 21–10 to sum the fields due to each strip to get the total field.]

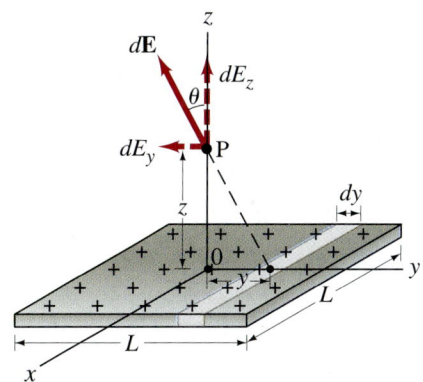

FIGURE 21–58 Problem 53.

Section 21–10

54. (II) An electron with speed $v_0 = 21.5 \times 10^6$ m/s is traveling parallel to an electric field $(\mathbf{v}_0 \| \mathbf{E})$ of magnitude $E = 11.4 \times 10^3$ N/C. (a) How far will it travel before it stops? (b) How much time will elapse before it returns to its starting point?

55. (II) An electron has an initial velocity $v_0 = 8.0 \times 10^4$ m/s $\mathbf{i}$. It enters a region where $\mathbf{E} = (2.0\mathbf{i} + 8.0\mathbf{j}) \times 10^4$ N/C. (a) Determine the vector acceleration of the electron as a function of time. (b) At what angle θ is it moving (relative to its initial direction) at $t = 1.0$ ns?

56. (II) A water droplet of radius 0.020 mm remains stationary, in the air. If the electric field of the Earth is 150 N/C downward, how many excess electron charges must the water droplet have?

57. (II) At what angle will the electrons in Example 21–15 leave the uniform electric field at the end of the parallel plates (point P in Fig. 21–40)? Assume the plates are 6.0 cm long and $E = 5.0 \times 10^3$ N/C. Ignore fringing of the field.

58. (II) Suppose electrons enter the uniform electric field midway between two plates, moving at an upward 45° angle as shown in Fig. 21–59. What maximum speed can the electrons have if they are to avoid striking the upper plate? Ignore fringing of the field.

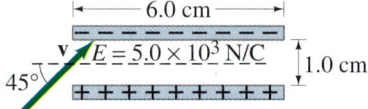

FIGURE 21–59 Problem 58.

59. (III) A positive charge q is placed at the center of a circular ring of radius R. The ring carries a uniformly distributed negative charge of total magnitude $-Q$. (a) If the charge q is displaced from the center a small distance x as shown in Fig. 21–60, show that it will undergo simple harmonic motion when released. (b) If its mass is m, what is its period?

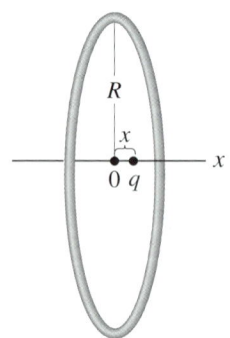

FIGURE 21–60 Problem 59.

Section 21–11

60. (II) A dipole consists of charges $+e$ and $-e$ separated by 0.68 nm. It is in an electric field $E = 2.7 \times 10^4$ N/C. (a) What is the value of the dipole moment? (b) What is the torque on the dipole when it is perpendicular to the field? (c) What is the torque on the dipole when it is at an angle of 45° to the field? (d) What is the work required to rotate the dipole from being oriented parallel to the field to being antiparallel to the field?

61. (II) The HCl molecule has a dipole moment of about 3.4×10^{-30} C·m. The two atoms are separated by about 1.0×10^{-10} m. (a) What is the net charge on each atom? (b) Is this equal to an integral multiple of e? If not, explain. (c) What maximum torque would this dipole experience in a 2.5×10^4 N/C electric field? (d) How much energy would be needed to rotate one molecule 45° from its equilibrium position of lowest potential energy?

62. (II) Suppose both charges in Fig. 21–44 were positive. (a) Show that the field on the perpendicular bisector, for $r \gg l$, is given by $(1/4\pi\epsilon_0)(2Q/r^2)$. (b) Explain why the field decreases as $1/r^2$ here whereas for a dipole it decreases as $1/r^3$.

63. (II) An electric dipole, of dipole moment p and moment of inertia I, is placed in a uniform electric field $\mathbf{E}$. (a) If displaced by an angle θ as shown in Fig. 21–43 and released, under what conditions will it oscillate in simple harmonic motion? (b) What will be its frequency?

64. (III) Suppose a dipole $\mathbf{p}$ is placed in a nonuniform electric field $\mathbf{E} = E\mathbf{i}$ that points along the x axis. If E depends only on x, show that the net force on the dipole is

$$\mathbf{F} = \left(\mathbf{p} \cdot \frac{d\mathbf{E}}{dx}\right)\mathbf{i},$$

where $d\mathbf{E}/dx$ is the gradient of the field in the x direction.

65. (III) (a) Show that at points along the axis of a dipole (along the same line that contains $+Q$ and $-Q$), the electric field has magnitude

$$E = \frac{1}{4\pi\epsilon_0}\frac{2p}{r^3}$$

for $r \gg l$ (Fig. 21–44), where r is the distance from the point to the center of the dipole. (b) In what direction does $\mathbf{E}$ point?

General Problems

66. How close must two electrons be if the electric force between them is equal to the weight of either at the Earth's surface?

67. Imagine that space invaders could deposit extra electrons in equal amounts on the Earth and on your car, which has a mass of 1050 kg. Note that the rubber tires would provide some insulation. How much charge Q would need to be placed on your car (same amount on the Earth) in order to levitate it (overcome gravity)? [Hint: Assume that the Earth's charge is spread uniformly so it acts as if it were located at the Earth's center, and then the separation distance is the radius of the Earth.]

68. A 3.0-g copper penny has a positive charge of 5.5 μC. What fraction of its electrons has it lost?

69. Suppose that electrical attraction, rather than gravity, were responsible for holding the Moon in orbit around the Earth. If equal and opposite charges Q were placed on the Earth and the Moon, what should be the value of Q to maintain the present orbit? Use these data: mass of Earth = 5.97×10^{24} kg, mass of Moon = 7.35×10^{22} kg, radius of orbit = 3.84×10^8 m. Treat the Earth and Moon as point particles.

70. One type of *electric quadrupole* consists of two dipoles placed end to end with their negative charges (say) overlapping; that is, in the center is $-2Q$ flanked (on a line) by a $+Q$ to either side (Fig. 21–61). Determine the electric field **E** at points along the perpendicular bisector and show that E decreases as $1/r^4$.

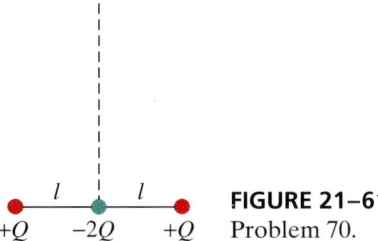

FIGURE 21–61 Problem 70.

71. Three charged particles are placed at the corners of an equilateral triangle of side 1.20 m (Fig. 21–62). The charges are $+4.0\,\mu\text{C}$, $-8.0\,\mu\text{C}$, and $-6.0\,\mu\text{C}$. Calculate the magnitude and direction of the net force on each due to the other two.

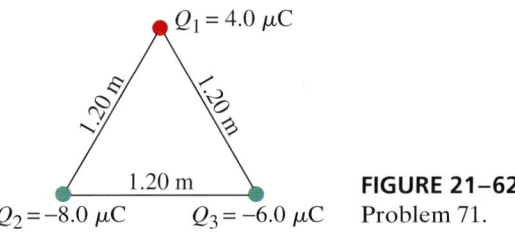

FIGURE 21–62 Problem 71.

72. A proton ($m = 1.67 \times 10^{-27}$ kg) is suspended at rest in a uniform electric field **E**. Take into account gravity and determine **E**.

73. Calculate the magnitude of the electric field at the center of a square with sides 35 cm long if the corners, taken in rotation, have charges of $1.0\,\mu\text{C}$, $2.0\,\mu\text{C}$, $3.0\,\mu\text{C}$, and $4.0\,\mu\text{C}$ (all positive).

74. In a simple model of the hydrogen atom, the electron revolves in a circular orbit around the proton with a speed of 1.1×10^6 m/s. What is the radius of the electron's orbit?

75. Two charges, $-Q_0$ and $-4Q_0$, are a distance l apart. These two charges are free to move but do not because there is a third charge nearby. What must be the magnitude of the third charge and its placement in order for the first two to be in equilibrium?

76. A point charge ($m = 1.0$ g) at the end of an insulating string of length 55 cm is observed to be in equilibrium in a uniform horizontal electric field of 10,000 N/C, when the pendulum's position is as shown in Fig. 21–63, with the charge 12 cm above the lowest (vertical) position. If the field points to the right in Fig. 21–63, determine the magnitude and sign of the point charge.

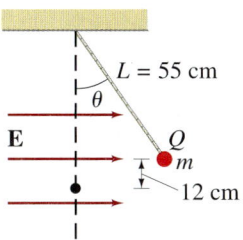

FIGURE 21–63 Problem 76.

77. A positive point charge $Q_1 = 1.85 \times 10^{-5}$ C is fixed at the origin of coordinates, and a negative charge $Q_2 = -7.65 \times 10^{-6}$ C is fixed to the x axis at $x = +2.00$ m. Find the location of the place(s) along the x axis where the electric field due to these two charges is zero.

78. Estimate the net force between the CO group and the HN group shown in Fig. 21–64. The C and O have effective charges $\pm 0.40e$ and the H and N have effective charges $\pm 0.20e$ where $e = 1.6 \times 10^{-19}$ C. [*Hint:* Do not include the "internal" forces between C and O, or between H and N.]

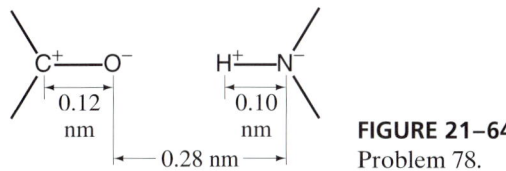

FIGURE 21–64 Problem 78.

79. An electron moving with a speed of 2.0×10^6 m/s to the right enters a uniform electric field region where the field is known to be parallel to its direction of motion. If the electron is to be brought to rest in the space of 5.4 cm, (*a*) in what direction is the electric field, and (*b*) what is the magnitude of the field?

80. The two strands of the helix-shaped DNA molecule (the genetic material in living cells) are held together by electrostatic forces as shown in Fig. 21–65. Assume that the net average charge indicated on H and N atoms is effectively $0.2e$, and on the indicated C and O atoms is $0.4e$, that atoms on each molecule are separated by 1.0×10^{-10} m, and that all relevant angles are 120°. Estimate the net force between: (*a*) a thymine and an adenine; and (*b*) a cytosine and a guanine. (*c*) Estimate the total force for a DNA molecule containing 10^5 pairs of such molecules.

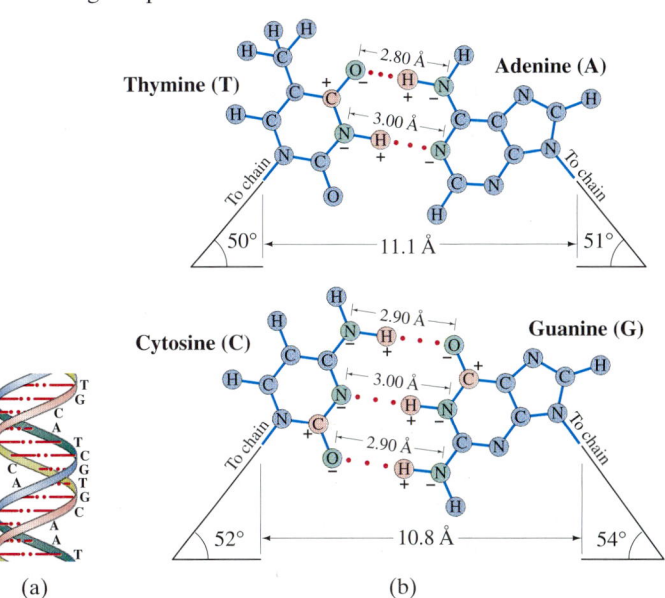

FIGURE 21–65 Problem 80. (a) Section of a DNA double helix. (b) "Close-up" view of the helix, showing how A and T attract each other and how G and C attract each other through electrostatic forces, to hold the double helix together. The red dots are used to indicate the electrostatic attraction (often called a "weak bond" or "hydrogen bond"). Note that there are two weak bonds between A and T, and three between C and G. The distance unit is the angstrom (1 Å = 10^{-10} m).

81. Suppose electrons enter the electric field midway between two plates at an angle θ_0 to the horizontal, as shown in Fig. 21–66. The path is symmetrical, so they leave at the same angle θ_0 and just barely miss the top plate. What is θ_0? Ignore fringing of the field.

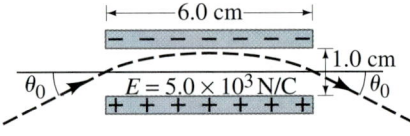

FIGURE 21–66 Problem 81.

82. Two point charges, $Q_1 = -6.7 \,\mu C$ and $Q_2 = 1.3 \,\mu C$, are located between two oppositely charged parallel plates, as shown in Fig. 21–67. The two point charges are separated by a distance of $x = 0.34$ m. Assume that the electric field produced by the charged plates is uniform and equal to $E = 73{,}000$ N/C. Calculate the net electrostatic force on Q_1 and give its direction.

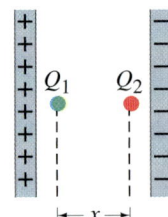

FIGURE 21–67 Problem 82.

83. A small lead ball is encased in insulating plastic and suspended vertically from an ideal spring $(k = 126\text{ N/m})$ above a lab table, Fig. 21–68. The total mass of the coated ball is 0.800 kg, and its center lies 15.0 cm above the tabletop when in equilibrium. The ball is pulled down 5.00 cm below equilibrium, an electric charge $Q = -3.00 \times 10^{-6}$ C is deposited on the ball, and the system is released. Using what you know about harmonic oscillation, write an expression for the electric field strength as a function of time that would be measured at the point on the tabletop (P) directly below the ball.

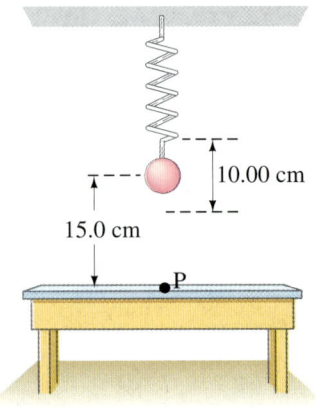

FIGURE 21–68 Problem 83.

84. A large electroscope is made with "leaves" that are 75-cm-long wires with 22-g balls at the ends. When charged, nearly all the charge resides on the balls. If the wires each make a 30° angle with the vertical (Fig. 21–69), what total charge Q must have been applied to the electroscope?

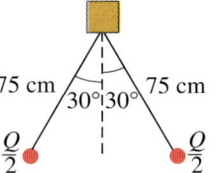

FIGURE 21–69 Problem 84.

85. Three very large square planes of charge are arranged as shown (on edge) in Fig. 21–70. From left to right, the planes have charge densities per unit area of $-0.50\,\mu C/m^2$, $+0.10\,\mu C/m^2$, and $-0.35\,\mu C/m^2$. Find the total electric field (direction and magnitude) at the points A, B, C, and D. Assume the plates are much larger than the distance AD.

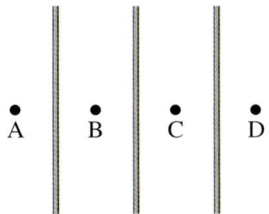

FIGURE 21–70 Problem 85.

86. What is the total charge of all the electrons in a 15-kg bar of aluminum? What is the net charge of the bar? (Aluminum has 13 electrons per atom and an atomic mass of 27 u.)

87. Given the two charges shown in Fig. 21–71, at what position(s) x is the electric field zero? Is the field zero at any other points, not on the x axis?

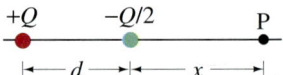

FIGURE 21–71 Problem 87.

88. An electron moves in a circle of radius r around a very long uniformly charged wire in a vacuum chamber, as shown in Fig. 21–72. The charge density on the wire is $\lambda = 0.14\,\mu C/m$. (a) What is the electric field at the electron (magnitude and direction)? (b) What is the speed of the electron?

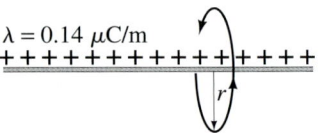

FIGURE 21–72 Problem 88.

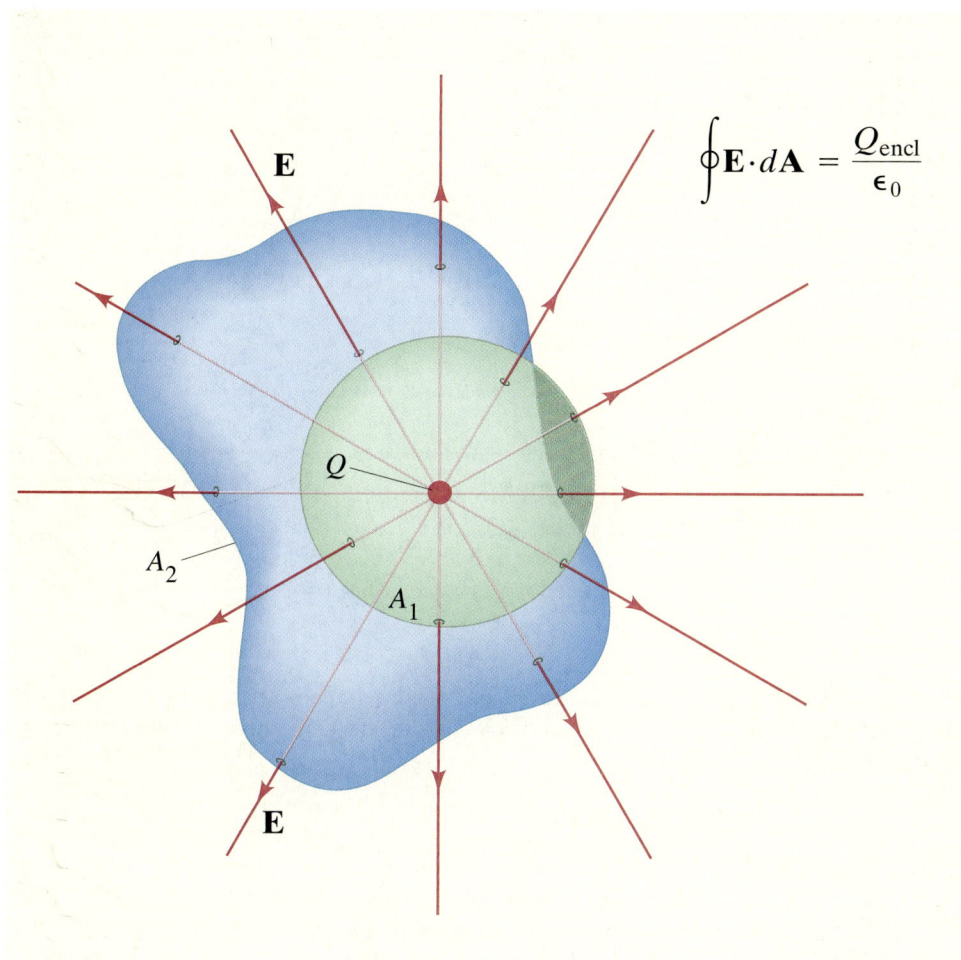

Gauss's law is an elegant relation between electric charge and electric field. It is more general than Coulomb's law. Gauss's law involves an integral of the electric field **E** at each point on a closed surface. The surface is only imaginary, and we choose the shape and placement of the surface so that we can figure out the integral. In this drawing, two different surfaces are shown, both enclosing a point charge Q. Gauss's law states that the product $\mathbf{E} \cdot d\mathbf{A}$, where $d\mathbf{A}$ is an infinitesimal area of the surface, integrated over the entire surface, equals the charge enclosed by the surface Q_{encl} divided by ϵ_0. Both surfaces here enclose the same charge Q. Hence $\oint \mathbf{E} \cdot d\mathbf{A}$ will give the same result for both.

CHAPTER 22

Gauss's Law

Gauss's law, which we develop and discuss in this chapter, is a statement of the relation between electric charge and electric field. It is a more general and elegant form of Coulomb's law.

We can, in principle, determine the electric field due to any given distribution of electric charge using Coulomb's law. The total electric field at any point will be the vector sum (or integral) of contributions from all charges present (see Eqs. 21–5 and 21–6). Except for some simple cases, the sum or integral can be quite complicated to evaluate. For situations in which an analytic solution (such as we carried out in the Examples of Sections 21–6 and 21–7) is not possible, a computer can be used.

In some cases, however, the electric field due to a given charge distribution can be calculated more easily or more elegantly using Gauss's law, as we shall see in this chapter. But the major importance of Gauss's law is that it gives us additional insight into the nature of electrostatic fields, and a more general relationship between charge and field.

Before discussing Gauss's law itself, we first discuss the concept of *flux*.

FIGURE 22–1 A uniform field **E** (indicated by the parallel field lines) passing through a surface of area A: (a) perpendicular to **E**; (b) not perpendicular to **E**. The dashed surface of area $A_\perp$ in (b) is the projection of **A** perpendicular to the field **E**.

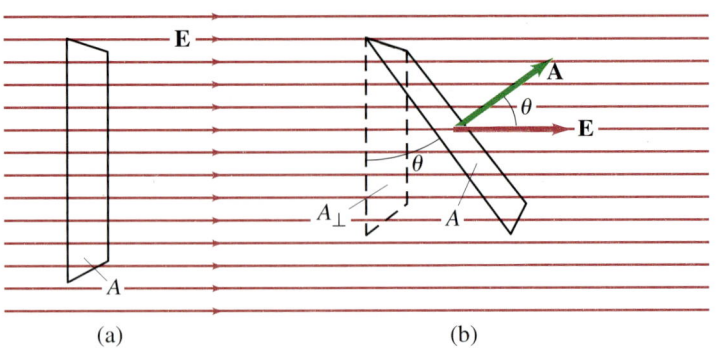

22–1 Electric Flux

Imagine a surface of area A through which a uniform electric field **E** passes, Fig. 22–1. The surface could be a rectangle (as shown), a circle, or any other shape. If the electric field direction is perpendicular to the surface, as in Fig. 22–1a, the **electric flux**, Φ_E, through this surface is defined as the product

$$\Phi_E = EA.$$

If the area A is not perpendicular to **E**, but rather makes an angle θ as shown in Fig. 22–1b, fewer field lines will pass through the area. In this case we define the electric flux through the surface as

$$\Phi_E = EA_\perp = EA\cos\theta, \qquad \text{[E uniform]} \quad (22\text{–}1\text{a})$$

where $A_\perp$ is the projection of the area A on a surface perpendicular to **E** as shown. The area A of a surface can be represented by a vector **A** whose magnitude is A and whose direction is perpendicular to the surface, as shown in Fig. 22–1b. The angle θ is the angle between **E** and **A**, so the electric flux can also be written

$$\Phi_E = \mathbf{E}\cdot\mathbf{A}. \qquad \text{[E uniform]} \quad (22\text{–}1\text{b})$$

Because of how we defined it, the electric flux has a simple intuitive interpretation in terms of field lines. We saw in Section 21–8 that field lines can always be drawn so that the number (N) passing through unit area perpendicular to the field ($A_\perp$) is proportional to the magnitude of the field (E): that is, $E \propto N/A_\perp$. Hence,

Electric flux is proportional to the number of field lines passing through the area

$$N \propto EA_\perp = \Phi_E,$$

so the flux through an area is proportional to the number of lines passing through that area.

EXAMPLE 22–1 Electric flux. (a) Calculate the electric flux through the rectangle shown in Fig. 22–1a. The rectangle is 10 cm by 20 cm and the electric field is uniform at 200 N/C. (b) What is the flux in Fig. 22–1b if the angle θ is 30°?

SOLUTION (a) The electric flux is

$$\Phi_E = EA\cos\theta$$
$$= (200\,\text{N/C})(0.10\,\text{m}\times 0.20\,\text{m})\cos 0° = 4.0\,\text{N}\cdot\text{m}^2/\text{C}.$$

(b) In this case the flux is

$$\Phi_E = (200\,\text{N/C})(0.10\,\text{m}\times 0.20\,\text{m})\cos 30° = 3.5\,\text{N}\cdot\text{m}^2/\text{C}.$$

Now let us consider the more general case, when the electric field **E** is not uniform and the surface is not flat, Fig. 22–2. We divide up the chosen surface into n small elements of surface whose areas are $\Delta A_1, \Delta A_2, \ldots \Delta A_n$. We choose the division so that each ΔA_i is small enough that (1) it can be considered flat, and (2) the electric field varies so little over this small area that it can be considered uniform over this tiny area. Then the electric flux through the entire surface is approximately

$$\Phi_E \approx \sum_{i=1}^{n} \mathbf{E}_i \cdot \Delta \mathbf{A}_i,$$

where $\mathbf{E}_i$ is the field passing through $\Delta \mathbf{A}_i$. In the limit as we let $\Delta A_i \to 0$, the sum becomes an integral over the entire surface and the relation becomes mathematically exact:

$$\Phi_E = \int \mathbf{E} \cdot d\mathbf{A}. \tag{22–2}$$

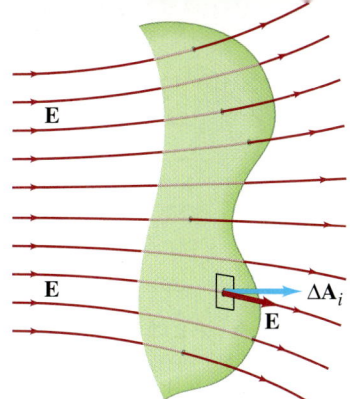

FIGURE 22–2 Electric flux through a curved surface. One small area of the surface, $\Delta \mathbf{A}_i$, is indicated.

Electric flux defined

In many cases (in particular, for Gauss's law) we deal with the flux through a *closed* surface—that is, a surface that completely encloses a volume (like a sphere or the surface of a football), Fig. 22–3. In this case, the net flux through the surface is given by

$$\Phi_E = \oint \mathbf{E} \cdot d\mathbf{A}, \tag{22–3}$$

Electric flux through a closed surface

where the integral sign is written $\oint$ to indicate that the integral is over the value of **E** on an enclosing surface.

Up to now we have not been concerned with the fact that there is an ambiguity in the direction of the vector **A** that represents a surface. For example, in Fig. 22–1, the vector **A** could point upward and to the right (as shown) or downward to the left and still be perpendicular to the surface. For a closed surface, we define (arbitrarily) the direction of **A**, or of $d\mathbf{A}$, to point *outward* from the enclosed volume, Fig. 22–4. For a line leaving the enclosed volume (on the right in Fig. 22–4), the angle θ between **E** and $d\mathbf{A}$ must be less than $\pi/2$ ($= 90°$), so $\cos\theta > 0$. For a line entering the volume (on the left in Fig. 22–4) $\theta > \pi/2$, so $\cos\theta < 0$. Hence, *flux entering the enclosed volume is negative* ($\int E \cos\theta \, dA < 0$), whereas *flux leaving the volume is positive*. Consequently, Eq. 22–3 gives the net flux *out* of the volume. If Φ_E is negative, there is a net flux *into* the volume.

Flux entering is negative
Flux leaving is positive

In Figs. 22–3 and 22–4, each field line that enters the volume also leaves the volume. Hence $\Phi_E = \oint \mathbf{E} \cdot d\mathbf{A} = 0$. There is no net flux into or out of this surface. The flux, $\oint \mathbf{E} \cdot d\mathbf{A}$, will be nonzero only if one or more lines start or end within the surface.

FIGURE 22–3 Electric flux through a closed surface.

FIGURE 22–4 The direction of an element of area $d\mathbf{A}$ is taken to point outward from an enclosed surface.

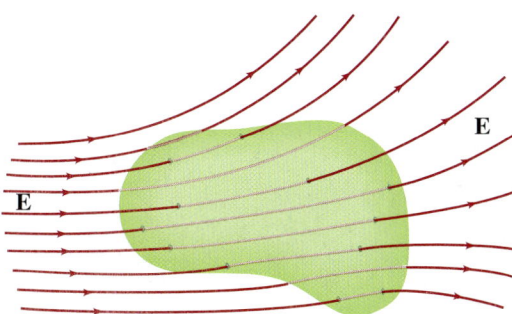
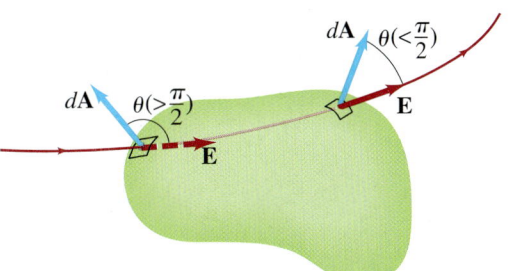

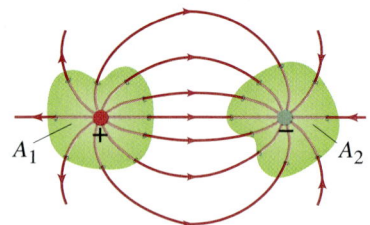

FIGURE 22–5 An electric dipole. Flux through surface A_1 is positive. Flux through A_2 is negative.

FIGURE 22–6 Net flux through surface A is negative.

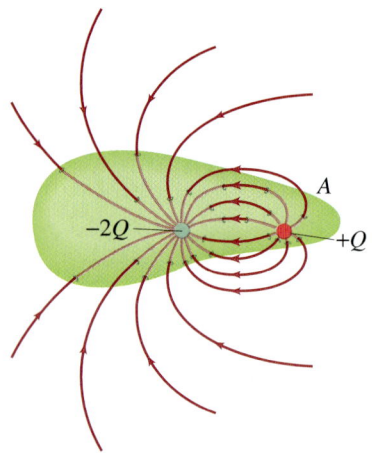

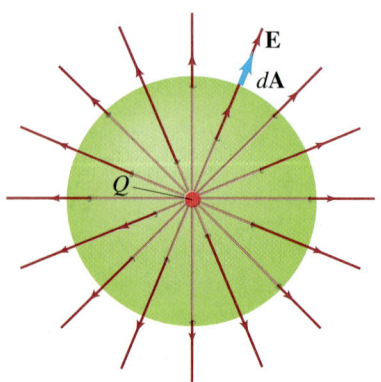

FIGURE 22–7 A single point charge Q at the center of an imaginary sphere of radius r (our "gaussian surface"—that is, the closed surface we choose to use for applying Gauss's law in this case).

Since electric field lines start and stop only on electric charges, the flux will be nonzero only if the surface encloses a net charge. For example, the surface labeled A_1 in Fig. 22–5 encloses a positive charge and there is a net outward flux through this surface ($\Phi_E > 0$). The surface A_2 encloses an equal magnitude negative charge and there is a net inward flux ($\Phi_E < 0$). For the configuration shown in Fig. 22–6, the flux through the surface shown is negative (count the lines). The value of Φ_E depends on the charge enclosed by the surface, and this is what Gauss's law is all about.

[The concept of flux applies equally well to fluid flow, and thus makes an interesting analogy. (Indeed, the word "flux" comes from the Latin word for "flow.") The electric field **E** at each point corresponds to the fluid flow velocity **v**, so the electric field lines correspond to streamlines of a fluid flow. The flux, Φ, through a surface for a fluid is the volume rate of flow and is given by $\Phi = \int \mathbf{v} \cdot d\mathbf{A}$. In Figs. 22–1, 22–2, and 22–3, the lines can correspond to the steady streamline flow of a fluid with no sources (such as a faucet) or sinks (such as a leak or drain). In this case, the net flux through a closed surface, as in Fig. 22–3, is zero: what flows in also flows out. In Figs. 22–5 and 22–6, there are sources (corresponding to a positive charge) where flow lines start, and also sinks (corresponding to negative charge) where flow lines end. Although this comparison between electric flux and fluid flux is interesting, and perhaps offers some insight, do not get them confused—an electric flux is not a flow of any substance. Flux can be defined for any vector field, and we will later use it also for the magnetic field.]

22–2 Gauss's Law

The precise relation between the electric flux through a closed surface and the net charge Q_{encl} enclosed within that surface is given by **Gauss's law**:

$$\oint \mathbf{E} \cdot d\mathbf{A} = \frac{Q_{\text{encl}}}{\epsilon_0}, \qquad (22\text{–}4)$$

where ϵ_0 is the same constant (permittivity of free space) that appears in Coulomb's law. The integral on the left is over the value of **E** on a closed surface, which we choose for our convenience in any given situation. The charge Q_{encl} is the net charge *enclosed* by that surface. It doesn't matter where or how the charge is distributed within the surface. Any charge outside this surface must not be included. A charge outside the chosen surface may affect the position of the electric field lines, but will not affect the net number of lines entering or leaving the surface. For example, Q_{encl} for the gaussian surface A_1 in Fig. 22–5 would be the positive charge enclosed by A_1; the negative charge does contribute to the electric field at A_1 but it is *not* enclosed by surface A_1 and so is not included in Q_{encl}.

Before we discuss the validity of Gauss's law, we note that the integral is often rather difficult to carry out in practice. We rarely need to do it except for some fairly simple situations that will be discussed shortly (Section 22–3).

Now let us see how Gauss's law is related to Coulomb's law.[†] First, we show that Coulomb's law follows from Gauss's law. In Fig. 22–7 we have a single charge Q.

[†]Note that Gauss's law would look more complicated in terms of the constant $k = 1/4\pi\epsilon_0$ that we originally used in Coulomb's law (Eq. 21–1 or 21–4a):

Coulomb's law	*Gauss's law*
$E = k\dfrac{Q}{r^2}$	$\oint \mathbf{E} \cdot d\mathbf{A} = 4\pi k Q$
$E = \dfrac{1}{4\pi\epsilon_0}\dfrac{Q}{r^2}$	$\oint \mathbf{E} \cdot d\mathbf{A} = \dfrac{Q}{\epsilon_0}.$

Gauss's law has a simpler form using ϵ_0; Coulomb's law is simpler using k. The normal convention is to use ϵ_0 rather than k because Gauss's law is considered more general and therefore it is preferable to have it in simpler form.

For our "gaussian surface," we choose an imaginary sphere of radius r centered on the charge. Because Gauss's law is supposed to be valid for any surface, we have chosen one that will make our calculation easy. Because of the symmetry of this (imaginary) sphere about the charge at its center, we know that $\mathbf{E}$ must have the same magnitude at any point on the surface, and that $\mathbf{E}$ points radially outward (or inward) parallel to $d\mathbf{A}$, an element of the surface area. Hence we write the integral in Gauss's law as

$$\oint \mathbf{E} \cdot d\mathbf{A} = \oint E \, dA = E \oint dA = E(4\pi r^2)$$

since the surface area of a sphere of radius r is $4\pi r^2$, and the magnitude of $\mathbf{E}$ is the same at all points on this spherical surface. Then Gauss's law becomes, with $Q_{\text{encl}} = Q$,

$$\frac{Q}{\epsilon_0} = \oint \mathbf{E} \cdot d\mathbf{A} = E(4\pi r^2).$$

Solving for E we obtain

$$E = \frac{Q}{4\pi \epsilon_0 r^2},$$

which is the electric field form of Coulomb's law, Eq. 21–4b.

Now let us do the reverse, and derive Gauss's law from Coulomb's law for static electric charges. First we consider a single point charge Q surrounded by an imaginary spherical surface as in Fig. 22–7. Coulomb's law tells us that the electric field at the spherical surface is $E = (1/4\pi\epsilon_0)(Q/r^2)$. Reversing the argument we just used, we have

$$\oint \mathbf{E} \cdot d\mathbf{A} = \oint \frac{1}{4\pi\epsilon_0} \frac{Q}{r^2} dA = \frac{Q}{4\pi\epsilon_0 r^2}(4\pi r^2) = \frac{Q}{\epsilon_0}.$$

This is Gauss's law, with $Q_{\text{encl}} = Q$, and we derived it for the special case of a spherical surface enclosing a point charge at its center. But what about some other surface, such as the irregular surface labeled A_2 in Fig. 22–8? The same number of field lines (due to our charge Q) pass through surface A_2, as pass through the spherical surface, A_1. Therefore, because the flux through a surface is proportional to the number of lines through it as we saw in Section 22–1, the flux through A_2 is the same as through A_1:

$$\oint_{A_2} \mathbf{E} \cdot d\mathbf{A} = \oint_{A_1} \mathbf{E} \cdot d\mathbf{A} = \frac{Q}{\epsilon_0}.$$

Hence we can expect that

$$\oint \mathbf{E} \cdot d\mathbf{A} = \frac{Q}{\epsilon_0}$$

would be valid for *any* surface surrounding a single point charge Q.

Finally, let us look at the case of more than one charge. For each charge, Q_i, enclosed by the chosen surface,

$$\oint \mathbf{E}_i \cdot d\mathbf{A} = \frac{Q_i}{\epsilon_0},$$

where $\mathbf{E}_i$ refers to the electric field produced by Q_i alone. By the superposition principle for electric fields (Eq. 21–5), the total field $\mathbf{E}$ is equal to the sum of the fields due to each separate charge, $\mathbf{E} = \Sigma \mathbf{E}_i$. Hence

$$\oint \mathbf{E} \cdot d\mathbf{A} = \oint (\Sigma \mathbf{E}_i) \cdot d\mathbf{A} = \Sigma \frac{Q_i}{\epsilon_0} = \frac{Q_{\text{encl}}}{\epsilon_0},$$

where $Q_{\text{encl}} = \Sigma Q_i$ is the total net charge enclosed within the surface. Thus we see, based on this simple argument, that Gauss's law follows from Coulomb's law for any distribution of electric charge enclosed within a closed surface of any shape.

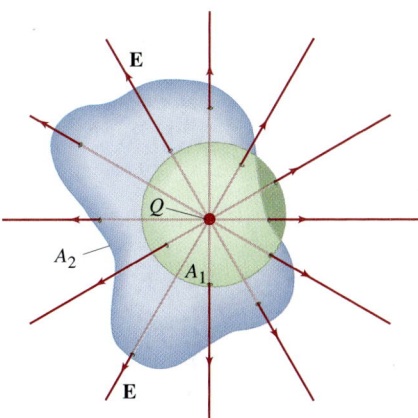

FIGURE 22–8 A single point charge surrounded by a spherical surface A_1, and an irregular surface, A_2.

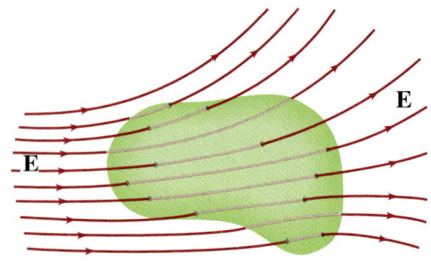

FIGURE 22–9 Electric flux through a closed surface. (Same as Fig. 22–3.) No electric charge is enclosed by this surface ($Q_{encl} = 0$).

The derivation of Gauss's law from Coulomb's law is valid for electric fields produced by static electric charges. We will see later that electric fields can also be produced by changing magnetic fields. Coulomb's law cannot be used to describe such electric fields. But Gauss's law *is* found to hold also for electric fields generated in this way. Hence *Gauss's law is a more general law than Coulomb's law.* It holds for any electric field whatsoever.

Even for the case of static electric fields we are treating in this chapter, it is important to recognize that **E** on the left side of Gauss's law is not necessarily due only to the charge Q_{encl} that appears on the right. For example, in Fig. 22–9 there is an electric field **E** at all points on the imaginary gaussian surface, but it is not due to the charge enclosed by the surface (which is $Q_{encl} = 0$ in this case). The electric field **E** which appears on the left side of Gauss's law is the *total* electric field at each point, on the gaussian surface chosen, not just that due to the charge Q_{encl}, which appears on the right side. Gauss's law has been found to be valid for the total field at any surface. It tells us that any *difference* between the input and output flux of the electric field over any surface is due to charge within that surface.

CONCEPTUAL EXAMPLE 22–2 **Flux from Gauss's law.** Consider the two gaussian surfaces, A_1 and A_2, shown in Fig. 22–10. The only charge present is the charge Q at the center of surface A_1. What is the net flux through each surface, A_1 and A_2?

RESPONSE The surface A_1 encloses the charge $+Q$. By Gauss's law, the net flux through A_1 is then Q/ϵ_0. For surface A_2, the charge $+Q$ is outside the surface. Surface A_2 encloses zero net charge, so the net electric flux through A_2 is zero, by Gauss's law. Note that all field lines that enter the volume enclosed by surface A_2 also leave it.

FIGURE 22–10 Example 22–2. Two gaussian surfaces.

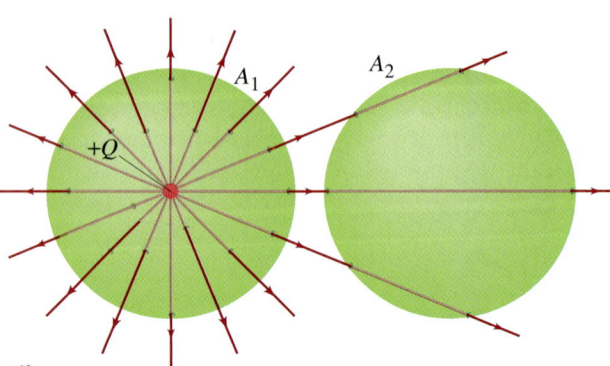

22–3 Applications of Gauss's Law

Gauss's law is a very compact and elegant way to write the relation between electric charge and electric field. It also offers a simple way to determine the electric field when the charge distribution is simple and/or possesses a high degree of symmetry. In order to apply Gauss's law however, we must choose the "gaussian" surface very carefully (for the integral on the left side of Gauss's law) so we can determine **E**. We normally try to think of a surface that has just the symmetry needed so that E will be constant on all or on parts of its surface. Sometimes we choose a surface so the flux through part of the surface is zero.

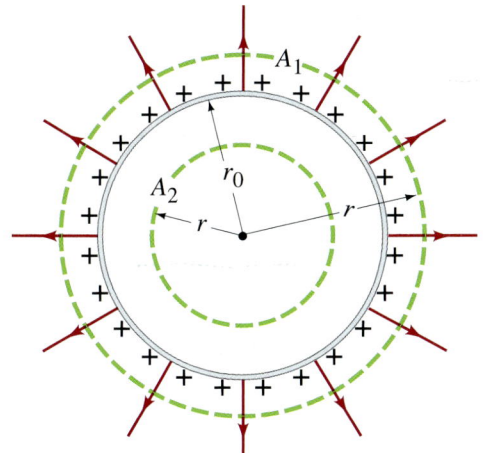

FIGURE 22–11 Cross sectional drawing of a thin spherical shell of radius r_0, carrying a net charge Q uniformly distributed. A_1 and A_2 represent two Gaussian surfaces we use to determine **E**.

EXAMPLE 22–3 Spherical conductor. A thin spherical shell of radius r_0 possesses a total net charge Q that is uniformly distributed on it, Fig. 22–11. Determine the electric field at points (*a*) outside the shell, and (*b*) inside the shell. (*c*) What if the conductor were a solid sphere?

SOLUTION (*a*) Because the charge is distributed symmetrically, the electric field must also be symmetric. Thus the field must be directed radially outward (inward if $Q < 0$) and must depend only on r, not on angle (spherical coordinates). First we want to find **E** outside the spherical shell, so we choose our imaginary gaussian surface to be a sphere of radius r ($r > r_0$) concentric with the shell, and shown in Fig. 22–11 as a dashed circle A_1 outside the shell. The electric field **E** then has the same magnitude at all points on the surface, and because **E** is perpendicular to this surface, the cosine of the angle between **E** and $d\mathbf{A}$ is always 1. Gauss's law then gives (with $Q_{\text{encl}} = Q$)

$$\oint \mathbf{E} \cdot d\mathbf{A} = E(4\pi r^2) = \frac{Q}{\epsilon_0}$$

or

$$E = \frac{1}{4\pi\epsilon_0} \frac{Q}{r^2}. \qquad [r > r_0]$$

Thus the field outside a uniform spherical shell of charge is the same as if all the charge were concentrated at the center as a point charge.

Field outside spherical shell is same as for point charge at center

(*b*) Inside the shell, the field must also be symmetric. So E must again have the same value at all points on a spherical gaussian surface (A_2 in Fig. 22–11) concentric with the shell. Thus E can be factored out of the integral and, with $Q_{\text{encl}} = 0$, we have

$$\oint \mathbf{E} \cdot d\mathbf{A} = E(4\pi r^2) = 0.$$

Hence

$$E = 0 \qquad [r < r_0]$$

E = 0 inside uniformly charged spherical shell

inside a uniform spherical shell of charge.

(*c*) These same results also apply to a uniformly charged solid spherical conductor, since all the charge would lie in a thin layer at the surface.

Above results apply also for solid conducting sphere

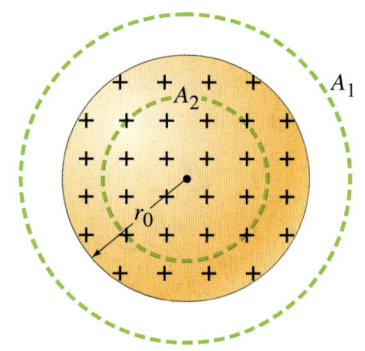

FIGURE 22–12 A solid sphere of uniform charge density.

Uniform sphere produces same field as a point charge at center

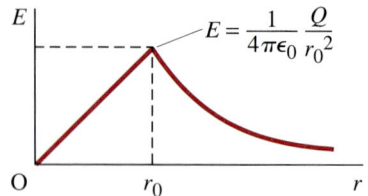

FIGURE 22–13 Magnitude of the electric field as a function of the distance r from the center of a uniformly charged solid sphere.

Electric field inside a uniformly charged nonconducting sphere

EXAMPLE 22–4 Solid sphere of charge. An electric charge Q is distributed uniformly throughout a nonconducting sphere of radius r_0, Fig. 22–12. Determine the electric field (a) outside the sphere $(r > r_0)$ and (b) inside the sphere $(r < r_0)$.

SOLUTION Since the charge is distributed symmetrically in the sphere, the electric field at all points must again be symmetric. **E** depends only on r and is directed radially outward (or inward if $Q < 0$).

(a) For our gaussian surface we choose a sphere of radius r $(r > r_0)$, labeled A_1 in Fig. 22–12. Since E depends only on r, Gauss's law gives, with $Q_{encl} = Q$,

$$\oint \mathbf{E} \cdot d\mathbf{A} = E(4\pi r^2) = \frac{Q}{\epsilon_0}$$

or

$$E = \frac{1}{4\pi\epsilon_0} \frac{Q}{r^2}.$$

Again, the field outside a spherically symmetric distribution of charge is the same as that for a point charge of the same magnitude located at the center of the sphere.

(b) Inside the sphere, we choose for our gaussian surface a concentric sphere of radius r $(r < r_0)$, labeled A_2 in Fig. 22–12. From symmetry, the magnitude of **E** is the same at all points on A_2, and **E** is perpendicular to the surface, so

$$\oint \mathbf{E} \cdot d\mathbf{A} = E(4\pi r^2).$$

We must equate this to Q_{encl}/ϵ_0 where Q_{encl} is the charge enclosed by A_2. Q_{encl} is not the total charge Q but only a portion of it. We define the **charge density**, ρ_E, as the charge per unit volume ($\rho_E = dQ/dV$), and here we are given that $\rho_E =$ constant. So the charge enclosed by the gaussian surface A_2, a sphere of radius r, is

$$Q_{encl} = \left(\frac{\frac{4}{3}\pi r^3 \rho_E}{\frac{4}{3}\pi r_0^3 \rho_E}\right) Q = \frac{r^3}{r_0^3} Q.$$

Hence, from Gauss's law,

$$E(4\pi r^2) = \frac{Q_{encl}}{\epsilon_0} = \frac{r^3}{r_0^3} \frac{Q}{\epsilon_0}$$

or

$$E = \frac{1}{4\pi\epsilon_0} \frac{Q}{r_0^3} r. \qquad [r < r_0]$$

Thus the field increases linearly with r, until $r = r_0$. It then decreases as $1/r^2$, as plotted in Fig. 22–13.

The results above would have been difficult to obtain from Coulomb's law by integrating over the sphere. Using Gauss's law and the symmetry of the situation, this result is obtained rather easily, and shows the great power of Gauss's law. However, its use in this way is limited mainly to cases where the charge distribution has a high degree of symmetry. In such cases, we *choose* a simple surface on which $E =$ constant, so the integration is simple. Gauss's law holds, of course, for any surface. The next two Examples are symmetric cases that we did treat before, using Coulomb's law, but we get the result more easily using Gauss's law.

EXAMPLE 22-5 Long uniform line of charge. A very long straight wire possesses a uniform positive charge per unit length, λ. Calculate the electric field at points near (but outside) the wire, far from the ends.

SOLUTION Because of the symmetry, we expect the field to be directed radially outward and to depend only on the perpendicular distance, r, from the wire. Because of the cylindrical symmetry, the field will be the same at all points on a gaussian surface that is a cylinder with the wire along its axis, Fig. 22–14. **E** is perpendicular to this surface at all points. For Gauss's law, we need a closed surface, so we include the flat ends of the cylinder. Since **E** is parallel to the ends, there is no flux through the ends (the cosine of the angle between **E** and $d\mathbf{A}$ on the ends is $\cos 90° = 0$). So Gauss's law tells us

$$\oint \mathbf{E} \cdot d\mathbf{A} = E(2\pi r l) = \frac{Q_{\text{encl}}}{\epsilon_0} = \frac{\lambda l}{\epsilon_0},$$

where l is the length of our chosen gaussian surface ($l \ll$ length of wire), and $2\pi r$ is its circumference. Hence

$$E = \frac{1}{2\pi\epsilon_0} \frac{\lambda}{r}.$$

This is the same result as we got in Example 21–10 using Coulomb's law (we used x there instead of r), but here it took much less effort. Again we see the great power of Gauss's law.[†]

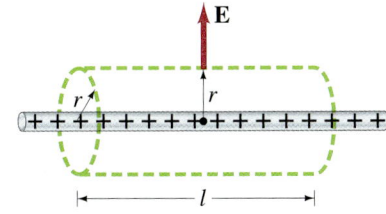

FIGURE 22–14 Calculation of **E** due to a very long line of charge.

EXAMPLE 22-6 Infinite plane of charge. Charge is distributed uniformly, with a surface charge density σ (σ = charge per unit area = dQ/dA), over a very large but very thin nonconducting flat plane surface. Determine the electric field at points near the plane.

SOLUTION We choose as our gaussian surface a small, closed cylinder whose axis is perpendicular to the plane and which extends through the plane as shown in Fig. 22–15. Because of the symmetry, we expect **E** to be directed perpendicular to the plane on both sides as shown, and to be uniform over the end caps of the cylinder, each of whose area is A. Since no flux passes through the curved sides of the cylinder, all the flux is through the two end caps. So Gauss's law gives

$$\oint \mathbf{E} \cdot d\mathbf{A} = 2EA = \frac{Q_{\text{encl}}}{\epsilon_0} = \frac{\sigma A}{\epsilon_0},$$

where $Q_{\text{encl}} = \sigma A$ is the charge enclosed by our gaussian cylinder. The electric field is then

$$E = \frac{\sigma}{2\epsilon_0}.$$

This is the same result we obtained much more laboriously in Chapter 21, Eq. 21–7. The field is uniform for points far from the ends of the plane, and close to its surface.

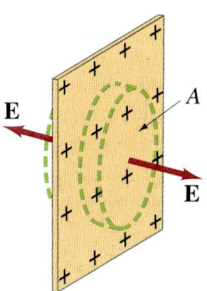

FIGURE 22–15 Calculation of the electric field outside a large uniformly charged nonconducting plane surface.

Electric field near a thin uniformly charged plane

[†] But note that the method of Example 21–10 allows calculation of E also for a short line of charge by using the appropriate limits for the integral, whereas Gauss's law is not readily adapted due to lack of symmetry.

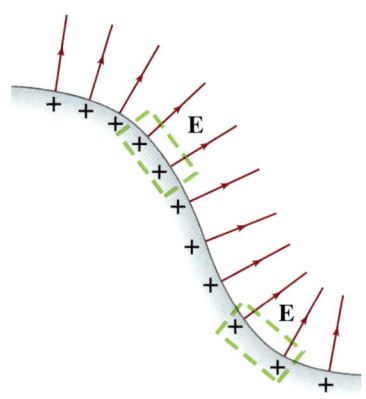

FIGURE 22–16 Electric field near surface of a conductor.

EXAMPLE 22–7 **Electric field near any conducting surface.** Show that the electric field just outside the surface of any good conductor of arbitrary shape is given by

$$E = \frac{\sigma}{\epsilon_0},$$

where σ is the surface charge density on the conductor's surface at that point.

SOLUTION We choose as our gaussian surface a small cylindrical box, as we did in the previous example. We choose the cylinder to be very small in height, so that one of its circular ends is just above the conductor (Fig. 22–16). The other end is just below the conductor's surface, and the sides are perpendicular to it. The electric field is zero inside a conductor and is perpendicular to the surface just outside it (Section 21–9), so electric flux passes only through the outside end of our cylindrical box. We choose the area A (of the flat cylinder end) small enough so that E is essentially uniform over it. Then Gauss's law gives

$$\oint \mathbf{E} \cdot d\mathbf{A} = EA = \frac{Q_{\text{encl}}}{\epsilon_0} = \frac{\sigma A}{\epsilon_0},$$

so that

Electric field at surface of a conductor

$$E = \frac{\sigma}{\epsilon_0}. \qquad \text{[at surface of conductor]} \quad (22\text{–}5)$$

This is a useful result which applies for a conductor of any shape.

When is $E = \sigma/\epsilon_0$ and when is $E = \sigma/2\epsilon_0$

Why is it that the field outside a large plane nonconductor is $E = \sigma/2\epsilon_0$ (Example 22–6) whereas outside a conductor it is $E = \sigma/\epsilon_0$ (Example 22–7)? The reason for the factor of 2 comes not so much from conductor versus nonconductor as from what we mean by the surface charge density σ. For a conductor, the charge lies at the surface and all the electric field lines leave on one side of the surface. For a thin plane nonconductor, the lines leave both sides, Fig. 22–15. If we had a large, thin, flat conducting plane, the charge would accumulate on both surfaces (Fig. 22–17) and the field would emanate from both sides. If we called σ' the surface charge for the plane as a *whole*, each face of the plane would have surface charge $\sigma = \sigma'/2$, and so the result of Example 22–7 would give $E = (\sigma'/2)/\epsilon_0 = \sigma'/2\epsilon_0$, the same as for a nonconducting plane. Normally, however, we use σ to apply to each face of a conducting plane and then we would get $E = \sigma/\epsilon_0$. Thus the factor of 2 between Examples 22–6 and 22–7 comes from two different ways of defining σ.

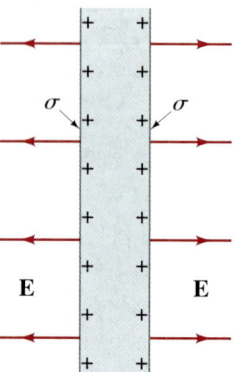

FIGURE 22–17 Thin flat charged conductor with surface charge density σ at each surface, but for the conductor as a whole, the charge density is $\sigma' = 2\sigma$.

We saw in Section 21–9 that in the static situation, the electric field inside any conductor must be zero even if it has a net charge. (Otherwise, the free charges in the conductor would move—until the net force on each, and hence **E**, were zero.) We also mentioned there that any net electric charge on a conductor must all reside on its outer surface. This is readily shown using Gauss's law. Consider any charged conductor of any shape, such as that shown in Fig. 22–18, which carries a net charge Q. We choose the gaussian surface, shown dashed in the diagram, so that it lies just below the surface of the conductor. Our gaussian surface can be arbitrarily close to the surface, but still *inside* the conductor. The electric field is zero at all points on this gaussian surface since it is inside the conductor. Hence, from Gauss's law, Eq. 22–4, the net charge within the surface must be zero. Hence, there can be no net charge within the conductor. Any net charge must lie on the surface of the conductor.

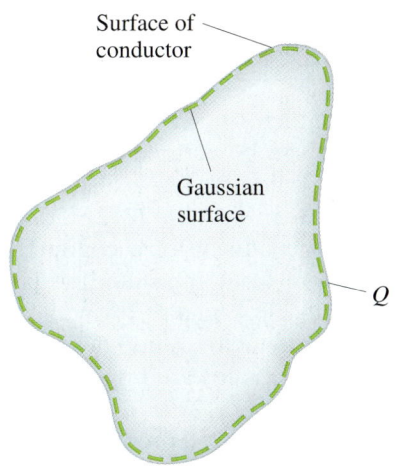

FIGURE 22–18 An insulated charged conductor of arbitrary shape, showing gaussian surface (dashed) just below the surface of the conductor.

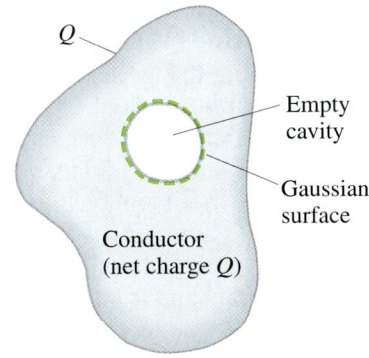

FIGURE 22–19 An empty cavity inside a charged conductor carries zero net charge.

If there is an empty cavity inside a conductor, can charge accumulate on that (inner) surface too? As shown in Fig. 22–19, if we imagine a gaussian surface (shown dashed) just inside the conductor above the cavity, we know that **E** must be zero everywhere on this surface since it is inside the conductor. Hence, by Gauss's law, *there can be no net charge at the surface of the cavity*.

But what if the cavity is not empty and there is a charge inside it?

CONCEPTUAL EXAMPLE 22–8 **Conductor with charge inside a cavity.** Suppose a conductor carries a net charge $+Q$ and contains a cavity, inside of which resides a point charge $+q$. What can you say about the charges on the inner and outer surfaces of the conductor?

RESPONSE As shown in Fig. 22–20, a gaussian surface just inside the conductor surrounding the cavity must contain zero net charge ($E = 0$ in a conductor). Thus a net charge of $-q$ must exist on the cavity surface. The conductor itself carries a net charge $+Q$, so its outer surface must carry a charge equal to $Q + q$.

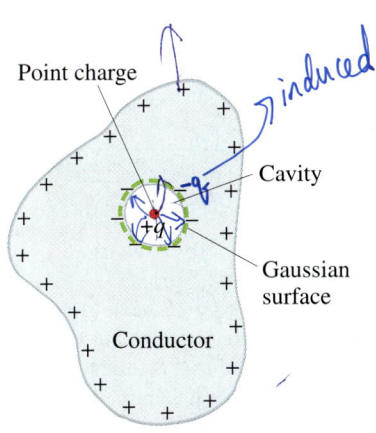

FIGURE 22–20 Example 22–8.

PROBLEM SOLVING — **Gauss's Law for Symmetric Charge Distributions**

1. First identify the symmetry of the charge distribution: spherical, cylindrical, planar. This identification should suggest a gaussian surface for which **E** will be constant and/or zero on all or on parts of the surface: a sphere for spherical symmetry, a cylinder for cylindrical symmetry and a small cylinder or "pillbox" for planar symmetry.

2. Draw the appropriate gaussian surface making sure it passes through the point where you want to know the electric field.

3. Use the symmetry of the charge distribution to determine the direction of **E** on the gaussian surface.

4. Evaluate the flux, $\oint \mathbf{E} \cdot d\mathbf{A}$. With the appropriate gaussian surface, the dot product $\mathbf{E} \cdot d\mathbf{A}$ should be zero or equal to $\pm E\, dA$, with the magnitude of E being constant.

5. Calculate the charge *enclosed* by the gaussian surface. Remember it's the enclosed charge that matters. Ignore all the charge outside the gaussian surface.

6. Equate the flux to the enclosed charge and solve for E.

*22–4 Experimental Basis of Gauss's and Coulomb's Law

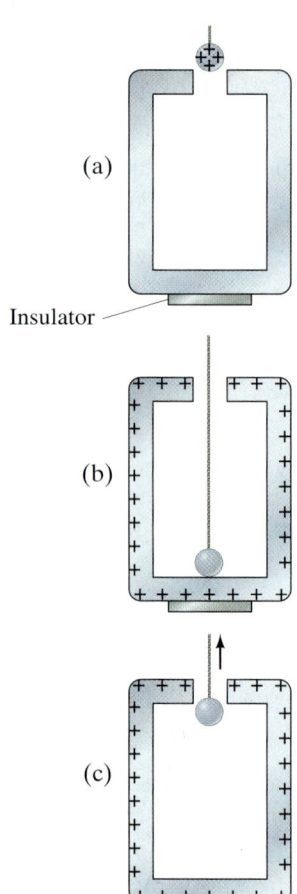

FIGURE 22–21 (a) A charged conductor (metal ball) is lowered into an insulated metal can (a good conductor) carrying zero net charge. (b) The charged ball is touched to the can and all of its charge quickly flows to the outer surface of the can. (c) When the ball is then removed, it is found to carry zero net charge.

Gauss's law predicts that any net charge on a conductor must lie only on its surface. But is this true in real life? Let us see how it can be verified experimentally. And in confirming this prediction of Gauss's law, Coulomb's law is also confirmed since the latter follows from Gauss's law, as we saw in Section 22–2. Indeed, the earliest observation that charge resides only on the outside of a conductor was recorded by Benjamin Franklin some 30 years before Coulomb stated his law.

A simple experiment is illustrated in Fig. 22–21. A metal can with a small opening at the top rests on an insulator. The can, a conductor, is initially uncharged (Fig. 22–21a). A charged metal ball (also a conductor) is lowered by an insulating thread into the can, and is allowed to touch the can (Fig. 22–21b). The ball and can now form a single conductor. Gauss's law, as discussed above, predicts that all the charge will flow to the outer surface of the can. (The flow of charge in such situations does not occur instantaneously, but the time involved is usually negligible.) These predictions are confirmed in experiments by (1) connecting an electroscope to the can, which will show that the can is charged, and (2) connecting an electroscope to the ball after it has been withdrawn from the can (Fig. 22–21c), which will show that the ball carries zero charge.

The precision with which Coulomb's and Gauss's laws hold can be stated quantitatively by writing Coulomb's law as

$$F = k\frac{Q_1 Q_2}{r^{2+\delta}}.$$

For a perfect inverse-square law, $\delta = 0$. The most recent and precise experiments (1971) give $\delta = (2.7 \pm 3.1) \times 10^{-16}$. Thus Coulomb's and Gauss's laws are found to be valid to an extremely high precision!

Summary

The **electric flux** passing through a flat area A for a uniform electric field $\mathbf{E}$ is

$$\Phi_E = \mathbf{E} \cdot \mathbf{A}.$$

If the field is not uniform, the flux is determined from the integral

$$\Phi_E = \int \mathbf{E} \cdot d\mathbf{A}.$$

The direction of the vector $\mathbf{A}$ or $d\mathbf{A}$ is chosen to be perpendicular to the surface whose area is A or dA, and points outward from an enclosed surface. The flux through a surface is proportional to the number of field lines passing through it.

Gauss's law states that the net flux passing out of any closed surface is equal to the net charge Q_{encl} enclosed by the surface divided by ϵ_0:

$$\oint \mathbf{E} \cdot d\mathbf{A} = \frac{Q_{\text{encl}}}{\epsilon_0}.$$

Gauss's law can in principle be used to determine the electric field due to a given charge distribution, but its usefulness is mainly limited to a small number of cases, usually where the charge distribution displays much symmetry. The real importance of Gauss's law is that it is a more general and elegant statement (than Coulomb's law) for the relation between electric charge and electric field. It is one of the basic equations of electromagnetism.

Questions

1. If the electric flux through a closed surface is zero, is the electric field necessarily zero at all points on the surface? What about the converse: If $\mathbf{E} = 0$ at all points on the surface is the flux through the surface zero?
2. Is the electric field $\mathbf{E}$ in Gauss's law, $\oint \mathbf{E} \cdot d\mathbf{A} = Q_{encl}/\epsilon_0$, that due only to the charge Q_{encl}?
3. A point charge is surrounded by a spherical gaussian surface of radius r. If the sphere is replaced by a cube of side r, will Φ_E be larger, smaller, or the same?
4. What can you say about the flux through a closed surface that encloses an electric dipole?
5. The electric field $\mathbf{E}$ is zero at all points on a closed surface; is there necessarily no net charge within the surface? If a surface encloses zero net charge, is the electric field necessarily zero at all points on the surface?
6. Define gravitational flux in analogy to electric flux. Are there "sources" and "sinks" for the gravitational field as there are for the electric field? Discuss.
7. Would Gauss's law be helpful in determining the electric field due to an electric dipole?
8. A spherical basketball (a nonconductor) is given a charge Q distributed uniformly over its surface. What can you say about the electric field inside the ball? A person now steps on the ball, collapsing it, and forcing most of the air out without altering the charge. What can you say about the field inside now?
9. In Example 22–5, it may seem that the electric field calculated is due only to the charge on the wire that is enclosed by the cylinder chosen as our gaussian surface. In fact, the entire charge along the whole length of the wire contributes to the field. Explain how the charge outside the cylindrical gaussian surface of Fig. 22–14 contributes to E at the gaussian surface. [*Hint*: Compare to what the field would be due to a short wire.]
10. Suppose the line of charge in Example 22–5 extended only a short way beyond the ends of the cylinder shown in Fig. 22–14. How would the result of Example 22–5 be altered?
11. A point charge Q is surrounded by a spherical surface of radius r_0, whose center is at Q. Later, the charge is moved to the right a distance $\frac{1}{2}r_0$, but the sphere remains where it was, Fig. 22–22. How is the electric flux Φ_E through the sphere changed? Is the electric field at the surface of the sphere changed? For each "yes" answer, describe the change.

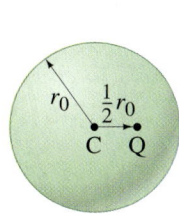
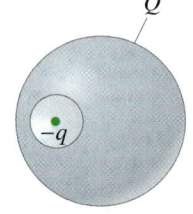

FIGURE 22–22 Question 11.

FIGURE 22–23 Question 12.

12. A conductor carries a net positive charge Q. There is a hollow cavity within the conductor, at whose center is a negative point charge $-q$ (Fig. 22–23). What is the charge on (*a*) the outer surface of the conductor and (*b*) the inner surface of the conductor?
13. A point charge q is placed at the center of the cavity of a thin metal shell which is neutral. Will a charge Q placed outside the shell feel an electric force? Explain.
14. In Fig. 22–24, two objects, O_1 and O_2, have charges $+1.0\,\mu C$ and $-2.0\,\mu C$ respectively, and a third object, O_3, is electrically neutral. (*a*) What is the electric flux through the surface A_1 that encloses all the three objects? (*b*) What is the electric flux through the surface A_2 that encloses the third object only?

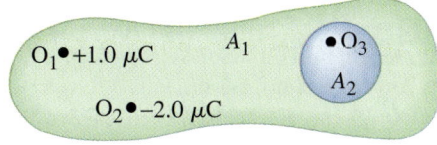

FIGURE 22–24 Question 14.

Problems

Section 22–1

1. (I) A flat circle of radius 15 cm is placed in a uniform electric field of magnitude $5.8 \times 10^2\,\text{N/C}$. What is the electric flux through the circle when its face is (*a*) perpendicular to the field lines, (*b*) at 45° to the field lines, and (*c*) parallel to the field lines.
2. (I) The Earth possesses an electric field of (average) magnitude 150 N/C near its surface. The field points radially inward. Calculate the net electric flux outward through a spherical surface surrounding, and just beyond, the Earth's surface.
3. (II) A cube of side l is placed in a uniform field $E = 6.50 \times 10^3\,\text{N/C}$ with edges parallel to the field lines. What is the net flux through the cube? What is the flux through each of its six faces?
4. (II) A uniform field $\mathbf{E}$ is parallel to the axis of a hollow hemisphere of radius R, Fig. 22–25. (*a*) What is the electric flux through the hemispherical surface? (*b*) What is the result if $\mathbf{E}$ is instead perpendicular to the axis?

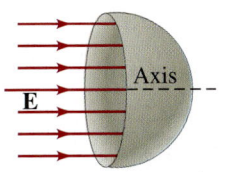

FIGURE 22–25 Problem 4.

Section 22-2

5. (I) The total electric flux from a cubical box 28.0 cm on a side is 1.45×10^3 N·m²/C. What charge is enclosed by the box?

6. (I) Figure 22-26 shows five closed surfaces that surround various charges as indicated. Determine the electric flux through each surface, $S_1, S_2, S_3, S_4,$ and S_5. The surfaces are flat "pill-box" surfaces that extend only slightly above and below the plane of the paper.

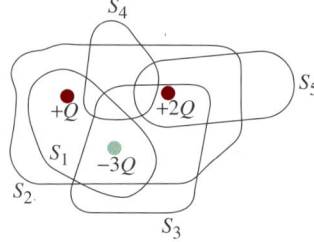

FIGURE 22-26 Problem 6.

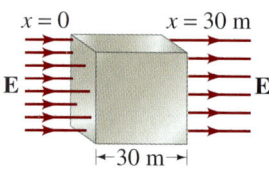

FIGURE 22-27 Problem 7.

7. (II) In a certain region of space, the electric field is constant in direction (say horizontally, in the x direction), but its magnitude decreases from $E = 560$ N/C at $x = 0$ to $E = 410$ N/C at $x = 30$ m. Determine the charge within a cubical box of side $l = 30$ m, where the box is oriented so that four of its sides are parallel to the field lines (Fig. 22-27).

8. (II) A point charge Q is placed at the center of a cube of side l. What is the flux through one face of the cube?

Section 22-3

9. (I) The field just outside a 3.50-cm-radius metal ball is 2.75×10^2 N/C and points toward the ball. What charge resides on the ball?

10. (I) Starting from the result of Example 22-3, show that the electric field just outside a uniformly charged spherical conductor is $E = \sigma/\epsilon_0$, consistent with Example 22-7.

11. (I) A long thin wire, hundreds of meters long, carries a uniformly distributed charge of $-2.8\,\mu\text{C}$ per meter of length. What are the magnitude and direction of the electric field at points (a) 5.0 m and (b) 2.0 m from the wire?

12. (II) A solid metal sphere of radius 3.00 m carries a total charge of $-3.50\,\mu\text{C}$. What is the magnitude of the electric field at a distance from the sphere's center of (a) 0.15 m, (b) 2.90 m, (c) 3.10 m, and (d) 6.00 m? How would the answers differ if the sphere were (e) a thin shell, or (f) a solid nonconductor uniformly charged throughout?

13. (II) A 15.0-cm-diameter nonconducting sphere carries a total charge of $12.0\,\mu\text{C}$ distributed uniformly throughout its volume. Graph the electric field E as a function of the distance r from the center of the sphere from $r = 0$ to $r = 30$ cm.

14. (II) A flat square sheet of thin aluminum foil, 25 cm on a side, carries a uniformly distributed 35 nC charge. What, approximately, is the electric field (a) 1.0 cm above the sheet and (b) 20 m above the sheet?

15. (II) A spherical cavity of radius 4.50 cm is at the center of a metal sphere of radius 18.0 cm. A point charge $Q = 5.50\,\mu\text{C}$ rests at the very center of the cavity, whereas the metal conductor carries no net charge. Determine the electric field at a point (a) 3.0 cm from the center of the cavity and (b) 6.0 cm from the center of the cavity.

16. (II) A point charge Q rests at the center of an uncharged thin spherical conducting shell. What is the electric field E as a function of r (a) for r less than the radius of the shell, (b) inside the shell, and (c) beyond the shell? (d) Does the shell affect the field due to Q alone? Does the charge Q affect the shell?

17. (II) A solid metal cube has a spherical cavity at its center as shown in Fig. 22-28. At the center of the cavity there is a point charge $Q = +8.00\,\mu\text{C}$. The metal cube carries a net charge $q = -7.00\,\mu\text{C}$ (not including Q). Determine (a) the total charge on the surface of the spherical cavity and (b) the total charge on the outer surface of the cube.

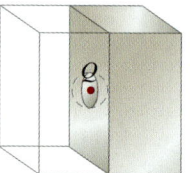

FIGURE 22-28 Problem 17.

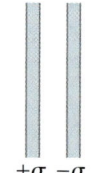

FIGURE 22-29 Problems 18, 19, and 20.

18. (II) Two large, flat metal plates are separated by a distance that is very small compared to their height and width. The conductors are given equal but opposite uniform surface charge densities $\pm\sigma$. Ignore edge effects and use Gauss's law to show that for points far from the edges, (a) the electric field between the plates is $E = \sigma/\epsilon_0$ and (b) that outside the plates on either side the field is zero. (c) How would your results be altered if the two plates were nonconductors? (See Fig. 22-29.)

19. (II) Suppose the two conducting plates in Problem 18 have the *same* sign of charge. What then will be the electric field (a) between them and (b) outside them on either side? (c) What if the planes are nonconducting?

20. (II) The electric field between two square metal plates is 100 N/C. The plates are 1.0 m on a side and are separated by 3.0 cm, as in Fig. 22-29. What is the charge on each plate? Neglect edge effects.

21. (II) Two thin concentric spherical shells of radii r_1 and r_2 $(r_1 < r_2)$ contain uniform surface charge densities σ_1 and σ_2, respectively (see Fig. 22-30). Determine the electric field for (a) $r < r_1$, (b) $r_1 < r < r_2$, and (c) $r > r_2$. (d) Under what conditions will $E = 0$ for $r > r_2$? (e) Under what conditions will $E = 0$ for $r_1 < r < r_2$?

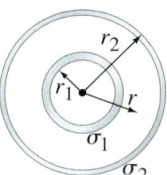

FIGURE 22-30 Two spherical shells (Problem 21).

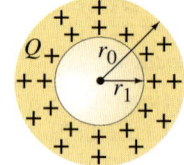

FIGURE 22-31 Problems 22, 23, and 34.

22. (II) Suppose the nonconducting sphere of Example 22-4 has a spherical cavity of radius r_1 centered at the sphere's center (Fig. 22-31). Assuming the charge Q is distributed uniformly in the "shell" (between $r = r_1$ and $r = r_0$), determine the electric field as a function of r for (a) $r < r_1$, (b) $r_1 < r < r_0$, and (c) $r > r_0$.

23. (II) Suppose in Fig. 22-31, Problem 22, there is also a charge q at the center of the cavity. Determine the electric field for (a) $0 < r < r_1$, (b) $r_1 < r < r_0$, and (c) $r > r_0$.

24. (II) Suppose the thick spherical shell of Problem 22 is a conductor. It carries a total net charge Q and at its center there is a point charge q. What total charge is found on (a) the inner surface of the shell and (b) the outer surface of the shell? Determine the electric field for (c) $0 < r < r_1$, (d) $r_1 < r < r_0$, and (e) $r > r_0$.

25. (II) Suppose that at the center of the cavity inside the shell (charge Q) of Fig. 22–11 (and Example 22–3), there is a point charge q ($\neq Q$). Determine the electric field for (a) $r < r_0$ and (b) $r > r_0$. What are your answers if (c) $q = Q$ and (d) $q = -Q$?

26. (II) A spherical rubber balloon carries a total charge Q uniformly distributed on its surface. At $t = 0$ the nonconducting balloon has radius r_0 and the balloon is then slowly blown up so that r increases linearly to $2r_0$ in a time T. Determine the electric field as a function of time (a) just outside the balloon surface and (b) at $r = 4r_0$.

27. (II) A long cylindrical shell of radius R_0 and length L ($R_0 \ll L$) possesses a uniform surface charge density (charge per unit area) σ (Fig. 22–32). Determine the electric field at points (a) outside the cylinder ($r > R_0$) and (b) inside the cylinder ($r < R_0$); assume the points are far from the ends and not too far from the shell ($r \ll L$). (c) Compare to the result for a long line of charge, Example 22–5.

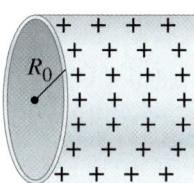

FIGURE 22–32 Problem 27.

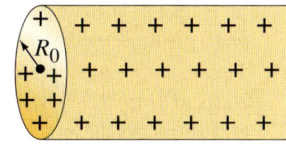

FIGURE 22–33 Problem 28.

28. (II) A very long solid nonconducting cylinder of radius R_0 and length L ($R_0 \ll L$) possesses a uniform volume charge density ρ_E (C/m^3), Fig. 22–33. Determine the electric field at points (a) outside the cylinder ($r > R_0$) and (b) inside the cylinder ($r < R_0$). Do only for points far from the ends and for which $r \ll L$.

29. (II) A thin cylindrical shell of radius R_1 is surrounded by a second concentric cylindrical shell of radius R_2 (Fig. 22–34). The inner shell has a total charge $+Q$ and the outer shell $-Q$. Assuming the length L of the shells is much greater than R_1 or R_2, determine the electric field as a function of r (the perpendicular distance from the common axis of the cylinders) for (a) $r < R_1$, (b) $R_1 < r < R_2$, and (c) $r > R_2$. (d) What is the kinetic energy of an electron if it moves between (and concentric with) the shells in a circular orbit of radius $(R_1 + R_2)/2$?

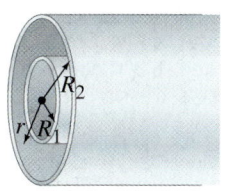

FIGURE 22–34 Problems 29, 30, 31, and 32.

30. (II) In Problem 29 (a) under what conditions will $E = 0$ for $r > R_2$? (b) Under what conditions will $E = 0$ for $R_1 < r < R_2$?

31. (II) A thin cylindrical shell of radius $R_1 = 5.0$ cm is surrounded by a second cylindrical shell of radius $R_2 = 9.0$ cm, as in Fig. 22–34. Both cylinders are 5.0 m long and the inner one carries a total charge $Q_1 = -3.8$ μC and the outer one $Q_2 = +3.2$ μC. For points far from the ends of the cylinders, determine the electric field at a radial distance r from the central axis of (a) 3.0 cm, (b) 6.0 cm, and (c) 12.0 cm.

32. (II) (a) If an electron ($m = 9.1 \times 10^{-31}$ kg) escaped from the surface of the inner cylinder in Problem 31 (Fig. 22–34) with negligible speed, what would be its speed when it reached the outer cylinder? (b) If a proton ($m = 1.67 \times 10^{-27}$ kg) revolves in a circular orbit of radius $r = 6.0$ cm about the axis (i.e., between the cylinders), what must be its speed?

33. (II) A very long solid nonconducting cylinder of radius R_1 is uniformly charged with a charge density ρ_E. It is surrounded by a concentric cylindrical tube of inner radius R_2 and outer radius R_3 as shown in Fig. 22–35, and it too carries a uniform charge density ρ_E. Determine the electric field as a function of the distance r from the center of the cylinders for (a) $0 < r < R_1$, (b) $R_1 < r < R_2$, (c) $R_2 < r < R_3$, and (d) $r > R_3$. (e) If $\rho_E = 15$ μC/m^3 and $R_1 = \frac{1}{2}R_2 = \frac{1}{3}R_3 = 5.0$ cm, plot E as a function of r from $r = 0$ to $r = 20.0$ cm. Assume the cylinders are very long compared to R_3.

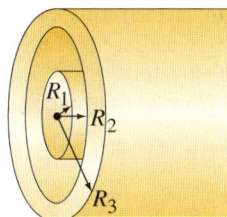

FIGURE 22–35 Problem 33.

34. (III) Suppose the density of charge between r_1 and r_0 of the hollow sphere of Problem 22 (Fig. 22–31) varies as $\rho_E = \rho_0 r_1/r$. Determine the electric field as a function of r for (a) $r < r_1$, (b) $r_1 < r < r_0$, and (c) $r > r_0$. (d) Plot E versus r from $r = 0$ to $r = 2r_0$.

35. (III) A point charge Q is on the axis of a cylinder at its center. The diameter of the cylinder is equal to its length L (Fig. 22–36). What is the total flux through the curved sides of the cylinder? [*Hint*: First calculate the flux through the ends.]

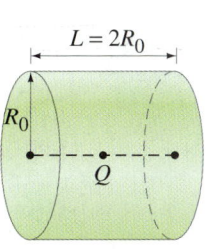

FIGURE 22–36 Problem 35.

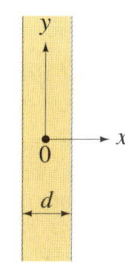

FIGURE 22–37 Problem 36.

36. (III) A flat slab of nonconducting material (Fig. 22–37) carries a uniform charge per unit volume, ρ_E. The slab has thickness d which is small compared to the height and breadth of the slab. Determine the electric field as a function of x (a) inside the slab and (b) outside the slab (at distances much less than the slab's height or breadth). Take the origin at the center of the slab.

General Problems

37. Write Gauss's law for the gravitational field **g** (see Section 6–6).

38. The Earth is surrounded by an electric field, pointing inward at every point, of magnitude $E \approx 150$ N/C near the surface. (a) What is the net charge on the Earth? (b) How many excess electrons per square meter on the Earth's surface does this correspond to?

39. A cube of side l has one corner at the origin of coordinates, and extends along the positive x, y, and z axes. Suppose the electric field in this region is given by $E = (a + by)\mathbf{j}$. Determine the charge inside the cube.

40. A solid nonconducting sphere of radius r_0 has a total charge Q which is distributed according to $\rho_E = br$, where ρ_E is the charge per unit volume, or charge density (C/m^3), and b is a constant. Determine (a) b in terms of Q, (b) the electric field at points inside the sphere, and (c) the electric field at points outside the sphere.

41. A point charge of 3.50 nC is located at the origin and a second charge of -5.00 nC is located on the x axis at $x = 1.50$ m. Calculate the electric flux through a sphere centered at the origin with radius 1.00 m. Repeat the calculation for a sphere of radius 2.00 m.

42. A point charge produces an electric flux of $+500$ N·m²/C through a gaussian sphere of radius 15.0 cm centered on the charge. (a) What is the flux through a gaussian sphere with a radius 35.0 cm? (b) What is the magnitude and sign of the charge?

43. A point charge Q is placed a distance $r_0/2$ above the surface of an imaginary spherical surface of radius r_0 (Fig. 22–38). (a) What is the electric flux through the sphere? (b) What range of values does E have at the surface of the sphere? (c) Is **E** perpendicular to the sphere at all points? (d) Is Gauss's law useful for obtaining E at the surface of the sphere?

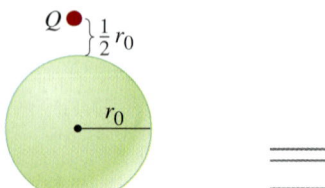

FIGURE 22–38 Problem 43. **FIGURE 22–39** Problem 44.

44. Three large but thin charged sheets are parallel to each other as shown in Fig. 22–39. Sheet I has a total surface charge density of 9.0 nC/m², sheet II a charge of -2.0 nC/m², and sheet III a charge of 5.0 nC/m². What is the force per unit area on each sheet, in N/m²?

45. Neutral hydrogen can be modeled as a positive point charge $+e$ surrounded by a distribution of negative charge with volume density given by $\rho_E(r) = -Ae^{-2r/a_0}$ where $a_0 = 0.53 \times 10^{-10}$ m is called the *Bohr radius*, and A is a constant such that the total amount of negative charge is $-e$. (a) What is the net charge inside a sphere of radius a_0? (b) What is the strength of the electric field at a distance a_0 from the nucleus?

46. A very large thin plane has uniform surface charge density σ. Touching it on the right (Fig. 22–40) is a long and wide slab of thickness d with uniform volume charge density ρ_E. Determine the electric field (a) to the left of the plane, (b) to the right of the slab, and (c) everywhere inside the slab.

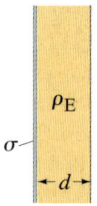 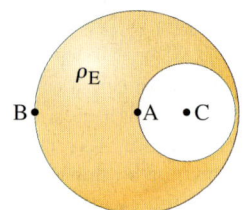

FIGURE 22–40 Problem 46. **FIGURE 22–41** Problem 47.

47. A sphere of radius r_0 carries a volume charge density ρ_E (Fig. 22–41). A spherical cavity of radius $r_0/2$ is then scooped out and left empty, as shown. (a) What is the magnitude and direction of the electric field at point A? (b) What is the direction and magnitude of the electric field at point B?

48. Dry air will break down and generate a spark if the electric field exceeds about 3×10^6 N/C. How much charge could be packed onto a green pea (diameter 0.75 cm) before the pea spontaneously discharges?

49. Three very large sheets are separated by equal distances of 20.0 cm (Fig. 22–42). The first and third sheets are very thin and nonconducting and have surface charge densities of $+5.00\ \mu C/m^2$ and $-5.00\ \mu C/m^2$ respectively. The middle sheet is conducting but has no net charge. (a) What is the electric field inside the middle sheet? What is the electric field (b) between the left and middle sheets, and (c) between the middle and right sheets? (d) What is the charge density on the surface of the middle sheet facing the left sheet, and (e) on the surface facing the right sheet?

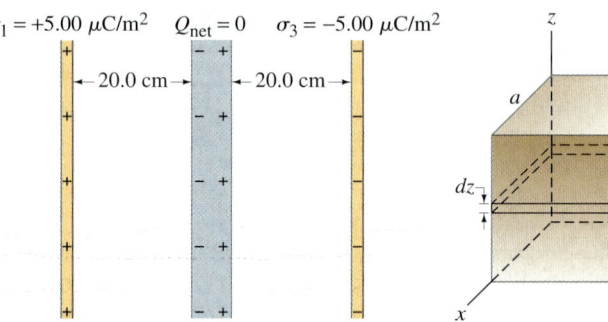

FIGURE 22–42 Problem 49. **FIGURE 22–43** Problem 50.

50. Careful measurement of the electric field in the ground can help provide useful information about charge. In a particular region, a technician has determined that the electric field in a cubical volume, 1.00 m on a side, is

$$E = E_0\left(1 + \frac{z}{a}\right)\hat{\mathbf{i}} + E_0\left(\frac{z}{a}\right)\hat{\mathbf{j}}$$

where $E_0 = 1.00$ N/C and $a = 1.00$ m. The cube has its sides parallel to the coordinate axes, Fig. 22–43. Determine the net charge within the cube.

Lightning: The potential difference (voltage) between clouds and the Earth can become so high that electrons are pulled off atoms of the air by the large electric field. The air becomes a conductor as the ionized atoms and freed electrons flow rapidly, colliding with more atoms, and causing more ionization. The massive flow of charge reduces the potential difference and the "discharge" quickly ceases. The light represents energy released when the ions and electrons recombine to form atoms.

CHAPTER 23

Electric Potential

We saw in Chapters 7 and 8 that the concept of energy was extremely valuable in dealing with mechanical problems. For one thing, energy is a conserved quantity and is thus an important tool for understanding nature. Furthermore, we saw that many problems could be solved using the energy concept even though a detailed knowledge of the forces involved was not possible, or when a calculation involving Newton's laws would have been too difficult.

The energy point of view can be used in electricity, and it is especially useful. It not only extends the law of conservation of energy, but it gives us another way to view electrical phenomena; and it is a tool in solving problems more easily, in many cases, than by using forces and electric fields.

23-1 Electric Potential and Potential Difference

To apply conservation of energy, we need to define electric potential energy as for other types of potential energy (Chapter 8). As we saw, potential energy can be defined only for a conservative force. Recall that the work done by a conservative force in moving an object between any two positions is independent of the path

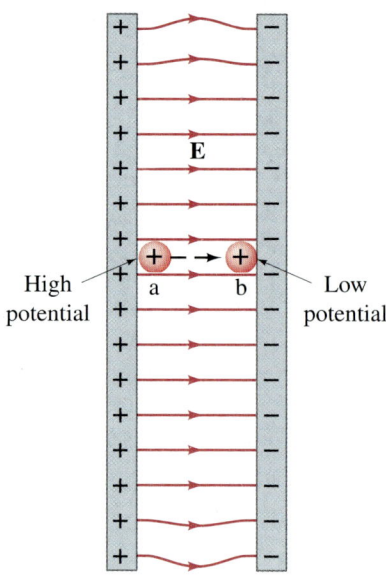

FIGURE 23–1 Work is done by the electric field in moving the positive charge from position a to position b.

Potential is potential energy per unit charge

Potential difference

The volt (1 V = 1 J/C)

Voltage = potential difference

V = 0 chosen arbitrarily

taken. It is easy to see that the electrostatic force between any two charges $(F = kQ_1Q_2/r^2)$ is conservative: the dependence on position is $1/r^2$ just as for the gravitational force, which we saw in Section 8–7 is conservative. Hence the electrostatic force given by Coulomb's law is conservative and we can define potential energy U for it.

We define the change in electric potential energy, $U_b - U_a$, when a charge q moves from some point a to a second point b, as the negative of the work done by the electric force to move the charge from a to b. For example, consider the electric field between two equally but oppositely charged parallel plates whose separation is small compared to their width and height, so the field $\mathbf{E}$ will be uniform over most of the region, Fig. 23–1. Now consider a tiny positive point charge q placed at point a very near the positive plate as shown. This charge q is so small it doesn't affect $\mathbf{E}$. If this charge q at point a is released, the electric force will do work on the charge and accelerate it toward the negative plate. In the process, the charged particle will have its kinetic energy K increased. By conservation of energy, the potential energy will decrease by an equal amount, equal to the negative of the work done by the electric force. In accord with the conservation of energy, electric potential energy is transformed into kinetic energy, and the total energy is conserved. Note that the positive charge q has its greatest potential energy at point a, near the positive plate,[†] so $(U_b - U_a) < 0$. The reverse is true for a negative charge: its potential energy is greatest near the negative plate.

We defined the electric field (Chapter 21) as the force per unit charge. Similarly, it is useful to define the **electric potential** (or simply the **potential** when "electric" is understood) as the *potential energy per unit charge*. Electric potential is given the symbol V. If a positive test charge q has electric potential energy U_a at some point a (relative to some zero potential energy), the electric potential V_a at this point is

$$V_a = \frac{U_a}{q}.$$

As we discussed in Chapter 8, only differences in potential energy are physically meaningful. Hence only the **difference in potential**, or the **potential difference**, between two points a and b (such as between a and b in Fig. 23–1) is measurable. When the electric force does positive work on a charge, the kinetic energy increases and the potential energy decreases. The difference in potential energy, $U_b - U_a$, is equal to the negative of the work, W_{ba}, done by the electric force to move the charge from point a to point b, so the potential difference V_{ba} is

$$V_{ba} = V_b - V_a = \frac{U_b - U_a}{q} = -\frac{W_{ba}}{q}. \quad (23\text{–}1)$$

Note that electric potential, like electric field, does not depend on our test charge q. V depends on the other charges that create the field, not on q; q acquires potential energy by being in the potential V due to the other charges.

We can see from our definition that the positive plate in Fig. 23–1 is at a higher potential than the negative plate. Thus a positively charged object moves naturally from a high potential to a low potential. A negative charge does the reverse.

The unit of electric potential, and of potential difference, is joules/coulomb and is given a special name, the **volt**, in honor of Alessandro Volta (1745–1827; he is best known for having invented the electric battery). The volt is abbreviated V, so 1 V = 1 J/C. Potential difference, since it is measured in volts, is often referred to as **voltage**.

If we wish to speak of the potential, V_a, at some point a, we must be aware that V_a depends on where the potential is chosen to be zero. The zero point for electric potential in a given situation, just as for potential energy, can be chosen arbitrarily

[†] At this point it has its greatest ability to do work (on some other object or system).

since only differences in potential energy can be measured. Often the ground, or a conductor connected directly to the ground (the Earth), is taken as zero potential, and other potentials are given with respect to ground. (Thus, a point where the voltage is 50 V is one where the difference of potential between it and ground is 50 V.) In other cases, as we shall see, we may choose the potential to be zero at an infinite distance ($r = \infty$).

CONCEPTUAL EXAMPLE 23–1 **A negative charge.** Suppose a negative charge, such as an electron, is placed at point b in Fig. 23–1. If the electron is free to move, will its electric potential energy increase or decrease? How will the electric potential change?

RESPONSE An electron placed at point b will move toward the positive plate. (An electron placed at point a would not move.) As the electron moves to the left, its potential energy *decreases* as its kinetic energy gets larger. But note that the electron moves from a point b at low potential to a point a at higher potential: $\Delta V = V_a - V_b > 0$. (The potentials V_a and V_b are due to the charges on the plates, not due to the electron.)

Since the electric potential difference is defined as the potential energy difference per unit charge, then the change in potential energy of a charge q when moved between two points a and b is

$$U_b - U_a = q(V_b - V_a) = qV_{ba}. \quad (23\text{-}2)$$

Electric potential and potential energy

That is, if an object with charge q moves through a potential difference V_{ba}, its potential energy changes by an amount qV_{ba}. For example, if the potential difference between the two plates in Fig. 23–1 is 6 V, then a +1 C charge moved (say by an external force) from b to a will gain $(1\,\text{C})(6\,\text{V}) = 6\,\text{J}$ of electric potential energy. (And it will lose 6 J of electric potential energy if it moves from a to b.) Similarly, a 2-C charge will gain 12 J, and so on. Thus, electric potential difference is a measure of how much energy an electric charge can acquire in a given situation. And, since energy is the ability to do work, the electric potential difference is also a measure of how much work a given charge can do. The exact amount depends both on the potential difference and on the charge.

To better understand electric potential, let's make a comparison to the gravitational case when a rock falls from the top of a cliff. The greater the height, h, of a cliff, the more potential energy ($=mgh$) the rock has at the top of the cliff, relative to the bottom, and the more kinetic energy it will have when it reaches the bottom. The actual amount of kinetic energy it will acquire, and the amount of work it can do, depends both on the height of the cliff and the mass m of the rock. A large rock and a small rock can be at the same height h (Fig. 23–2a) and thus have the same "gravitational potential," but the larger rock has the greater potential energy. The electrical case is similar (Fig. 23–2b): the potential energy change, or the work that can be done, depends both on the potential difference (corresponding to the height of the cliff) and on the charge (corresponding to mass), Eq. 23–2. [But note a significant difference: electric charge comes in two types, + and −, whereas gravitational mass is always +.]

Potential likened to height of a cliff

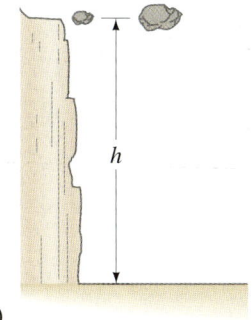
(a)

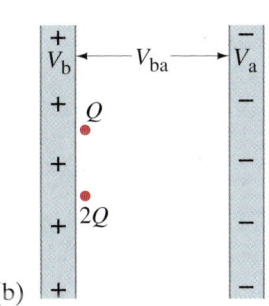
(b)

FIGURE 23–2 (a) Two rocks are at the same height. The larger rock has more potential energy. (b) Two charges have the same electric potential. The $2Q$ charge has more potential energy.

SECTION 23–1   593

TABLE 23–1 Some Typical Voltages

Source	Voltage (approx.)
Thundercloud to ground	10^8 V
High-voltage power line	10^6 V
Power supply for TV tube	10^4 V
Automobile ignition	10^4 V
Household outlet	10^2 V
Automobile battery	12 V
Flashlight battery	1.5 V
Resting potential across nerve membrane	10^{-1} V
Potential changes on skin (EKG and EEG)	10^{-4} V

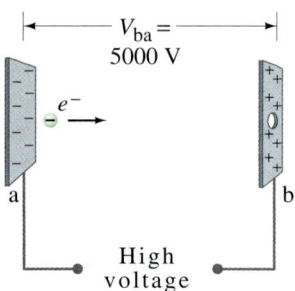

FIGURE 23–3 Electron accelerated in TV picture tube. Example 23–2.

Practical sources of electrical energy such as batteries and electric generators are meant to maintain a potential difference. The actual amount of energy used or transformed depends on how much charge flows. For example, consider an automobile headlight connected to a 12.0-V battery. The amount of energy transformed (into light and thermal energy) is proportional to how much charge flows, which in turn depends on how long the light is on. If over a given period of time 5.0 C of charge flows through the light, the total energy transformed is $(5.0\,\text{C})(12.0\,\text{V}) = 60\,\text{J}$. If the headlight is left on twice as long, 10.0 C of charge will flow and the energy transformed is $(10.0\,\text{C})(12.0\,\text{V}) = 120\,\text{J}$.

Table 23–1 presents some typical voltages.

EXAMPLE 23–2 **Electron in TV tube.** Suppose an electron in the picture tube of a television set is accelerated from rest through a potential difference $V_{ba} = +5000\,\text{V}$ (Fig. 23–3). (a) What is the change in potential energy of the electron? (b) What is the speed of the electron ($m = 9.1 \times 10^{-31}$ kg) as a result of this acceleration? (c) Repeat for a proton ($m = 1.67 \times 10^{-27}$ kg) that accelerates through a potential difference of $V_{ba} = -5000\,\text{V}$.

SOLUTION (a) The charge on an electron is $e = -1.6 \times 10^{-19}$ C. Therefore its change in potential energy (Eq. 23–2) is equal to

$$\Delta U = qV_{ba} = (-1.6 \times 10^{-19}\,\text{C})(+5000\,\text{V})$$
$$= -8.0 \times 10^{-16}\,\text{J}.$$

The minus sign indicates that the potential energy decreases. (The potential difference, V_{ba}, has a positive sign since the final potential is higher than the initial potential; that is, negative electrons are attracted from a negative electrode to a positive one.)

(b) The potential energy lost by the electron becomes kinetic energy ($= K$). From conservation of energy (Eq. 8–9), $\Delta K + \Delta U = 0$, so

$$\Delta K = -\Delta U$$
$$\tfrac{1}{2}mv^2 - 0 = -qV_{ba},$$

where the initial kinetic energy is zero since we assume the electron started from rest. We solve for v and put in the mass of the electron $m = 9.1 \times 10^{-31}$ kg:

$$v = \sqrt{-\frac{2qV_{ba}}{m}}$$
$$= \sqrt{-\frac{2(-1.6 \times 10^{-19}\,\text{C})(5000\,\text{V})}{9.1 \times 10^{-31}\,\text{kg}}}$$
$$= 4.2 \times 10^7\,\text{m/s}.$$

[Note: For such a high speed, which is $\tfrac{1}{7}$ the speed of light, we really should use the theory of relativity, Chapter 37, to get a more precise result.]

(c) The proton has the same magnitude of charge as the electron, though of opposite sign. Hence for the same magnitude of V_{ba} we expect the same change in U, but a lesser speed since the proton's mass is greater. Thus:

$$\Delta U = qV_{ba} = (+1.6 \times 10^{-19}\,\text{C})(-5000\,\text{V}) = -8.0 \times 10^{-16}\,\text{J},$$

and

$$v = \sqrt{-\frac{2qV_{ba}}{m}} = \sqrt{-\frac{2(1.6 \times 10^{-19}\,\text{C})(-5000\,\text{V})}{(1.67 \times 10^{-27}\,\text{kg})}}$$
$$= 9.8 \times 10^5\,\text{m/s}.$$

Note that the energy doesn't depend on the mass, only on the charge and voltage. The speed *does* depend on m.

23–2 Relation Between Electric Potential and Electric Field

The effects of any charge distribution can be described either in terms of electric field or in terms of electric potential. Electric potential is often easier to use because it is a scalar, as compared to electric field which is a vector. There is a crucial connection between the electric potential produced by a given arrangement of charges and the electric field due to those charges, which we now examine.

We start by recalling the relation between a conservative force **F** and the potential energy U associated with that force. As discussed in Section 8–2, the difference in potential energy between any two points in space, a and b, is given by Eq. 8–4:

$$U_b - U_a = -\int_a^b \mathbf{F} \cdot d\mathbf{l},$$

where $d\mathbf{l}$ is an infinitesimal increment of displacement, and the integral is taken along any path in space from point a to point b. For the electrical case, we are more interested in the potential difference, given by Eq. 23–1, $V_{ba} = V_b - V_a = (U_b - U_a)/q$, rather than in the potential energy itself. Also, the electric field **E** at any point in space is defined as the force per unit charge (Eq. 21–3): $\mathbf{E} = \mathbf{F}/q$. Putting these two relations in the above equation gives us

$$V_{ba} = V_b - V_a = -\int_a^b \mathbf{E} \cdot d\mathbf{l}. \tag{23–3}$$

V related to **E**

This is the general relation between electric field and potential difference. See Fig. 23–4. If we are given the electric field due to some arrangement of electric charge, we can use Eq. 23–3 to determine V_{ba}.

A simple special case is when the field is uniform. In Fig. 23–1, for example, a path parallel to the electric field lines from point a at the positive plate to point b at the negative plate gives (since **E** and $d\mathbf{l}$ are in the same direction at each point),

$$V_b - V_a = -\int_a^b \mathbf{E} \cdot d\mathbf{l} = -E\int_a^b dl = -Ed$$

or

$$V_{ba} = -Ed \qquad \text{[only if } E \text{ is uniform]} \tag{23–4}$$

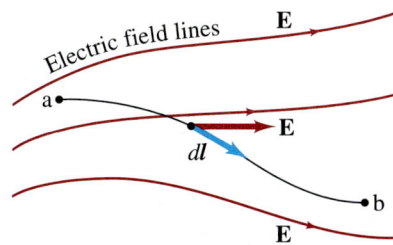

FIGURE 23–4 Integrating $\mathbf{E} \cdot d\mathbf{l}$ from point a to point b in a nonuniform electric field **E**.

where d is the distance, parallel to the field lines, between points a and b. Be careful not to use Eq. 23–4 unless you are sure the electric field is uniform.

From either Eq. 23–3 or 23–4 we can see that the units for electric field intensity can be written as volts per meter (V/m) as well as newtons per coulomb (N/C). These are equivalent in general, since $1\,\text{N/C} = 1\,\text{N·m/C·m} = 1\,\text{J/C·m} = 1\,\text{V/m}$.

EXAMPLE 23–3 Special case: Uniform electric field obtained from voltage. Two parallel plates are charged to a voltage of 50 V. If the separation between the plates is 5.0 cm, calculate the electric field between them, ignoring any fringing.

SOLUTION We have from Eq. 23–4, dealing only with magnitudes for convenience,

$$E = \frac{V_{ba}}{d} = \frac{50\,\text{V}}{0.050\,\text{m}} = 1000\,\text{V/m}.$$

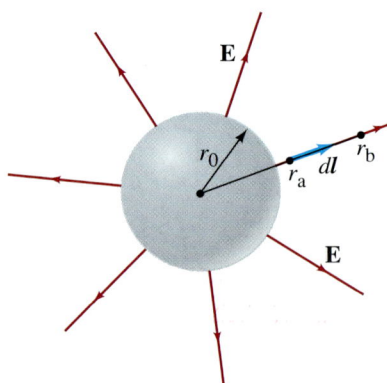

FIGURE 23–5 Example 23–4: integrating $\mathbf{E} \cdot d\mathbf{l}$ for the field outside a spherical conductor.

FIGURE 23–6 (a) E versus r, and (b) V versus r, for a uniformly charged solid conducting sphere of radius r_0 (the charge distributes itself on the surface); r is the distance from the center of the sphere.

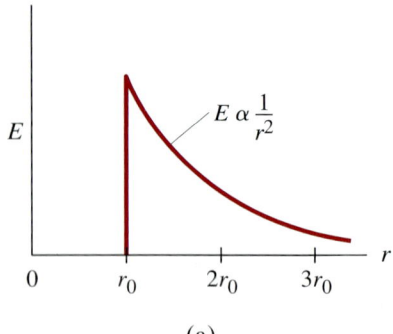

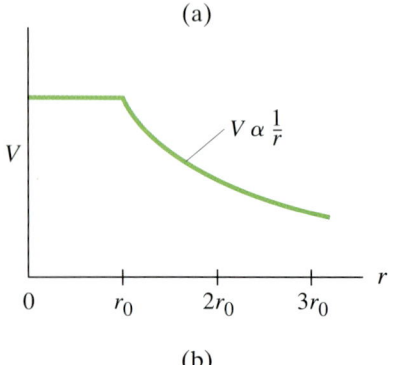

EXAMPLE 23–4 Charged conducting sphere. Determine the potential at a distance r from the center of a uniformly charged conducting sphere of radius r_0 for (a) $r > r_0$, (b) $r = r_0$, (c) $r < r_0$. The total charge on the sphere is Q.

SOLUTION (a) The charge Q is distributed over the surface of the sphere since it is a conductor. We saw in Example 22–3 that the electric field outside a conducting sphere is

$$E = \frac{1}{4\pi\epsilon_0} \frac{Q}{r^2} \qquad [r > r_0]$$

and points radially outward (inward if $Q < 0$). Since we know $\mathbf{E}$, we can use Eq. 23–3 and integrate along a radial line with $d\mathbf{l}$ parallel to $\mathbf{E}$ (Fig. 23–5) between two points which are distances r_a and r_b from the sphere's center:

$$V_b - V_a = -\int_{r_a}^{r_b} \mathbf{E} \cdot d\mathbf{l} = -\frac{Q}{4\pi\epsilon_0} \int_{r_a}^{r_b} \frac{dr}{r^2} = \frac{Q}{4\pi\epsilon_0}\left(\frac{1}{r_b} - \frac{1}{r_a}\right).$$

If we let $V = 0$ for $r = \infty$ (say $V_b = 0$ at $r_b = \infty$), then at any other point r (for $r > r_0$) we have

$$V = \frac{1}{4\pi\epsilon_0}\frac{Q}{r}. \qquad [r > r_0]$$

We will see in the next Section that this same equation applies for the potential a distance r from a single point charge. Thus the electric potential outside a spherical conductor with a uniformly distributed charge is the same as if all the charge were at its center.

(b) As r approaches r_0, we see that

$$V = \frac{1}{4\pi\epsilon_0}\frac{Q}{r_0} \qquad [r = r_0]$$

at the surface of the conductor.

(c) For points within the conductor, $E = 0$. Thus the integral, $\int \mathbf{E} \cdot d\mathbf{l}$, between $r = r_0$ and any point within the conductor gives zero change in V. Hence V is constant within the conductor:

$$V = \frac{1}{4\pi\epsilon_0}\frac{Q}{r_0}. \qquad [r \leq r_0]$$

The whole conductor, not just its surface, is at this same potential. Plots of both E and V as a function of r are shown in Fig. 23–6 for a conducting sphere.

EXAMPLE 23–5 Breakdown voltage. In many kinds of equipment, very high voltages are used. A problem with high voltage is that the air can become ionized due to the high electric fields: free electrons in the air (produced by cosmic rays, for example) can be accelerated by such high fields to speeds sufficient to ionize O_2 and N_2 molecules by collision, knocking out one or more of their electrons. The air then becomes conducting and the high voltage cannot be maintained as charge flows. The breakdown of air occurs for electric fields of about 3×10^6 V/m. (a) Show that the breakdown voltage for a spherical conductor in air is proportional to the radius of the sphere, and (b) estimate the breakdown voltage in air for a sphere of diameter 1.0 cm.

SOLUTION (a) The electric potential at the surface of a spherical conductor of radius r_0 (Example 23–4), and the electric field just outside its surface, are

$$V = \frac{1}{4\pi\epsilon_0}\frac{Q}{r_0} \quad \text{and} \quad E = \frac{1}{4\pi\epsilon_0}\frac{Q}{r_0^2}.$$

Combining these we obtain

$$V = r_0 E. \qquad \text{[at surface of spherical conductor]}$$

(b) For $r_0 = 5 \times 10^{-3}$ m, the breakdown voltage in air is

$$V = (5 \times 10^{-3}\,\text{m})(3 \times 10^6\,\text{V/m}) \approx 15{,}000\,\text{V}.$$

Example 23–5 makes clear why large rounded terminals without sharp edges are used for high-voltage equipment. It also explains why breakdown, or sparks, occur at rough edges or points (regions with small radius of curvature) on a conductor, and why conductors are usually made very smooth.

23–3 Electric Potential Due to Point Charges

The electric potential at a distance r from a single point charge Q can be derived directly from Eq. 23–3, $V_b - V_a = -\int \mathbf{E} \cdot d\mathbf{l}$. The electric field due to a single point charge has magnitude (Eq. 21–4)

$$E = \frac{1}{4\pi\epsilon_0}\frac{Q}{r^2} \quad \text{or} \quad E = k\frac{Q}{r^2}$$

(where $k = 1/4\pi\epsilon_0 = 8.99 \times 10^9 \text{ N}\cdot\text{m}^2/\text{C}^2$), and is directed radially outward from the charge (inward if $Q < 0$). We take the integral in Eq. 23–3 along a (straight) field line (Fig. 23–7) from point a, a distance r_a from Q, to point b, a distance r_b from Q. Then $d\mathbf{l}$ will be parallel to $\mathbf{E}$ and $dl = dr$. Thus

$$V_b - V_a = -\int_{r_a}^{r_b} \mathbf{E} \cdot d\mathbf{l} = -\frac{Q}{4\pi\epsilon_0}\int_{r_a}^{r_b} \frac{1}{r^2}dr = \frac{1}{4\pi\epsilon_0}\left(\frac{Q}{r_b} - \frac{Q}{r_a}\right).$$

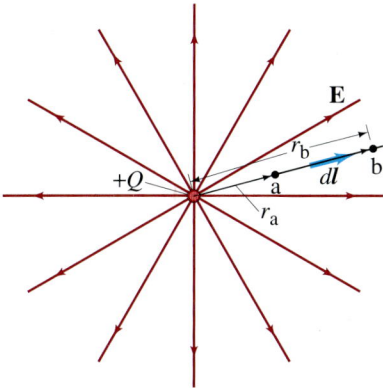

FIGURE 23–7 We integrate Eq. 23–3 along the straight line (shown in black) from point a to point b. The line ab is parallel to a field line.

As mentioned earlier, only differences in potential have physical meaning. We are free, therefore, to choose the value of the potential at some one point to be whatever we please. It is common to choose the potential to be zero at infinity (let $V_b = 0$ at $r_b = \infty$). Then the electric potential V at a distance r from a single point charge is

$$V = \frac{1}{4\pi\epsilon_0}\frac{Q}{r}. \qquad \begin{bmatrix}\text{single point charge;}\\V = 0 \text{ at } r = \infty\end{bmatrix} \quad (23\text{–}5)$$

We can think of V here as representing the absolute potential, where $V = 0$ at $r = \infty$, or we can think of V as the potential difference between r and infinity. Notice that the potential V decreases with the first power of the distance, whereas the electric field (Eq. 21–4) decreases as the *square* of the distance. The potential near a positive charge is large, and it decreases toward zero at very large distances (Fig. 23–8). For a negative charge, the potential is negative and increases toward zero at large distances (Fig. 23–9).

In Example 23–4 we found that the potential due to a uniformly charged sphere is given by the same relation, Eq. 23–5, for points outside the sphere. Thus we see that the potential outside a uniformly charged sphere is the same as if all the charge were concentrated at its center.

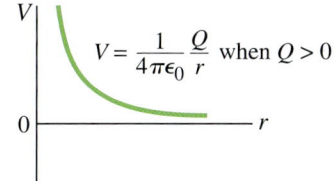

FIGURE 23–8 Potential V as a function of distance r from a single point charge Q when the charge is positive.

FIGURE 23–9 Potential V as a function of distance r from a single point charge Q when the charge is negative.

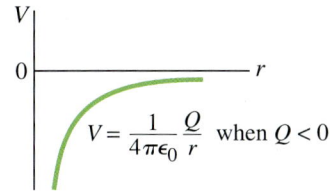

EXAMPLE 23–6 Work to force two + charges close together. What minimum work is required by an external force to bring a charge $q = 3.00\,\mu\text{C}$ from a great distance away (take $r = \infty$) to a point 0.500 m from a charge $Q = 20.0\,\mu\text{C}$?

SOLUTION The work done by the electric field is equal to the negative of the change in potential energy:

$$W = -qV_{ba} = -\frac{q}{4\pi\epsilon_0}\left(\frac{Q}{r_b} - \frac{Q}{r_a}\right),$$

where $r_b = 0.500$ m and $r_a = \infty$. The second term is zero ($1/\infty = 0$), so

$$W = -(3.00 \times 10^{-6}\,\text{C})\frac{(8.99 \times 10^9\,\text{N}\cdot\text{m}^2/\text{C}^2)(2.00 \times 10^{-5}\,\text{C})}{(0.500\,\text{m})} = -1.08\,\text{J}.$$

The electric field does negative work in this case. In order to bring the charge to this point, an *external* force would have to do work $W = +1.08\,\text{J}$, assuming no acceleration of the charges.

Potentials add as scalars (Fields add as vectors)

To determine the electric field surrounding a collection of two or more point charges requires adding up the electric fields due to each charge. Since the electric field is a vector, this can often be a chore. To find the electric potential due to a collection of point charges is far easier, since the electric potential is a scalar, and hence you only need to add numbers together without concern for direction. This is a major advantage in using electric potential. We do have to include the signs of charges, however.

EXAMPLE 23–7 Potential above two charges. Calculate the electric potential at points A and B in Fig. 23–10 due to the two charges shown. (This is the same situation as Example 21–8, Fig. 21–26, where we calculated the electric field at these points.) Assume $V = 0$ at $r = \infty$.

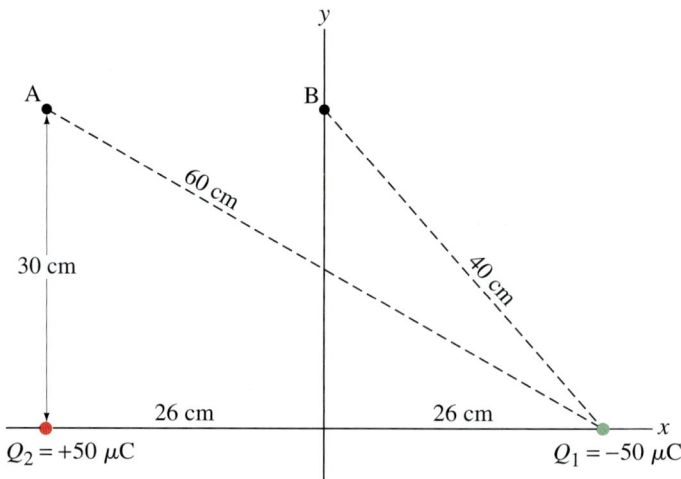

FIGURE 23–10 Example 23–7. (See also Example 21–8, Fig. 21–26).

SOLUTION The potential at point A is the sum of the potentials due to the $+$ and $-$ charges, and we use Eq. 23–5 for each:

$$V_A = V_{A2} + V_{A1}$$
$$= \frac{(9.0 \times 10^9 \, \text{N} \cdot \text{m}^2/\text{C}^2)(5.0 \times 10^{-5} \, \text{C})}{0.30 \, \text{m}}$$
$$+ \frac{(9.0 \times 10^9 \, \text{N} \cdot \text{m}^2/\text{C}^2)(-5.0 \times 10^{-5} \, \text{C})}{0.60 \, \text{m}}$$
$$= 1.50 \times 10^6 \, \text{V} - 0.75 \times 10^6 \, \text{V}$$
$$= 7.5 \times 10^5 \, \text{V}.$$

At point B:
$$V_B = V_{B2} + V_{B1}$$
$$= \frac{(9.0 \times 10^9 \, \text{N} \cdot \text{m}^2/\text{C}^2)(5.0 \times 10^{-5} \, \text{C})}{0.40 \, \text{m}}$$
$$+ \frac{(9.0 \times 10^9 \, \text{N} \cdot \text{m}^2/\text{C}^2)(-5.0 \times 10^{-5} \, \text{C})}{0.40 \, \text{m}}$$
$$= 0 \, \text{V}.$$

It should be clear that the potential will be zero everywhere on the plane equidistant between the two charges. Thus this plane is an equipotential surface with $V = 0$.

A summation like these can be easily performed for any number of point charges.

23–4 Potential Due to Any Charge Distribution

If we know the electric field in a region of space due to any distribution of electric charge, we can determine the difference in potential between two points in the region using Eq. 23-3, $V_{ba} = -\int_a^b \mathbf{E} \cdot d\mathbf{l}$. In many cases we don't know $\mathbf{E}$ as a function of position, and it may be difficult to calculate. We can calculate the potential V due to a given charge distribution in another way, often easier, using the potential due to a single point charge, Eq. 23-5:

$$V = \frac{1}{4\pi\epsilon_0} \frac{Q}{r},$$

where $V = 0$ at $r = \infty$. Then we can sum over all the charges. If we have n individual point charges, the potential at some point a (relative to $V = 0$ at $r = \infty$) is

$$V_a = \sum_{i=1}^n V_i = \frac{1}{4\pi\epsilon_0} \sum_{i=1}^n \frac{Q_i}{r_{ia}}, \qquad (23\text{–}6a)$$

Potential due to many point charges

where r_{ia} is the distance from the ith charge (Q_i) to the point a. (We already used this approach in Example 23-7.) If the charge distribution can be considered continuous, then

$$V = \frac{1}{4\pi\epsilon_0} \int \frac{dq}{r}, \qquad (23\text{–}6b)$$

Potential due to a continuous charge distribution

where r is the distance from a tiny element of charge, dq, to the point where V is being determined.

EXAMPLE 23–8 Potential due to a ring of charge. A thin circular ring of radius R carries a uniformly distributed charge Q. Determine the electric potential at a point P on the axis of the ring a distance x from its center, Fig. 23-11.

SOLUTION Each point on the ring is equidistant from point P, and this distance is $(x^2 + R^2)^{\frac{1}{2}}$. So the potential at P is:

$$V = \frac{1}{4\pi\epsilon_0} \int \frac{dq}{r} = \frac{1}{4\pi\epsilon_0} \frac{1}{(x^2 + R^2)^{\frac{1}{2}}} \int dq = \frac{1}{4\pi\epsilon_0} \frac{Q}{(x^2 + R^2)^{\frac{1}{2}}}.$$

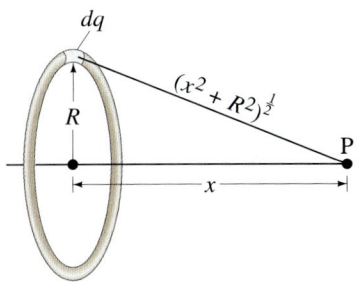

FIGURE 23–11 Calculating the potential at point P, a distance x from the center of a uniform ring of charge (Example 23-8).

Note that for points very far away from the ring, $x \gg R$, this result reduces to $(1/4\pi\epsilon_0)(Q/x)$, the potential of a point charge, as we should expect.

EXAMPLE 23–9 Potential due to a charged disk. A thin flat disk, of radius R, carries a uniformly distributed charge Q, Fig. 23-12. Determine the potential at a point P on the axis of the disk, a distance x from its center.

SOLUTION Divide the disk into thin rings of radius r and thickness dr. The charge Q is distributed uniformly, so the charge contained in each ring is proportional to its area. The disk has area πR^2 and each thin ring has area $dA = (2\pi r)(dr)$. Hence

$$\frac{dq}{Q} = \frac{2\pi r\, dr}{\pi R^2}$$

so

$$dq = Q\frac{(2\pi r)(dr)}{\pi R^2} = \frac{2Qr\, dr}{R^2}.$$

FIGURE 23–12 Calculating the electric potential at point P on the axis of a uniformly charged thin disk. Example 23-9.

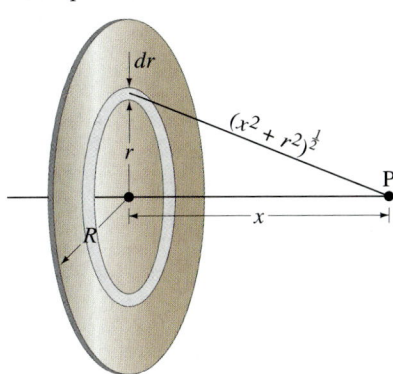

Then the potential at P, using Eq. 23-6b in which r is replaced by $(x^2 + r^2)^{\frac{1}{2}}$, is

$$V = \frac{1}{4\pi\epsilon_0} \int \frac{dq}{(x^2 + r^2)^{\frac{1}{2}}} = \frac{2Q}{4\pi\epsilon_0 R^2} \int_0^R \frac{r\, dr}{(x^2 + r^2)^{\frac{1}{2}}} = \frac{Q}{2\pi\epsilon_0 R^2}(x^2 + r^2)^{\frac{1}{2}}\Big|_{r=0}^{r=R}$$

$$= \frac{Q}{2\pi\epsilon_0 R^2}\left[(x^2 + R^2)^{\frac{1}{2}} - x\right].$$

23–5 Equipotential Surfaces

The electric potential can be represented graphically by drawing **equipotential lines** or, in three dimensions, **equipotential surfaces**. An equipotential surface is one on which all points are at the same potential. That is, the potential difference between any two points on the surface is zero, and no work is required to move a charge from one point to the other. An *equipotential surface must be perpendicular to the electric field* at any point. If this were not so—that is, if there were a component of **E** parallel to the surface—it would require work to move the charge along the surface against this component of **E**; and this would contradict the idea that it is an equipotential surface. This can also be seen from Eq. 23-3, $\Delta V = -\int \mathbf{E} \cdot d\mathbf{l}$. On a surface where V is constant, $\Delta V = 0$, so we must have either $\mathbf{E} = 0$, $d\mathbf{l} = 0$, or $\cos\theta = 0$ where θ is the angle between **E** and $d\mathbf{l}$. Thus in a region where **E** is not zero, the path $d\mathbf{l}$ along an equipotential must have $\cos\theta = 0$, meaning $\theta = 90°$ and **E** is perpendicular to the equipotential.

The fact that the electric field lines and equipotential surfaces are mutually perpendicular helps us locate the equipotentials when the electric field lines are known. In a normal two-dimensional drawing, we show equipotential *lines*, which are the intersections of equipotential surfaces with the plane of the drawing. In Fig. 23–13, a few of the equipotential lines are drawn (dashed green lines) for the electric field (red lines) between two parallel plates at a potential difference of 20 V. The negative plate is arbitrarily chosen to be zero volts and the potential of each equipotential line is indicated. Note that **E** points toward lower values of V. The equipotential lines for the case of two equal but oppositely charged particles are shown in Fig. 23–14 as green dashed lines. Equipotential lines and surfaces, unlike field lines, are always continuous and never end, and so continue beyond the borders of Figs. 23–13 and 23–14.

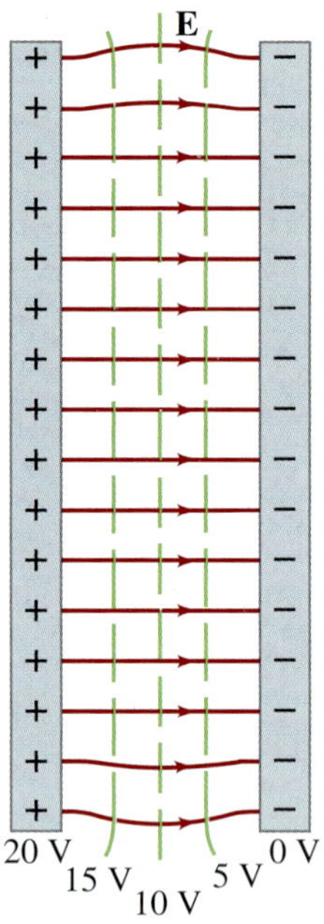

FIGURE 23–13 Equipotential lines (the green dashed lines) between two oppositely charged parallel plates. Note that they are perpendicular to the electric field lines (solid red lines).

We saw in Section 21–9 that there can be no electric field within a conductor in the static case, for otherwise the free electrons would feel a force and would move. Indeed, the entire volume of *a conductor must be entirely at the same potential in the static case*, and the surface of a conductor is then an equipotential surface. (If it weren't, the free electrons at the surface would move, since whenever there is a potential difference between two points, free charges will move.) This is fully consistent with our result, discussed earlier, that the electric field at the surface of a conductor must be perpendicular to the surface.

Conductors are equipotential surfaces

FIGURE 23–14 Equipotential lines (green, dashed) are always perpendicular to the electric field lines (solid red) shown here for two equal but oppositely charged particles.

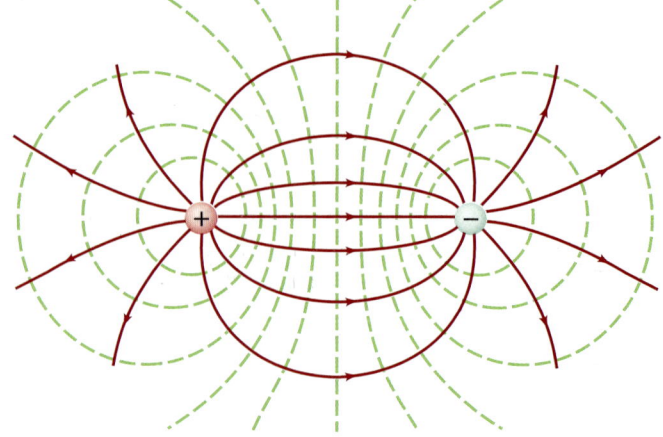

FIGURE 23–15 A topographic map (here, a portion of the Sierra Nevada in California) shows continuous contour lines, each of which is at a fixed height above sea level. Here they are at 80 ft (25 m) intervals. If you walk along one contour line, you neither climb nor descend. If you cross lines, and especially (maximally), if you climb perpendicular to the lines, you will be changing your gravitational potential (rapidly, if the lines are close together).

A useful analogy for equipotential lines is a topographic map: the contour lines are essentially gravitational equipotential lines (Fig. 23–15).

23–6 Electric Dipoles

Two equal point charges Q, of opposite sign, separated by a distance l, are called an **electric dipole**, as we already saw in Section 21–11. Also, the two charges we saw in Figs. 23–10 and 23–14 constitute an electric dipole, and the latter shows the electric field lines and equipotential surfaces for a dipole. Because electric dipoles occur often in physics, as well as in other fields, it is useful to examine them more closely.

Let us calculate the electric potential at an arbitrary point P due to a dipole, as shown in Fig. 23–16. As usual, we take $V = 0$ at $r = \infty$. Since V is the sum of the potentials due to each of the two charges, we have

$$V = \frac{1}{4\pi\epsilon_0}\frac{Q}{r} + \frac{1}{4\pi\epsilon_0}\frac{(-Q)}{r + \Delta r} = \frac{1}{4\pi\epsilon_0}Q\left(\frac{1}{r} - \frac{1}{r + \Delta r}\right) = \frac{Q}{4\pi\epsilon_0}\frac{\Delta r}{r(r + \Delta r)},$$

where r is the distance from P to the positive charge and $r + \Delta r$ is the distance to the negative charge. This equation becomes simpler if we consider points P whose distance from the dipole is much larger than the separation of the two charges—that is, for $r \gg l$. From the diagram we can see that in this case, $\Delta r \approx l \cos\theta$. Since $r \gg \Delta r$, we can neglect Δr in the denominator as compared to r. Therefore, we obtain

$$V = \frac{1}{4\pi\epsilon_0}\frac{Ql\cos\theta}{r^2} = \frac{1}{4\pi\epsilon_0}\frac{p\cos\theta}{r^2} \qquad \text{[dipole; } r \gg l\text{]} \quad (23\text{–}7)$$

where $p = Ql$ is called the **dipole moment**. When θ is between $0°$ and $90°$, V is positive. If θ is between $90°$ and $180°$, V is negative (since $\cos\theta$ is then negative). This makes sense since in the first case P is closer to the positive charge and in the second case it is closer to the negative charge. At $\theta = 90°$, the potential is zero ($\cos 90° = 0$), in agreement with the result of Example 23–7 (point B). From Eq. 23–7, we see that the potential decreases as the *square* of the distance from the dipole, whereas for a single point charge the potential decreases with the first power of the distance (Eq. 23–5). It is not surprising that the potential should fall off faster for a dipole; for when you are far from a dipole, the two equal but opposite charges appear so close together as to tend to neutralize each other.

Table 23–2 gives the dipole moments for several molecules. The $+$ and $-$ signs indicate on which atoms these charges lie. The last two entries are a part of many organic molecules and play an important role in molecular biology.

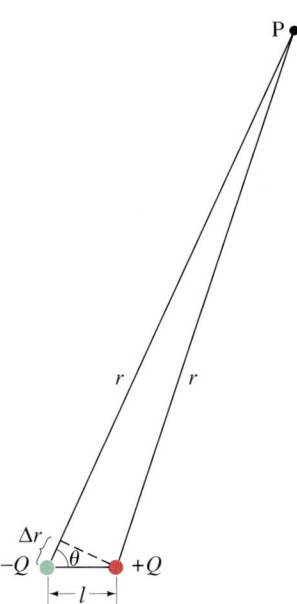

FIGURE 23–16 Electric dipole. Calculation of potential V at point P.

TABLE 23–2 Dipole Moments of Selected Molecules

Molecule	Dipole Moment (C·m)
$H_2^{(+)}O^{(-)}$	6.1×10^{-30}
$H^{(+)}Cl^{(-)}$	3.4×10^{-30}
$N^{(-)}H_3^{(+)}$	5.0×10^{-30}
$>N^{(-)}\!-\!H^{(+)\ddagger}$	$\approx 3.0 \times 10^{-30}$
$>C^{(+)}\!=\!O^{(-)\ddagger}$	$\approx 8.0 \times 10^{-30}$

‡These groups often appear on larger molecules; hence the value for the dipole moment will vary somewhat, depending on the rest of the molecule.

EXAMPLE 23–10 **The C=O group dipole.** The distance between the carbon (+) and oxygen (−) atoms in the group C=O which occurs in many organic molecules is about 1.2×10^{-10} m and the dipole moment of this group is about 8.0×10^{-30} C·m. Calculate (a) the effective charge Q on the C (carbon) and O (oxygen) atoms, and (b) the potential 9.0×10^{-10} m from the dipole along its axis, with the oxygen being the nearer atom (that is, to the left in Fig. 23–16, so $\theta = 180°$). (c) What would the potential be at this point if only the oxygen (O) were charged?

SOLUTION (a) The dipole moment $p = Ql$. Therefore

$$Q = \frac{p}{l} = \frac{8.0 \times 10^{-30}\,\text{C·m}}{1.2 \times 10^{-10}\,\text{m}} = 6.7 \times 10^{-20}\,\text{C}.$$

Although this charge is less than e, the smallest known charge, it is not a charge that can be isolated, but is the effective charge that results from unequal sharing of the electrons, Fig. 23–17.

(b) Since $\theta = 180°$, we have, using Eq. 23–7:

$$V = \frac{1}{4\pi\epsilon_0} \frac{p \cos\theta}{r^2}$$

$$= \frac{(9.0 \times 10^9\,\text{N·m}^2/\text{C}^2)(8.0 \times 10^{-30}\,\text{C·m})(-1.00)}{(9.0 \times 10^{-10}\,\text{m})^2} = -0.089\,\text{V}.$$

(c) If we assume that the oxygen has charge $Q = -6.7 \times 10^{-20}$ C, as in part (a) above, and that the carbon is not charged, we use the formula for a single charge:

$$V = \frac{1}{4\pi\epsilon_0} \frac{Q}{r} = \frac{(9.0 \times 10^9\,\text{N·m}^2/\text{C}^2)(-6.7 \times 10^{-20}\,\text{C})}{9.0 \times 10^{-10}\,\text{m}} = -0.67\,\text{V}.$$

Of course, we expect the potential of a single charge to have greater magnitude than that of a dipole of equal charge at the same distance.

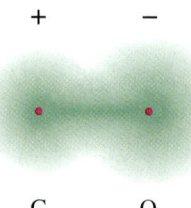

FIGURE 23–17 Electron "cloud" around C and O in the C=O group. The C=O group has a dipole moment because two electrons originally on the carbon atom spend some of their time in the vicinity of the oxygen atom.

23–7 E Determined from V

We can use Eq. 23–3, $V_b - V_a = -\int_a^b \mathbf{E} \cdot d\mathbf{l}$, to determine the difference in potential between two points if the electric field is known in the region between those two points. By inverting Eq. 23–3, we can write the electric field in terms of the potential. Then the electric field can be determined from a knowledge of V. Let us see how to do this.

We write Equation 23–3 in differential form as

$$dV = -\mathbf{E} \cdot d\mathbf{l} = -E_l\,dl,$$

where dV is the infinitesimal difference in potential between two points a distance dl apart, and E_l is the component of the electric field in the direction of the infinitesimal displacement $d\mathbf{l}$. We can then write

$$E_l = -\frac{dV}{dl}. \qquad (23\text{–}8)$$

Thus *the component of the electric field in any direction is equal to the negative of the rate of change of the electric potential with distance in that direction.* The quantity dV/dl is called the **gradient** of V in a particular direction. If the direction is not specified, the term *gradient* refers to that direction in which V changes most rapidly; this would be the direction of $\mathbf{E}$ at that point, so we can write

$$E = -\frac{dV}{dl}. \qquad [\text{if } d\mathbf{l} \parallel \mathbf{E}]$$

If **E** is written as a function of x, y, and z, and we let l refer to the x, y, and z axes, then Eq. 23–8 becomes

$$E_x = -\frac{\partial V}{\partial x}, \qquad E_y = -\frac{\partial V}{\partial y}, \qquad E_z = -\frac{\partial V}{\partial z}. \qquad (23\text{–}9)$$

E related to V

Here, $\partial V/\partial x$ is the "partial derivative" of V with respect to x, with y and z held constant.[†]

EXAMPLE 23–11 E for ring and disk. Determine the electric field at point P on the axis of (a) a circular ring of charge (Fig. 23–11) and (b) a uniformly charged disk (Fig. 23–12).

SOLUTION (a) From Example 23–8,

$$V = \frac{1}{4\pi\epsilon_0} \frac{Q}{(x^2 + R^2)^{\frac{1}{2}}}.$$

Then

$$E_x = -\frac{\partial V}{\partial x} = -\frac{1}{4\pi\epsilon_0} \frac{Qx}{(x^2 + R^2)^{\frac{3}{2}}}$$

$$E_y = E_z = 0.$$

This is the same result we obtained in Example 21–9, but here we didn't have to break the vector electric field into components and then integrate.

(b) From Example 23–9,

$$V = \frac{Q}{2\pi\epsilon_0 R^2}\left[(x^2 + R^2)^{\frac{1}{2}} - x\right],$$

so

$$E_x = -\frac{\partial V}{\partial x} = \frac{Q}{2\pi\epsilon_0 R^2}\left[1 - \frac{x}{(x^2 + R^2)^{\frac{1}{2}}}\right]$$

$$E_y = E_z = 0.$$

For points very close to the disk, $x \ll R$, this can be approximated by

$$E_x \approx \frac{Q}{2\pi\epsilon_0 R^2} = \frac{\sigma}{2\epsilon_0}$$

where $\sigma = Q/\pi R^2$ is the surface charge density. We also obtained these results in Chapter 21, Example 21–11 and Eq. 21–7.

If we compare this last Example with Examples 21–9 and 21–11, we see that here, as for many charge distributions, it is easier to calculate V first, and then **E** from Eq. 23–9, than to calculate **E** due to each charge from Coulomb's law. This is because V due to many charges is a scalar sum, whereas **E** is a vector sum.

23–8 Electrostatic Potential Energy; the Electron Volt

Suppose a point charge q is moved between two points in space, a and b, where the electric potential due to other charges is V_a and V_b, respectively. The change in electrostatic potential energy of q in the field of these other charges is, according to Eq. 23–2,

$$\Delta U = U_b - U_a = q(V_b - V_a) = qV_{ba}.$$

[†]Equation 23–9 can be written as a vector equation,

$$\mathbf{E} = -\text{grad } V = -\boldsymbol{\nabla}V = -\left(\mathbf{i}\frac{\partial}{\partial x} + \mathbf{j}\frac{\partial}{\partial y} + \mathbf{k}\frac{\partial}{\partial z}\right)V$$

where the symbol $\boldsymbol{\nabla}$ is called the *del* or *gradient operator*: $\boldsymbol{\nabla} = \mathbf{i}\frac{\partial}{\partial x} + \mathbf{j}\frac{\partial}{\partial y} + \mathbf{k}\frac{\partial}{\partial z}.$

Now suppose we have a system of several point charges. What is the electrostatic potential energy of the system? It is most convenient to choose the electric potential energy to be zero when the charges are very far (ideally infinitely far) apart. A single point charge, Q_1, in isolation, has no potential energy, because if there are no other charges around, no electric force can be exerted on it. If a second point charge Q_2 is brought close to Q_1, the potential due to Q_1 at the position of this second charge is

$$V = \frac{1}{4\pi\epsilon_0} \frac{Q_1}{r_{12}},$$

where r_{12} is the distance between the two. The potential energy of the two charges, relative to $V = 0$ at $r = \infty$, is

Potential energy, 2 point charges

$$U = Q_2 V = \frac{1}{4\pi\epsilon_0} \frac{Q_1 Q_2}{r_{12}}. \qquad (23\text{–}10)$$

This represents the work that needs to be done by an external force to bring Q_2 from infinity $(V = 0)$ to a distance r_{12} from Q_1. It is also the negative of the work needed to separate them to infinity.

If the system consists of three charges, the total potential energy will be the work needed to bring all three together. Equation 23–10 represents the work needed to bring Q_2 close to Q_1; to bring a third charge Q_3 so that it is a distance r_{13} from Q_1 and r_{23} from Q_2 requires work equal to

$$\frac{1}{4\pi\epsilon_0} \frac{Q_1 Q_3}{r_{13}} + \frac{1}{4\pi\epsilon_0} \frac{Q_2 Q_3}{r_{23}}.$$

So the potential energy of a system of three point charges is

Potential energy, 3 point charges

$$U = \frac{1}{4\pi\epsilon_0} \left(\frac{Q_1 Q_2}{r_{12}} + \frac{Q_1 Q_3}{r_{13}} + \frac{Q_2 Q_3}{r_{23}} \right). \qquad [V = 0 \text{ at } r = \infty]$$

For a system of four charges, the potential energy would contain six such terms, and so on. (Caution must be used when making such sums to avoid double counting of the different pairs.)

The Electron Volt Unit

The joule is a very large unit for dealing with energies of electrons, atoms, or molecules (see Example 23–2), and for this purpose, the unit **electron volt** (eV) is used. One electron volt is defined as the energy acquired by a particle carrying a charge equal to that on an electron $(q = e)$ when it moves through a potential difference of 1 V. Since $e = 1.6 \times 10^{-19}$ C, and since the change in potential energy equals qV, 1 eV is equal to $(1.6 \times 10^{-19} \text{ C})(1.0 \text{ V}) = 1.6 \times 10^{-19}$ J:

Electron volt (energy unit)

$$1 \text{ eV} = 1.6 \times 10^{-19} \text{ J}.$$

An electron that accelerates through a potential difference of 1000 V will lose 1000 eV of potential energy and will thus gain 1000 eV or 1 keV (kilo-electron volt) of kinetic energy. On the other hand, if a particle has a charge equal to twice the charge on the electron $(= 2e = 3.2 \times 10^{-19} \text{ C})$, when it moves through a potential difference of 1000 V its energy will change by 2000 eV.

Although the electron volt is handy for *stating* the energies of molecules and elementary particles, it is not a proper SI unit. For calculations it should be converted to joules using the conversion factor given above. In Example 23–2, for example, the electron acquired a kinetic energy of 8.0×10^{-16} J. We normally would quote this energy as 5000 eV $(= 8.0 \times 10^{-16} \text{ J}/1.6 \times 10^{-19} \text{ J/eV})$. But in determining its speed in SI units we had to use the kinetic energy in J.

EXAMPLE 23–12 Disassembling a hydrogen atom. Calculate the work needed to "disassemble" a hydrogen atom. Assume that the proton and electron are initially separated by a distance equal to the "average" radius of the hydrogen atom in its ground state, 0.529×10^{-10} m, and that they end up an infinite distance apart from each other.

SOLUTION From Eq. 23-10 we have initially

$$U = \frac{1}{4\pi\epsilon_0}\frac{Q_1Q_2}{r} = -\frac{1}{4\pi\epsilon_0}\frac{e^2}{r} = \frac{-(8.99 \times 10^9 \text{ N}\cdot\text{m}^2/\text{C}^2)(1.60 \times 10^{-19}\text{ C})^2}{(0.529 \times 10^{-10}\text{ m})}$$
$$= -27.2(1.60 \times 10^{-19})\text{ J} = -27.2 \text{ eV}.$$

This represents the potential energy. The total energy must include also the kinetic energy of the electron moving in an orbit of radius $r = 0.529 \times 10^{-10}$ m. From $F = ma$ for centripetal acceleration, we have $(1/4\pi\epsilon_0)(e^2/r^2) = mv^2/r$. Then

$$K = \tfrac{1}{2}mv^2 = \tfrac{1}{2}\left(\frac{1}{4\pi\epsilon_0}\right)\frac{e^2}{r}$$

which equals $-\tfrac{1}{2}U$ (as calculated above), so $K = +13.6$ eV. The total energy initially is $E = K + U = 13.6$ eV $- 27.2$ eV $= -13.6$ eV. To separate a stable hydrogen atom into a proton and an electron at rest very far apart ($U = 0$ at $r = \infty$, $K = 0$ because $v = 0$) requires $+13.6$ eV. This is, in fact, the measured ionization energy for hydrogen.

*23-9 Cathode Ray Tube: TV and Computer Monitors, Oscilloscope

An important device that makes use of voltage, and that allows us to "visualize" voltages in the sense of displaying graphically how a voltage changes in time, is the **cathode ray tube (CRT)**. A CRT used in this way is an *oscilloscope*—but an even more common use of a CRT is as the picture tube of television sets and computer monitors.

The operation of a CRT depends first of all on the phenomenon of **thermionic emission**, discovered by Thomas Edison (1847–1931) in the course of experiments on developing the electric light bulb. To understand how thermionic emission occurs, consider two small plates (electrodes) inside an evacuated "bulb" or "tube" as shown in Fig. 23–18, to which is applied a potential difference (by a battery, say). The negative electrode is called the **cathode**, the positive one the **anode**. If the negative cathode is heated (usually by an electric current, as in a lightbulb) so that it becomes hot and glowing, it is found that negative charge leaves the cathode and flows to the positive anode. These negative charges are now called electrons, but originally they were called **cathode rays** since they seemed to come from the cathode.

We can understand how electrons might be "boiled off" a hot metal plate if we treat electrons like molecules in a gas. This makes sense if electrons are relatively free to move about inside a metal, which is consistent with metals being good conductors. However, electrons don't readily escape from the metal. If an electron were to escape outside the metal surface, a net positive charge would remain behind, and this would attract the electron back. To escape, an electron needs a certain minimum kinetic energy, just as molecules in a liquid must have a minimum kinetic energy to "evaporate" into the gaseous state. We saw in Chapter 18 that the average kinetic energy ($\overline{K}$) of molecules in a gas is proportional to the absolute temperature T. We can apply this idea, but only very roughly, to free electrons in a metal as if they made up an "electron gas." Of course, some electrons have more kinetic energy than average and others less. At room temperature, very few electrons would have sufficient energy to escape. At high temperature, $\overline{K}$ is larger and many electrons escape—just as molecules evaporate from liquids, which occurs more readily at high temperatures. Thus, significant thermionic emission occurs only at elevated temperatures.

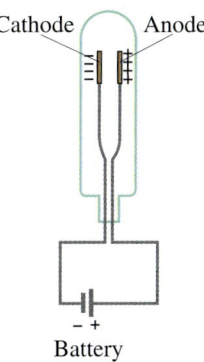

FIGURE 23–18 If the cathode inside the evacuated glass tube is heated to glowing, negatively charged "cathode rays" (electrons) are "boiled off" and flow across to the anode (+) to which they are attracted.

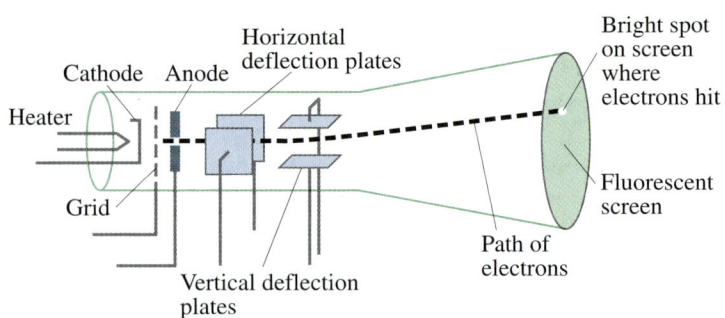

FIGURE 23–19 A cathode-ray tube. Magnetic deflection coils are often used in place of the electric deflection plates. The relative positions of the elements have been exaggerated for clarity.

PHYSICS APPLIED
CRT

PHYSICS APPLIED
TV and computer monitors

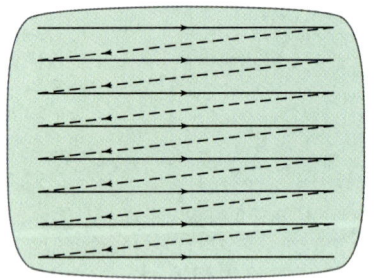

FIGURE 23–20 Electron beam sweeps across a television screen in a succession of horizontal lines.

FIGURE 23–21 An electrocardiogram (ECG) trace displayed on a CRT.

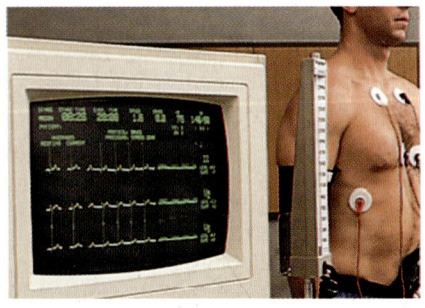

The **cathode-ray tube** (CRT) derives its name from the fact that inside an evacuated glass tube, a beam of cathode rays (electrons) is directed to various parts of a screen to produce a "picture." A simple CRT is diagrammed in Fig. 23–19. Electrons emitted by the heated cathode are accelerated by a high voltage (5,000–50,000 V) applied to the anode. The electrons pass out of this "electron gun" through a small hole in the anode. The inside of the tube face is coated with a fluorescent material that glows when struck by electrons. A tiny bright spot is thus visible where the electron beam strikes the screen. Two horizontal and two vertical plates deflect the beam of electrons when a voltage is applied to them. The electrons are deflected toward whichever plate is positive. By varying the voltage on the deflection plates, the bright spot can be placed at any point on the screen. Today many CRTs use magnetic deflection coils (Chapter 27) instead of electric plates.

In the picture tube or monitor for a computer or television set, the electron beam is made to sweep over the screen in the manner shown in Fig. 23–20. The beam is swept horizontally by the horizontal deflection plates or coils. When the horizontal deflecting field is maximum in one direction, the beam is at one edge of the screen. As the field decreases to zero, the beam moves to the center; and as the field increases to a maximum in the opposite direction, the beam approaches the opposite edge. When the beam reaches this edge, the voltage or current abruptly changes to return the beam to the opposite side of the screen. Simultaneously, the beam is deflected downward slightly by the vertical deflection plates (or coils), and then another horizontal sweep is made. For standard television in the United States, 525 lines constitutes a complete sweep over the entire screen. (High-definition TV provides more than double this number of lines, giving greater picture sharpness.) The complete picture of 525 lines is swept out in $\frac{1}{30}$ s. Actually, a single vertical sweep takes $\frac{1}{60}$ s and involves every other line. The lines in between are then swept out over the next $\frac{1}{60}$ s (called interlacing). We see a picture because the image is retained by the fluorescent screen and by our eyes for about $\frac{1}{20}$ s. The picture we see consists of the varied brightness of the spots on the screen. The brightness at any point is controlled by the grid (a "porous" electrode, such as a wire grid, that allows passage of electrons) which can limit the flow of electrons by means of the voltage applied to it: the more negative this voltage, the more electrons are repelled and the fewer pass through. The voltage on the grid is determined by the video signal (a voltage) sent out by the TV station and received by the TV set. Accompanying this signal are signals that synchronize the grid voltage to the horizontal and vertical sweeps.

An **oscilloscope** is a device for amplifying, measuring, and visually observing an electrical signal (a "signal" is usually a time-varying voltage), especially rapidly changing signals. The signal is displayed on the screen of a CRT. In normal operation, the electron beam is swept horizontally at a uniform rate in time by using a time varying potential difference applied to the horizontal deflection plates. The signal to be displayed is applied, after amplification, to the vertical deflection plates. The visible "trace" on the screen, which could be an ECG (Fig. 23–21), a voltage in an electronic device being repaired, or a signal from an experiment, is thus a plot of the signal voltage (vertically) versus time (horizontally).

Summary

Electric potential is defined as electric potential energy per unit charge. That is, the **electric potential difference** between any two points in space is defined as the difference in potential energy of a test charge q placed at those two points, divided by the charge q:

$$V_{ba} = \frac{U_b - U_a}{q}.$$

Potential difference is measured in volts (1 V = 1 J/C) and is sometimes referred to as **voltage**.

The change in potential energy of a charge q when it moves through a potential difference V_{ba} is

$$\Delta U = qV_{ba}.$$

The potential difference V_{ba} between two points, a and b, is given by the relation

$$V_{ba} = -\int_a^b \mathbf{E} \cdot d\mathbf{l}.$$

Thus V_{ba} can be found in any region where $\mathbf{E}$ is known. If the electric field is uniform, the integral is easy: $V_{ba} = -Ed$, where d is the distance (parallel to the field lines) between the two points.

When V is known, the components of $\mathbf{E}$ can be found from the inverse of the above relation, namely

$$E_x = -\frac{\partial V}{\partial x}, \quad E_y = -\frac{\partial V}{\partial y}, \quad E_z = -\frac{\partial V}{\partial z}.$$

An **equipotential line** or **surface** is all at the same potential, and is perpendicular to the electric field at all points.

The electric potential due to a single point charge Q, relative to zero potential at infinity, is given by

$$V = \frac{1}{4\pi\epsilon_0}\frac{Q}{r}.$$

The potential due to any charge distribution can be obtained by summing (or integrating) over the potentials for all the charges.

Questions

1. If two points are at the same potential, does this mean that no work is done in moving a test charge from one point to the other? Does this imply that no force must be exerted?

2. If a negative charge is initially at rest in an electric field, will it move toward a region of higher potential or lower potential? What about a positive charge? How does the potential energy of the charge change in each of these two instances?

3. State clearly the difference (a) between electric potential and electric field, (b) between electric potential and electric potential energy.

4. An electron is accelerated by a potential difference of, say, 0.10 V. How much greater would its final speed be if it were accelerated with four times as much voltage?

5. Can a particle ever move from a region of low electric potential to one of high potential and yet have its electric potential energy decrease? Explain.

6. If $V = 0$ at a point in space, must $\mathbf{E} = 0$? If $\mathbf{E} = 0$ at some point, must $V = 0$ at that point? Explain. Give examples for each.

7. When dealing with practical devices, we often take the ground (the Earth) to be 0 V. (a) If instead we said the ground was −10 V, how would this affect V and E at other points? (b) Does the fact that the Earth carries a net charge affect the choice of V at its surface?

8. Can two equipotential lines cross? Explain.

9. Draw in a few equipotential lines in Fig. 21–33b and c.

10. What can you say about the electric field in a region of space that has the same potential throughout?

11. A satellite orbits the Earth along a gravitational equipotential line. What shape must the orbit be?

12. Suppose the charged ring of Example 23–8 was not uniformly charged, so that the density of charge was twice as great near the top as near the bottom. Assuming the total charge Q is unchanged, would this affect the potential at point P on the axis (Fig. 23–11)? Would it affect the value of E at that point? Is there a discrepancy here? Explain.

13. Consider a metal conductor in the shape of a football. If it carries a total charge Q, where would you expect the charge density σ to be greatest, at the ends or along the flatter sides? Explain. (*Hint*: Near the surface of a conductor, $E = \sigma/\epsilon_0$.)

14. A conducting sphere carries a charge Q and a second identical conducting sphere is neutral. The two are initially insulated, but then they are placed in contact. (a) What can you say about the potential of each when they are in contact? (b) Will charge flow from one to the other? If so, how much? (c) If the spheres do not have the same radius, how are your answers to parts (a) and (b) altered?

15. At a particular point, the electric field points due north. In what direction(s) will the rate of change of potential be (a) greatest, (b) least, and (c) zero?

16. If you know V at a point in space, can you calculate $\mathbf{E}$ at that point? If you know $\mathbf{E}$ at a point can you calculate V at that point? If not, what else must be known in each case?

17. Equipotential lines are spaced 1.00 V apart. Does the distance between the lines in different regions of space tell you anything about the relative strengths of $\mathbf{E}$ in those regions? If so, what?

18. If the electric field $\mathbf{E}$ is uniform in a region, what can you infer about the electric potential V? If V is uniform in a region of space, what can you infer about $\mathbf{E}$?

19. Is the electric potential energy of two unlike charges positive or negative? What about two like charges? What is the significance of the sign of the potential energy in each case?

Problems

Section 23–1

1. (I) How much work is needed to move a $-7.0\text{-}\mu\text{C}$ charge from ground to a point whose potential is $+6.00$ V higher?

2. (I) How much work is needed to move a proton from a point with a potential of $+100$ V to a point where it is -50 V?

3. (I) How much kinetic energy will an electron gain (in joules) if it falls through a potential difference of 21,000 V in a TV picture tube?

4. (I) An electron acquires 16.4×10^{-16} J of kinetic energy when it is accelerated by an electric field from plate A to plate B. What is the potential difference between the plates, and which plate is at the higher potential?

5. (II) The work done by an external force to move a $-8.10\text{-}\mu\text{C}$ charge from point a to point b is 8.00×10^{-4} J. If the charge was started from rest and had 2.10×10^{-4} J of kinetic energy when it reached point b, what must be the potential difference between a and b?

Section 23–2

6. (I) The electric field between two parallel plates connected to a 45-V battery is 1500 V/m. How far apart are the plates?

7. (I) An electric field of 640 V/m is desired between two parallel plates 11.0 mm apart. How large a voltage should be applied?

8. (I) How strong is the electric field between two parallel plates 5.0 mm apart if the potential difference between them is 110 V?

9. (I) What is the maximum amount of charge that a spherical conductor of radius 5.0 cm can hold in air?

10. (I) What minimum radius must a large conducting sphere of an electrostatic generating machine have if it is to carry 30,000 V without discharge into the air? How much charge will it carry?

11. (II) A uniform electric field $\mathbf{E} = -300\,\text{N/C}\,\hat{\mathbf{i}}$ points in the negative x direction as shown in Fig. 23–22. The x and y coordinates of points A, B, and C are given on the diagram (in meters). Determine the differences in potential (a) V_{BA}, (b) V_{CB}, and (c) V_{CA}.

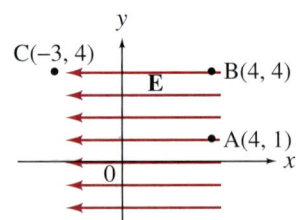

FIGURE 23–22 Problem 11.

12. (II) The electric potential of a very large flat metal plate is V_0. It carries a uniform distribution of charge of surface density $\sigma(\text{C/m}^2)$. Determine V at a distance x from the plate. Consider the point x to be far from the edges and assume x is much smaller than the plate dimensions.

13. (II) The Earth produces an inwardly directed electric field of magnitude 150 V/m near its surface. (a) What is the potential of the Earth's surface relative to $V = 0$ at $r = \infty$? (b) If the potential of the Earth is chosen to be zero, what is the potential at infinity? (Ignore the fact that positive charge in the ionosphere approximately cancels the Earth's net charge; how would this affect your answer?)

14. (II) A 32-cm-diameter conducting sphere is charged to 500 V relative to $V = 0$ at $r = \infty$. (a) What is the surface charge density σ? (b) At what distance will the potential due to the sphere be only 10 V?

15. (II) An insulated spherical conductor of radius r_1 carries a charge Q. A second conducting sphere of radius r_2 and initially uncharged is then connected to the first by a long conducting wire. (a) After the connection, what can you say about the electric potential of each sphere? (b) How much charge is transferred to the second sphere? Assume the connected spheres are far apart compared to their radii. (Why make this assumption?)

16. (II) Determine the difference in potential between two points that are distances R_a and R_b from a very long ($\gg R_a$ or R_b) straight wire carrying a uniform charge per unit length λ.

17. (II) A nonconducting sphere of radius r_0 carries a total charge Q distributed uniformly throughout its volume. Determine the electric potential as a function of the distance r from the center of the sphere for (a) $r > r_0$ and (b) $r < r_0$. Take $V = 0$ at $r = \infty$. (c) Plot V versus r and E versus r.

18. (III) Repeat Problem 17 assuming the charge density ρ_E increases as the square of the distance from the center of the sphere, and $\rho_E = 0$ at the center.

19. (III) A very long conducting cylinder (length L) of radius R_0 ($R_0 \ll L$) carries a uniform surface charge density $\sigma(\text{C/m}^2)$. The cylinder is at an electric potential V_0. What is the potential, at points far from the end, at a distance r from the center of the cylinder? Determine for (a) $r > R_0$ and (b) $r < R_0$. (c) Is $V = 0$ at $r = \infty$ (assume $L = \infty$)? Explain.

20. (III) A hollow spherical conductor, carrying a net charge $+Q$, has inner radius r_1 and outer radius $r_2 = 2r_1$ (Fig. 23–23). At the center of the sphere is a point charge $+Q/2$. (a) Write the electric field strength E in all three regions as a function of r. Then determine the potential as a function of r, the distance from the center, for (b) $r > r_2$, (c) $r_1 < r < r_2$, and (d) $0 < r < r_1$. (e) Plot both V and E as a function of r from $r = 0$ to $r = 2r_2$.

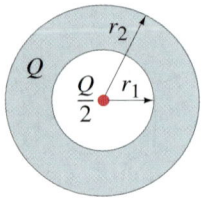

FIGURE 23–23 Problem 20.

Section 23–3

21. (I) (a) What is the electric potential 0.50×10^{-10} m from a proton (charge $+e$)? Let $V = 0$ at $r = \infty$. (b) What is the potential energy of an electron at this point?

22. (I) A charge Q creates an electric potential of $+125$ V at a distance of 15 cm. What is Q?

23. (II) A $+25$-μC charge is placed 6.0 cm from an identical $+25$-μC charge. How much work would be required by an external force to move a $+0.10$-μC test charge from a point midway between them to a point 1.0 cm closer to either of the charges?

24. (II) A 3.0-μC and a -2.0-μC charge are placed 4.0 cm apart. At what points along the line joining them is (a) the electric field zero and (b) the potential zero? Let $V = 0$ at $r = \infty$.

25. (II) How much voltage must be used to accelerate a proton (radius 1.2×10^{-15} m) so that it has sufficient energy to just penetrate a silicon nucleus? A silicon nucleus has a charge of $+14e$ and its radius is about 3.6×10^{-15} m. Assume the potential is that for point charges.

26. (II) Consider point a which is 70 cm north of a -3.8-μC point charge, and point b which is 80 cm west of the charge (Fig. 23–24). Determine (a) $V_{ba} = V_b - V_a$, and (b) $\mathbf{E}_b - \mathbf{E}_a$ (magnitude and direction).

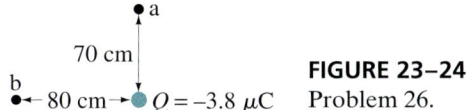

FIGURE 23–24 Problem 26.

27. (II) An electron starts from rest 72.5 cm from a fixed point charge with $Q = -0.125$ μC. How fast will the electron be moving when it is very far away?

28. (II) Two identical $+7.5$-μC point charges are initially spaced 5.5 cm from each other. If they are released at the same instant from rest, how fast will they be moving when they are very far away from each other? Assume they have identical masses of 1.0 mg.

29. (II) Two equal but opposite charges are separated by a distance d, as shown in Fig. 23–25. Determine a formula for $V_{BA} = V_B - V_A$ for points B and A on the line between the charges situated as shown.

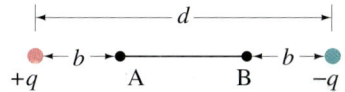

FIGURE 23–25 Problem 29.

Section 23–4

30. (II) Three point charges are arranged at the corners of a square of side L as shown in Fig. 23–26. What is the potential at the fourth corner (point A), taking $V = 0$ at a great distance.

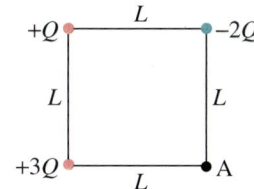

FIGURE 23–26 Problem 30.

31. (II) A flat ring of inner radius R_1 and outer radius R_2, Fig. 23–27, carries a uniform surface charge density σ. Determine the electric potential at points along the axis (the x axis).

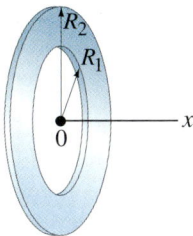

FIGURE 23–27 Problem 31.

32. (II) A thin rod of length $2L$ is centered on the x axis as shown in Fig. 23–28. The rod carries a uniformly distributed charge Q. Determine the potential V as a function of y for points along the y axis. Let $V = 0$ at infinity.

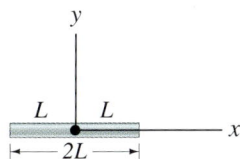

FIGURE 23–28 Problems 32, 33, 34, and 48.

33. (III) Determine the potential $V(x)$ for points along the x axis outside the rod of Fig. 23–28 (Problem 32).

34. (III) The charge on the rod of Fig. 23–28 has a nonuniform linear charge distribution, $\lambda = ax$. Determine the potential V for (a) points along the y axis and (b) points along the x axis outside the rod.

35. (III) Suppose the flat circular disk of Fig. 23–12 (Example 23–9) has a nonuniform surface charge density $\sigma = ar^2$, where r is measured from the center of the disk. Find the potential $V(x)$ at points along the x axis, relative to $V = 0$ at $x = \infty$.

Section 23–5

36. (I) Draw a conductor in the shape of a football. This conductor carries a net negative charge, $-Q$. Draw in a dozen or so electric field lines and equipotential lines.

37. (II) Equipotential surfaces are to be drawn 100 V apart near a very large uniformly charged plate carrying a charge density $\sigma = 0.55$ μC/m^2. How far apart (in space) are the equipotential surfaces?

38. (II) A metal sphere of radius $r_0 = 0.30$ m carries a charge $Q = 0.50$ μC. Equipotential surfaces are to be drawn for 100-V intervals outside the sphere. Determine the radius r of (a) the first, (b) the tenth, and (c) the 100th equipotential from the surface.

Section 23–6

39. (I) An electron and a proton are 0.53×10^{-10} m apart. (a) What is their dipole moment if they are at rest? (b) What is the average dipole moment if the electron revolves about the proton in a circular orbit?

40. (II) Calculate the electric potential due to a dipole whose dipole moment is 4.8×10^{-30} C·m at a point 1.1×10^{-9} m away if this point is far from the dipole, and: (a) along the axis of the dipole nearer the positive charge; (b) 45° above the axis but nearer the positive charge; (c) 45° above the axis but nearer the negative charge.

41. (II) (a) In Example 23–10 part b, calculate the electric potential without using the dipole approximation, Eq. 23–7; that is, don't assume $r \gg l$. (b) What is the percent error in this case when the dipole approximation is used?

42. (III) Show that if an electric dipole is placed in a uniform electric field, then a torque is exerted on it equal to $pE \sin \phi$, where ϕ is the angle between the dipole moment vector and the direction of the electric field as shown in Fig. 23–29. What is the net force on the dipole? How are your answers affected if the field is nonuniform? Note that the dipole moment vector $\mathbf{p}$ is defined so that its magnitude is Ql and its direction is pointing from the negative end to the positive end as shown.

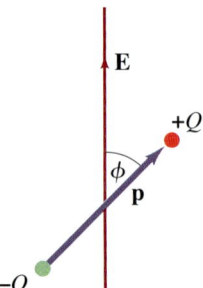

FIGURE 23–29
Problem 42.

43. (III) The dipole moment, considered as a vector, points from the negative to the positive charge. The water molecule, Fig. 23–30, has a dipole moment $\mathbf{p}$ which can be considered as the vector sum of the two dipole moments $\mathbf{p}_1$ and $\mathbf{p}_2$ as shown. The distance between each H and the O is about 0.96×10^{-10} m; the lines joining the center of the O atom with each H atom make an angle of 104° as shown, and the net dipole moment has been measured to be $p = 6.1 \times 10^{-30}$ C·m. (a) Determine the effective charge q on each H atom. (b) Determine the electric potential, far from the molecule, due to each dipole, $\mathbf{p}_1$ and $\mathbf{p}_2$, and show that

$$V = \frac{1}{4\pi\epsilon_0} \frac{p \cos \theta}{r^2},$$

where p is the magnitude of the net dipole moment, $\mathbf{p} = \mathbf{p}_1 + \mathbf{p}_2$, and V is the total potential due to both $\mathbf{p}_1$ and $\mathbf{p}_2$. Take $V = 0$ at $r = \infty$.

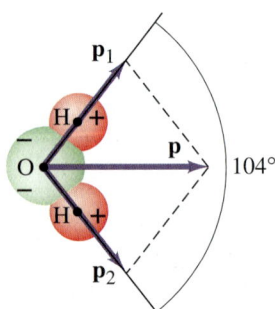

FIGURE 23–30
Problem 43.

Section 23–7

44. (I) What is the potential gradient just outside the surface of a uranium nucleus ($Q = +92e$) whose diameter is about 15×10^{-15} m?

45. (I) Show that the electric field of a single point charge (Eq. 21–4) follows from Eq. 23–5, $V = (1/4\pi\epsilon_0)(Q/r)$.

46. (II) The electric potential in a region of space varies as $V = ay/(b^2 + y^2)$. Determine $\mathbf{E}$.

47. (II) In a certain region of space, the electric potential is given by $V = y^2 + 2xy - 4xyz$. Determine the electric field vector, $\mathbf{E}$, in this region.

48. (III) Use the results of Problems 32 and 33 to determine the electric field due to the uniformly charged rod of Fig. 23–28 for points (a) along the y axis and (b) along the x axis.

Section 23–8

49. (I) Determine the mutual electrostatic potential energy (in electron volts) of two protons in a uranium (^{235}U) nucleus (a) if they are at the surface, on opposite sides of the nucleus, and (b) if one is at the center and the other is at the surface. The diameter of a ^{235}U nucleus is about 15×10^{-15} m. Ignore the other protons for this calculation.

50. (I) How much work must be done to bring three electrons from a great distance apart to within 1.0×10^{-10} m from one another?

51. (I) What potential difference is needed to give a helium nucleus ($Q = 3.2 \times 10^{-19}$ C) 48 keV of kinetic energy?

52. (I) What is the speed of (a) a 3.5 keV (kinetic energy) electron and (b) a 3.5 keV proton?

53. (II) Write the total electrostatic potential energy, U, for (a) four point charges and (b) five point charges. Draw a diagram defining all quantities.

54. (II) An alpha particle (which is a helium nucleus, $Q = +2e$, $m = 6.64 \times 10^{-27}$ kg) is emitted in a radioactive decay with kinetic energy 5.53 MeV. What is its speed?

55. (II) An electron starting from rest acquires 2.0 keV of kinetic energy in moving from point A to point B. (a) How much kinetic energy would a proton acquire, starting from rest at B and moving to point A? (b) Determine the ratio of their speeds at the end of their respective trajectories.

56. (II) Four equal point charges, Q, are fixed at the corners of a square of side b. (a) What is their total electrostatic potential energy? (b) How much potential energy will a fifth charge, Q, have at the center of the square (relative to $V = 0$ at $r = \infty$)? (c) If constrained to remain in that plane, is the fifth charge in stable or unstable equilibrium? If unstable, what maximum kinetic energy could it acquire? (d) Repeat part (c) for a negative ($-Q$) charge.

57. (II) Repeat Problem 56, parts a and b, assuming that two of the charges, on diagonally opposite corners, are replaced by $-Q$ charges.

58. (II) Determine the total electrostatic potential energy of a conducting sphere of radius r_0 that carries a total charge Q distributed uniformly on its surface.

59. (III) Determine the total electrostatic potential energy of a nonconducting sphere of radius r_0 carrying a total charge Q distributed uniformly throughout its volume.

* Section 23–9

* 60. (I) Use the ideal gas as a model to estimate the rms speed of a free electron in a metal at 300 K, and at 2500 K (the typical temperature of the cathode in a CRT).

* 61. (III) In a given CRT, electrons are accelerated horizontally by 15 kV. They then pass through a uniform electric field E for a distance of 2.8 cm which deflects them upward so they reach the top of the screen 22 cm away, 11 cm above the center. Estimate the value of E.

General Problems

62. Sketch the electric field and equipotential lines for two charges of the same sign and magnitude separated by a distance d.

63. By rubbing a nonconducting material, a charge of 10^{-8} C can readily be produced. If this is done to a sphere of radius 10 cm, estimate the potential produced at the surface. Let $V = 0$ at $r = \infty$.

64. If the electrons in a single raindrop, 3 mm in diameter, could be removed from the Earth (without removing the atomic nuclei), by how much would the potential of the Earth increase?

65. A lightning flash transfers 4.0 C of charge and 4.2 MJ of energy to the Earth. (a) Between what potential difference did it travel? (b) How much water could this boil, starting from room temperature?

66. At each corner of a cube of side l there is a point charge Q, Fig. 23–31. (a) What is the potential at the center of the cube ($V = 0$ at $r = \infty$)? (b) What is the potential at each corner due to the other seven charges? (c) What is the total potential energy of this system?

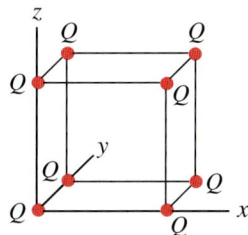

FIGURE 23–31 Problem 66.

67. How much voltage must be used to accelerate a proton so that it has just sufficient energy to touch the surface of an iron nucleus? An iron nucleus has a charge 26 times that of the proton ($= e$) and its radius is about 4.0×10^{-15} m, whereas the proton has a radius of about 1.2×10^{-15} m. Assume the nucleus is spherical and uniformly charged.

* 68. Electrons are accelerated by 14 kV in a CRT. The screen is 30 cm wide and is 34 cm from the 2.6-cm-long deflection plates. Over what range must the horizontally deflecting electric field vary to sweep the beam fully across the screen?

69. Suppose a uniform layer of electrons was held near the surface of the Earth by the Earth's gravity. What is the maximum number of electrons that could be held in this manner? Ignore all other electric charges and fields except those of the electrons themselves.

70. Four point charges are located at the corners of a square that is 8.0 cm on a side. The charges, going in rotation around the square, are Q, $2Q$, $-3Q$, and $2Q$, where $Q = 4.8\,\mu\text{C}$ (Fig. 23–32). What is the total electric potential energy stored in the system, relative to $U = 0$ at infinite separation?

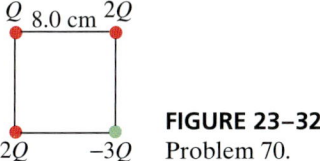

FIGURE 23–32 Problem 70.

71. In a television picture tube, electrons are accelerated by thousands of volts through a vacuum. If a television set were laid on its back, would electrons be able to move upward against the force of gravity? What potential difference, acting over a distance of 3.0 cm, would be needed to balance the downward force of gravity so that an electron would remain stationary? Assume that the electric field is uniform.

72. A proton starting from rest acquires 5.2 keV of kinetic energy in moving from point P to point Q. (a) How much kinetic energy would an electron acquire, starting from rest at Q and moving to point P? (b) Determine the ratio of their speeds at the end of their respective trajectories.

73. In a photocell, ultraviolet (UV) light provides enough energy to some electrons in barium metal to eject them from the surface at high speed. See Fig. 23–33. To measure the maximum energy of the electrons, another plate above the barium surface is kept at a negative enough potential that the emitted electrons are slowed down and stopped, and return to the barium surface. If the plate voltage is -3.02 V (compared to the barium) when the fastest electrons are stopped, what was the speed of these electrons when they were emitted?

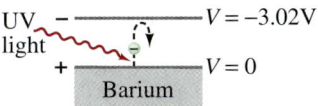

FIGURE 23–33 Problem 73.

74. Near the surface of the Earth there is an electric field of about 150 V/m which points downward. Two identical balls with mass $m = 0.540$ kg are dropped from a height of 2.00 m, but one of the balls is positively charged with $q_1 = 550\,\mu\text{C}$, and the second is negatively charged with $q_2 = -550\,\mu\text{C}$. Use conservation of energy to determine the difference in the speed of the two balls when they hit the ground. (Neglect air resistance.)

75. Three charges are at the corners of an equilateral triangle (side L) as shown in Fig. 23–34. Determine the potential at the midpoint of each of the sides.

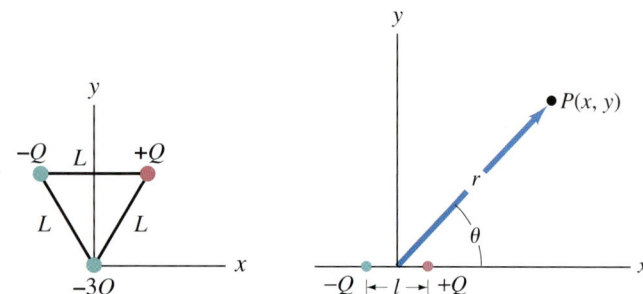

FIGURE 23–34 Problem 75.   FIGURE 23–35 Problem 76.

76. Determine the components of the electric field, E_x and E_y, at any point P in the xy plane due to a dipole, Fig. 23–35, starting with Eq. 23–7. Assume $r = (x^2 + y^2)^{\frac{1}{2}} \gg l$.

77. (a) What is the electric potential a distance of 2.5×10^{-15} m away from a proton? Let $V = 0$ at $r = \infty$. (b) What is the electric potential energy of a system that consists of two protons 2.5×10^{-15} m apart—as might occur inside a typical nucleus?

78. A thin flat nonconducting disk, with radius R and charge Q, has a hole with a radius $R/2$ in its center. Find the electric potential $V(x)$ at points along the symmetry (x) axis of the disk (a line perpendicular to the disk, passing through its center). Let $V = 0$ at $x = \infty$.

79. A Geiger counter is used to detect charged particles emitted by radioactive nuclei. It consists of a thin, positively charged central wire of radius R_a surrounded by a concentric conducting cylinder of radius R_b with an equal negative charge (Fig. 23–36). The charge per unit length on the inner wire is λ (units C/m). The cylindrical assembly is filled with low-pressure inert gas. Charged particles ionize some of these gas atoms; the resulting free electrons are attracted toward the central wire. If the radial electric field is strong enough, the freed electrons gain enough energy to ionize other atoms, causing an "avalanche" of electrons to strike the central wire, generating an electric "signal." Find the expression for the electric field between the wire and the cylinder, and show that the potential difference between R_a and R_b is

$$V_a - V_b = \Delta V = \frac{\lambda}{2\pi\epsilon_0 \ln(R_b/R_a)}.$$

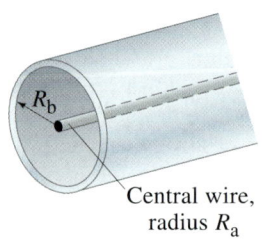

FIGURE 23–36 Problem 79.

80. A Van de Graaff generator (Fig. 23–37) can develop a very large potential difference, even millions of volts. Electrons are pulled off the belt by the high voltage pointed electrode at A, leaving the belt positively charged. The belt carries the positive charge up inside the spherical shell where it passes to the pointed conductor at B, and races to the outer surface of the conducting sphere. As more charge is brought up, the sphere reaches extremely high voltage. Consider a Van de Graaff generator with a sphere of radius 0.15 m. (a) What is the electric potential on the surface of the sphere when electrical breakdown occurs? (Assume that $V = 0$ at $r = \infty$.) (b) What is the charge on the sphere for the potential found in part (a)?

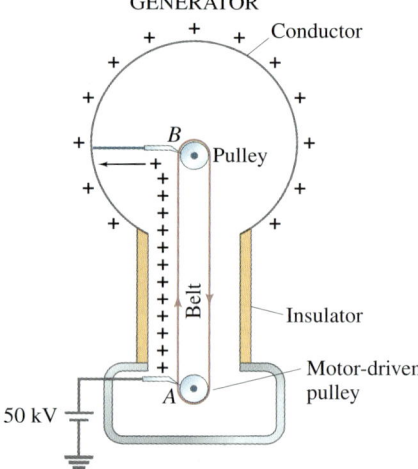

FIGURE 23–37 Problem 80.

81. Show that if two dipoles with dipole moments p_1 and p_2 are in line with one another (Fig. 23–38), the potential energy of one in the presence of the other (their "interaction energy") is given by

$$U = -\frac{1}{2\pi\epsilon_0} \frac{p_1 p_2}{r^3}$$

where r is the distance between the two dipoles. Assume that r is much greater than the length of either dipole.

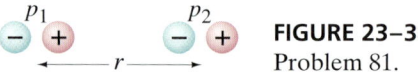

FIGURE 23–38 Problem 81.

82. Show that the electrostatic potential energy of two dipoles in a plane as shown in Fig. 23–39 is given by

$$U = \frac{1}{4\pi\epsilon_0} \frac{p_1 p_2}{r^3} \left[\cos(\theta_1 - \theta_2) - 3\cos\theta_1 \cos\theta_2 \right].$$

Assume r is much larger than the length of each dipole. The vector dipole moments, $\mathbf{p}_1$ and $\mathbf{p}_2$, point from the negative charge toward the positive charge of the dipole.

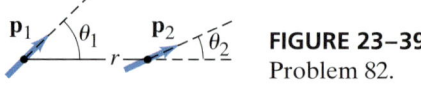

FIGURE 23–39 Problem 82.

83. A nonconducting sphere of radius r_2 contains a concentric spherical cavity of radius r_1. The material between r_1 and r_2 carries a uniform charge density ρ_E (C/m³). Determine the electric potential V, relative to $V = 0$ at $r = \infty$, as a function of the distance r from the center for (a) $r > r_2$, (b) $r_1 < r < r_2$, and (c) $r < r_1$. Is V continuous at r_1 and r_2?

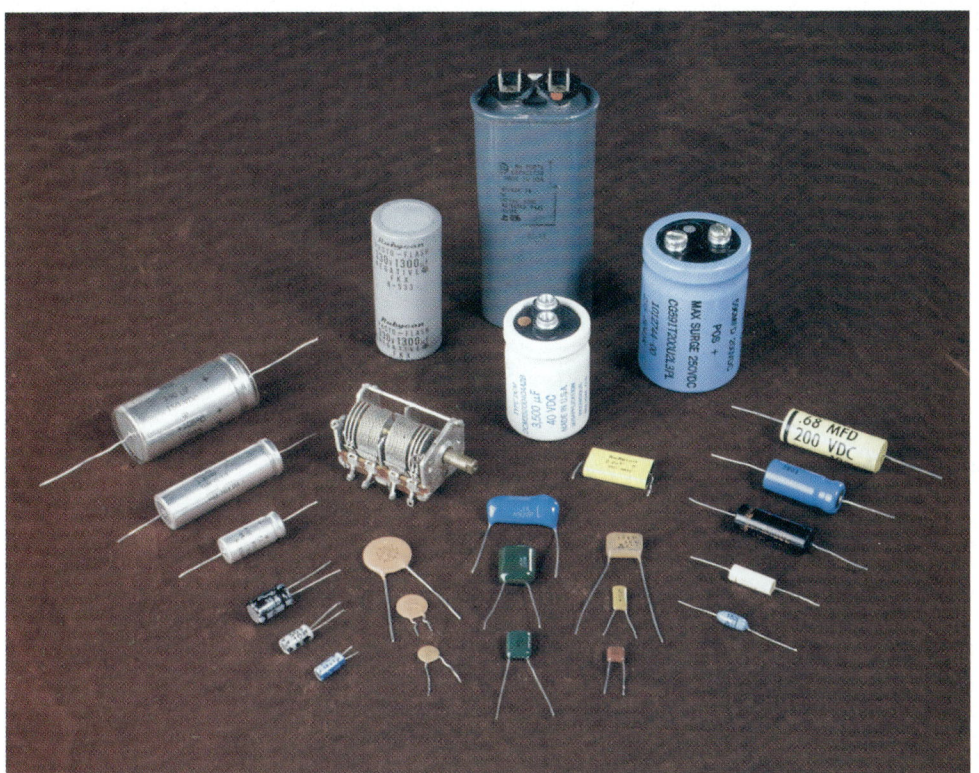

Capacitors come in a wide range of sizes and shapes, only a few of which are shown here. A capacitor is basically two conductors that do not touch, and which therefore can store charge of opposite sign on its two conductors. Capacitors are used in a wide variety of circuits, as we shall see in this and later chapters.

Capacitance, Dielectrics, Electric Energy Storage

This chapter will complete our study of electrostatics. It deals first of all with an important device, the capacitor, which is used in nearly all electronic circuits. We will also discuss electric energy storage and the effects of an insulator, or dielectric, on electric fields and potential differences.

24–1 Capacitors

A **capacitor**, sometimes called a *condenser*, is a device that can store electric charge, and usually consists of two conducting objects (usually plates or sheets) placed near each other but not touching. Capacitors are widely used in electronic circuits: they store charge for later use, as in a camera flash, and as energy backup in computers if the power fails; capacitors block surges of charge and energy to protect circuits; they form part of the tuner of a radio; very tiny capacitors serve as memory for the "ones" and "zeroes" of the binary code in the random access memory (RAM) of computers; and capacitors serve many other applications, some of which we will discuss.

→ **PHYSICS APPLIED**

Uses of capacitors

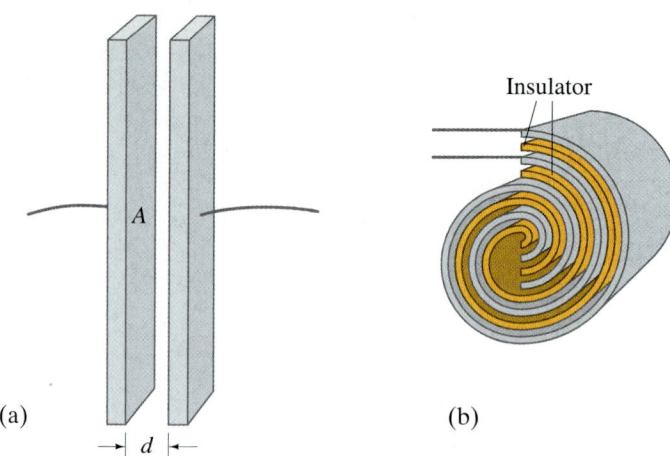

FIGURE 24–1 Capacitors: Diagrams of (a) parallel plate, (b) cylindrically shaped (rolled up parallel plate).

A simple capacitor consists of a pair of parallel plates of area A separated by a small distance d (Fig. 24–1a). Often the two plates are rolled into the form of a cylinder with paper or other insulator separating the plates, Fig. 24–1b. On the previous page is a photo of some actual capacitors used for various applications. In a diagram, a capacitor is represented by the symbol

   [capacitor symbol]

Another symbol for a capacitor you may encounter is ⊣⊦. A battery, which is a source of voltage, is indicated by the symbol:

⊣⊦   [battery symbol]

with unequal arms.

If a voltage is applied to a capacitor, say by connecting the capacitor to a battery as in Fig. 24–2, it quickly becomes charged. One plate acquires a negative charge, the other an equal amount of positive charge. Each battery terminal, connecting wire, and plate of the capacitor are conductors and at the same potential; hence the full battery voltage appears across the capacitor. For a given capacitor, it is found that the amount of charge Q acquired by each plate is proportional to the magnitude of the potential difference V_{ba} between them:

$$Q = CV_{ba}. \tag{24–1}$$

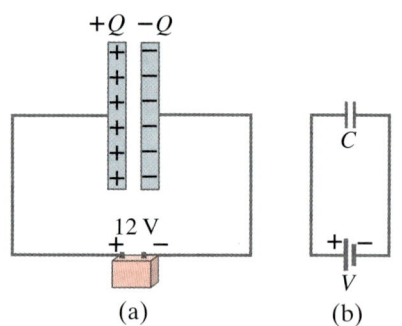

FIGURE 24–2 (a) Parallel-plate capacitor connected to a battery. (b) Same circuit shown using symbols.

Capacitance

Unit is farad $(1\,\text{F} = 1\,\text{C/V})$

The constant of proportionality, C, in the above relation is called the **capacitance** of the capacitor. The unit of capacitance is coulombs per volt and this unit is called a **farad** (F). Most capacitors have capacitance in the range of 1 pF (picofarad $= 10^{-12}\,\text{F}$) to 1 µF (microfarad $= 10^{-6}\,\text{F}$). The relation, Eq. 24–1, was first suggested by Volta in the late eighteenth century. The capacitance C does not in general depend on Q or V. Its value depends only on the size, shape, and relative position of the two conductors, and also on the material that separates them.

24–2 Determination of Capacitance

The capacitance of a given capacitor can be determined experimentally directly from Eq. 24–1, by measuring the charge Q on either conductor for a given potential difference V_{ba}.

For capacitors whose geometry is simple, we can determine C analytically, and in this Section we assume the conductors are separated by a vacuum or air. To illustrate this, we now determine C for a parallel-plate capacitor, Fig. 24–3. Each plate has area A and the two plates are separated by a distance d. We assume d is small compared to the dimensions of each plate so that the electric field $\mathbf{E}$ is uniform between them and we can ignore fringing (lines of $\mathbf{E}$ not straight) at the edges. We saw earlier (Example 21–12) that the electric field between two closely spaced parallel plates has magnitude $E = \sigma/\epsilon_0$ and its direction is perpendicular to the plates. Since σ is the charge per unit area, $\sigma = Q/A$, then the field between the plates is

$$E = \frac{Q}{\epsilon_0 A}.$$

The relation between electric field and electric potential, as given by Eq. 23–4, is

$$V_{ba} = -\int_a^b \mathbf{E} \cdot d\mathbf{l}.$$

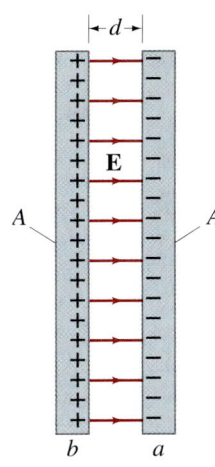

FIGURE 24–3 Parallel-plate capacitor, each of whose plates has area A. Fringing of the field is ignored.

We can take the line integral along a path antiparallel to the field lines, from one plate to the other; then $\theta = 180°$ and $\cos 180° = -1$, so

$$V_{ba} = V_b - V_a = -\int_a^b E\, dl \cos 180° = +\int_a^b E\, dl = \frac{Q}{\epsilon_0 A}\int_a^b dl = \frac{Qd}{\epsilon_0 A}.$$

This relates Q to V_{ba}, and from it we can get the capacitance C in terms of the geometry of the plates:

$$C = \frac{Q}{V_{ba}} = \epsilon_0 \frac{A}{d}. \qquad \text{[parallel-plate capacitor]} \quad (24\text{--}2)$$

Parallel plate capacitance formula

This relation makes sense intuitively: a larger area A means that for a given number of charges (electrons), there will be less repulsion between them (they're farther apart), so we expect that more charge can be held on each plate. And a greater separation d means the charge on each plate exerts less attractive force on the other plate, so less charge is drawn from the battery, and the capacitance is less.

Note also from Eq. 24–2 that the value of C does not depend on Q or V_{ba}, so Q is predicted to be proportional to V_{ba} as is found experimentally.

EXAMPLE 24–1 Capacitor calculations. (a) Calculate the capacitance of a capacitor whose plates are 20 cm × 3.0 cm and are separated by a 1.0-mm air gap. (b) What is the charge on each plate if the capacitor is connected to a 12-V battery? (c) What is the electric field between the plates? (d) Estimate the area of the plates needed to achieve a capacitance of 1 F, given the same air gap d.

SOLUTION (a) The area $A = (20 \times 10^{-2}\,\text{m})(3.0 \times 10^{-2}\,\text{m}) = 6.0 \times 10^{-3}\,\text{m}^2$. The capacitance C is then

$$C = \epsilon_0 \frac{A}{d} = (8.85 \times 10^{-12}\,\text{C}^2/\text{N}\cdot\text{m}^2)\frac{6.0 \times 10^{-3}\,\text{m}^2}{1.0 \times 10^{-3}\,\text{m}} = 53\,\text{pF}.$$

(b) The charge on each plate (where we write simply V in place of V_{ba}) is:

$$Q = CV = (53 \times 10^{-12}\,\text{F})(12\,\text{V}) = 6.4 \times 10^{-10}\,\text{C}.$$

(c) From Eq. 23–4 for a uniform electric field, the magnitude of E is

$$E = \frac{V}{d} = \frac{12\,\text{V}}{1.0 \times 10^{-3}\,\text{m}} = 1.2 \times 10^4\,\text{V/m}.$$

(d) We solve for A in Eq. 24–2 and find that to make a 1-F capacitor with a 1.0-mm air gap requires plates with an area

$$A = \frac{Cd}{\epsilon_0} \approx \frac{(1\,\text{F})(1.0 \times 10^{-3}\,\text{m})}{(9 \times 10^{-12}\,\text{C}^2/\text{N}\cdot\text{m}^2)} \approx 10^8\,\text{m}^2.$$

This is the area of a square $10^4\,\text{m} = 10\,\text{km}$ on a side: the size of a city like San Francisco or Boston!

> **PHYSICS APPLIED**
> *Very high capacitance*

Ten or fifteen years ago, a capacitance greater than $1\,\mu\text{F}$ was unusual. Today there are capacitors available that are 1 or 2 F, yet they are physically small, a few cm on a side. Such capacitors are used as power backups for low voltage applications, such as computer memory and VCR's where the time and date can be maintained through tiny charge flow. [Capacitors are superior to rechargable batteries for this purpose because they can be recharged more than 10^5 times with no degradation.] How are these high capacitance capacitors made? One type uses activated carbon which has very high porosity, so that the surface area is very large; one tenth of a gram of activated carbon can have a surface area of $100\,\text{m}^2$. Furthermore, the equal and opposite charges exist in an electric "double layer" which is a layer of charge at the interface between the carbon particles and the sulfuric acid surrounding them. The positive charge resides at the edge of the carbon and the negative charge at the edge of the acid with a spacing between them of about $10^{-9}\,\text{m}$. Thus, the capacitance of 0.1 g of activated carbon, whose internal area can be $10^2\,\text{m}^2$, is $C = \epsilon_0 A/d = (8.85 \times 10^{-12}\,\text{C}^2/\text{N}\cdot\text{m}^2)(10^2\,\text{m}^2)/(10^{-9}\,\text{m}) \approx 1\,\text{F}$.

One type of computer keyboard operates by capacitance. As shown in Fig. 24–4, each key is connected to the upper plate of a capacitor. The upper plate moves down when the key is pressed, reducing the spacing between the capacitor plates, and increasing the capacitance (Eq. 24–2: smaller d, larger C). The *change* in capacitance becomes an electric signal that is detected by an electronic circuit.

> *Computer keys*

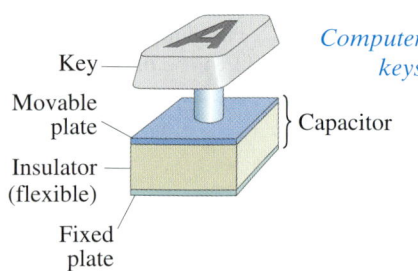

FIGURE 24–4 Key on a computer keyboard. Pressing the key reduces the capacitor spacing thus increasing the capacitance which can be detected electronically.

The proportionality, $C \propto A/d$ in Eq. 24–2, is valid also for a parallel-plate capacitor that is rolled up into a spiral cylinder, as in Fig. 24–1b. However, the constant factor, ϵ_0, must be replaced if an insulator such as paper separates the plates, as is usual, and this is discussed in Section 24–5. For a true cylindrical capacitor—consisting of two long coaxial cylinders—the result is somewhat different as the next Example shows.

EXAMPLE 24–2 Cylindrical capacitor. A cylindrical capacitor consists of a cylinder (or wire) of radius R_b surrounded by a coaxial cylindrical shell of inner radius R_a, Fig. 24–5a. Both cylinders have length L which we assume is much greater than the separation of the cylinders, $R_a - R_b$, so we can neglect end effects. The capacitor is charged (say, by connecting it to a battery) so that one cylinder has a charge $+Q$ (say, the inner one) and the other one a charge $-Q$. Determine a formula for the capacitance.

SOLUTION To obtain $C = Q/V_{ba}$, we need to determine the potential difference between the cylinders, V_{ba}, in terms of Q. We can use our earlier result (Example 21–10 or 22–5) that the electric field outside a long wire is directed radially outward and has magnitude $E = (1/2\pi\epsilon_0)(\lambda/r)$, where r is the distance from the axis and λ is the charge per unit length, Q/L; then $E = (1/2\pi\epsilon_0)(Q/Lr)$ for points between the cylinders.

To obtain V_{ba} in terms of Q, we use this result in Eq. 23–3, $V_{ba} = -\int_a^b \mathbf{E}\cdot d\mathbf{l}$, and write the line integral from the outer cylinder to the inner one (so $V_{ba} > 0$) along a radial line[†]:

FIGURE 24–5 (a) Cylindrical capacitor consists of two coaxial cylindrical conductors. (b) The electric field lines are shown in cross-sectional view.

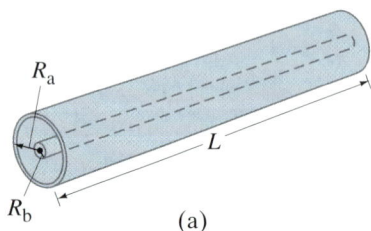

$$V_{ba} = -\int_a^b \mathbf{E}\cdot d\mathbf{l} = -\frac{Q}{2\pi\epsilon_0 L}\int_{R_a}^{R_b}\frac{dr}{r} = -\frac{Q}{2\pi\epsilon_0 L}\ln\frac{R_b}{R_a} = \frac{Q}{2\pi\epsilon_0 L}\ln\frac{R_a}{R_b}.$$

Q and V_{ba} are proportional, and the capacitance C is

$$C = \frac{Q}{V_{ba}} = \frac{2\pi\epsilon_0 L}{\ln(R_a/R_b)}. \qquad \text{[cylindrical capacitor]}$$

Does the dependence on L, R_a, and R_b make sense intuitively? (See discussion immediately after Eq. 24–2.)

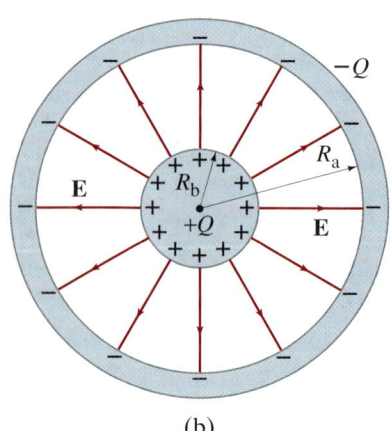

[†]Note that $\mathbf{E}$ points outward in Fig. 24–5b, but $d\mathbf{l}$ points inward for our chosen direction of integration; the angle between $\mathbf{E}$ and $d\mathbf{l}$ is 180° and cos 180° = −1. Also, $dl = -dr$ because dr increases outward. These two minus signs cancel.

EXAMPLE 24-3 Spherical capacitor. A spherical capacitor consists of two thin concentric spherical conducting shells, of radius r_a and r_b as shown in Fig. 24–6. The inner shell carries a uniformly distributed charge Q on its surface, and the outer shell an equal but opposite charge $-Q$. Determine the capacitance of the two shells.

SOLUTION In Example 22–3 we used Gauss's law to show that the electric field outside a uniformly charged conducting sphere is $E = Q/4\pi\epsilon_0 r^2$ as if all the charge were concentrated at the center. Now we use Eq. 23–3, $V_{ba} = -\int_a^b \mathbf{E} \cdot d\mathbf{l}$, and integrate along a radial line to obtain the potential difference between the two conducting shells:

$$V_{ba} = -\int_a^b \mathbf{E} \cdot d\mathbf{l} = -\frac{Q}{4\pi\epsilon_0}\int_{r_a}^{r_b} \frac{1}{r^2} dr$$

$$= \frac{Q}{4\pi\epsilon_0}\left(\frac{1}{r_b} - \frac{1}{r_a}\right) = \frac{Q}{4\pi\epsilon_0}\left(\frac{r_a - r_b}{r_a r_b}\right).$$

Finally,

$$C = \frac{Q}{V_{ba}} = 4\pi\epsilon_0\left(\frac{r_a r_b}{r_a - r_b}\right).$$

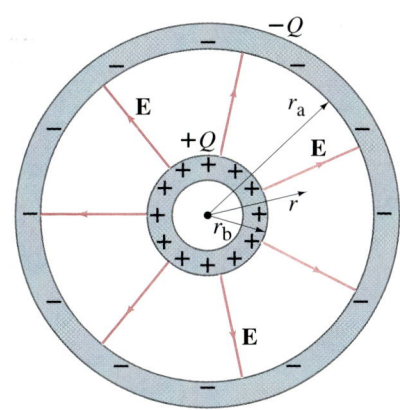

FIGURE 24–6 Cross section of a spherical capacitor. The thin inner shell has radius r_b and the thin outer shell has radius r_a.

A single isolated conductor can also be said to have a capacitance, C. In this case, C can still be defined as the ratio of the charge to absolute potential V on the conductor (relative to $V = 0$ at $r = \infty$), so that the relation

$$Q = CV$$

remains valid. For example, the potential of a single conducting sphere of radius r_b can be obtained from our results in Example 24–3 by letting r_a become infinitely large. As $r_a \to \infty$, then

$$V = \frac{Q}{4\pi\epsilon_0}\left(\frac{1}{r_b} - \frac{1}{r_a}\right) = \frac{1}{4\pi\epsilon_0}\frac{Q}{r_b};$$

so its capacitance is

$$C = \frac{Q}{V} = 4\pi\epsilon_0 r_b.$$

But note that a single conductor alone is not considered a capacitor. In practical cases, a single conductor may be near other conductors or the Earth (which can be thought of as the other "plate" of a capacitor), and these will affect the value of the capacitance.

24–3 Capacitors in Series and Parallel

Capacitors are found in many electric circuits. By electric circuit we mean a closed path of conductors, usually wires connecting capacitors and/or other devices, in which charge can flow and which includes a source of voltage such as a battery. The battery voltage is usually given the symbol V, which means that V represents a potential *difference*. Capacitors can be connected together in various ways. Two common ways are in *series*, or in *parallel*, and we now discuss both.

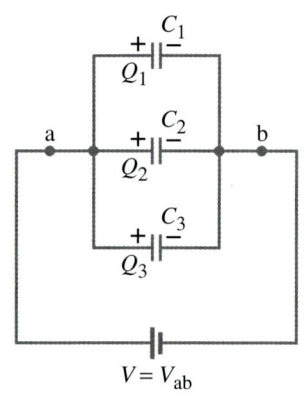

FIGURE 24–7 Capacitors in parallel: $C = C_1 + C_2 + C_3$.

A circuit containing three capacitors connected in **parallel** is shown in Fig. 24–7. They are in "parallel" because when a battery of voltage V is connected to points a and b, this voltage $V = V_{ab}$ exists across each of the capacitors. That is, since the left-hand plates of all the capacitors are connected by conductors, they all reach the same potential V_a when connected to the battery; and the right-hand plates each reach potential V_b. Each capacitor plate acquires a charge given by $Q_1 = C_1 V$, $Q_2 = C_2 V$, and $Q_3 = C_3 V$. The total charge Q that must leave the battery is then

$$Q = Q_1 + Q_2 + Q_3 = C_1 V + C_2 V + C_3 V.$$

Let us try to find a single equivalent capacitor that will hold the same charge Q at the same voltage $V = V_{ab}$. It will have a capacitance C_{eq} given by

$$Q = C_{eq} V.$$

Combining the two previous equations, we have

$$C_{eq} V = C_1 V + C_2 V + C_3 V = (C_1 + C_2 + C_3) V$$

or

Capacitors in parallel

$$C_{eq} = C_1 + C_2 + C_3. \qquad \text{[parallel]} \quad (24\text{–}3)$$

The net effect of connecting capacitors in parallel is thus to *increase* the capacitance. This makes sense because we are essentially increasing the area of the plates where charge can accumulate (see, for example, Eq. 24–2).

Capacitors can also be connected in **series**. That is, end to end, as shown in Fig. 24–8. A charge $+Q$ flows from the battery to one plate of C_1, and $-Q$ flows to one plate of C_3. The regions A and B between the capacitors were originally neutral; so the net charge there must still be zero. The $+Q$ on the left plate of C_1 attracts a charge of $-Q$ on the opposite plate. Because region A must have a zero net charge, there is thus $+Q$ on the left plate of C_2. The same considerations apply to the other capacitors, so we see the charge on each capacitor is the same value Q. A single capacitor that could replace these three in series without affecting the circuit (that is, Q and V the same) would have a capacitance C_{eq} where

$$Q = C_{eq} V.$$

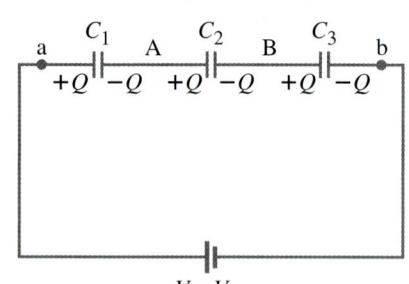

FIGURE 24–8 Capacitors in series: $\dfrac{1}{C} = \dfrac{1}{C_1} + \dfrac{1}{C_2} + \dfrac{1}{C_3}$.

Now the total voltage V across the three capacitors in series must equal the sum of the voltages across each capacitor:

$$V = V_1 + V_2 + V_3.$$

We also have $Q = C_1 V_1$, $Q = C_2 V_2$, and $Q = C_3 V_3$, so we substitute for V, V_1, V_2, and V_3 into the last equation and get

$$\frac{Q}{C_{eq}} = \frac{Q}{C_1} + \frac{Q}{C_2} + \frac{Q}{C_3} = Q\left(\frac{1}{C_1} + \frac{1}{C_2} + \frac{1}{C_3}\right)$$

or

Capacitors in series

$$\frac{1}{C_{eq}} = \frac{1}{C_1} + \frac{1}{C_2} + \frac{1}{C_3}. \qquad \text{[series]} \quad (24\text{–}4)$$

Note that the equivalent capacitance C_{eq} is smaller than the smallest contributing capacitance.

Other connections of capacitors can be analyzed similarly using charge conservation, and often simply in terms of series and parallel connections.

EXAMPLE 24-4 Equivalent capacitance. Determine the capacitance of a single capacitor that will have the same effect as the combination shown in Fig. 24–9. Take $C_1 = C_2 = C_3 = C$.

SOLUTION C_2 and C_3 are connected in parallel, so they are equivalent to a single capacitor having capacitance

$$C_{23} = C_2 + C_3 = 2C.$$

C_{23} is in series with C_1, so the equivalent capacitance, C_{eq}, is given by

$$\frac{1}{C_{eq}} = \frac{1}{C_1} + \frac{1}{C_{23}} = \frac{1}{C} + \frac{1}{2C} = \frac{3}{2C}.$$

Hence $C_{eq} = \frac{2}{3}C$, which is the equivalent capacitance of the entire combination.

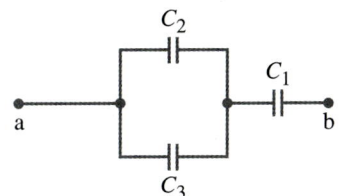

FIGURE 24–9 Example 24–4.

→ **PROBLEM SOLVING**

Remember to take the reciprocal

EXAMPLE 24-5 Capacitor combination. (a) Determine the equivalent capacitance of the combination shown in Fig. 24–10a (that is, the capacitance between points a and b). Take $C_1 = 6.0\,\mu\text{F}$, $C_2 = 4.0\,\mu\text{F}$, and $C_3 = 8.0\,\mu\text{F}$. (b) If the capacitors are charged by a 12-V battery placed between a and b, determine the charge on each capacitor and the potential difference across each.

SOLUTION (a) C_2 and C_3 are connected in parallel, so they are equivalent to a single capacitor of capacitance (Eq. 24–3)

$$C_{23} = C_2 + C_3 = 4.0\,\mu\text{F} + 8.0\,\mu\text{F} = 12.0\,\mu\text{F}.$$

C_{23} is in series with C_1 as shown in Fig. 24–10b. So the equivalent capacitance C of the entire combination is given by (Eq. 24–4)

$$\frac{1}{C_{eq}} = \frac{1}{C_1} + \frac{1}{C_{23}} = \frac{1}{6.0\,\mu\text{F}} + \frac{1}{12.0\,\mu\text{F}} = \frac{3}{12.0\,\mu\text{F}}.$$

Hence $C = 12.0\,\mu\text{F}/3 = 4.0\,\mu\text{F}$.

(b) The total charge that flows from the battery is

$$Q = CV = (4.0 \times 10^{-6}\,\text{F})(12\,\text{V}) = 4.8 \times 10^{-5}\,\text{C}.$$

Both C_1 and C_{23} carry this charge Q. The voltage across C_1 is then

$$V_1 = \frac{Q}{C_1} = \frac{4.8 \times 10^{-5}\,\text{C}}{6.0 \times 10^{-6}\,\text{F}} = 8.0\,\text{V}.$$

The voltage across the combination C_{23} is

$$V_{23} = \frac{Q}{C_{23}} = \frac{4.8 \times 10^{-5}\,\text{C}}{12.0 \times 10^{-6}\,\text{F}} = 4.0\,\text{V}.$$

Since the actual capacitors C_2 and C_3 are in parallel, this represents the voltage across each of them:

$$V_2 = V_3 = 4.0\,\text{V}.$$

The charges on C_2 and C_3 are

$$Q_2 = C_2 V_2 = (4.0 \times 10^{-6}\,\text{F})(4.0\,\text{V}) = 1.6 \times 10^{-5}\,\text{C}$$
$$Q_3 = C_3 V_3 = (8.0 \times 10^{-6}\,\text{F})(4.0\,\text{V}) = 3.2 \times 10^{-5}\,\text{C}.$$

To summarize:

$$V_1 = 8.0\,\text{V} \quad Q_1 = 48\,\mu\text{C}$$
$$V_2 = 4.0\,\text{V} \quad Q_2 = 16\,\mu\text{C}$$
$$V_3 = 4.0\,\text{V} \quad Q_3 = 32\,\mu\text{C}.$$

Note that $Q_2 + Q_3 = Q_1 = Q$, as it should.

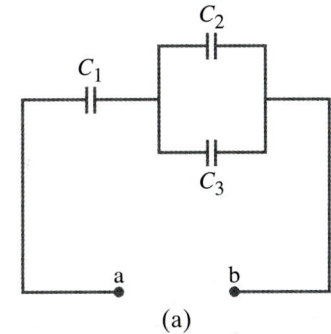

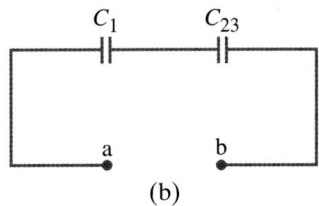

FIGURE 24–10 Example 24–5.

SECTION 24–3 Capacitors in Series and Parallel

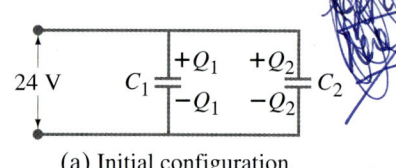

(a) Initial configuration.

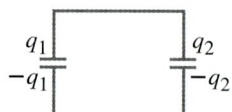

(b) At the instant of reconnection only.

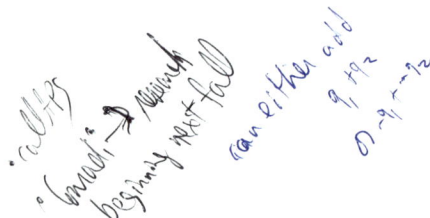

(c) A short time later.

FIGURE 24–11 Example 24–6.

EXAMPLE 24–6 **Capacitors reconnected.** Two capacitors, $C_1 = 2.2\,\mu F$ and $C_2 = 1.2\,\mu F$ are connected in parallel to a 24-volt source as shown in Fig. 24–11a. After they are charged they are disconnected from the source and from each other and reconnected directly to each other, with plates of opposite sign connected together (see Fig. 24–11b). Find the charge on each capacitor and the potential across each after equilibrium is established.

SOLUTION First we need to calculate how much charge has been placed on each capacitor after the power source has charged them fully. Using Eq. 24–1 for each capacitor we find:

$$Q_1 = C_1 V = (2.2\,\mu F)(24\,V) = 52.8\,\mu C,$$
$$Q_2 = C_2 V = (1.2\,\mu F)(24\,V) = 28.8\,\mu C.$$

Now we examine Fig. 24–11b. The capacitors are connected in parallel, and the potential difference across each must quickly reach the same value. Thus, the charge cannot remain as shown in Fig. 24–11b, but the charge must rearrange itself so that the upper plates at least have the same sign of charge, with the lower plates having the opposite charge as shown in Fig. 24–11c. Equation 24–1 applies for each:

$$q_1 = C_1 V' \quad \text{and} \quad q_2 = C_2 V'$$

where V' is the voltage across each capacitor after the charges have rearranged themselves. We don't know q_1, q_2, or V', so we need a third equation. This is provided by charge conservation. The charges have rearranged themselves between Figs. 24–11b and c. The total charge on the upper plates in those two figures must be the same, so we have

$$q_1 + q_2 = Q_1 - Q_2 = 24.0\,\mu C.$$

Combining the last three equations we find:

$$V' = (q_1 + q_2)/(C_1 + C_2) = 24.0\,\mu C/3.4\,\mu F = 7.06\,V \approx 7.1\,V$$
$$q_1 = C_1 V' = (2.2\,\mu F)(7.06\,V) = 15.5\,\mu F \approx 16\,\mu F$$
$$q_2 = C_2 V' = (1.2\,\mu F)(7.06\,V) = 8.5\,\mu F$$

where we have kept only two significant figures in our final answers.

24–4 Electric Energy Storage

A charged capacitor stores electrical energy. The energy stored in a capacitor will be equal to the work done to charge it. The net effect of charging a capacitor is to remove charge from one plate and add it to the other plate. This is what a battery does when it is connected to a capacitor. A capacitor does not become charged instantly. It takes time (Section 26–4). Initially, when the capacitor is uncharged, it requires no work to move the first bit of charge over. When some charge is on each plate, it requires work to add more charge of the same sign because of the electric repulsion. The more charge already on a plate, the more work is required to add additional charge. The work needed to add a small amount of charge dq, when a potential difference V is across the plates, is $dW = V\,dq$. Since $V = q/C$ at any moment (Eq. 24–1), where C is the capacitance, the work needed to store a total charge Q is

$$W = \int_0^Q V\,dq = \frac{1}{C}\int_0^Q q\,dq = \frac{1}{2}\frac{Q^2}{C}.$$

Thus we can say that the energy "stored" in a capacitor is

$$U = \frac{1}{2}\frac{Q^2}{C}$$

when the capacitor C carries charges $+Q$ and $-Q$ on its two conductors. Since $Q = CV$, where V is the potential difference across the capacitor, we can also write

Energy stored in a capacitor

$$U = \frac{1}{2}\frac{Q^2}{C} = \frac{1}{2}CV^2 = \frac{1}{2}QV. \tag{24–5}$$

EXAMPLE 24–7 Energy stored in a capacitor. A camera flash unit stores energy in a 150 μF capacitor at 200 V. How much electric energy can be stored?

SOLUTION From Eq. 24–5, we have

$$U = \text{energy} = \tfrac{1}{2}CV^2 = \tfrac{1}{2}(150 \times 10^{-6}\,\text{F})(200\,\text{V})^2 = 3.0\,\text{J}.$$

Notice how the units work out: $FV^2 = \left(\dfrac{C}{V}\right)(V^2) = CV = C\left(\dfrac{J}{C}\right) = J.$

If this energy could be released in $\tfrac{1}{1000}$ of a second $(10^{-3}\,\text{s})$ the power output would be equivalent to 3000 W.

▶ **PHYSICS APPLIED**
Camera flash

It is useful to think of the energy stored in a capacitor as being stored in the electric field between the plates. As an example let us calculate the energy stored in a parallel-plate capacitor in terms of the electric field.

We have seen (Eq. 23–4) that the electric field **E** between two close parallel plates is (approximately) uniform and its magnitude is related to the potential difference by $V = Ed$ where d is the separation. Also, Eq. 24–2 tells us $C = \epsilon_0 A/d$ for a parallel-plate capacitor. Thus

$$U = \tfrac{1}{2}CV^2 = \tfrac{1}{2}\left(\dfrac{\epsilon_0 A}{d}\right)(E^2 d^2)$$
$$= \tfrac{1}{2}\epsilon_0 E^2\, Ad.$$

The quantity Ad is the volume between the plates in which the electric field E exists. If we divide both sides by the volume, we obtain an expression for the energy per unit volume or **energy density**, u:

$$u = \text{energy density} = \tfrac{1}{2}\epsilon_0 E^2. \qquad (24\text{–}6)$$

Energy stored per unit volume in electric field

The *electric energy stored per unit volume in any region of space is proportional to the square of the electric field* in that region. We derived Eq. 24–6 for the special case of a parallel-plate capacitor. But it can be shown to be true for any region of space where there is an electric field.

24–5 Dielectrics

In most capacitors there is an insulating sheet of material (such as paper or plastic) called a **dielectric** between the plates. This serves several purposes. First of all, dielectrics break down (allowing electric charge to flow) less readily than air, so higher voltages can be applied without charge passing across the gap. Furthermore, a dielectric allows the plates to be placed closer together without touching, thus allowing an increased capacitance because d is less in Eq. 24–2. Finally, it is found experimentally that if the dielectric fills the space between the two conductors, it increases the capacitance by a factor K which is known as the **dielectric constant**. Thus

$$C = KC_0, \qquad (24\text{–}7)$$

Dielectric constant (defined)

where C_0 is the capacitance when the space between the two conductors of the capacitor is a vacuum and C is the capacitance when the space is filled with a material whose dielectric constant is K.

TABLE 24–1
Dielectric constants (at 20° C)

Material	Dielectric constant K	Dielectric strength (V/m)
Vacuum	1.0000	
Air (1 atm)	1.0006	3×10^6
Paraffin	2.2	10×10^6
Polystyrene	2.6	24×10^6
Rubber, neoprene	6.7	12×10^6
Vinyl (plastic)	2–4	50×10^6
Paper	3.7	15×10^6
Quartz	4.3	8×10^6
Oil	4	12×10^6
Glass, Pyrex	5	14×10^6
Porcelain	6–8	5×10^6
Mica	7	150×10^6
Water (liquid)	80	
Strontium titanate	300	8×10^6

The values of the dielectric constant for various materials are given in Table 24–1. Note that for air (at 1 atm pressure) $K = 1.0006$, which differs so little from 1.0000 that the capacitance for an air-filled capacitor hardly differs from that for a vacuum. Also shown in Table 24–1 is the **dielectric strength**, the maximum electric field before breakdown (charge flow) occurs.

For a parallel-plate capacitor (see Eq. 24–2),

$$C = K\epsilon_0 \frac{A}{d} \qquad \text{[parallel-plate capacitor]} \quad (24\text{–}8)$$

when the space between the plates is completely filled with a dielectric whose dielectric constant is K. (The situation when the dielectric only partially fills the space will be discussed shortly in Example 24–9.) The quantity $K\epsilon_0$ appears so often in formulas that we define a new quantity

$$\epsilon = K\epsilon_0 \qquad (24\text{–}9)$$

called the **permittivity** of a material. Then the capacitance of a parallel-plate capacitor becomes

$$C = \epsilon \frac{A}{d}.$$

Note that ϵ_0 represents the permittivity of free space (a vacuum).

The energy density stored in an electric field E (Section 24–4) in a dielectric is given by (see Eq. 24–6)

$$u = \tfrac{1}{2} K\epsilon_0 E^2 = \tfrac{1}{2} \epsilon E^2.$$

Energy density in electric field inside a dielectric

Two simple experiments illustrate the effect of a dielectric. In the first, Fig. 24–12a, a battery of voltage V_0 is kept connected to a capacitor as a dielectric is inserted between the plates. If the charge on the plates without dielectric is Q_0, then when the dielectric is inserted, it is found experimentally (first by Faraday) that the charge Q on the plates is increased by a factor K,

$$Q = KQ_0. \qquad \text{[voltage constant]}$$

The capacitance has increased to $C = Q/V_0 = KQ_0/V_0 = KC_0$, which is Eq. 24–7. In a second experiment, Fig. 24–12b, a battery V_0 is connected to a capacitor C_0 which then holds a charge $Q_0 = C_0V_0$. The battery is then disconnected, leaving the capacitor isolated with charge Q_0 and still at voltage V_0. Next a dielectric is inserted between the plates of the capacitor. The charge remains Q_0 (there is nowhere for the charge to go) but the voltage is found experimentally to drop by a factor K:

$$V = \frac{V_0}{K}. \qquad \text{[charge constant]}$$

Note that the capacitance changes to $C = Q_0/V = Q_0/(V_0/K) = KQ_0/V_0 = KC_0$, so this experiment too confirms Eq. 24–7.

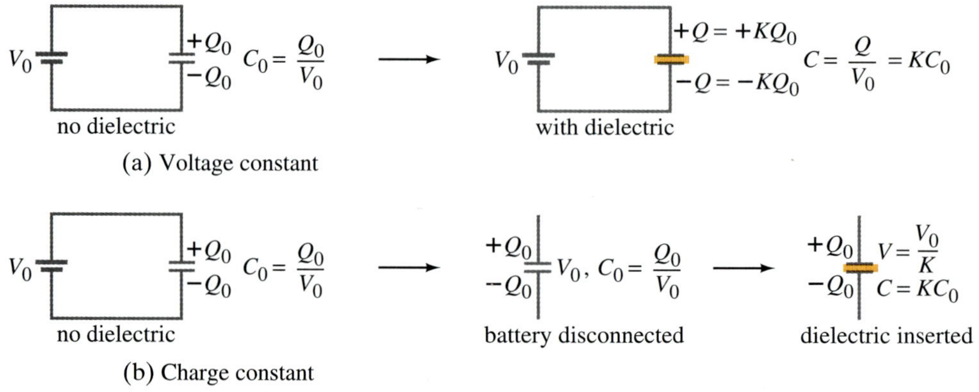

FIGURE 24–12 Two experiments with a capacitor. Dielectric inserted with (a) voltage held constant, (b) charge held constant.

The electric field within a dielectric is also altered. When no dielectric is present, the electric field between the plates of a parallel-plate capacitor is given by Eq. 23–4:

$$E_0 = \frac{V_0}{d},$$

where V_0 is the potential difference between the plates and d is their separation. If the capacitor is isolated so that the charge remains fixed on the plates when a dielectric is inserted, filling the space between the plates, the potential difference drops to $V = V_0/K$. So the electric field in the dielectric is now

$$E = E_D = \frac{V}{d} = \frac{V_0}{Kd}$$

or

$$E_D = \frac{E_0}{K}. \qquad \text{[in a dielectric]} \quad (24\text{–}10)$$

The electric field within a dielectric is thus also reduced by a factor equal to the dielectric constant. Although the field is reduced in a dielectric (or insulator), it is not reduced all the way to zero as in a conductor.

EXAMPLE 24–8 Dielectric removal. A parallel-plate capacitor, filled with a dielectric with $K = 3.4$, is connected to a 100-V battery (Fig. 24–13a). After the capacitor is fully charged, the battery is disconnected. The plates have area $A = 4.0\,\text{m}^2$, and are separated by $d = 4.0\,\text{mm}$. (a) Find the capacitance, the charge on the capacitor, the electric field strength, and the energy stored in the capacitor. (b) The dielectric is carefully removed, without changing the plate separation nor does any charge leave the capacitor (Fig. 24–13b). Find the new values of capacitance, electric field strength, voltage between the plates, and the energy stored on the capacitor.

SOLUTION (a) First we find the capacitance, with dielectric:

$$C = \frac{K\epsilon_0 A}{d} = \frac{3.4(8.85 \times 10^{-12}\,\text{C}^2/\text{N}\cdot\text{m}^2)(4.0\,\text{m}^2)}{4.0 \times 10^{-3}\,\text{m}} = 3.0 \times 10^{-8}\,\text{F}.$$

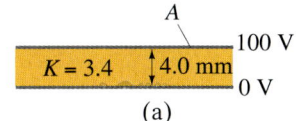

FIGURE 24–13 Example 24–8.

The charge Q on the plates is

$$Q = CV = (3.0 \times 10^{-8}\,\text{F})(100\,\text{V}) = 3.0 \times 10^{-6}\,\text{C}.$$

The electric field between the plates is

$$E = \frac{V}{d} = \frac{100\,\text{V}}{4.0 \times 10^{-3}\,\text{m}} = 25\,\text{kV/m}.$$

Finally, the total energy stored in the capacitor is

$$U = \tfrac{1}{2}CV^2 = \tfrac{1}{2}(3.0 \times 10^{-8}\,\text{F})(100\,\text{V})^2 = 1.5 \times 10^{-4}\,\text{J}.$$

(b) The capacitance without dielectric is

$$C_0 = \frac{C}{K} = \frac{(3.0 \times 10^{-8}\,\text{F})}{(3.4)} = 8.8 \times 10^{-9}\,\text{F}.$$

The charge Q doesn't change, so $V = Q/C$ increases by a factor $K = 3.4$ to 340 V. The electric field is

$$E = \frac{V}{d} = \frac{340\,\text{V}}{4.0 \times 10^{-3}\,\text{m}} = 85\,\text{kV/m}.$$

The energy stored is

$$U = \tfrac{1}{2}CV^2 = \tfrac{1}{2}(8.8 \times 10^{-9}\,\text{F})(340\,\text{V})^2 = 5.1 \times 10^{-4}\,\text{J}.$$

Where did all this extra energy come from? The energy increased because work had to be done to remove the dielectric. The work required was $W = 5.1 \times 10^{-4}\,\text{J} - 1.5 \times 10^{-4}\,\text{J} = 3.6 \times 10^{-4}\,\text{J}$. (We will see in the next Section that work is required because of the force of attraction between induced charge on the dielectric and the charges on the plates—see Fig. 24–14c.)

*24–6 Molecular Description of Dielectrics

Molecular description of dielectrics

Let us now examine, from the molecular point of view, why the capacitance of a capacitor should increase when a dielectric is inserted between the plates. Consider a capacitor whose plates are separated by an air gap. This capacitor has a charge $+Q$ on one plate and $-Q$ on the other (Fig. 24–14a). The capacitor is isolated (not connected to a battery) so charge cannot flow to or from the plates. The potential difference between the plates, V_0, is given by Eq. 24–1: $Q = C_0 V_0$; the subscripts $(_0)$ refer to the situation when only air is between the plates. Now we insert a dielectric between the plates (Fig. 24–14b). The molecules of the dielectric may be *polar*. That is, although the molecules are neutral, they may have a permanent dipole moment (as water does). Because of the electric field between the plates, the molecules will tend to become oriented as shown in Fig. 24–14b; they won't be perfectly aligned because of thermal motion (Chapter 18), but they will usually be at least partially aligned (the stronger the electric field the more alignment).

Even if the molecules are not polar, the electric field between the plates will induce some separation of charge in the molecules (induced dipole moment). Although the electrons do not leave the molecules, they will move slightly within the molecules toward the positive plate. So the situation is still as illustrated in Fig. 24–14b.

The net effect in either case is as if there were a net negative charge on the outer edge of the dielectric facing the positive plate, and a net positive charge on the opposite side, as shown in Fig. 24–14c.

We can visualize that some of the electric field lines do not pass through the dielectric but instead end on charges induced on the surface of the dielectric as shown. Hence the electric field within the dielectric is less than in air. Now imagine a positive test charge within the dielectric. Because the electric field is less, the force a test charge feels is reduced by some factor K (equal, as we shall see, to the dielectric constant). Because the force on our test charge is reduced by a factor K, the work needed to move it from one plate to the other is reduced by a factor K. (We assume

FIGURE 24–14 Molecular view of the effects of a dielectric.

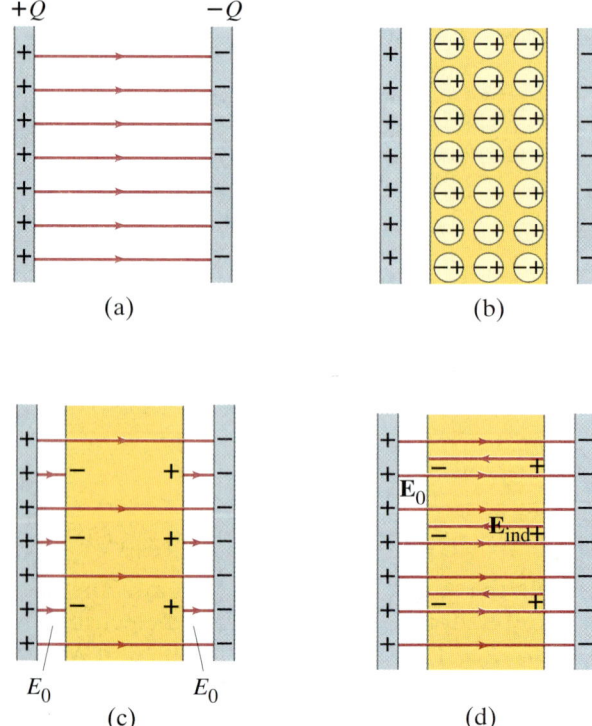

that the dielectric fills all the space between the plates, even though Fig. 24–14 leaves a space so we can show the field there.) The voltage, which is the work done per unit charge, must therefore also have decreased by the factor K. That is, the voltage between the plates is now

$$V = \frac{V_0}{K}.$$

Now the charge Q on the plates has not changed, because they are isolated. So we have

$$Q = CV,$$

where C is the capacitance when the dielectric is present. When we combine this with the relation, $V = V_0/K$, we obtain

$$C = \frac{Q}{V} = \frac{Q}{V_0/K} = \frac{QK}{V_0} = KC_0,$$

since $C_0 = Q/V_0$. Thus we see why the capacitance is increased by a factor K.

As shown in Fig. 24–14d, the electric field within the dielectric E_D can be considered as the vector sum of the electric field $\mathbf{E}_0$ due to the "free" charges on the conducting plates, and the field $\mathbf{E}_{ind}$ due to the induced charge on the surfaces of the dielectric. Since these two fields are in opposite directions, the net field within the dielectric, $E_0 - E_{ind}$, is less than E_0. The precise relationship is given by Eq. 24–10:

$$E_D = E_0 - E_{ind} = \frac{E_0}{K},$$

so

$$E_{ind} = E_0\left(1 - \frac{1}{K}\right).$$

The electric field between two parallel plates is related to the surface charge density, σ (see Section 22–3 and Example 22–7), by $E = \sigma/\epsilon_0$. Thus

$$E_0 = \frac{\sigma}{\epsilon_0}$$

where $\sigma = Q/A$ is the surface charge density on the conductor; Q is the net charge on the conductor and is often called the **free charge** (since charge is free to move in a conductor). Similarly, we define an equivalent induced surface charge density σ_{ind} on the dielectric; then

$$E_{ind} = \frac{\sigma_{ind}}{\epsilon_0}$$

where E_{ind} is the electric field due to the induced charge $Q_{ind} = \sigma_{ind} A$ on the surface of the dielectric, Fig. 24–14d. Q_{ind} is often called the **bound charge**, since it is on an insulator and is not free to move. Since $E_{ind} = E_0(1 - 1/K)$ as shown above, we now have

$$\sigma_{ind} = \sigma\left(1 - \frac{1}{K}\right) \tag{24–11a}$$

and

$$Q_{ind} = Q\left(1 - \frac{1}{K}\right). \tag{24–11b}$$

Since K is always greater than 1, we see that the charge induced on the dielectric is always less than the free charge on each of the capacitor plates.

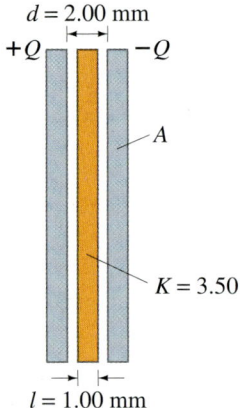

FIGURE 24–15 Example 24–9.

EXAMPLE 24–9 **Dielectric partially fills capacitor.** A parallel-plate capacitor has plates of area $A = 250 \text{ cm}^2$ and separation $d = 2.00 \text{ mm}$. The capacitor is charged to a potential difference $V_0 = 150 \text{ V}$. Then the battery is disconnected (the charge Q on the plates then won't change), and a dielectric sheet ($K = 3.50$) of the same area A but thickness $l = 1.00 \text{ mm}$ is placed between the plates as shown in Fig. 24–15. Determine (a) the initial capacitance of the air-filled capacitor, (b) the charge on each plate before the dielectric is inserted, (c) the charge induced on each face of the dielectric after it is inserted, (d) the electric field in the space between each plate and the dielectric, (e) the electric field in the dielectric, (f) the potential difference between the plates after the dielectric is added, and (g) the capacitance after the dielectric is in place.

SOLUTION (a) Before the dielectric is in place, the capacitance is

$$C_0 = \epsilon_0 \frac{A}{d} = (8.85 \times 10^{-12} \text{ C}^2/\text{N} \cdot \text{m}^2)\left(\frac{2.50 \times 10^{-2} \text{ m}^2}{2.00 \times 10^{-3} \text{ m}}\right) = 111 \text{ pF}.$$

(b) The charge on each plate is

$$Q = C_0 V_0 = (1.11 \times 10^{-10} \text{ F})(150 \text{ V}) = 1.66 \times 10^{-8} \text{ C}.$$

(c) From Eq. 24–11b,

$$Q_{\text{ind}} = Q\left(1 - \frac{1}{K}\right) = (1.66 \times 10^{-8} \text{ C})\left(1 - \frac{1}{3.50}\right) = 1.19 \times 10^{-8} \text{ C}.$$

(d) The electric field in the gaps between the plates and the dielectric (see Fig. 24–14c) is the same as in the absence of the dielectric since the charge on the plates has not been altered. Gauss's law, as applied in Example 22–7, could be used here, which gives $E_0 = \sigma/\epsilon_0$. Or we can note that, in the absence of the dielectric, $E_0 = V_0/d = Q/C_0 d$ (since $V_0 = Q/C_0$) $= Q/\epsilon_0 A$ (since $C_0 = \epsilon_0 A/d$) which is the same result. Thus

$$E_0 = \frac{Q}{\epsilon_0 A} = \frac{1.66 \times 10^{-8} \text{ C}}{(8.85 \times 10^{-12} \text{ C}^2/\text{N} \cdot \text{m}^2)(2.50 \times 10^{-2} \text{ m}^2)} = 7.50 \times 10^4 \text{ V/m}.$$

(e) In the dielectric the electric field is (Eq. 24–10)

$$E_D = \frac{E_0}{K} = \frac{7.50 \times 10^4 \text{ V/m}}{3.50} = 2.14 \times 10^4 \text{ V/m}.$$

(f) To obtain the potential difference in the presence of the dielectric we use Eq. 23–3, and integrate along a straight line parallel to the field lines:

$$V = -\int \mathbf{E} \cdot d\mathbf{l} = E_0(d - l) + E_D l,$$

which can be simplified to

$$V = E_0\left(d - l + \frac{l}{K}\right)$$
$$= (7.50 \times 10^4 \text{ V/m})\left(1.00 \times 10^{-3} \text{ m} + \frac{1.00 \times 10^{-3} \text{ m}}{3.50}\right)$$
$$= 96.4 \text{ V}.$$

(g) In the presence of the dielectric, the capacitance is

$$C = \frac{Q}{V} = \frac{1.66 \times 10^{-8} \text{ C}}{96.4 \text{ V}} = 172 \text{ pF}.$$

Note that if the dielectric filled the space between the plates, the answers to (f) and (g) would be 42.9 V and 387 pF, respectively.

Summary

A **capacitor** is a device used to store charge and consists of two separated conductors. The two conductors generally carry equal and opposite charges, Q, and the ratio of this charge to the potential difference V between the conductors is called the **capacitance**, C; so

$$Q = CV.$$

The capacitance of a parallel-plate capacitor is proportional to the area A of each plate and inversely proportional to their separation d:

$$C = \epsilon_0 \frac{A}{d}.$$

The space between the conductors contains a nonconducting material such as air, paper, or plastic. The latter materials are referred to as **dielectrics**, and the capacitance is proportional to a property of dielectrics called the **dielectric constant**, K (nearly equal to 1 for air). For a parallel-plate capacitor

$$C = K\epsilon_0 \frac{A}{d} = \epsilon \frac{A}{d}$$

where $\epsilon = K\epsilon_0$ is called the **permittivity** of the dielectric material.

When capacitors are connected in **parallel**, the equivalent capacitance is the sum of the individual capacitances:

$$C_{eq} = C_1 + C_2 + \cdots.$$

When capacitors are connected in **series**, the reciprocal of the equivalent capacitance equals the sum of the reciprocals of the individual capacitances:

$$\frac{1}{C_{eq}} = \frac{1}{C_1} + \frac{1}{C_2} + \cdots.$$

A charged capacitor stores an amount of electric energy given by

$$U = \tfrac{1}{2} QV = \tfrac{1}{2} CV^2 = \tfrac{1}{2} \frac{Q^2}{C}.$$

This energy can be thought of as stored in the electric field between the plates. In any electric field $\mathbf{E}$ in free space the **energy density** u (energy per unit volume) is

$$u = \tfrac{1}{2} \epsilon_0 E^2.$$

If a dielectric is present, the energy density is

$$u = \tfrac{1}{2} K\epsilon_0 E^2 = \tfrac{1}{2} \epsilon E^2.$$

Questions

1. Suppose two nearby conductors carry the same negative charge. Can there be a potential difference between them? If so, can the definition of capacitance, $C = Q/V$, be used here?
2. Suppose the separation of plates d in a parallel-plate capacitor is not very small compared to the dimensions of the plates. Would you expect Eq. 24–2 to give an overestimate or underestimate of the true capacitance? Explain.
3. Suppose one of the plates of a parallel-plate capacitor was moved so that the area of overlap was reduced by half, but they are still parallel. How would this affect the capacitance?
4. Explain how the relation for the capacitance of a cylindrical capacitor, Example 24–2, makes sense intuitively. Use arguments such as those just after Eq. 24–2.
5. Describe a simple method of measuring ϵ_0 using a capacitor.
6. When a battery is connected to a capacitor, why do the two plates acquire charges of the same magnitude? Will this be true if the two conductors are different sizes or shapes?
7. A large copper sheet of thickness l is placed between the parallel plates of a capacitor, but does not touch the plates. How will this affect the capacitance?
8. Suppose three identical capacitors are connected to a battery. Will they store more energy if connected in series or in parallel?
9. The parallel plates of an isolated capacitor carry opposite charges, Q. If the separation of the plates is increased, is a force required? Is the potential difference changed? What happens to the work done in the pulling process?
10. How does the energy in a capacitor change if (a) the potential difference is doubled, (b) the charge on each plate is doubled, and (c) the separation of the plates is doubled, as the capacitor remains connected to a battery?
11. For dielectrics consisting of polar molecules, how would you expect the dielectric constant to change with temperature?
12. An isolated charged capacitor has horizontal plates. If a thin dielectric is inserted a short way between the plates, Fig. 24–16, how will it move when it is then released?

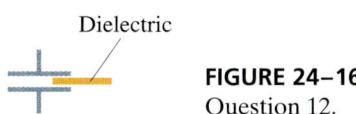

FIGURE 24–16
Question 12.

13. Suppose a battery remains connected to the capacitor in Question 12. What then will happen when the dielectric is released?
14. A dielectric is pulled from between the plates of a capacitor which remains connected to a battery. What changes occur to the capacitance, charge on the plates, potential difference, energy stored, and electric field?
15. How does the energy stored in a capacitor change when a dielectric is inserted if (a) the capacitor is isolated so Q doesn't change; (b) the capacitor remains connected to a battery so V doesn't change?

16. We have seen that the capacitance C depends on the size, shape, and position of the two conductors, as well as on the dielectric constant K. What then did we mean when we said that C is a constant in Eq. 24–1?

17. What value might we assign to the dielectric constant for a good conductor? Explain.

18. *Dissolving Power of Water.* The very high dielectric constant of water, $K = 80$ (Table 24–1), has a profound effect on materials in that it allows many of them to be dissolved in water. For example, ordinary table salt, NaCl (sodium chloride), whose crystal structure (Fig. 24–17a) is held together by the attractive forces between the ions Na$^+$ and Cl$^-$, is easily dissolved when placed in water. Explain why we would expect that the electric field produced by each ion would be reduced by a factor equal to the dielectric constant; that is, discuss the extension of Eq. 24–10 to the field of a point charge in a dielectric, and thus explain (using this simple model) how salt is dissolved (see Fig. 24–17b).

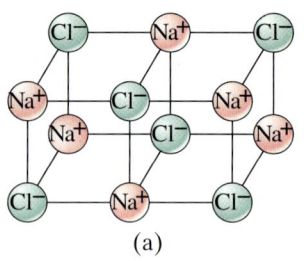

(a)

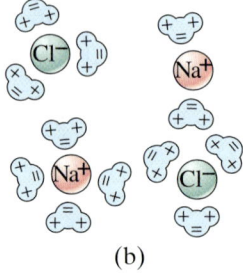

(b)

FIGURE 24–17 (a) Sodium chloride crystal; (b) sodium chloride dissolving in water. Question 18.

Problems

Section 24–1

1. (I) The two plates of a capacitor hold $+2500\,\mu\text{C}$ and $-2500\,\mu\text{C}$ of charge, respectively, when the potential difference is 950 V. What is the capacitance?

2. (I) A 12,000-pF capacitor holds $28.0 \times 10^{-8}\,\text{C}$ of charge. What is the voltage across the capacitor?

3. (I) The potential difference between two parallel wires in air is 12.0 V. They carry equal and opposite charge of magnitude 75 pC. What is the capacitance of the two wires?

4. (I) How much charge flows from a 12-V battery when it is connected to a 15.6-μF capacitor?

5. (I) The charge on a capacitor increases by $16\,\mu\text{C}$ when the voltage across it increases from 28 V to 48 V. What is the capacitance of the capacitor?

6. (II) A capacitor C_1 carries a charge Q_0. It is then connected directly to a second, initially uncharged, capacitor C_2. What charge will each carry now? What will be the potential difference across each?

7. (II) It takes 25 J of energy to move a 0.20-mC charge from one plate of a 16-μF capacitor to the other. How much charge is on each plate?

8. (II) A 2.40-μF capacitor is charged to 880 V and a 4.00-μF capacitor is charged to 560 V. (*a*) These capacitors are then disconnected from their batteries, and the positive plates are now connected to each other and the negative plates are connected to each other. What will be the potential difference across each capacitor and the charge on each? (*b*) What is the voltage and charge for each capacitor if plates of opposite sign are connected?

Section 24–2

9. (I) A 0.40-μF capacitor is desired. What area must the plates have if they are to be separated by a 4.0-mm air gap?

10. (I) What is the capacitance per unit length (F/m) of a coaxial cable whose inner conductor has a 1.0-mm diameter and the outer cylindrical sheath has a 5.0-mm diameter? Assume the space between is filled with air.

11. (I) Determine the capacitance of the Earth, assuming it to be a spherical conductor.

12. (II) Use Gauss's law to show that $\mathbf{E} = 0$ inside the inner conductor of a cylindrical capacitor (see Fig. 24–5 and Example 24–2) as well as outside the outer cylinder.

13. (II) Dry air will break down if the electric field exceeds about 3.0×10^6 V/m. What amount of charge can be placed on a capacitor if the area of each plate is 8.5 cm^2?

14. (II) An electric field of 2.80×10^5 V/m is desired between two parallel plates each of area 21.0 cm^2 and separated by 0.250 cm of air. What charge must be on each plate?

15. (II) In the limit of a very small separation of the two cylinders of a cylindrical capacitor $(R_a - R_b \ll R_a$ in Fig. 24–5) show that the relation derived in Example 24–2 reduces to that of a parallel-plate capacitor (Eq. 24–2).

16. (II) Suppose a capacitor carries a charge of $\pm 4.2\,\mu\text{C}$, and an electric field of 2.0 kV/mm is desired between the plates which are separated by 4.0 mm of air. What must each plate's area be?

17. (II) How strong is the electric field between the plates of a 0.80-μF air-gap capacitor if they are 2.0 mm apart and each has a charge of $72\,\mu\text{C}$?

18. (II) Show that for a spherical capacitor (Example 24–3), if the spacing between shells is very small $(r_a - r_b \ll r_a)$, the formula reduces to that for a parallel-plate capacitor.

19. (II) A large metal sheet of thickness l is placed between, and parallel to, the plates of the parallel-plate capacitor of Fig. 24–3. It does not touch the plates, and extends beyond their edges. (a) What is now the net capacitance in terms of A, d, and l? (b) If $l = \frac{2}{3}d$, by what factor does the capacitance change when the sheet is inserted?

Section 24–3

20. (I) Six 1.8-μF capacitors are connected in parallel. What is the equivalent capacitance? What is their equivalent capacitance if connected in series?

21. (I) The capacitance of a portion of a circuit is to be reduced from 3600 pF to 1600 pF. What capacitance can be added to the circuit to produce this effect without removing existing circuit elements? Must any existing connections be broken to accomplish this?

22. (II) Suppose three parallel-plate capacitors, whose plates have areas A_1, A_2, and A_3 and separations d_1, d_2, and d_3, are connected in parallel. Show, using only Eq. 24–2, that Eq. 24–3 is valid.

23. (II) (a) Determine the equivalent capacitance of the circuit shown in Fig. 24–18. (b) If $C_1 = C_2 = 2C_3 = 14.0\,\mu\text{F}$, how much charge is stored on each capacitor when $V = 25.0\,\text{V}$?

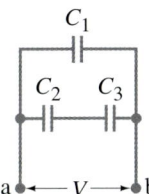

FIGURE 24–18 Problems 23, 24, and 43.

24. (II) In Fig. 24–18, suppose $C_1 = C_2 = C_3 = 16.0\,\mu\text{F}$. If the charge on C_2 is $Q_2 = 24.0\,\mu\text{C}$, determine the charge on each of the other capacitors, the voltage across each capacitor, and the voltage V_{ab} across the entire combination.

25. (II) A 3.00-μF and a 4.00-μF capacitor are connected in series and this combination is connected in parallel with a 2.00-μF capacitor (see Fig. 24–19). (a) What is the net capacitance? (b) If 26.0 V is applied across the whole network, calculate the voltage across each capacitor.

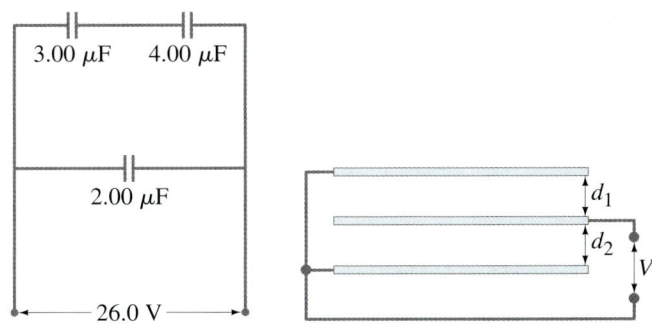

FIGURE 24–19 Problem 25. **FIGURE 24–20** Problem 26.

26. (II) Three conducting plates, each of area A, are connected as shown in Fig. 24–20. (a) Are the two capacitors thus formed connected in series or in parallel? (b) Determine C as a function of d_1, d_2, and A. Assume $d_1 + d_2$ is much less than the dimensions of the plates. (c) The middle plate can be moved (changing the values of d_1 and d_2), so as to vary the capacitance. What are the minimum and maximum values of the net capacitance?

27. (II) Consider three capacitors, of capacitance 3000 pF, 5000 pF, and 0.010 μF. What are the maximum and minimum capacitances that you can form from these? How do you make the connection in each case?

28. (II) A 0.20-μF and a 0.30-μF capacitor are connected in series to a 9.0-V battery. Calculate (a) the potential difference across each capacitor and (b) the charge on each. (c) Repeat parts (a) and (b) assuming the two capacitors are in parallel.

29. (II) In Fig. 24–21, suppose $C_1 = C_2 = C_3 = C_4 = C$. (a) Determine the equivalent capacitance between points a and b. (b) Determine the charge on each capacitor and the potential difference across each if $V_{ba} = V$.

30. (II) Suppose in Fig. 24–21 that $C_1 = C_2 = C_3 = 16.0\,\mu\text{F}$ and $C_4 = 36.0\,\mu\text{F}$. If the charge on C_2 is $Q_2 = 12.4\,\mu\text{C}$, determine the charge on each of the other capacitors, the voltage across each capacitor, and the voltage V_{ab} across the entire combination.

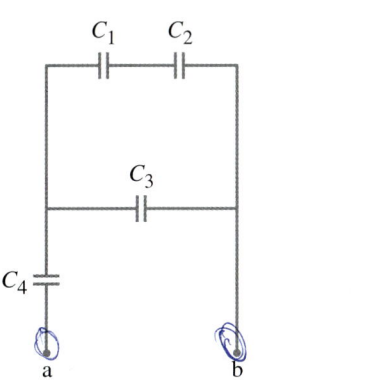

FIGURE 24–21 Problems 29, 30, and 44. **FIGURE 24–22** Problem 31.

31. (II) The switch S in Fig. 24–22 is connected downward so that capacitor C_2 becomes fully charged by the battery of voltage V_0. If the switch is then connected upward, determine the charge on each capacitor after the switching.

32. (II) (a) Determine the equivalent capacitance between points a and b for the combination of capacitors shown in Fig. 24–23. (b) Determine the charge on each capacitor and the voltage across each if $V_{ab} = V$.

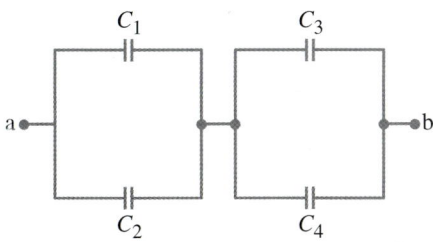

FIGURE 24–23 Problems 32 and 33.

33. (II) Suppose in Problem 32, Fig. 24–23, that $C_1 = C_3 = 8.0\,\mu\text{F}$, $C_2 = C_4 = 16\,\mu\text{F}$, and $Q_3 = 30\,\mu\text{C}$. Determine (a) the charge on each of the other capacitors, (b) the voltage across each capacitor, and (c) the voltage V_{ba} across the combination.

34. (II) Two capacitors connected in parallel produce an equivalent capacitance of 35.0 μF but when connected in series the equivalent capacitance is only 4.0 μF. What is the individual capacitance of each capacitor?

35. (II) In the *capacitance bridge* shown in Fig. 24–24, a voltage V_0 is applied and the variable capacitor C_1 is adjusted until there is zero voltage between points a and b as measured on the voltmeter (●—Ⓥ—●). Determine the unknown capacitance C_x if $C_1 = 8.9\,\mu\text{F}$ and the fixed capacitors have $C_2 = 18.0\,\mu\text{F}$ and $C_3 = 6.0\,\mu\text{F}$.

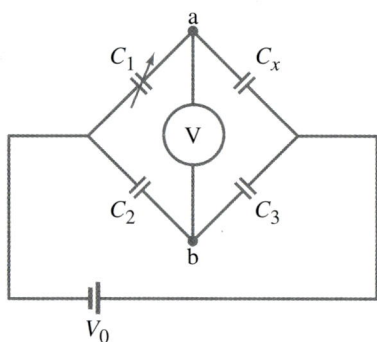

FIGURE 24–24 Problem 35.

36. (II) Two capacitors, $C_1 = 3200\,\text{pF}$ and $C_2 = 2200\,\text{pF}$, are connected in series to a 12.0-V battery. The capacitors are later disconnected from the battery and connected directly to each other, positive plate to positive plate, and negative plate to negative plate. What then will be the charge on each capacitor?

37. (III) Suppose one plate of a parallel-plate capacitor were tilted so it made a small angle θ with the other plate, as shown in Fig. 24–25. Determine a formula for C in terms of A, d, and θ, where A is the area of each plate and θ is small. Assume the plates are square. [*Hint*: Imagine the capacitor as many infinitesimal capacitors in parallel.]

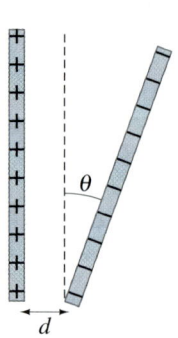

FIGURE 24–25 Problem 37.

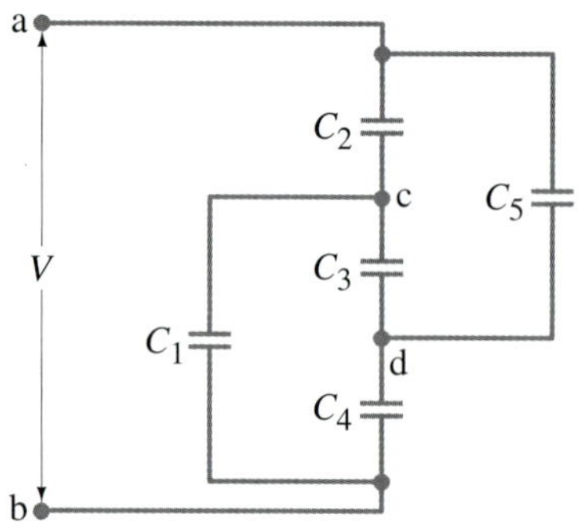

FIGURE 24–26 Problem 38.

38. (III) A voltage V is applied to the capacitor network shown in Fig. 24–26. (*a*) What is the equivalent capacitance? [*Hint*: Assume a potential difference V_{ab} exists across the network as shown; write potential differences for various pathways through the network from a to b in terms of the charges on the capacitors and the capacitances.] (*b*) Determine the equivalent capacitance if $C_2 = C_4 = 8.0\,\mu\text{F}$ and $C_1 = C_3 = C_5 = 6.0\,\mu\text{F}$.

Section 24–4

39. (I) 1200 V is applied to a 2800-pF capacitor. How much electric energy is stored?

40. (I) There is an electric field near the Earth's surface whose intensity is about 150 V/m. How much energy is stored per cubic meter in this field?

41. (I) How much energy is stored by the electric field between two square plates, 8.0 cm on a side, separated by a 1.5 mm air gap? The charges on the plates are equal and opposite and of magnitude 420 μC.

42. (II) A parallel-plate capacitor has fixed charges $+Q$ and $-Q$. The separation of the plates is then doubled. (*a*) By what factor does the energy stored in the electric field change? (*b*) How much work must be done if the separation of the plates is doubled from d to $2d$? The area of each plate is A.

43. (II) In Fig. 24–18, let $V = 10.0\,\text{V}$ and $C_1 = C_2 = C_3 = 2200\,\text{pF}$. How much energy is stored in the capacitor network?

44. (II) What is the total energy stored in the network of capacitors in Fig. 24–21, Problem 29?

45. (II) How much energy must a 12-V battery expend to fully charge a 0.15-μF and a 0.20-μF capacitor when they are placed (*a*) in parallel, (*b*) in series? (*c*) How much charge flowed from the battery in each case?

46. (II) (*a*) Suppose the outer radius R_a of a cylindrical capacitor was doubled, but the charge was kept constant. By what factor would the stored energy change? Where would the energy come from? (*b*) Repeat, assuming the voltage remains constant.

47. (II) A 3.0-μF capacitor is charged by a 12-V battery. It is disconnected from the battery and then connected to an uncharged 5.0-μF capacitor. Determine the total stored energy (*a*) before the two capacitors are connected and (*b*) after they are connected. (*c*) What is the change in energy? (*d*) Is energy conserved? Explain why.

48. (II) How much work would be required to remove the metal sheet from between the plates of the capacitor in Problem 19, assuming: (*a*) the battery remains connected so the voltage remains constant; (*b*) the battery is disconnected so the charge remains constant?

49. (II) (*a*) Show that each plate of a parallel-plate capacitor exerts a force

$$F = \frac{1}{2}\frac{Q^2}{\epsilon_0 A}$$

on the other, by calculating dW/dx where dW is the work needed to increase the separation by dx. (*b*) Why does using $F = QE$ give the wrong answer?

50. (II) Show that the electrostatic energy stored in the electric field outside an isolated spherical conductor of radius R carrying a net charge Q is

$$U = \frac{1}{8\pi\epsilon_0}\frac{Q^2}{R}.$$

Do this in three ways: (*a*) Use Eq. 24–6 for the energy density in an electric field [*Hint*: Consider spherical shells of thickness dr]; (*b*) use Eq. 24–5 together with the capacitance of an isolated sphere (Section 24–2); (*c*) by calculating the work needed to bring all the charge Q up from infinity in infinitesimal bits dq.

Section 24–5

51. (I) What is the capacitance of a pair of circular plates with a radius of 5.0 cm separated by 3.2 mm of mica?

52. (I) What is the capacitance of two square parallel plates 5.5 cm on a side that are separated by 1.8 mm of paraffin?

53. (II) A 3500-pF air-gap capacitor is connected to a 22-V battery. If a piece of mica is placed between the plates, how much charge will flow from the battery?

54. (II) Suppose the capacitor in Example 24–8 remains connected to the battery as the dielectric is removed. What will be the work required to remove the dielectric in this case?

55. (II) How much energy would be stored in the capacitor of Problem 41 if a mica dielectric is placed between the plates? Assume the mica is 1.5 mm thick (and therefore fills the space between the plates).

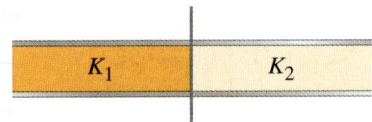

FIGURE 24–27 Problem 56.

56. (II) Two different dielectrics each fill half the space between the plates of a parallel-plate capacitor as shown in Fig. 24–27. Determine a formula for the capacitance in terms of K_1, K_2, the area A of the plates, and the separation d. [*Hint*: Can you consider this capacitor as two capacitors in series or in parallel?]

57. (II) Two different dielectrics fill the space between the plates of a parallel-plate capacitor as shown in Fig. 24–28. Determine a formula for the capacitance in terms of K_1, K_2, the area A, of the plates, and the separation $d_1 = d_2 = d/2$. [*Hint*: Can you consider this capacitor as two capacitors in series or in parallel?]

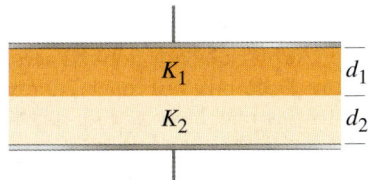

FIGURE 24–28 Problems 57 and 58.

58. (II) Repeat Problem 57 (Fig. 24–28) but assume the separation $d_1 \neq d_2$.

59. (II) Two identical capacitors are connected in parallel and each acquires a charge Q_0 when connected to a source of voltage V_0. The voltage source is disconnected and then a dielectric ($K = 4.0$) is inserted to fill the space between the plates of one of the capacitors. Determine (*a*) the charge now on each capacitor, and (*b*) the voltage now across each capacitor.

60. (III) A slab of width d and dielectric constant K is inserted a distance x between the square parallel plates (of side l) of a capacitor as shown in Fig. 24–29. Determine, as a function of x, (*a*) the capacitance, (*b*) the energy stored if the potential difference is V_0, and (*c*) the magnitude and direction of the force exerted on the slab (assume V_0 is constant).

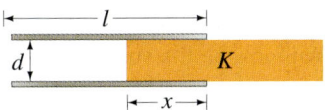

FIGURE 24–29 Problem 60.

*Section 24–6

***61.** (II) Repeat Example 24–9 assuming the battery remains connected when the dielectric is inserted. Also, what is the free charge on the plates after the dielectric is added (let this be part (*h*) of the problem)?

***62.** (II) Show that the capacitor in Example 24–9 with dielectric inserted can be considered as equivalent to three capacitors in series, and using this assumption show that the same value for the capacitance is obtained as was obtained in part (*g*) of the Example.

***63.** (II) In Example 24–9 what percent of the stored energy is stored in the electric field in the dielectric?

***64.** (II) Using Example 24–9 as a model, derive a formula for the capacitance of a parallel-plate capacitor whose plates have area A, separation d, with a dielectric of thickness $l(l < d)$ with dielectric constant K placed between the plates.

***65.** (III) The capacitor shown in Fig. 24–30 is connected to a 90.0-V battery. Calculate (and sketch) the electric field everywhere between the capacitor plates. Find both the free charge on the capacitor plate and the induced charge on the faces of the glass dielectric plate.

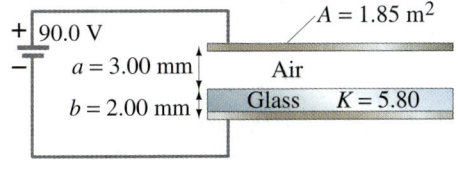

FIGURE 24–30 Problem 65.

General Problems

66. An electric circuit was accidentally constructed using a 5.0-μF capacitor instead of the required 16-μF value. What can a technician add to correct this circuit?

67. A *cardiac defibrillator* is used to shock a heart that is beating erratically. A capacitor in this device is charged to 6000 V and stores 200 J of energy. What is its capacitance?

68. A homemade capacitor is assembled by placing two 9-inch pie pans 10 cm apart and connecting them to the opposite terminals of a 9-V battery. Estimate (a) the capacitance, (b) the charge on each plate, (c) the electric field halfway between the plates, (d) the work done by the battery to charge the plates. (e) Which of the above values change if a dielectric is inserted?

69. How does the energy stored in a capacitor change if (a) the potential difference is doubled, (b) the charge on each plate is doubled, and (c) the separation of the plates is doubled, as the capacitor remains connected to a battery?

70. A huge 7.0-F capacitor has enough stored energy to heat 2.5 kg of water from 20°C to 95°C. What is the potential difference across the plates?

71. An uncharged capacitor is connected to a 24.0-V battery until it is fully charged, after which it is disconnected from the battery. A slab of paraffin is then inserted between the plates. What will now be the voltage between the plates?

72. It takes 18.5 J of energy to move a 13.0-mC charge from one plate of a 12.0-μF capacitor to the other. How much charge is on each plate?

73. A coaxial cable, Fig. 24–31, consists of an inner cylindrical conducting wire of radius R_b surrounded by a dielectric insulator. Surrounding the dielectric insulator is an outer conducting sheath of radius R_a, which is usually "grounded." (a) Determine an expression for the capacitance per unit length of a cable whose insulator has dielectric constant K. (b) For a given cable, $R_b = 3.5$ mm and $R_a = 9.0$ mm. The dielectric constant of the dielectric insulator is $K = 2.6$. Suppose that there is a potential of 1.0 kV between the inner conducting wire and the outer conducting sheath. Find the capacitance per meter of the cable.

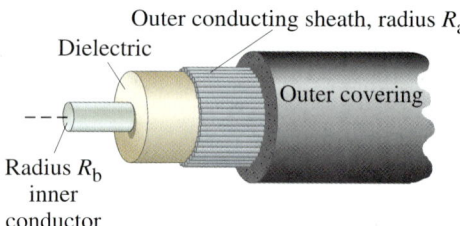

FIGURE 24–31 Problem 73.

74. The electric field between the plates of a paper-separated ($K = 3.75$) capacitor is 9.21×10^4 V/m. The plates are 1.95 mm apart and the charge on each plate is $0.475\ \mu$C. Determine the capacitance of this capacitor and the area of each plate.

75. A parallel-plate capacitor is isolated with a charge $\pm Q$ on each plate. If the separation of the plates is halved and a dielectric (constant K) is inserted in place of air, by what factor does the energy storage change? To what do you attribute the change in stored potential energy? How does the new value of the electric field between the plates compare with the original value?

76. A 3.5-μF capacitor is charged by a 12.4-V battery and then is disconnected from the battery. When this capacitor (C_1) is then connected to a second (initially uncharged) capacitor, C_2, the voltage on the first drops to 5.2 V. What is the value of C_2?

77. The power supply for a pulsed nitrogen laser has a 0.060-μF capacitor with a maximum voltage rating of 25 kilovolts. (a) Estimate how much energy could be stored in this capacitor. (b) If 10 percent of this stored electrical energy is converted to light energy in a pulse that is 10 microseconds long, what is the power of the laser pulse?

78. The variable capacitance of an old radio tuner consists of four plates connected together placed alternately between four other plates, also connected together (Fig. 24–32). Each plate is separated from its neighbor by 1.0 mm of air. One set of plates can move so that the area of overlap of each plate varies from 2.0 cm² to 9.0 cm². (a) Are these seven capacitors connected in series or in parallel? (b) Determine the range of capacitance values.

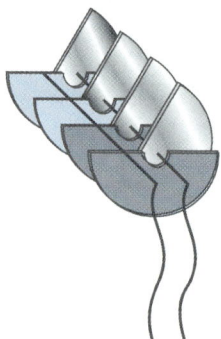

FIGURE 24–32 Problems 78 and 79.

79. A high-voltage supply can be constructed from a variable capacitor with interleaving plates which can be rotated as in Fig. 24–32. A version of this type of capacitor with more plates has a capacitance which can be varied from 10 pF to 1 pF. (a) Initially, this capacitor is charged by a 10,000-volt power supply when the capacitance is 10 pF. It is then disconnected from the power supply and the capacitance reduced to 1.0 pF by rotating the plates. What is the voltage across the capacitor now? (b) What is a major disadvantage of this as a high-voltage power supply?

80. A 150-pF capacitor is connected in series with an unknown capacitor, and as a series combination they are connected to a 25.0-V battery. If the 150-pF capacitor stores 125 pC of charge on its plates, what is the unknown capacitance?

81. A circuit contains a single 330-pF capacitor hooked across a battery. It is desired to store three times as much energy by adding a single capacitor to this one. How would you hook it up and what would its value be?

82. What are the values of effective capacitance which can be obtained by connecting four identical capacitors, each having a capacitance C.

83. In the circuit shown in Fig. 24–33, $C_1 = 1.0\,\mu\text{F}$, $C_2 = 2.0\,\mu\text{F}$, $C_3 = 3.0\,\mu\text{F}$, and a voltage $V_{ab} = 24\,\text{V}$ is applied across points a and b. After C_1 is fully charged the switch is thrown to the right. What is the final charge and potential difference on each capacitor?

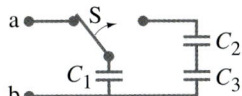

FIGURE 24–33 Problem 83.

84. The long cylindrical capacitor shown in Fig. 24–34 consists of 4 concentric cylinders, with respective radii R_a, R_b, R_c, and R_d. The cylinders b and c are joined by metal strips, as indicated. Determine the capacitance per unit length of this arrangement.

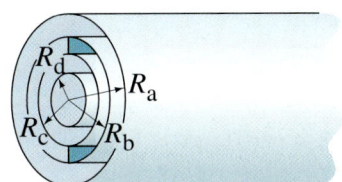

FIGURE 24–34 Problem 84.

85. A parallel plate capacitor has plate area A, plate separation x, and has a charge Q stored on its plates (Fig. 24–35). Find the amount of work required to double the plate separation to $2x$, assuming the charge remains constant at Q. Show that your answer is consistent with the change in energy stored by the capacitor.

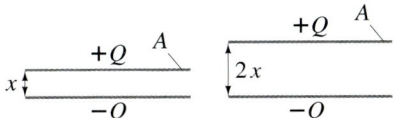

FIGURE 24–35 Problem 85.

86. Consider the use of capacitors as memory cells. A charged capacitor would represent a one and an uncharged capacitor a zero. Suppose these capacitors were fabricated on a silicon chip and had a capacitance of 30 femto-farads each $(1\,\text{fF} = 10^{-15}\,\text{F}.)$ The dielectric filling the space between the parallel plates has dielectric constant $K = 1.00 \times 10^4$ and a dielectric strength of $5.0 \times 10^7\,\text{V/m}$. (a) If the operating voltage is 3.0 volts, how many electrons would be stored on one of these capacitors when charged? (b) If no safety factor is allowed, how thin a dielectric layer could we use for operation at 3.0 volts? (c) Using the layer thickness from your answer to part (b), what would be the area of the capacitor plates?

87. A parallel plate capacitor with plate area $A = 2.5\,\text{m}^2$ and plate separation $d = 3.0\,\text{mm}$ is connected to a 45 V battery (Fig. 24–36). (a) Determine the charge on the capacitor, the electric field, the capacitance, and the energy stored in the capacitor. (b) With the capacitor still connected to the battery, a slab of plastic with dielectric strength $K = 3.6$ is placed between the plates of the capacitor, so that the gap is completely filled with the dielectric. What are the new values of charge, electric field, capacitance, and the energy U stored in the capacitor?

FIGURE 24–36 Problem 87.

88. A smooth conducting sphere of radius r_0 carries a charge Q. Half of the energy stored in its electric field is contained in a volume of what radius?

89. Paper has a dielectric constant of $K = 3.7$ and a dielectric strength of $15 \times 10^6\,\text{V/m}$. Suppose that a typical sheet of paper has a thickness of 0.030 mm. You make a "home made" capacitor by placing a sheet of 8.5×11 inch paper between two aluminum foil sheets (Fig. 24–37). The thickness of the aluminum foil is 0.040 mm. (a) What is the capacitance C_0 of your device? (b) About how much charge could you store on your capacitor before it would break down? (c) Show in a sketch how you could overlay sheets of paper and aluminum for a parallel combination. If you made 100 such capacitors, and connected the ends of the sheets in parallel, so that you have a single large capacitor of capacitance $100\,C_0$, how thick would your new large capacitor be? (d) What is the maximum voltage you can apply to this $100\,C_0$ capacitor without breakdown?

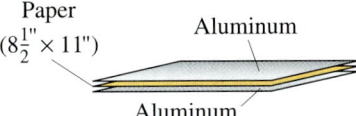

FIGURE 24–37 Problem 89.

90. In lightning storms, the potential difference between the Earth and the bottom of the thunderclouds can be as high as 35,000,000 V. The bottoms of the thunderclouds are typically 1500 m above the Earth, and can have an area of 110 km². For the purpose of this problem, model the Earth-cloud system as a huge capacitor and calculate (a) the capacitance of the Earth-cloud system, (b) the charge stored in the "capacitor," and (c) the energy stored in the "capacitor."

The glow of the thin wire filament of a light bulb is caused by the electric current passing through it. Electric energy is transformed to thermal energy (via collisions between moving electrons and atoms of the wire), which causes the wire's temperature to become so high that it glows. Electric current and electric power in electric circuits are of basic importance in everyday life. We examine both dc and ac in this chapter, and include the microscopic analysis of electric current, as well as a look at electric hazards.

CHAPTER 25

Electric Currents and Resistance

FIGURE 25–1 Alessandro Volta. In this portrait, Volta exhibits his battery to Napoleon in 1801.

In the previous four chapters we have been studying static electricity: electric charges at rest. In this chapter we begin our study of charges in motion, and we call a flow of charge an electric current.

In everyday life we are familiar with electric currents flowing in wires and other conductors. Indeed, most practical electrical devices depend on electric current: current flowing through a light bulb, current in the heating element of a stove or electric heater, and of course currents in electronic devices. Electric currents can exist in conductors such as wires and also in other devices, such as the CRT of a television or computer monitor whose charged electrons flow through space (Section 23–9).

In electrostatic situations, we saw (Sections 21–9 and 22–3) that the electric field must be zero inside a conductor (if it weren't, the charges would move). But when charges are moving in a conductor, there can be an electric field in the conductor. Indeed, an electric field is needed to get charges into motion, and to keep them in motion in any normal conductor.

We first look at electric current from a macroscopic point of view: that is, current as measured in a laboratory. Later in the chapter we look at currents from a microscopic (theoretical) point of view as a flow of electrons in a wire.

We can control the flow of charge using electric fields and electric potential (voltage), concepts we have just been discussing. In order to have a current in a wire, a potential difference is needed, which can be provided by a battery.

Until the year 1800, the technical development of electricity consisted mainly of producing a static charge by friction. It all changed in 1800 when Alessandro Volta (1745–1827; Fig. 25–1) invented the electric battery, and with it produced the first steady flow of electric charge—that is, a steady electric current.

25–1 The Electric Battery

The events that led to the discovery of the battery are interesting. For not only was this an important discovery, but it also gave rise to a famous scientific debate.

In the 1780s, Luigi Galvani (1737–1798), professor at the University of Bologna, carried out a series of experiments on the contraction of a frog's leg muscle through electricity produced by static electricity. Galvani found that contraction of the muscle could also be produced when dissimilar metals were inserted into the frog. Galvani believed that the source of the electric charge was in the frog muscle or nerve itself, and that the metal merely transmitted the charge to the proper points. When he published his work in 1791, he termed this charge "animal electricity." Many wondered, including Galvani himself, if he had discovered the long-sought "life-force."

Volta, at the University of Pavia 200 km away, was skeptical of Galvani's results, and came to believe that the source of the electricity was not in the animal itself, but rather in the *contact between the dissimilar metals*. Volta realized that a moist conductor, such as a frog muscle or moisture at the contact point of two dissimilar metals, was necessary in the circuit if it was to be effective. He also saw that the contracting frog muscle was a sensitive instrument for detecting electric "tension" or "electromotive force" (his words for what we now call potential), in fact more sensitive than the best available electroscopes that he and others had developed.[†]

Volta's research found that certain combinations of metals produced a greater effect than others, and, using his measurements, he listed them in order of effectiveness. (This "electrochemical series" is still used by chemists today.) He also found that carbon could be used in place of one of the metals.

Volta then conceived his greatest contribution to science. Between a disc of zinc and one of silver, he placed a piece of cloth or paper soaked in salt solution or dilute acid and piled a "battery" of such couplings, one on top of another, as shown in Fig. 25–2. This "pile" or "battery" produced a much increased potential difference. Indeed, when strips of metal connected to the two ends of the pile were brought close, a spark was produced. Volta had designed and built the first electric battery.

FIGURE 25–2 A voltaic battery: Taken from Volta's original publication.

Cells and batteries

A battery produces electricity by transforming chemical energy into electrical energy. Today a great variety of electric cells and batteries are available, from flashlight batteries to the storage battery of a car. The simplest batteries contain two plates or rods made of dissimilar metals (one can be carbon) called **electrodes**. The electrodes are immersed in a solution, such as a dilute acid, called the **electrolyte**. Such a device is properly called an **electric cell**, and several cells connected together is a **battery**, although today even a single cell is called a battery. The chemical reactions involved in most electric cells are quite complicated. Here we describe how one very simple cell works, emphasizing the physical aspects.

The cell shown in Fig. 25–3 uses dilute sulfuric acid as the electrolyte. One of the electrodes is made of carbon, the other of zinc. That part of each electrode outside the solution is called the **terminal**, and connections to wires and circuits are made here. The acid reacts with the zinc electrode and tends to dissolve it. Each zinc atom leaves two electrons behind and enters the solution as a positive ion. The zinc electrode thus acquires a negative charge. As the electrolyte becomes positively charged, electrons are pulled off the carbon electrode. Thus the carbon electrode becomes positively charged. Because there is an opposite charge on the two electrodes, there is a potential difference between the two terminals. In a cell

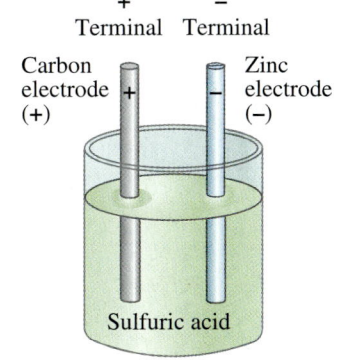

FIGURE 25–3 Simple electric cell.

[†]Volta's most sensitive electroscope measured about 40 V per degree (angle of leaf separation). Nonetheless, he was able to estimate the potential differences produced by dissimilar metals in contact: for a silver-zinc contact he got about 0.7 V, remarkably close to today's value of 0.78 V.

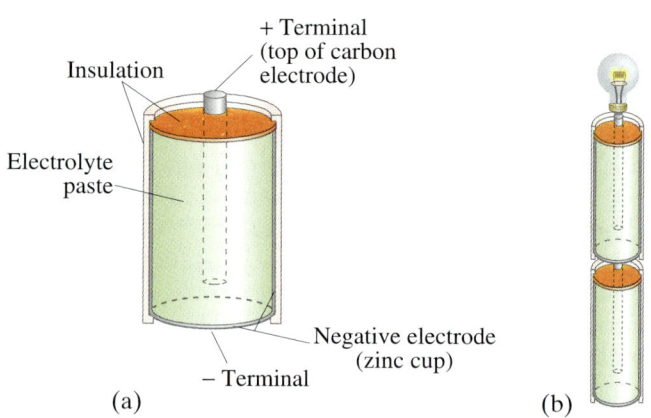

FIGURE 25–4 (a) Diagram of an ordinary dry cell (like a D-cell or AA). The cylindrical zinc cup is covered on the sides; its flat bottom is the negative terminal. (b) Two dry cells (AA type) connected in series. Note that the positive terminal of one cell pushes against the negative terminal of the other.

whose terminals are not connected, only a small amount of the zinc is dissolved, for as the zinc electrode becomes increasingly negative, any new positive zinc ions produced are attracted back to the electrode. Thus, a particular potential difference or voltage is maintained between the two terminals. If charge is allowed to flow between the terminals, say, through a wire (or a lightbulb), then more zinc can be dissolved. After a time, one or the other electrode is used up and the cell becomes "dead."

The voltage that exists between the terminals of a battery depends on what the electrodes are made of and their relative ability to be dissolved or give up electrons.

When two or more cells are connected so that the positive terminal of one is connected to the negative terminal of the next, they are said to be connected in *series* and their voltages add up. Thus, the voltage between the ends of two 1.5-V flashlight batteries connected in series is 3.0 V, whereas the six 2-V cells of an automobile storage battery give 12 V. Figure 25–4 shows (a) a diagram of a common "dry cell" or "flashlight battery" used in portable radios, Walkmans, flashlights, etc., and (b) shows two of them in series.

25–2 Electric Current

Electric circuit

Battery

When a continuous conducting path is connected between the terminals of a battery, we have an electric **circuit**, Fig. 25–5a. On any diagram of a circuit, as in Fig. 25–5b, we represent a battery by the symbol

$$-\!\!|\!\!|\!\!-$$
 $\;+\;\;-$. [battery symbol]

The device powered by the battery could be a lightbulb (which is just a fine wire inside an evacuated glass bulb), a heater, a radio, or whatever. When such a circuit is formed, charge can flow through the wires of the circuit, from one terminal of the battery to the other. Any flow of charge such as this is called an **electric current**. Electric current can flow whenever there is a potential difference between the ends of a conductor—or, more simply, if there are opposite charges at the two ends of the conductor, or even in empty space.

FIGURE 25–5 (a) A simple electric circuit. (b) Schematic drawing of the same circuit.

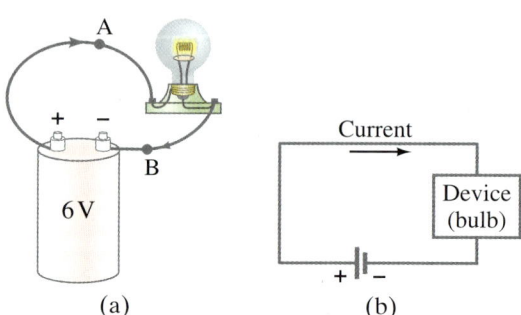

More precisely, the electric current in a wire is defined as the net amount of charge that passes through the wire's full cross section at any point per unit time. Thus, the average current $\overline{I}$ is defined as

$$\overline{I} = \frac{\Delta Q}{\Delta t}, \qquad (25\text{-}1\text{a})$$

Electric current

where ΔQ is the amount of charge that passes through the conductor at any location during the time interval Δt. The instantaneous current is defined by the differential limit

$$I = \frac{dQ}{dt}. \qquad (25\text{-}1\text{b})$$

Electric current is measured in coulombs per second; this is given a special name, the **ampere** (abbreviated amp or A), after the French physicist André Ampère (1775–1836). Thus, $1\,\text{A} = 1\,\text{C/s}$. Smaller units of current are often used, such as the milliampere $(1\,\text{mA} = 10^{-3}\,\text{A})$ and microampere $(1\,\mu\text{A} = 10^{-6}\,\text{A})$.

The ampere $(1\,\text{A} = 1\,\text{C/s})$

In any single circuit, with only a single path for current to follow such as in Fig. 25–5, a steady current at any instant is the same at one point (say point A) as at any other point (such as B). This follows from the conservation of electric charge (charge doesn't disappear).

EXAMPLE 25–1 **Current is flow of charge.** A steady current of 2.5 A flows in a wire for 4.0 min. (*a*) How much charge passed by any point in the circuit? (*b*) How many electrons would this be?

SOLUTION (*a*) Since the current was 2.5 A, or 2.5 C/s, then in 4.0 minutes ($= 240$ seconds) the total charge that flowed was, from Eq. 25–1,

$$\Delta Q = I\,\Delta t$$
$$= (2.5\,\text{C/s})(240\,\text{s}) = 600\,\text{C}.$$

(*b*) The charge on one electron is $1.60 \times 10^{-19}\,\text{C}$, so 600 C would consist of

$$\frac{600\,\text{C}}{1.6 \times 10^{-19}\,\text{C/electron}} = 3.8 \times 10^{21}\,\text{electrons}.$$

We saw in Chapter 21 that conductors contain many free electrons. Thus, if a continuous conducting wire is connected to the terminals of a battery, negatively charged electrons flow in the wire. When the wire is first connected, the potential difference between the terminals of the battery sets up an electric field inside the wire and parallel to it. Thus free electrons at one end of the wire are attracted into the positive terminal, and at the same time, electrons leave the negative terminal of the battery and enter the wire at the other end. There is a continuous flow of electrons through the wire that begins as soon as the wire is connected to *both* terminals. However, when the conventions of positive and negative charge were devised two centuries ago, it was assumed that positive charge flowed in a wire. For nearly all purposes, positive charge flowing in one direction is exactly equivalent to negative charge flowing in the opposite direction,[†] as shown in Fig. 25–6. Today, we still use the historical convention of positive current when discussing the direction of a current. So when we speak of the current in a circuit, we mean the direction positive charge would flow. This is sometimes referred to as **conventional current**. When we want to speak of the direction of electron flow, we will specifically state it is the electron current. In liquids and gases, both positive and negative charges (ions) can move.

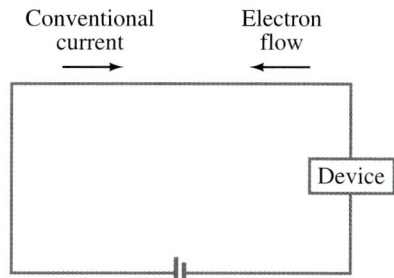

FIGURE 25–6 Conventional current from + to − is equivalent to a negative (electron) current flowing from − to +.

Conventional current

[†]An exception is discussed in Section 27–8.

25–3 Ohm's Law: Resistance and Resistors

To produce an electric current in a circuit, a difference in potential is required. One way of producing a potential difference is by a battery. It was Georg Simon Ohm (1787–1854) who established experimentally that the current in a metal wire is proportional to the potential difference V applied to its two ends:

$$I \propto V.$$

If, for example, we connect a wire to a 6-V battery, the current flow will be twice what it would be if the wire were connected to a 3-V battery. For simplicity, we are using V, rather than V_{ba}, to represent potential difference or voltage. It is also found that reversing the sign of the voltage does not affect the magnitude of the current.

Water analogy

It is helpful to compare an electric current to the flow of water in a river or a pipe acted on by gravity. If the pipe (or river) is nearly level, the flow rate is small. But if one end is somewhat higher than the other, the flow rate—or current—is greater. The greater the difference in height, the swifter the current. We saw in Chapter 23 that electric potential is analogous, in the gravitational case, to the height of a cliff. This applies in the present case to the height through which the fluid flows. Just as an increase in height can cause a greater flow of water, so a greater electric potential difference, or voltage, causes a greater electric current flow.

Exactly how much current flows in a wire depends not only on the voltage, but also on the resistance the wire offers to the flow of electrons. The walls of a pipe, or the banks of a river and rocks in the middle, offer resistance to the flow of current. Similarly, electrons are slowed down because of interactions with the atoms of the wire. The higher this resistance, the less the current for a given voltage V. We then define *resistance* so that the current is inversely proportional to the resistance: that is,

$$R = \frac{V}{I} \tag{25–2a}$$

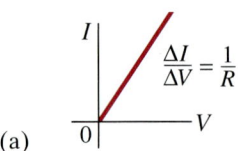

(a)

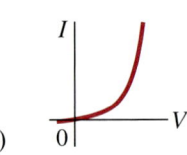

(b)

FIGURE 25–7 Graphs of current vs. voltage for (a) a metal conductor which obeys Ohm's law, and (b) for a nonohmic device, in this case a semiconductor diode.

where R is the **resistance** of a wire or other device, V is the potential difference across the device, and I is the current that flows through it. Eq. 25–2a is often written as

$$V = IR. \tag{25–2b}$$

As mentioned above, Ohm found experimentally that in metal conductors R is a constant independent of V, a result known as **Ohm's law**. Also, Eq. 25–2b, $V = IR$, is itself sometimes called Ohm's law, but only when referring to materials or devices for which R is a constant independent of V. But R is not a constant for many substances, nor for devices such as diodes, vacuum tubes, transistors, and so on. Thus "Ohm's law" is not a fundamental law, but rather a description of a certain class of materials (metal conductors). Materials or devices that do not follow Ohm's law (R = constant) are said to be *nonohmic*. See Fig. 25–7.

The ohm $(1 \, \Omega = 1 \, \text{V/A})$

The unit for resistance is called the **ohm** and is abbreviated Ω (Greek capital omega). Because $R = V/I$, we see that $1.0 \, \Omega$ is equivalent to $1.0 \, \text{V/A}$.

FIGURE 25–8 Example 25–2.

CONCEPTUAL EXAMPLE 25–2 **Current and potential.** Current I enters a resistor R as shown in Fig. 25–8. (*a*) Is the potential higher at point A or at point B? (*b*) Is the current greater at point A or at point B?

RESPONSE (*a*) Positive charge always flows from + to −, from high potential to low potential. Think again of the gravitational analogy: a mass will fall down from high gravitational potential to low. So for positive current I, point A is at a higher potential than point B. (*b*) Conservation of charge requires that whatever current flows into the resistor at point A, emerges at point B. Current does not get "used up" by a resistor, just as an object that falls through a gravitational potential difference does not gain or lose mass. So the current is the same at A and B.

EXAMPLE 25-3 Flashlight bulb resistance. A small flashlight bulb (Fig. 25–9) draws 300 mA from its 1.5-V battery. (*a*) What is the resistance of the bulb? (*b*) If the voltage dropped to 1.2 V, how would the current change?

SOLUTION (*a*) From Eq. 25–2,

$$R = \frac{V}{I} = \frac{1.5 \text{ V}}{0.30 \text{ A}} = 5.0 \text{ }\Omega.$$

(*b*) If the resistance didn't change, the current would be

$$I = \frac{V}{R} = \frac{1.2 \text{ V}}{5.0 \text{ }\Omega} = 0.24 \text{ A},$$

or a drop of 60 mA. Actually, resistance does depend on temperature (Section 25–4), so this is only a rough approximation.

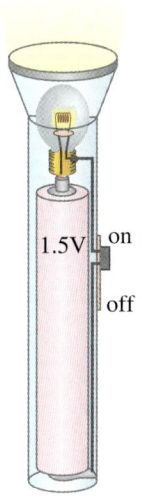

FIGURE 25–9 Flashlight (Example 25–3). Note how circuit is completed along the side strip.

All electric devices, from heaters to lightbulbs to stereo amplifiers, offer resistance to the flow of current. The filaments of lightbulbs and electric heaters are special types of wires whose resistance results in their becoming very hot. Generally, the connecting wires have very low resistance in comparison to the resistance of the wire filaments or coils. In many circuits, particularly in electronic devices, **resistors** are used to control the amount of current. Resistors have resistances from less than an ohm to millions of ohms (see Figs. 25–10 and 25–11). The main types are "wire-wound" resistors which consist of a coil of fine wire; "composition" resistors which are usually made of the semiconductor carbon; and thin metal films.

When we draw a diagram of a circuit, we indicate a resistance with the symbol

─⋀⋀⋀─ .   [resistor symbol]

Wires whose resistance is negligible, however, are shown simply as straight lines.

FIGURE 25–10 Photo of resistors (mostly).

Resistor Color Code			
Color	Number	Multiplier	Tolerance (%)
Black	0	1	
Brown	1	10^1	
Red	2	10^2	
Orange	3	10^3	
Yellow	4	10^4	
Green	5	10^5	
Blue	6	10^6	
Violet	7	10^7	
Gray	8	10^8	
White	9	10^9	
Gold		10^{-1}	5%
Silver		10^{-2}	10%
No color			20%

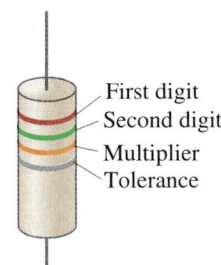

FIGURE 25–11 The resistance value of a given resistor is written on the exterior, or may be given as a color code, as shown above and in the table: the first two colors represent the first two digits in the value of the resistance, the third color represents the power of ten that it must be multiplied by, and the fourth is the manufactured tolerance. For example, a resistor whose four colors are red, green, orange, and silver has a resistance of 25,000 Ω (25 kΩ), give or take 10 percent.

TABLE 25–1 Resistivity and Temperature Coefficients (at 20°C)

Material	Resistivity, ρ ($\Omega \cdot$m)	Temperature Coefficient, α (C°)$^{-1}$
Conductors		
Silver	1.59×10^{-8}	0.0061
Copper	1.68×10^{-8}	0.0068
Gold	2.44×10^{-8}	0.0034
Aluminum	2.65×10^{-8}	0.00429
Tungsten	5.6×10^{-8}	0.0045
Iron	9.71×10^{-8}	0.00651
Platinum	10.6×10^{-8}	0.003927
Mercury	98×10^{-8}	0.0009
Nichrome (alloy of Ni, Fe, Cr)	100×10^{-8}	0.0004
Semiconductors†		
Carbon (graphite)	$(3-60) \times 10^{-5}$	-0.0005
Germanium	$(1-500) \times 10^{-3}$	-0.05
Silicon	$0.1-60$	-0.07
Insulators		
Glass	$10^9 - 10^{12}$	
Hard rubber	$10^{13} - 10^{15}$	

†Values depend strongly on presence of even slight amounts of impurities.

25–4 Resistivity

It is found experimentally that the resistance R of a metal wire is directly proportional to its length l and inversely proportional to its cross-sectional area A. That is,

$$R = \rho \frac{l}{A}, \tag{25-3}$$

Resistivity

where ρ, the constant of proportionality, is called the **resistivity** and depends on the material used. Typical values of ρ, whose units are $\Omega \cdot$m (see Eq. 25–3), are given for various materials in the middle column of Table 25–1. The values depend somewhat on purity, heat treatment, temperature, and other factors. Notice that silver has the lowest resistivity and is thus the best conductor (although it is expensive). Copper is close and much less expensive, so it is clear why most wires are made of copper. Aluminum, although it has a higher resistivity, is much less dense than copper; it is thus preferable to copper in some situations, such as transmission lines, because its resistance for the same weight is less than that for copper.

The reciprocal of the resistivity, called the **conductivity** σ, is

Conductivity

$$\sigma = \frac{1}{\rho}, \tag{25-4}$$

and has units of $(\Omega \cdot m)^{-1}$.

EXAMPLE 25–4 Speaker wires. Suppose you want to connect your stereo to remote speakers (Fig. 25–12). (*a*) If each wire must be 20 m long, what diameter copper wire should you use to keep the resistance less than 0.10 Ω per wire? (*b*) If the current to each speaker is 4.0 A, what is the voltage drop across each wire?

SOLUTION (a) We solve Eq. 25–3 for the area A and use Table 25–1:

$$A = \rho\frac{l}{R} = \frac{(1.68 \times 10^{-8}\,\Omega\cdot m)(20\,m)}{(0.10\,\Omega)} = 3.4 \times 10^{-6}\,m^2.$$

The cross-sectional area A of a circular wire is related to its diameter d by $A = \pi d^2/4$. The diameter must then be at least

$$d = \sqrt{\frac{4A}{\pi}} = 2.1 \times 10^{-3}\,m = 2.1\,mm.$$

(b) From $V = IR$ we have

$$V = IR = (4.0\,A)(0.10\,\Omega) = 0.40\,V.$$

CONCEPTUAL EXAMPLE 25–5 **Stretching changes resistance.** A wire of resistance R is stretched uniformly until it is twice its original length. What happens to its resistance?

RESPONSE If the length l doubles then the cross-sectional area A halves, so that the volume ($V = Al$) of the wire remains the same. From Eq. 25–3 we see that the resistance would increase by a factor of four ($2/\frac{1}{2} = 4$).

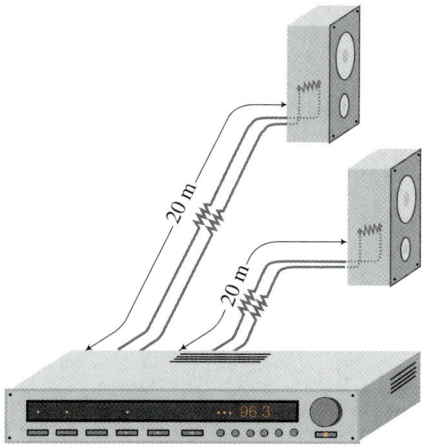

FIGURE 25–12 Example 25–4.

Temperature Dependence of Resistivity

The resistivity of a material depends somewhat on temperature. In general, the resistance of metals increases with temperature. This is not surprising, for at higher temperatures, the atoms are moving more rapidly and are arranged in a less orderly fashion. So they might be expected to interfere more with the flow of electrons. If the temperature change is not too great, the resistivity of metals usually increases nearly linearly with temperature. That is,

$$\rho_T = \rho_0[1 + \alpha(T - T_0)] \qquad (25\text{–}5)$$

Effect of temperature

where ρ_0 is the resistivity at some reference temperature T_0 (such as 0°C or 20°C), ρ_T is the resistivity at a temperature T and α is the *temperature coefficient of resistivity*. Values for α are given in Table 25–1. Note that the temperature coefficient for semiconductors can be negative. Why? It seems that at higher temperatures, some of the electrons that are not normally free in a semiconductor become free and can contribute to the current. Thus, the resistance of a semiconductor can decrease with an increase in temperature, although this is not always the case.

EXAMPLE 25–6 **Resistance thermometer.** The variation in electrical resistance with temperature can be used to make precise temperature measurements. Platinum is commonly used since it is relatively free from corrosive effects and has a high melting point. Suppose at 20°C the resistance of a platinum resistance thermometer is 164.2 Ω. When placed in a particular solution, the resistance is 187.4 Ω. What is the temperature of this solution?

→ **PHYSICS APPLIED**
Resistance thermometer

SOLUTION Since the resistance R is directly proportional to the resistivity ρ, we can combine Eq. 25–3 with Eq. 25–5:

$$R = R_0[1 + \alpha(T - T_0)].$$

Here $R_0 = \rho_0 L/A$ is the resistance of the wire at $T_0 = 20°C$. We solve this equation for T and find (see Table 25–1 for α)

$$T = T_0 + \frac{R - R_0}{\alpha R_0} = 20°C + \frac{187.4\,\Omega - 164.2\,\Omega}{(3.927 \times 10^{-3}(C°)^{-1})(164.2\,\Omega)} = 56.0°C.$$

More convenient for some applications is a *thermistor* (Fig. 25–13), which consists of a metal oxide or semiconductor whose resistance also varies in a repeatable way with temperature. They can be made quite small and respond very quickly to temperature changes. Resistance thermometers have another advantage in that they can be used at very high or low temperatures where gas or liquid thermometers would be useless.

FIGURE 25–13 A thermistor shown next to a millimeter ruler for scale.

The value of α in Eq. 25–5 itself can depend on temperature, so it is important to check the temperature range of validity of any value (say, in a handbook of physical data). If the temperature range is wide, Eq. 25–5 is not adequate and terms proportional to the square and cube of the temperature are needed, but they are generally very small except when $T - T_0$ is large.

25–5 Electric Power

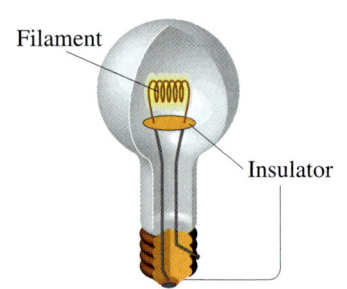

FIGURE 25–14 Incandescent lightbulb.

Electric energy is useful to us because it can be easily transformed into other forms of energy. Motors, whose operation we will examine in Chapter 27, transform electric energy into mechanical work.

In other devices such as electric heaters, stoves, toasters, and hair dryers, electric energy is transformed into thermal energy in a wire resistance known as a "heating element." And in an ordinary lightbulb, the tiny wire filament (Fig. 25–14) becomes so hot it glows; only a few percent of the energy is transformed into visible light, and the rest, over 90 percent, into thermal energy. Lightbulb filaments and heating elements in household appliances have resistances typically of a few ohms to a few hundred ohms.

Electric energy is transformed into thermal energy or light in such devices, and there are many collisions between the moving electrons and the atoms of the wire. In each collision, part of the electron's kinetic energy is transferred to the atom with which it collides. As a result, the kinetic energy of the wire's atoms increases and hence the temperature of the wire element increases. The increased thermal energy can be transferred as heat by conduction and convection to the air in a heater or to food in a pan, by radiation to bread in a toaster, or radiated as light.

To find the power transformed by an electric device, recall that the energy transformed when an infinitesimal charge dq moves through a potential difference V is $dU = dq\,V$ (Eq. 23–2). If dt is the time required for an amount of charge dq to move through the potential difference V, the power P, which is the rate energy is transformed, is

$$P = \frac{dU}{dt} = \frac{dq}{dt}V.$$

The charge that flows per second, dq/dt, is the electric current I. Thus we have

Electric power (general)

$$P = IV. \quad (25\text{–}6)$$

This general relation gives us the power transformed by any device, where I is the current passing through it and V is the potential difference across it. It also gives the power delivered by a source such as a battery. The SI unit of electric power is the same as for any kind of power, the **watt** (1 W = 1 J/s).

The rate of energy transformation in a resistance R can be written, using $V = IR$, in two other ways:

Electric power (in resistance R)

$$P = IV \quad (25\text{–}7a)$$
$$= I(IR) = I^2R \quad (25\text{–}7b)$$
$$= \left(\frac{V}{R}\right)V = \frac{V^2}{R}. \quad (25\text{–}7c)$$

Equations 25–7b and c apply only to resistors, whereas Eq. 25–7a, $P = IV$, applies to any device.

EXAMPLE 25–7 **Headlights.** Calculate the resistance of a 40-W automobile headlight designed for 12 V (Fig 25–15).

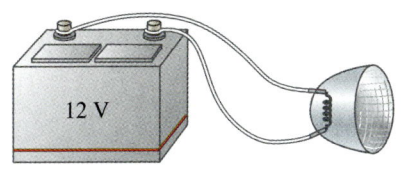

SOLUTION Since we are given $P = 40\,\text{W}$ and $V = 12\,\text{V}$, we can use Eq. 25–7c and solve for R:

$$R = \frac{V^2}{P} = \frac{(12\,\text{V})^2}{(40\,\text{W})} = 3.6\,\Omega.$$

40-W Headlight

FIGURE 25–15 Example 25–7.

This is the resistance when the bulb is burning brightly at 40 W. When the bulb is cold, the resistance is much lower, as we saw in Eq. 25–5. Since the current is high when the resistance is low, lightbulbs burn out most often when first turned on.

It is energy, not power, that you pay for on your electric bill. Since power is the *rate* energy is transformed, the total energy used by any device is simply its power consumption multiplied by the time it is on. If the power is in watts and the time is in seconds, the energy will be in joules since $1\,\text{W} = 1\,\text{J/s}$. Electric companies usually specify the energy with a much larger unit, the **kilowatt-hour** (kWh). One kWh = $(1000\,\text{W})(3600\,\text{s}) = 3.60 \times 10^6\,\text{J}$.

You pay for energy

Kilowatt-hour (unit of energy)

EXAMPLE 25–8 **Electric heater.** An electric heater draws a steady 15.0 A on a 120-V line. How much power does it require and how much does it cost per month (30 days) if it operates 3.0 h per day and the electric company charges 10.5 cents per kWh?

SOLUTION The power is

$$P = IV = (15.0\,\text{A})(120\,\text{V}) = 1800\,\text{W}$$

or 1.80 kW. To operate it for $(3.0\,\text{h/d})(30\,\text{d}) = 90\,\text{h}$ would cost $(1.80\,\text{kW})(90\,\text{h})(\$0.105) = \$17$.

EXAMPLE 25–9 **ESTIMATE** **Lightning bolt.** Lightning is a spectacular example of electric current in a natural phenomenon (Fig. 25–16). There is much variability to lightning bolts, but a typical event can transfer $10^9\,\text{J}$ of energy across a potential difference of perhaps $5 \times 10^7\,\text{V}$ during a time interval of about 0.2 s. Use this information to estimate the total amount of charge transferred, the current, and the average power over the 0.2 s.

▶ **PHYSICS APPLIED**
Lightning

SOLUTION From Eq. 23–2, energy = QV, so

$$Q \approx \frac{10^9\,\text{J}}{5 \times 10^7\,\text{V}} = 20\,\text{C}.$$

The current over the 0.2 s is about

$$I = \frac{Q}{t} \approx \frac{20\,\text{C}}{0.2\,\text{sec}} = 100\,\text{A}.$$

FIGURE 25–16 Example 25–9: a lightning bolt. See caption for photo at start of Chapter 23.

Since most lightning bolts consist of several stages, it is possible that individual parts could carry currents much higher than this. The average power delivered is

$$\overline{P} = \frac{\text{energy}}{\text{time}} = \frac{10^9\,\text{J}}{0.2\,\text{sec}} = 5 \times 10^9\,\text{W} = 5\,\text{GW}.$$

We can also use Eq. 25–7a:

$$P = IV = (100\,\text{A})(5 \times 10^7\,\text{V}) = 5\,\text{GW}.$$

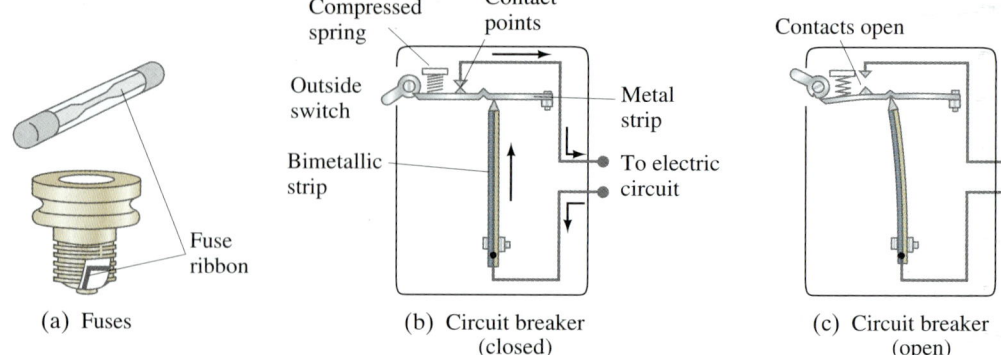

(a) Fuses (b) Circuit breaker (closed) (c) Circuit breaker (open)

FIGURE 25–17 (a) Fuses. When the current exceeds a certain value, the metallic ribbon melts and the circuit opens. Then the fuse must be replaced. (b) A circuit breaker. Electric current passes to a circuit via a bimetallic strip. When the current is great enough (i.e., too great to be safe), the bimetallic strip is heated and bends so far to the left that the notch in the spring-loaded metal strip drops down over the end of the bimetallic strip; (c) the circuit then opens at the contact points (one is attached to the metal strip) and the outside switch is also flipped. As soon as the bimetallic strip cools down, it can be reset using the outside switch.

25–6 Power in Household Circuits

The electric wires that carry electricity to lights and other electric appliances have some resistance, although usually it is quite small. Nonetheless, if the current is large enough, the wires will heat up and produce thermal energy at a rate equal to I^2R, where R is the wire's resistance. One possible hazard is that the current-carrying wires in the wall of a building may become so hot as to start a fire. Thicker wires, of course, have less resistance (see Eq. 25–3) and thus can carry more current without becoming too hot. When a wire carries more current than is safe, it is said to be "overloaded." To prevent overloading, fuses or circuit breakers are installed in circuits. They are basically switches (Fig. 25–17) that open the circuit when the current exceeds some particular value. A 20-A fuse or circuit breaker, for example, opens when the current passing through it exceeds 20 A. If a circuit repeatedly burns out a fuse or opens a circuit breaker, there are two possibilities: there may be too many devices drawing current in that circuit; or there is a fault somewhere, such as a "short." A short, or "short circuit," means that two wires have crossed (perhaps because the insulation has worn down) so the path of the current is shortened. The resistance of the circuit is then very small, so the current will be very large. Short circuits, of course, should be remedied immediately.

Household circuits are designed with the various devices connected so that each receives the standard voltage (usually 120 V in the United States) from the electric company (Fig. 25–18). Circuits with the devices arranged as in Fig. 25–18 are called *parallel circuits*, as we will discuss more fully in the next chapter. When a fuse blows or circuit breaker opens, the total current being drawn on that circuit should be checked.

→ **PHYSICS APPLIED**
Safety—wires getting hot

Fuses, circuit breakers, and shorts

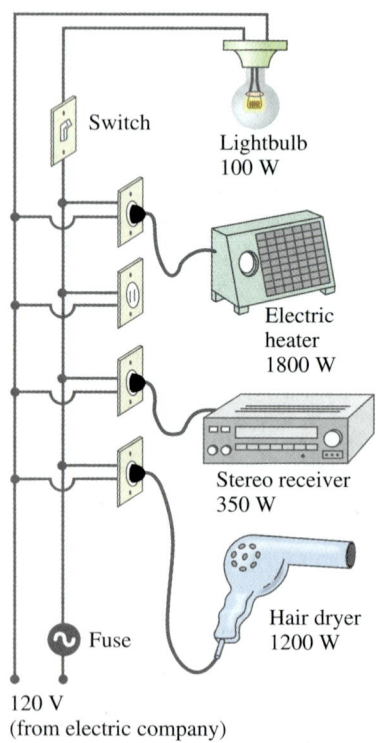

FIGURE 25–18 Connection of household appliances.

EXAMPLE 25–10 **Will a fuse blow?** Determine the total current drawn by all the devices in the circuit of Fig. 25–18.

SOLUTION The circuit in Fig. 25–18 draws the following currents: the lightbulb draws $I = P/V = 100\,\text{W}/120\,\text{V} = 0.8\,\text{A}$; the heater draws $1800\,\text{W}/120\,\text{V} = 15.0\,\text{A}$; the stereo draws a maximum of $350\,\text{W}/120\,\text{V} = 2.9\,\text{A}$; and the hair dryer draws $1200\,\text{W}/120\,\text{V} = 10.0\,\text{A}$. The total current drawn, if all devices are used at the same time, is

$$0.8\,\text{A} + 15.0\,\text{A} + 2.9\,\text{A} + 10.0\,\text{A} = 28.7\,\text{A}.$$

If the circuit in Fig. 25–18 is designed for a 20-A fuse, the fuse should blow, and we hope it will, to prevent overloaded wires from getting hot enough to start a fire. Something will have to be turned off to get this circuit below 20 A. (Houses and apartments usually have several circuits, each with its own fuse or circuit breaker; try moving one of the devices to another circuit.) If the circuit is designed for a 30-A fuse, it shouldn't blow, so if it does, a short may be the problem. (The most likely place is in the cord of one of the devices.) Proper fuse size is selected according to the wire used to supply the current; a properly rated fuse should *never* be replaced by a higher-rated one. A fuse blowing or a circuit breaker opening is acting like a switch, making an "open circuit." By an open circuit, we mean that there is no longer a complete conducting path, so no current can flow; it is as if $R = \infty$.

25–7 Alternating Current

When a battery is connected to a circuit, the current flows steadily in one direction. This is called a **direct current**, or **dc**. Electric generators at electric power plants, however, produce **alternating current**, or **ac**. (Sometimes capital letters are used, DC and AC.) An alternating current reverses direction many times per second and is commonly sinusoidal, as shown in Fig. 25–19. The electrons in a wire first move in one direction and then in the other. The current supplied to homes and businesses by electric companies is ac throughout virtually the entire world. We will discuss and analyze ac circuits in detail in Chapter 31. But because ac circuits are so common in real life, we will discuss some of their simpler aspects here.

The voltage produced by an ac electric generator is sinusoidal, as we shall see later. The current it produces is thus sinusoidal (Fig. 25–19b). We can write the voltage as a function of time as

$$V = V_0 \sin 2\pi f t = V_0 \sin \omega t.$$

The potential V oscillates between $+V_0$ and $-V_0$. V_0 is referred to as the **peak voltage**. The frequency f is the number of complete oscillations made per second, and $\omega = 2\pi f$. In most areas of the United States and Canada, f is 60 Hz (the unit "hertz," as we saw in Chapter 14, means cycles per second). In many countries, 50 Hz is used.

From Eq. 25–2, $V = IR$, if a voltage V exists across a resistance R, then the current I is

$$I = \frac{V}{R} = \frac{V_0}{R} \sin \omega t = I_0 \sin \omega t. \qquad (25\text{–}8)$$

The quantity $I_0 = V_0/R$ is the **peak current**. The current is considered positive when the electrons flow in one direction and negative when they flow in the opposite direction. It is clear from Fig. 25–19b that an alternating current is as often positive as it is negative. Thus, the average current is zero. This does not mean, however, that no power is needed or that no heat is produced in a resistor. Electrons do move back and forth, and do produce heat. Indeed, the power delivered to a resistance R at any instant is

$$P = I^2 R = I_0^2 R \sin^2 \omega t.$$

Because the current is squared, we see that the power is always positive, Fig. 25–20. The quantity $\sin^2 \omega t$ varies between 0 and 1; and it is not too difficult to show[†] that its average value is $\frac{1}{2}$ as can be seen graphically in the figure. Thus, the *average power* developed, $\overline{P}$, is

$$\overline{P} = \tfrac{1}{2} I_0^2 R.$$

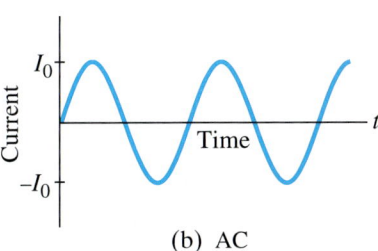

FIGURE 25–19 (a) Direct current. (b) Alternating current.

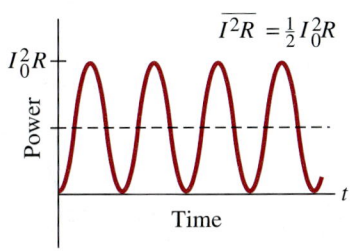

FIGURE 25–20 Power delivered to a resistor in an ac circuit.

[†]A graph of $\cos^2 \omega t$ versus t is identical to that for $\sin^2 \omega t$ in Fig. 25–20 except that the points are shifted (by $\frac{1}{4}$ cycle) on the time axis. Hence the average value of $\sin^2$ and $\cos^2$, averaged over one or more full cycles, will be the same: $\overline{\sin^2 \omega t} = \overline{\cos^2 \omega t}$. From the trigonometric identity $\sin^2 \theta + \cos^2 \theta = 1$, we can write

$$\overline{(\sin^2 \omega t)} + \overline{(\cos^2 \omega t)} = 2\overline{(\sin^2 \omega t)} = 1.$$

Hence the average value of $\sin^2 \omega t$ is $\frac{1}{2}$.

Since power can also be written $P = V^2/R = (V_0^2/R)\sin^2 \omega t$, we also have that the average power is

$$\overline{P} = \frac{1}{2}\frac{V_0^2}{R}.$$

The average or mean value of the *square* of the current or voltage is thus what is important for calculating average power: $\overline{I^2} = \frac{1}{2}I_0^2$ and $\overline{V^2} = \frac{1}{2}V_0^2$. The square root of each of these is the **rms** (root-mean-square) value of the current or voltage:

rms current
$$I_{\text{rms}} = \sqrt{\overline{I^2}} = \frac{I_0}{\sqrt{2}} = 0.707 I_0, \tag{25-9a}$$

rms voltage
$$V_{\text{rms}} = \sqrt{\overline{V^2}} = \frac{V_0}{\sqrt{2}} = 0.707 V_0. \tag{25-9b}$$

The rms values of V and I are sometimes called the "effective values." They are useful because they can be substituted directly into the power formulas, Eqs. 25–7, to get the average power:

$$\overline{P} = \frac{1}{2}I_0^2 R = I_{\text{rms}}^2 R \tag{25-10a}$$

$$\overline{P} = \frac{1}{2}\frac{V_0^2}{R} = \frac{V_{\text{rms}}^2}{R}. \tag{25-10b}$$

Thus, a direct current whose values of I and V equal the rms values of I and V for an alternating current will produce the same power. Hence it is usually the rms value of current that is specified or measured. For example, in the United States and Canada, standard line voltage† is 120 V ac. The 120 V is V_{rms}; the peak voltage V_0 is

$$V_0 = \sqrt{2}\, V_{\text{rms}} = 170 \text{ V}.$$

In most of Europe the rms voltage is 240 V, so the peak voltage is 340 V.

EXAMPLE 25–11 Hair dryer. (a) Calculate the resistance and the peak current in a 1000-W hair dryer (Fig. 25–21) connected to a 120-V line. (b) What happens if it is connected to a 240-V line in Britain?

SOLUTION (a) We can apply Eq. 25–7a using rms values. Then the rms current is

$$I_{\text{rms}} = \frac{\overline{P}}{V_{\text{rms}}} = \frac{1000 \text{ W}}{120 \text{ V}} = 8.33 \text{ A}.$$

Thus $I_0 = \sqrt{2}\, I_{\text{rms}} = 11.8 \text{ A}$. Then the resistance is

$$R = \frac{V_{\text{rms}}}{I_{\text{rms}}} = \frac{120 \text{ V}}{8.33 \text{ A}} = 14.4 \, \Omega.$$

The resistance could equally well be calculated using peak values: $R = V_0/I_0 = 170 \text{ V}/11.8 \text{ A} = 14.4 \, \Omega$.
(b) When connected to a 240-V line, more current would flow and the resistance would change with the increased temperature (Section 25–4). But let us make an estimate based on the same 14.4 Ω resistance. The average power delivered would be

$$\overline{P} = \frac{V_{\text{rms}}^2}{R} = \frac{(240 \text{ V})^2}{(14.4 \, \Omega)} = 4000 \text{ W}.$$

This is four times the dryer's power rating and would undoubtedly melt the heating element or the wire coils of the motor. Be sure your hair dryer (or electric shaver) has a 120/240 V switch before traveling too far from home, or carry a transformer (Section 29–6) for travelers.

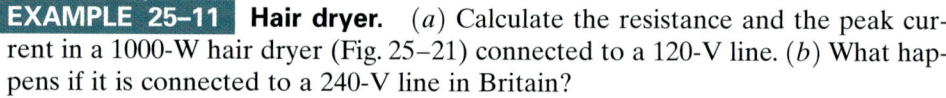

FIGURE 25–21 A hair dryer. Most of the current goes through the heating coils, a pure resistance; a small part goes to the motor to turn the fan. Example 25–11.

†The line voltage can vary, depending on the total load; the frequency of 60 Hz, however, remains extremely steady.

25-8 Microscopic View of Electric Current: Current Density and Drift Velocity

Up to now in this chapter we have dealt mainly with a macroscopic, everyday-world, view of electric current. We did mention, however, that according to atomic theory, the electric current in metal wires is carried by negatively charged electrons, and that in liquid solutions current can also be carried by positive and/or negative ions. Let us now look at this microscopic picture in more detail.

When a potential difference is applied to the two ends of a wire of uniform cross-section, the direction of the electric field **E** is parallel to the walls of the wire (Fig. 25–22). The existence of **E** within the conducting wire does not contradict our earlier result that $\mathbf{E} = 0$ inside a conductor in the electrostatic case. We are no longer dealing with the static case. Charges are free to move in a conductor, and hence can move under the action of the electric field. If all the charges are at rest, then **E** must be zero (electrostatics).

FIGURE 25–22 Electric field **E** in a uniform wire of cross-sectional area A carrying a current I. The current density $j = I/A$.

We now define a new microscopic quantity, the **current density**, **j**. It is defined as the *electric current per unit cross-sectional area* at any point in space. If the current density **j** in a wire of cross-sectional area A is uniform over the cross section, then j is related to the electric current by

$$j = \frac{I}{A} \quad \text{or} \quad I = jA. \qquad (25\text{--}11)$$

Current

If the current density is not uniform, then the general relation is

$$I = \int \mathbf{j} \cdot d\mathbf{A}, \qquad (25\text{--}12)$$

density

defined

where $d\mathbf{A}$ is an element of surface and I is the current through the surface over which the integration is taken. The direction of the current density at any point is the direction that a positive charge would move when placed at that point—that is, the direction of **j** at any point is generally the same as the direction of **E**, Fig. 25–22. (Inertial effects can usually be ignored.) The current density exists for any *point* in space. The current I, on the other hand, refers to a conductor as a whole, and hence is a macroscopic quantity.

The direction of **j** is chosen to represent the direction of flow of positive charge. In a conductor, it is negatively charged electrons that move, so they move in the direction of −**j**, or −**E** (to the left in Fig. 25–22). We can imagine the free electrons as moving about randomly at high speeds, bouncing off the atoms of the wire (somewhat like the molecules of a gas—Chapter 18). When an electric field exists in the wire, Fig. 25–23, the electrons feel a force and initially begin to accelerate. But they soon reach a more or less steady average speed (due to collisions with atoms in the wire), known as their **drift velocity**, v_d. The drift velocity is normally very much smaller than the electrons' average random speed.

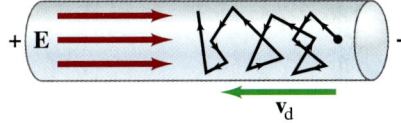

FIGURE 25–23 Electric field **E** in a wire gives electrons in random motion a drift velocity v_d.

Drift velocity

We can relate v_d to the macroscopic current I in the wire. In a time Δt, the electrons will travel a distance $l = v_\text{d} \Delta t$ on average. Suppose the wire has cross-sectional area A. Then in time Δt, electrons in a volume $V = Al = Av_\text{d} \Delta t$ will pass through the cross section A of wire, as shown in Fig. 25–24. If there are n free electrons (each of charge $-e$) per unit volume ($n = N/V$), then the total charge ΔQ that passes through the area A in a time Δt is

$$\begin{aligned}\Delta Q &= (\text{no. of charges, } N) \times (\text{charge per particle}) \\ &= (nV)(-e) = -(nAv_\text{d}\Delta t)(e).\end{aligned}$$

FIGURE 25–24 Electrons in the volume Al will all pass through the cross-section indicated in a time Δt, where $l = v_\text{d} \Delta t$.

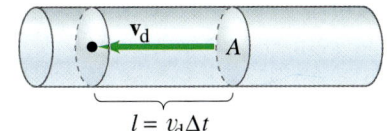

The current I in the wire is thus

$$I = \frac{\Delta Q}{\Delta t} = -neAv_\text{d}. \qquad (25\text{--}13)$$

Current (microscopic variables)

The current density, $j = I/A$, is

Current density in terms of drift velocity

$$j = -nev_d. \quad (25\text{–}14)$$

In vector form, this is written

$$\mathbf{j} = -ne\mathbf{v}_d, \quad (25\text{–}15)$$

where the minus sign indicates that the direction of (positive) current flow is opposite to the drift velocity of electrons.

We can generalize Eq. 25–15 to any type of charge flow, such as flow of ions in an electrolyte. If there are several types of ions (which can include free electrons), each of density n_i (number per unit volume), charge q_i ($q_i = -e$ for electrons) and drift velocity $\mathbf{v}_{di}$, then the net current density at any point is

$$\mathbf{j} = \sum_i n_i q_i \mathbf{v}_{di}. \quad (25\text{–}16)$$

The total current I passing through an area A perpendicular to a uniform $\mathbf{j}$ is then

$$I = \sum_i n_i q_i v_{di} A.$$

EXAMPLE 25–12 Electron speeds in a wire. A copper wire 3.2 mm in diameter, carries a 5.0-A current. Determine (a) the current density in the wire, and (b) the drift velocity of the free electrons. (c) Estimate the rms speed of electrons assuming they behave like an ideal gas at 20°C. Assume that one electron per Cu atom is free to move (the others remain bound to the atom).

SOLUTION (a) The cross-sectional area of the wire is

$$A = \pi r^2 = (3.14)(1.60 \times 10^{-3}\,\text{m})^2 = 8.0 \times 10^{-6}\,\text{m}^2.$$

The current density is then

$$j = \frac{I}{A} = \frac{5.0\,\text{A}}{8.0 \times 10^{-6}\,\text{m}^2} = 6.2 \times 10^5\,\text{A/m}^2.$$

(b) Since we assume there is one free electron per atom, the density of free electrons, n, is the same as the density of Cu atoms. The atomic mass of Cu is 63.5 u (see Periodic Table inside the back cover), so 63.5 g of Cu contains one mole or 6.02×10^{23} free electrons. The mass density of copper (Table 13–1) is $\rho_D = 8.9 \times 10^3\,\text{kg/m}^3$, where $\rho_D = m/V$. (We use ρ_D to distinguish it here from ρ for resistivity.) So the number of free electrons per unit volume is

$$n = \frac{N}{V} = \frac{N}{m/\rho_D} = \frac{N(1\,\text{mole})}{m(1\,\text{mole})}\rho_D$$

$$n = \left(\frac{6.02 \times 10^{23}\,\text{electrons}}{63.5 \times 10^{-3}\,\text{kg}}\right)(8.9 \times 10^3\,\text{kg/m}^3) = 8.4 \times 10^{28}\,\text{m}^{-3}.$$

Then, by Eq. 25–14, the drift velocity is

$$v_d = \frac{j}{ne} = \frac{6.2 \times 10^5\,\text{A/m}^2}{(8.4 \times 10^{28}\,\text{m}^{-3})(1.6 \times 10^{-19}\,\text{C})} = 4.6 \times 10^{-5}\,\text{m/s},$$

which is only about 0.05 mm/s.

(c) If we model the free electrons as an ideal gas (a rather rough approximation), we use Eq. 18–5 to estimate the random rms speed of an electron as it darts around:

$$v_{\text{rms}} = \sqrt{\frac{3kT}{m}} = \sqrt{\frac{3(1.38 \times 10^{-23}\,\text{J/K})(293\,\text{K})}{9.11 \times 10^{-31}\,\text{kg}}} = 1.2 \times 10^5\,\text{m/s}.$$

Thus we see that the drift velocity (average speed in the direction of the current) is very much less than the rms thermal speed of the electrons, by a factor of about 10^9. [Note: The result in (c) is an underestimate. Quantum theory calculations, and experiments, give the rms speed in copper to be about $1.6 \times 10^6\,\text{m/s}$.]

The drift velocity of electrons in a wire is clearly very slow, only about 0.05 mm/s for the example above, which means it takes an electron 20×10^3 s, or $5\frac{1}{2}$ h, to travel only 1 m. This is not, of course, how fast "electricity travels": when you flip a light switch, the light—even if many meters away—goes on nearly instantaneously, for electric fields travel essentially at the speed of light (3×10^8 m/s). We can think of electrons in a wire as being like a pipe full of water: when a little water enters one end of the pipe, almost immediately some water issues forth at the other end.

Electricity's "speed"

Equation 25–2b, $V = IR$, can be written in terms of microscopic quantities as follows. We write the resistance R in terms of the resistivity ρ:

$$R = \rho \frac{l}{A};$$

and we write V and I as

$$I = jA$$

and

$$V = El.$$

The last relation follows from Eq. 23–3, where we assume the electric field is uniform within the wire and l is the length of the wire (or a portion of the wire) between whose ends the potential difference is V. Thus, from $V = IR$, we have

$$V = IR$$

$$El = (jA)\left(\rho \frac{l}{A}\right) = j\rho l$$

so

$$j = \frac{1}{\rho}E = \sigma E, \tag{25–17}$$

where $\sigma = 1/\rho$ is the *conductivity* (Eq. 25–4). For a metal conductor, ρ and σ do not depend on V (and hence not on E). Therefore the current density **j** is proportional to the electrical field **E** in the conductor. This is the "microscopic" statement of Ohm's law. Equation 25–17, which can be written in vector form as

$$\mathbf{j} = \sigma \mathbf{E} = \frac{1}{\rho}\mathbf{E},$$

is sometimes taken as the definition of conductivity σ and resistivity ρ.

EXAMPLE 25–13 Electric field inside a wire. What is the electric field inside the wire of Example 25–12?

SOLUTION Table 25–1 gives $\rho = 1.68 \times 10^{-8}\,\Omega\cdot$m for copper. Since $j = 6.2 \times 10^5$ A/m²,

$$E = \rho j = (1.68 \times 10^{-8}\,\Omega\cdot\text{m})(6.2 \times 10^5\,\text{A/m}^2)$$

$$= 1.0 \times 10^{-2}\,\text{V/m}.$$

For comparison, the electric field between the plates of a capacitor is often much larger; in Example 24–1, for example, E is on the order of 10^4 V/m. Thus we see that only a modest electric field is needed for current flow in practical cases.

*25–9 Superconductivity

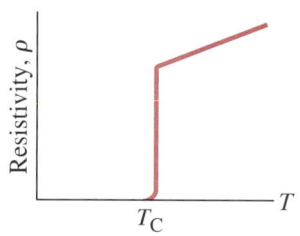

FIGURE 25–25 A superconducting material has zero resistivity when its temperature is below T_C, its "transition temperature." At T_C, the resistivity jumps to a "normal" non-zero value and increases with temperature as most materials do (Eq. 25–5).

High-temperature superconductors

FIGURE 25–26 An experimental train in Japan, supported by the magnetic field produced by current in coils beneath the tracks (in the red containers).

At very low temperatures, near absolute zero, the resistivity (Section 25–4) of certain metals and certain compounds or alloys becomes zero as measured by the highest-precision techniques. Materials in such a state are said to be **superconducting**. This phenomenon was first observed by H. K. Onnes (1853–1926) in 1911 when he cooled mercury below 4.2 K (−269°C). He found that at this temperature, the resistance of mercury suddenly dropped to zero. In general, superconductors become superconducting only below a certain *transition temperature* T_C, which is usually within a few degrees of absolute zero. Current in a ring-shaped superconducting material has been observed to flow for years in the absence of a potential difference, with no measurable decrease. Measurements show that the resistivity ρ of superconductors is less than $4 \times 10^{-25}\ \Omega\cdot\text{m}$, which is over 10^{16} times smaller than that for copper, and is considered to be zero in practice. See Fig. 25–25.

Much research has been done on superconductivity in recent years to try to understand why it occurs, and to find materials that superconduct at more reasonable temperatures to reduce the cost and inconvenience of refrigeration at the required very low temperature. Before 1986 the highest temperature at which a material was found to superconduct was 23 K, and this required liquid helium to keep the material cold. In 1987, a compound of yttrium, barium, copper, and oxygen was developed that can be superconducting at 90 K. Since this is above the temperature of liquid nitrogen, 77 K, boiling liquid nitrogen is sufficiently cold to keep the material superconducting. This was an important breakthrough since liquid nitrogen is much more easily and cheaply obtained than is the liquid helium needed for previous superconductors. Since then, superconductivity at temperatures in the vicinity of 160 K have been reported, though in fragile compounds.

Considerable research is being done to develop high-T_C superconductors as wires that can carry currents strong enough to be practical. Most applications today use a bismuth-strontium-calcium-copper oxide, known (for short) as BSCCO. A major use today of superconductors is for carrying the current in electromagnets (we shall see in Chapter 27 that electric currents produce magnetic fields). In large non-superconducting magnets, a great amount of energy is needed just to maintain the current, and this energy is wasted as heat.

A major problem is how to make a useable, bendable wire out of the BSCCO, which is very brittle. One solution is to embed tiny filaments of the high-T_C superconductor in a metal alloy matrix. The first major commercial use of high-T_C superconductors embeds the filaments in silver; the wires are formed into a cable to carry very high currents for the city of Detroit's electric distribution grid. The superconducting wire is wrapped around a tube carrying liquid nitrogen to keep the BSCCO below T_C. The wire is not resistanceless, because of the silver connections, but the resistance is much less than that of a conventional copper cable. Indeed, the 100 kg of this 130-m-long superconducting cable can carry as much current as 8000 kg of the copper cable it replaces.

Electric motors, generators, and transformers using superconductors are also being worked on, and will be much smaller and lighter than conventional ones. Prototype motors under development are half the size and weight of non-superconducting motors.

Superconductors could make electric cars more practical, make computers much faster, and have great potential in devices to store energy for use at peak demand. Superconductors are being studied for use in high-speed ground transportation: the magnetic fields produced by superconducting magnets would be used to "levitate" vehicles over tracks so there is essentially no friction (Fig. 25–26). The levitation arises from the repulsive force between the magnet (say, on the train) and the eddy currents produced in the track below (or, vice versa).

25–10 Electric Hazards; Leakage Currents

An electric shock can damage the body or may even be fatal. The severity of a shock depends on the magnitude of the current, how long it acts, and through what part of the body it passes. A current passing through vital organs such as the heart or brain is especially serious. Electric current heats tissue and can cause burns. A current also stimulates nerves and muscles, and we feel a "shock."

Most people can "feel" a current of about 1 mA. Currents of a few mA cause pain but rarely much damage in a healthy person. However, currents above 10 mA cause severe contraction of the muscles, and a person may not be able to release the source of the current (say, a faulty appliance or wire). Death from paralysis of the respiratory system can occur. Artificial respiration, however, can sometimes revive a victim. If a current above about 70 mA passes across the torso so that a portion passes through the heart for a second or more, the heart muscles will begin to contract irregularly and blood will not be properly pumped. This condition is called "ventricular fibrillation." If it lasts for long, death results. Strangely enough, if the current is much larger, on the order of 1 A, the damage may be less and death by heart failure may be less likely[†] under some conditions.

The seriousness of a shock depends on the effective resistance of the body. Living tissue has quite low resistance since the fluid of cells contains ions that can conduct quite well. However, the outer layers of skin, when dry, offer much resistance. The effective resistance between two points on opposite sides of the body when the skin is dry is in the range of 10^4 to 10^6 Ω. However, when the skin is wet, the resistance may be 10^3 Ω or less. A person in good contact with the ground who touches a 120-V dc line with wet hands can suffer a current

$$I = \frac{120 \text{ V}}{1000 \, \Omega} = 120 \text{ mA}.$$

As we saw above, this could be lethal.

Figure 25–27 shows how the circuit is completed when a person touches an electric wire. One side of a 120-V source is connected to ground by a wire connected to a buried conductor, such as a water pipe. Thus the current passes from the high-voltage wire, through the person, to the ground; it passes through the ground back to the other terminal of the source to complete the circuit. If the person in Fig. 25–27 stands on a good insulator—thick-soled shoes or a dry wood floor—there will be much more resistance in the circuit and consequently much less current will flow. However, if the person stands with bare feet on the ground, or is sitting in a bathtub, there is considerable danger because the resistance is much less. In a bathtub, not only are you wet, but the water is in contact with the drain pipe that leads to the ground. That is why it is strongly recommended not to touch anything electrical in such a situation.

A principal danger comes from touching a bare wire whose insulation has worn off, or from a bare wire inside an appliance when you're tinkering with it. (Always unplug an electrical device before investigating its insides!) Sometimes a wire inside a device breaks or loses its insulation and comes in contact with the case. If the case is metal, it will conduct electricity. A person could then suffer a

→ **PHYSICS APPLIED**
Electric shock

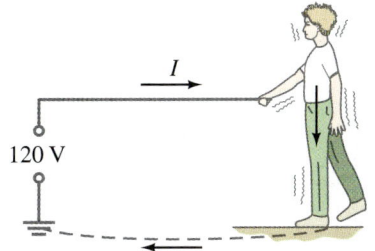

FIGURE 25–27 A person receives an electric shock when the circuit is completed.

→ **PHYSICS APPLIED**
Grounding and shocks

[†]Apparently, larger currents bring the entire heart to a standstill. Upon release of the current, the heart returns to its normal rhythm. This may not happen when fibrillation occurs since it is often hard to stop once it starts. Fibrillation may also occur as a result of a heart attack or during heart surgery. A device known as a *defibrillator* can apply a brief high current to the heart; this causes complete heart stoppage and is often followed by resumption of normal beating.

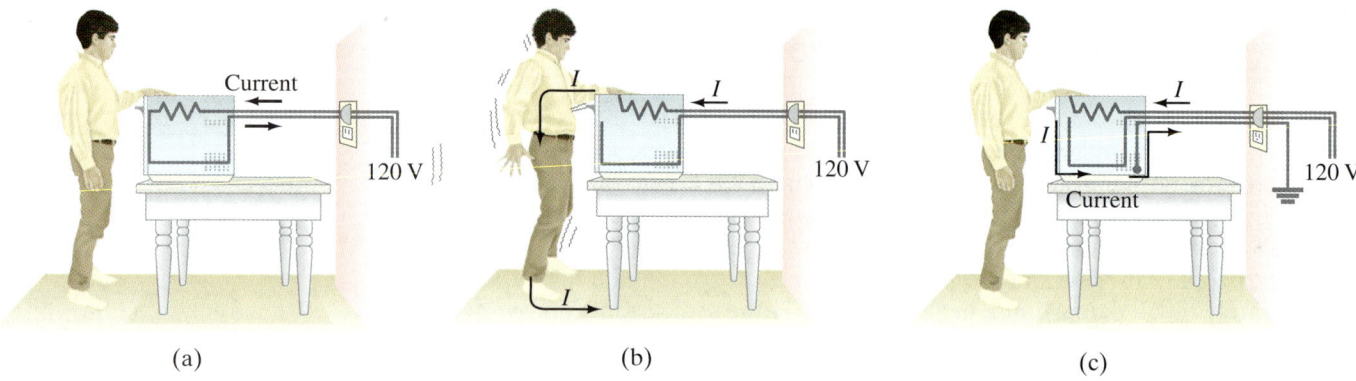

FIGURE 25–28 (a) An electric appliance operating normally with a two-prong plug. (b) Short to the case with ungrounded case: shock. (c) Short to the case with the case grounded with third prong.

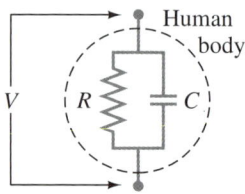

FIGURE 25–29 Human body modeled electrically as resistance and capacitor in parallel when a voltage is applied.

Leakage current

severe shock merely by touching the case, as shown in Fig. 25–28b. To prevent an accident, metal cases are supposed to be connected directly to ground, so they cannot become "hot." Then if a "hot" wire touches the grounded case, a short circuit to ground immediately occurs internally, as shown in Fig. 25–28c; most of the current passes through the low resistance ground wire rather than through the person. Furthermore, the high current immediately opens the fuse or circuit breaker in the circuit. Grounding a metal case is best done by a separate ground wire connected to the third (round) prong of a 3-prong plug. It can also be done by connecting the case to the larger prong of a so-called "polarized" 2-prong plug. Of course not only the device, but also the outlets must be wired correctly to ground.

The human body acts as if it had capacitance in parallel with its resistance (Fig. 25–29). A dc current can pass through the resistance, but not the capacitance. An ac current, like the changing currents discussed in Section 25–7, can exist also in the capacitive branch. Because of the additional path allowing current flow, the ac current for a given V_{rms} will be greater than for the same dc voltage. Thus an ac voltage is more dangerous than an equal dc voltage.

Another danger is *leakage current*, by which we mean a current along an unintended path. Leakage currents are often capacitively coupled. For example, a wire in a lamp forms a capacitor with the metal case; charges moving in one conductor attract or repel charge in the other, so there is a current. Typical electrical codes limit leakage currents to 1 mA for any device. A 1-mA leakage current is usually harmless. It can be very dangerous, however, to a hospital patient with implanted electrodes connected to ground through the apparatus. This is because the current can pass directly through the heart as compared to the usual situation where the current enters at the hands and spreads out through the body. Although 70 mA may be needed to cause heart fibrillation when entering through the hands (very little of it actually passes through the heart), as little as 0.02 mA has been known to cause fibrillation when passing directly to the heart. Thus, a "wired" patient is in considerable danger from leakage current even from as simple an act as touching a lamp.

Summary

An electric battery serves as a source of potential difference by transforming chemical energy into electric energy. A simple battery consists of two electrodes made of different metals immersed in a solution or paste known as an electrolyte.

Electric current, I, refers to the rate of flow of electric charge and is measured in **amperes** (A): 1 A equals a flow of 1 C/s past a given point.

The direction of **conventional current** flow is that of positive charge. In a wire, it is actually negatively charged electrons that move, so they flow in a direction opposite to the direction of the conventional current. Positive conventional current always flows from a high potential to a low potential.

The **resistance** R of a device is defined by the relation

$$V = IR,$$

where I is the current in the device when a potential difference V is applied across it. For materials such as metals, R is a constant independent of V (thus $I \propto V$), a result known as Ohm's law.

The unit of resistance is the **ohm** (Ω), where $1\,\Omega = 1\,\text{V/A}$. See Table 25–2.

TABLE 25–2	Summary of Units
Current	$1\,\text{A} = 1\,\text{C/s}$
Potential difference	$1\,\text{V} = 1\,\text{J/C}$
Power	$1\,\text{W} = 1\,\text{J/s}$
Resistance	$1\,\Omega = 1\,\text{V/A}$

The resistance R of a wire is inversely proportional to its cross-sectional area A, and directly proportional to its length l and to a property of the material called its resistivity: $R = \rho l / A$. The **resistivity**, ρ, increases with temperature for metals, but for semiconductors it may decrease.

The rate at which energy is transformed in a resistance R from electric to other forms of energy (such as heat and light) is equal to the product of current and voltage. That is, the **power** transformed, measured in watts, is given by

$$P = IV$$

and for resistors can be written as

$$P = I^2 R = \frac{V^2}{R}.$$

The total electric energy transformed in any device equals the product of power and the time during which the device is operated. In SI units, energy is given in joules ($1\,\text{J} = 1\,\text{W}\cdot\text{s}$), but electric companies use a larger unit, the **kilowatt-hour** ($1\,\text{kWh} = 3.6 \times 10^6\,\text{J}$).

Electric current can be **direct current (dc)**, in which the current is steady in one direction; or it can be **alternating current (ac)**, in which the current reverses direction at a particular frequency f, typically 60 Hz. Alternating currents are often sinusoidal in time, $I = I_0 \sin \omega t$, where $\omega = 2\pi f$, and are produced by an alternating voltage.

The rms values of sinusoidally alternating currents and voltages are given by

$$I_{\text{rms}} = \frac{I_0}{\sqrt{2}} \quad \text{and} \quad V_{\text{rms}} = \frac{V_0}{\sqrt{2}},$$

respectively, where I_0 and V_0 are the peak values. The power relationship, $P = IV = I^2 R = V^2/R$, is valid for the average power in alternating currents when the rms values of V and I are used.

Current density $\mathbf{j}$ is the current per cross-sectional area. From a microscopic point of view, the current density is related to the number of charge carriers per unit volume, n, their charge, q, and their drift velocity, $\mathbf{v}_d$, by $\mathbf{j} = nq\mathbf{v}_d$. The electric field within a wire is related to $\mathbf{j}$ by $\mathbf{j} = \sigma \mathbf{E}$ where $\sigma = 1/\rho$ is the **conductivity**.

At very low temperatures certain materials become **superconducting**, which means their electrical resistance becomes zero.

Electric shocks are caused by current passing through the body. To avoid shocks, the body must not become part of a circuit by allowing different parts of the body to touch objects at different potentials.

Questions

1. Car batteries can be rated in ampere-hours (A·h). What aspect of the battery is being rated?

2. When an electric cell is connected to a circuit, electrons flow away from the negative terminal in the circuit. But within the cell, electrons flow *to* the negative terminal. Explain.

3. Develop an analogy between blood circulation and an electrical circuit. Discuss what plays the role of the heart for the electric case, and so on.

4. In a car, one terminal of the battery is said to be connected to "ground." Since it is not really connected to the ground, what is meant by this expression?

5. When you turn on a water faucet, the water usually flows immediately. You don't have to wait for water to flow from the faucet valve to the spout. Why not? Is the same thing true when you connect a wire to the terminals of a battery?

6. Can a copper wire and an aluminum wire of the same length have the same resistance? Explain.

7. If a rectangular solid made of carbon has sides of length a, $2a$, $3a$, how would you connect the wires from a battery so as to obtain (a) the least resistance, (b) the greatest resistance?

8. The equation $P = V^2/R$ indicates that the power dissipated in a resistor decreases if the resistance is increased, whereas the equation $P = I^2R$ implies the opposite. Is there a contradiction here? Explain.

9. What happens when a lightbulb burns out?

10. Explain why lightbulbs almost always burn out just as they are turned on and not after they have been on for some time.

11. Which draws more current, a 100-W lightbulb or a 75-W bulb? Which has the higher resistance?

12. Electric power is transferred over large distances at very high voltages. Explain how the high voltage reduces power losses in the transmission lines.

13. Why is it dangerous to replace a 15-A fuse that blows repeatedly with a 25-A fuse?

14. When electric lights are operated on low-frequency ac (say, 10 Hz) they flicker noticeably. Why?

15. Driven by ac power, the same electrons pass back and forth through your reading lamp over and over again. Explain why the light stays lit instead of going out after the first pass of electrons.

16. The heating element in a toaster is made of Nichrome wire. Immediately after the toaster is turned on, is the current (I_{rms}) in the wire increasing, decreasing, or staying constant? Explain.

17. Is current used up in a resistor?

18. A voltage V is connected across a wire of length l and radius r. How is the electron drift velocity affected if (a) l is doubled, (b) r is doubled, (c) V is doubled.

19. Compare the drift velocities and electric currents in two wires that are geometrically identical and the density of atoms is similar, but the number of free electrons per atom in the material of one wire is twice that in the other.

20. Why is it more dangerous to turn on an electric appliance when you are standing outside in bare feet than when you are inside wearing shoes with thick soles?

Problems

Sections 25–2 and 25–3

1. (I) A current of 1.50 A flows in a wire. How many electrons are flowing past any point in the wire per second? The charge on one electron is 1.60×10^{-19} C.

2. (I) A service station charges a battery using a current of 5.7 A for 7.0 h. How much charge passes through the battery?

3. (I) What is the current in amperes if 1000 Na$^+$ ions were to flow across a cell membrane in 7.5 μs? The charge on the sodium is the same as on an electron, but positive.

4. (I) What is the resistance of a toaster if 110 V produces a current of 4.2 A?

5. (I) What voltage will produce 0.25 A of current through a 3000-Ω resistor?

6. (II) An electric device draws 5.50 A at 110 V. (a) If the voltage drops by 10 percent, what will be the current, assuming nothing else changes? (b) If the resistance of the device were reduced by 10 percent, what current would be drawn at 110 V?

7. (II) A 9.0-V battery is connected to a bulb whose resistance is 1.6 Ω. How many electrons leave the battery per minute?

8. (II) If a 12-V battery pushes a current of 0.50 A through a resistor, what is its resistance, and how many joules of energy does the battery lose in a minute?

9. (II) A hair dryer draws 7.5 A when plugged into a 120-V line. (a) What is its resistance? (b) How much charge passes through it in 15 min? (Assume direct current.)

10. (II) A bird stands on a dc electric transmission line carrying 2500 A (Fig. 25–30). The line has $2.5 \times 10^{-5}\,\Omega$ resistance per meter and the bird's feet are 4.0 cm apart. What potential difference does the bird feel?

FIGURE 25–30 Problem 10.

Section 25–4

11. (I) What is the diameter of a 1.00-m length of tungsten wire whose resistance is 0.22 Ω?

12. (I) What is the resistance of a 3.5-m length of copper wire 1.5 mm in diameter?

13. (II) Compare the resistance of 10.0 m of aluminum wire 2.0 mm in diameter with 20.0 m of copper filament wire 2.5 mm in diameter.

14. (II) Can a 2.5-mm-diameter copper wire have the same resistance as a tungsten wire of the same length? Give numerical details.

15. (II) A certain copper wire has a resistance of 10.0 Ω. At what point along the length must the wire be cut so that the resistance of one piece is 5.0 times the resistance of the other? What is the resistance of each piece.

16. (II) How much would you have to raise the temperature of a copper wire (originally at 20°C) to increase its resistance by 20 percent?

17. (II) A length of aluminum wire is connected to a precision 10.00-V power supply, and a current of 0.4212 A is precisely measured at 20.0°C. The wire is placed in a new environment of unknown temperature where the measured current is 0.3618 A. What is the unknown temperature?

18. (II) Estimate at what temperature copper will have the same resistivity as tungsten does at 20°C.

19. (II) A 100-W lightbulb has a resistance of about 12 Ω when off (20°C) and 140 Ω when on (hot). Estimate the temperature of the filament when "on" assuming an average temperature coefficient of resistivity $\alpha = 0.0060\ (C°)^{-1}$.

20. (II) A rectangular solid made of carbon has sides lying along the x, y, and z axes, whose lengths are 1.0 cm, 2.0 cm, and 4.0 cm, respectively (Fig. 25–31). Determine the resistance for current that flows through the solid in (a) the x direction, (b) the y direction, and (c) the z direction. Assume the resistivity is $\rho = 3.0 \times 10^{-5}\ \Omega \cdot m$.

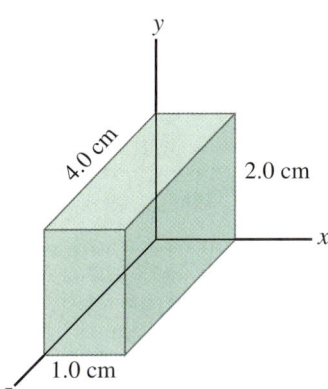

FIGURE 25–31 Problem 20.

21. (II) A length of wire is cut in half and the two lengths are wrapped together side by side to make a thicker wire. How does the resistance of this new combination compare to the resistance of the original wire?

22. (II) For some applications, it is important that the value of a resistance not change with temperature. For example, suppose you made a 4.70-kΩ resistor from a carbon resistor and a Nichrome wire-wound resistor connected together so the total resistance is the sum of their separate resistances. What value should each of these resistors have (at 0°C) so that the combination is temperature independent?

23. (II) (a) Show that if a straight wire of cross-sectional area A lies along the x axis, the rate at which charge flows is given by

$$\frac{dq}{dt} = -\sigma A \frac{dV}{dx},$$

where dV/dx is the potential gradient, and σ is the conductivity. (b) Make an analogy to heat conduction (Chapter 19). Would you expect σ and k (thermal conductivity) to be related?

24. (III) A hollow cylindrical resistor with inner radius r_1 and outer radius r_2, and length l, is made of a material whose resistivity is ρ (Fig. 25–32). (a) Show that the resistance is given by

$$R = \frac{\rho}{2\pi l} \ln \frac{r_2}{r_1}$$

for current that flows radially outward. [*Hint*: Divide the resistor into concentric cylindrical shells and integrate.] (b) Evaluate the resistance R for such a resistor made of carbon whose inner and outer radii are 1.0 mm and 1.8 mm and whose length is 3.0 cm. (c) What is the resistance in part (b) for current flowing *parallel* to the axis?

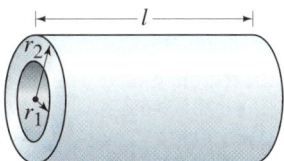

FIGURE 25–32 Problem 24.

25. (III) Determine a formula for the total resistance of a spherical shell made of material whose conductivity is σ and whose inner and outer radii are r_1 and r_2. Assume the current flows radially outward.

26. (III) The filament of a light bulb has a resistance of 12 Ω at 20°C and 140 Ω when hot (as in Problem 19). (a) Calculate the temperature of the filament when it is hot, and take into account the change in length and area of the filament due to thermal expansion (assume tungsten for which the thermal expansion coefficient is $\approx 5 \times 10^{-6}\ C°^{-1}$). (b) In this temperature range, what is the percentage change in resistance due to thermal expansion, and what is the percentage change in resistance due solely to the change in ρ? Use Eq. 25–5.

Sections 25–5 and 25–6

27. (I) What is the maximum power consumption of a 9.0-V portable cassette player that draws a maximum of 350 mA of current?

28. (I) The element of an electric oven is designed to produce 3.1 kW of heat when connected to a 240-V source. What must be the resistance of the element?

29. (I) What is the maximum voltage that can be applied to a 5.4-kΩ resistor rated at $\frac{1}{4}$ watt?

30. (I) A hair dryer has two settings: 600 W and 1200 W. (a) At which setting do you expect the resistance to be higher? After making a guess, determine the resistance at (b) the lower setting, (c) the higher setting.

31. (I) (a) What is the resistance and current through a 60-W lightbulb if it is connected to its proper source voltage of 120 V? (b) Repeat for a 150-W bulb.

32. (I) You buy a 60-W lightbulb in Europe, where electricity is delivered to homes at 240 V. If you use the lightbulb in the United States at 120 V (assume its resistance doesn't change), how bright will it be relative to 60-W 120-V bulbs? Estimate how much power it will consume.

33. (II) How many kWh of energy does a 550-W toaster use in the morning if it is in operation for a total of 10 min? At a cost of 12 cents/kWh, how much would this add to your monthly electric energy bill if you made toast five mornings per week?

34. (II) At $0.110 per kWh, what does it cost to leave a 60-W porch light on day and night for a year?

35. (II) What is the total amount of energy stored in a 12-V, 90-A·h car battery when it is fully charged?

36. (II) A transistor to be used in a circuit is rated for a maximum current of 28 mA if operated at 9.0 V. (a) What is the maximum acceptable power to the transistor, and (b) what would be the limiting current if the voltage applied were actually only 7.0 V?

37. (II) How many 100-W lightbulbs, connected to 120 V as in Fig. 25–18, can be used without blowing a 2.5-A fuse?

38. (II) What is the efficiency of a 0.50-hp electric motor that draws 4.6 A from a 120-V line?

39. (II) A power station delivers 520 kW of power to a factory through wires of total resistance of 3.0 Ω. How much less power is wasted if the electricity is delivered at 50,000 V rather than 12,000 V?

40. (II) A 2800-W oven is hooked to a 240-V source. (a) What is the resistance of the oven? (b) How long will it take to boil 100 mL of water assuming 80 percent efficiency? (c) How much will this cost at 10 cents/kWh?

41. (III) The current in an electromagnet connected to a 240-V line is 14.5 A. At what rate must cooling water pass over the coils if the water temperature is to rise by no more than 6.50 C°?

42. (III) A small immersion heater can be used in a car to heat a cup of water for coffee. If the heater can heat 150 mL of water from 5°C to 95°C in 6.0 min, approximately how much current does it draw from the 12-V battery, and what is its resistance? Assume the manufacturer's claim of 60 percent efficiency.

Section 25–7

43. (I) Calculate the peak current in a 1.8-kΩ resistor connected to a 120-V rms ac source.

44. (I) An ac voltage, whose peak value is 180 V, is across a 330-Ω resistor. What is the value of the peak and rms currents in the resistor?

45. (I) What is the resistance of the circuits in your house as seen by the power company, when (a) everything electrical is turned off, and (b) there is a lone 75-W lightbulb burning on the porch?

46. (II) The peak value of an alternating current passing through a 1500-W device is 6.0 A. What is the rms voltage across it?

47. (II) Calculate the peak voltage across, and peak current through, an 1800-W arc welder connected to a 450-V ac line.

48. (II) What is the maximum instantaneous power dissipated by, and maximum current passing through, a 3.0-hp pump connected to a 240-V ac power source?

49. (II) A heater coil connected to a 240-V ac line has a resistance of 38 Ω. (a) What is the average power used? (b) What are the maximum and minimum values of the instantaneous power?

50. (II) Suppose a current is given by the equation $I = 1.80 \sin 210t$, where I is in amperes, and t in seconds. (a) What is the frequency? (b) What is the rms value of the current? (c) If this is the current through a 42.0-Ω resistor, what is the equation that describes the voltage as a function of time?

Section 25–8

51. (II) A 0.55-mm-diameter copper wire carries a tiny current of 2.5 μA. Estimate (a) the electron drift velocity in the wire, (b) the current density, and (c) the electric field.

52. (II) A 5.00-m length of 2.0-mm-diameter wire carries a 750-mA current when 22.0 mV is applied to its ends. If the drift velocity has been measured (by the Hall effect—Section 27–8) to be 1.7×10^{-5} m/s, determine (a) the resistance R of the wire, (b) the resistivity ρ, (c) the current density j, (d) the electric field inside the wire, and (e) the number n of free electrons per unit volume.

53. (III) At a point high in the Earth's atmosphere, He^{2+} ions in a concentration of $2.8 \times 10^{12}/m^3$ are moving due north at a speed of 2.0×10^6 m/s. Also, an $8.0 \times 10^{11}/m^3$ concentration of O_2^- ions is moving due south at a speed of 7.2×10^6 m/s. Determine the magnitude and direction of the current density **j** at this point.

General Problems

54. How many coulombs are there in 1.00 ampere-hour?

55. A person accidentally leaves a car with the lights on. If each of the two front lights uses 40 W and each of the two rear lights 6 W, for a total of 92 W, how long will a fresh 12-V battery last if it is rated at 90 A·h? Assume the full 12 V appears across each bulb.

56. What is the average current drawn by a 1.5-hp 120-V motor?

57. The *conductance* G of an object is defined as the reciprocal of the resistance R: that is, $G = 1/R$. The unit of conductance is the mho (= ohm^{-1}), which is also called the siemens (S). What is the conductance (in siemens) of an object that draws 800 mA of current at 12.0 V?

58. 10.0 m of wire consists of 5.0 m of copper followed by 5.0 m of aluminum of equal diameter (both 1.0 mm). A potential difference of 25 V is placed across the composite wire. (a) What is the total resistance of the wire? (b) What is the current flow through the wire? (c) What are the voltages across the aluminum part and across the copper part?

59. An ordinary flashlight uses 2 D-cell 1.5 V batteries connected in series as in Fig. 25–4b. The bulb draws 350 mA when turned on. (a) Calculate the resistance of the bulb and the power dissipated. (b) By what factor would the power increase if 4 D-cells in series were used with the same bulb? (Neglect heating effects of the filament.) Why shouldn't you try this?

60. The heating element of a 110-V, 900-W heater is 5.4 m long. If it is made of iron, what must its diameter be?

61. (a) A particular household uses a 1.8-kW heater 3.0 h/day ("on" time), four 100-W lightbulbs 6.0 h/day, a 3.0-kW electric stove element for a total of 1.4 h/day, and miscellaneous power amounting to 2.0 kWh/day. If electricity costs $0.105 per kWh, what will be their monthly bill (30 days)? (b) How much coal (which produces 7000 kcal/kg) must be burned by a 35-percent-efficient power plant to provide the yearly needs of this household?

62. A small city requires about 10 MW of power. Suppose that instead of using high-voltage lines to supply the power, the power is delivered at 120 V. Assuming a two-wire line of 0.50-cm-diameter copper wire, estimate the cost of the energy lost to heat per hour per meter. Assume the cost of electricity is about 10 cents per kWh.

63. A 1200-W hair dryer is designed for 117 V. (a) What will be the percentage change in power output if the voltage drops to 105 V? Assume no change in resistance. (b) How would the actual change in resistivity with temperature affect your answer?

64. The wiring in a house must be thick enough so it doesn't become so hot as to start a fire. What diameter must a copper wire be if it is to carry a maximum current of 30 A and produce no more than 1.6 W of heat per meter of length?

65. An air conditioner draws 12 A at 120 V ac. The connecting cord is copper wire which has a diameter of 1.628 mm. (a) How much power does the air conditioner draw? (b) If the total length of wire is 15 m, how much power is dissipated in the wiring? (c) If no. 12 wire, with a diameter of 2.053 mm were used instead, how much power would be dissipated? (d) Assuming that the air conditioner is run 12 hours per day, how much money per day would be saved by using no. 12 wire? Assume that the cost of electricity is 10 cents per kWh.

66. In a "brownout" situation, the voltage supplied by the electric company falls. Assuming the percent drop is small, show that the power output of a given appliance falls by approximately twice that percent, assuming the resistance does not change. How much of a voltage drop does it take for a 60-W lightbulb to begin acting like a 50-W bulb?

67. A microwave oven running at 60 percent efficiency delivers 900 W of energy per second to the interior. Determine (a) the power drawn from the source, and (b) the rms current drawn. Assume a source voltage of 120 Vrms.

68. A 1.00-Ω wire is drawn out to 3.00 times its original length. What is its resistance now?

69. 220 V is applied to two different conductors made of the same material. One conductor is twice as long and twice as thick as the second. What is the ratio of the power transformed in the first relative to the second?

70. An electric heater is used to heat a room of volume 68 m³. Air is brought into the room at 5°C and is changed completely twice per hour. Heat loss through the walls amounts to approximately 850 kcal/h. If the air is to be maintained at 20°C, what minimum wattage must the heater have? (The specific heat of air is about 0.17 kcal/kg·C°.)

71. A 6.50-Ω resistor is made from a coil of copper wire whose total mass is 18.0 g. What is the diameter of the wire and how long is it?

72. The new EV-1 electric car makes use of storage batteries as its source of energy. Its mass is 1300 kg and it is powered by 26 batteries, each 12 V, 52 A·h. Assume that the car is driven on the level at an average speed of 40 km/h, and the average retarding force is 240 N. Assume 100 percent efficiency and neglect energy used for acceleration. Note that no energy is consumed when the vehicle is stopped since the engine doesn't need to idle. (a) Determine the horsepower required. (b) After approximately how many kilometers must the batteries be recharged?

73. A 100-W, 120-V lightbulb has a resistance of 12 Ω when cold (20°C) and 140 Ω when on (hot). Calculate its power consumption at (a) the instant it is turned on, and (b) after a few moments when it is hot.

74. A capacitor is often used in electronics to keep energy flowing even if there is a momentary loss of power from the electric company. What capacitance would be required for a television, operating at an internal dc voltage of 120 V at 150 W, to provide sufficient energy during a 0.10 s lapse in power?

75. The Tevatron accelerator at Fermilab (Illinois) is designed to carry an 11 mA beam of protons traveling at very nearly the speed of light $(3.0 \times 10^8 \text{ m/s})$ around a ring 6300 m in circumference. How many protons are stored in the beam?

76. How far does an average electron move along the wires of a 500-W toaster during an alternating current cycle? The power cord has copper wires of diameter 1.8 mm and is plugged into a standard 60 Hz 120 V ac outlet. [Hint: The maximum current in the cycle is related to the maximum drift velocity. The maximum velocity in an oscillation is related to the maximum displacement; see Chapter 14.]

77. For the wire in Fig. 25–33, whose diameter varies uniformly from a to b as shown, suppose a current $I = 2.0$ A enters at a. If $a = 3.0$ mm and $b = 4.0$ mm, what is the current density (assume uniform) at each end?

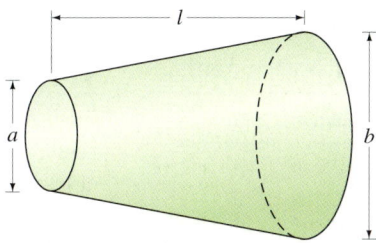

FIGURE 25–33 Problems 77 and 78.

78. The cross section of a portion of wire increases uniformly as shown in Fig. 25–34 so it has the shape of a truncated cone. The diameter at one end is a and at the other it is b, and the total length along the axis is l. If the material has resistivity ρ, determine the resistance R between the two ends in terms of a, b, l, and ρ. Assume that the current flows uniformly through each section, and that the taper is small, i.e., $(b - a) \ll l$.

This "Walkman" contains circuits within it that are dc, at least in part. (The audio signal is ac.) The circuit diagram below shows a possible amplifier circuit (actually, two identical circuits are used, one for each stereo channel). Although the large triangle is an amplifier circuit board containing transistors (not discussed in this chapter), the other circuit elements are ones we have met, resistors and capacitors, and we discuss them in circuits in this chapter. We also discuss voltmeters and ammeters, and how they are built and used.

CHAPTER 26

DC Circuits

TABLE 26–1 Symbols for Circuit Elements	
Symbol	**Device**
⊣⊢	Battery
⊣⊦	Capacitor
─⋀⋀⋀─	Resistor
────	Wire with negligible resistance
⏚ or ↧	Ground

Electric circuits are basic parts of all electronic gear from radio and TV sets to computers and automobiles. Scientific measurements, from physics to biology and medicine, make use of electric circuits. In Chapter 25, we discussed the basic principles of electric current. Now we will apply these principles to analyze dc circuits and to understand the operation of a number of useful instruments.[†]

When we draw a diagram for a circuit, we represent batteries, capacitors, and resistors by the symbols shown in Table 26–1. Wires whose resistance is negligible compared to other resistance in the circuit are drawn simply as straight lines. Some circuit diagrams show a ground symbol (⏚ or ↧) which may mean a real connection to the ground, perhaps via a metal pipe, or it may simply mean a common connection, such as the frame of a car.

For the most part in this chapter, except in Section 26–4, we will be interested in circuits operating in their steady state—that is, we won't be looking at a circuit at the moment a change is made in it, such as when a battery or resistor is connected or disconnected, but rather a short time later when the currents have reached their steady values.

[†] Ac circuits that contain only a voltage source and resistors can be analyzed like the dc circuits in this chapter. However, ac circuits that contain capacitors and other circuit elements are more complicated, and we discuss them in Chapter 31.

26–1 EMF and Terminal Voltage

To have current in an electric circuit, we need a device such as a battery or an electric generator that transforms one type of energy (chemical, mechanical, light, and so on) into electric energy. Such a device is called a *source* of *electromotive force* or of *emf*. (The term "electromotive force" is a misnomer since it does not refer to a "force" that is measured in newtons. Hence, to avoid confusion, we prefer to use the abbreviation, emf.) The potential difference between the terminals of such a source, when no current flows to an external circuit, is called the **emf** of the source. The symbol $\mathscr{E}$ is usually used for emf (don't confuse it with E for electric field).

emf

You may have noticed in your own experience that when a current is drawn from a battery, the voltage across its terminals drops below its rated emf. For example, if you start a car with the headlights on, you may notice the headlights dim. This happens because the starter draws a large current, and the battery voltage drops as a result. The voltage drop occurs because the chemical reactions in a battery cannot supply charge fast enough to maintain the full emf. For one thing, charge must flow (within the electrolyte) between the electrodes of the battery, and there is always some hindrance to completely free flow. Thus, a battery itself has some resistance, which is called its **internal resistance**; it is usually designated r. A real battery is then modeled as if it were a perfect emf $\mathscr{E}$ in series with a resistor r, as shown in Fig. 26–1. Since this resistance r is inside the battery, we can never separate it from the battery. The two points a and b in the diagram represent the two terminals of the battery. What we measure is the **terminal voltage** $V_{ab} = V_a - V_b$. When no current is drawn from the battery, the terminal voltage equals the emf, which is determined by the chemical reactions in the battery: $V_{ab} = \mathscr{E}$. However, when a current I flows naturally from the battery there is an internal drop in voltage equal to Ir. Thus the terminal voltage (the actual voltage delivered) is[†]

Why battery voltage isn't constant

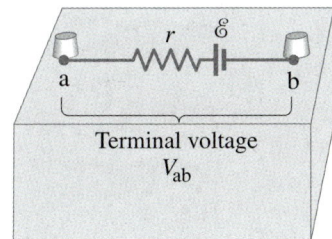

FIGURE 26–1 Diagram for an electric cell or battery.

Terminal voltage

$$V_{ab} = \mathscr{E} - Ir.$$

For example, if a 12-V battery has an internal resistance of 0.1 Ω, then when 10 A flows from the battery, the terminal voltage is 12 V − (10 A)(0.1 Ω) = 11 V. The internal resistance of a battery is usually small. For example, an ordinary flashlight battery when fresh may have an internal resistance of perhaps 0.05 Ω. (However, as it ages and the electrolyte dries out, the internal resistance increases to many ohms.) Car batteries have even lower internal resistance.

EXAMPLE 26–1 Battery with internal resistance. A 65.0-Ω resistor is connected to the terminals of a battery whose emf is 12.0 V and whose internal resistance is 0.5 Ω, Fig. 26–2. Calculate (a) the current in the circuit, (b) the terminal voltage of the battery, V_{ab}, and (c) the power dissipated in the resistor R and in the battery's internal resistance r.

SOLUTION (a) From the equation above relating emf $\mathscr{E}$ to terminal voltage, we have
$$V_{ab} = \mathscr{E} - Ir,$$
where $V_{ab} = IR$ (Eq. 25–2). Hence $IR = \mathscr{E} - Ir$ or $\mathscr{E} = I(R + r)$, and so
$$I = \frac{\mathscr{E}}{R + r} = \frac{12.0 \text{ V}}{65.5 \text{ Ω}} = 0.183 \text{ A}.$$

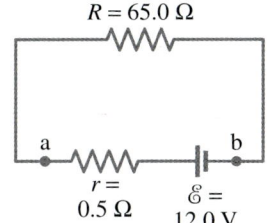

FIGURE 26–2 Example 26–1.

(b) The terminal voltage is
$$V_{ab} = \mathscr{E} - Ir = 12.0 \text{ V} - (0.183 \text{ A})(0.5 \text{ Ω}) = 11.9 \text{ V}.$$

(c) The power dissipated is
$$P_R = I^2R = (0.183 \text{ A})^2(65.0 \text{ Ω}) = 2.18 \text{ W}$$
$$P_r = I^2r = (0.183 \text{ A})^2(0.5 \text{ Ω}) = 0.02 \text{ W}.$$

[†]When a battery is being charged, a current is forced to pass through it and we have to write $V_{ab} = \mathscr{E} + Ir$. See Example 26–10.

In much of what follows, unless stated otherwise, we assume that the battery's internal resistance is negligible, and that the battery voltage given is its terminal voltage, which we will usually write simply as V rather than V_{ab}.

26–2 Resistors in Series and in Parallel

When two or more resistors are connected end to end as shown in Fig. 26–3, they are said to be connected in **series**. The resistors could be simple resistors as were pictured in Fig. 25–10, or they could be lightbulbs, heating elements, or other resistive devices. Any charge that passes through R_1 in Fig. 26–3a will also pass through R_2 and then R_3. Hence the same current I passes through each resistor. (If it did not, this would imply that charge was accumulating at some point in the circuit, which does not happen in the steady state.) We let V represent the voltage across all three resistors. We assume all other resistance in the circuit can be ignored, and so V equals the terminal voltage of the battery. We let V_1, V_2, and V_3 be the potential differences across each of the resistors, R_1, R_2, and R_3, respectively, as shown in Fig. 26–3a. From $V = IR$, we can write $V_1 = IR_1$, $V_2 = IR_2$, and $V_3 = IR_3$. Because the resistors are connected end to end, energy conservation tells us that the total voltage V is equal to the sum of the voltages across each resistor:

Series circuit: voltages add; current the same in each R

$$V = V_1 + V_2 + V_3 = IR_1 + IR_2 + IR_3. \qquad \text{[series]} \quad (26\text{–}1)$$

To see in more detail why this is true, note that an electric charge q passing through R loses potential energy by qV_1. In passing through R_2 and R_3, the potential energy U decreases by qV_2 and qV_3, for a total $\Delta U = qV_1 + qV_2 + qV_3$; this sum must equal the energy given to q by the battery, qV, so that energy is conserved. Hence $qV = q(V_1 + V_2 + V_3)$, and so $V = V_1 + V_2 + V_3$, which is Eq. 26–1.

Now let us determine the equivalent single resistance R_{eq} that would draw the same current as our combination; see Fig. 26–3c. Such a single resistance R_{eq} would be related to V by

$$V = IR_{eq}.$$

We equate this expression with Eq. 26–1, $V = I(R_1 + R_2 + R_3)$, and find

Resistances in series

$$R_{eq} = R_1 + R_2 + R_3. \qquad \text{[series]} \quad (26\text{–}2)$$

This is, in fact, what we expect. When we put several resistances in series, the total resistance is the sum of the separate resistances. This applies to any number of resistances in series. Note that when you add more resistance to the circuit, the current will decrease. For example, if a 12-V battery is connected to a 4-Ω resistor, the current will be 3 A. But if the 12-V battery is connected to three 4-Ω resistors in series, the total resistance is 12 Ω and the current will be only 1 A.

FIGURE 26–3 (a) Resistances connected in series: $R_{eq} = R_1 + R_2 + R_3$. (b) Resistances could be lightbulbs, or any other type of resistance. (c) Equivalent single resistance R_{eq} that draws the same current.

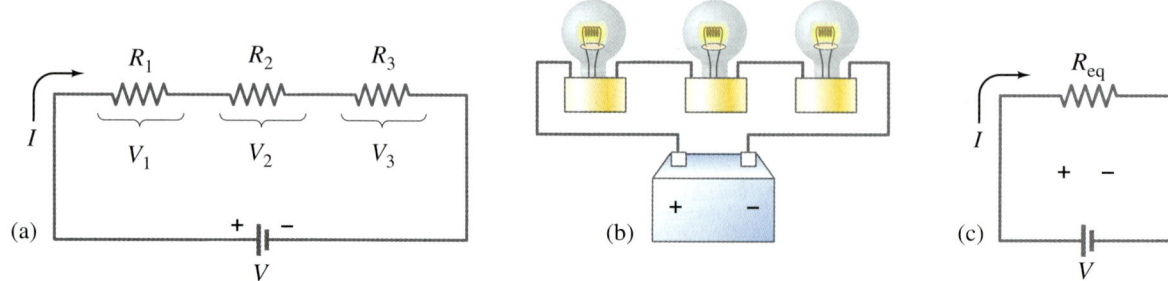

Another simple way to connect resistors is in **parallel**, so that the current from the source splits into separate branches, as shown in Fig. 26–4. The wiring in houses and buildings is arranged so all electric devices are in parallel, as we already saw in Chapter 25, Fig. 25–18. With parallel wiring, if you disconnect one device (say R_1 in Fig. 26–4), the current to the others is not interrupted. But in a series circuit, if one device (say R_1 in Fig. 26–3) is disconnected, the current *is* stopped to all the others.

In a parallel circuit, Fig. 26–4a, the total current I that leaves the battery breaks into three branches. We let I_1, I_2, and I_3 be the currents through each of the resistors, R_1, R_2, and R_3, respectively. Because electric charge is conserved, the current flowing into a junction (where different wires or conductors meet) must equal the current flowing out of the junction. Thus, in Fig. 26–4a,

$$I = I_1 + I_2 + I_3. \quad \text{[parallel]}$$

Parallel circuit: currents add; voltage the same across each R

When resistors are connected in parallel, each experiences the same voltage. (Indeed, any two points in a circuit connected by a wire of negligible resistance are at the same potential.) Hence the full voltage of the battery is applied to each resistor in Fig. 26–4a, so

$$I_1 = \frac{V}{R_1}, \quad I_2 = \frac{V}{R_2}, \quad \text{and} \quad I_3 = \frac{V}{R_3}.$$

Let us now determine what single resistor R_{eq} (Fig. 26–4c) will draw the same current I as these three resistances in parallel. This equivalent resistance R_{eq} must satisfy

$$I = \frac{V}{R_{eq}}.$$

We now combine the equations above:

$$I = I_1 + I_2 + I_3,$$

$$\frac{V}{R_{eq}} = \frac{V}{R_1} + \frac{V}{R_2} + \frac{V}{R_3}.$$

When we divide out the V from each term, we have

$$\frac{1}{R_{eq}} = \frac{1}{R_1} + \frac{1}{R_2} + \frac{1}{R_3}. \quad \text{[parallel]} \quad \text{(26–3)}$$

Resistances in parallel

For example, suppose you connect two 4-Ω loudspeakers to a single set of output terminals of your stereo amplifier or receiver. (Ignore the other channel for a moment—our two speakers are both connected to the left channel, say.) The equivalent resistance will then be found from

$$\frac{1}{R_{eq}} = \frac{1}{4\,\Omega} + \frac{1}{4\,\Omega} = \frac{2}{4\,\Omega} = \frac{1}{2\,\Omega}$$

and so $R_{eq} = 2\,\Omega$. Thus the net resistance is *less* than that of each single resistance. This may at first seem surprising. But remember that when you put resistors in parallel, you are giving the current additional paths to follow. Hence the net resistance will be less.

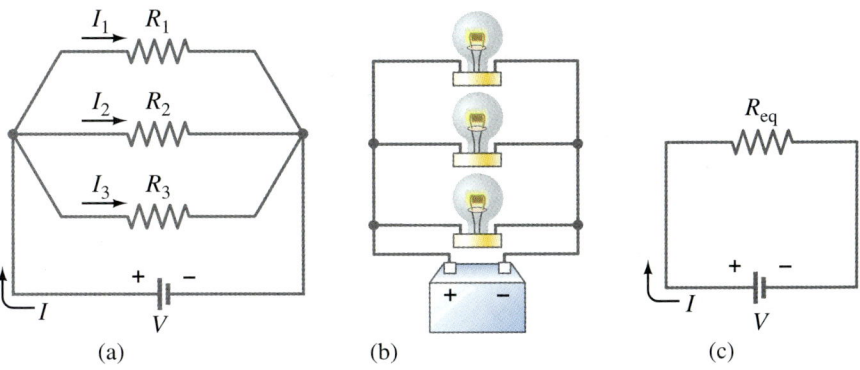

FIGURE 26–4 (a) Resistances connected in parallel: $1/R_{eq} = 1/R_1 + 1/R_2 + 1/R_3$, which could be (b) lightbulbs; (c) shows the equivalent circuit with R_{eq} obtained from Eq. 26–3.

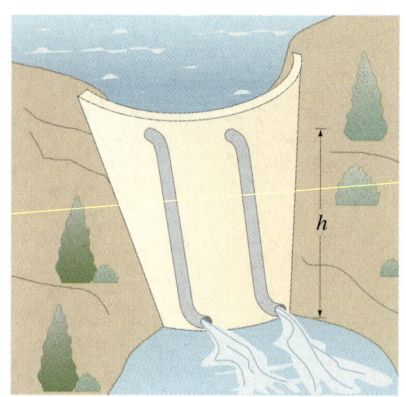

FIGURE 26–5 Water pipes in parallel—analogy to electric currents in parallel.

An analogy may help here. Consider two pipes taking in water near the top of a dam and releasing it below as shown in Fig. 26–5. The gravitational potential difference, proportional to the height h, is the same for both pipes, just as in the electrical case of parallel resistors. If both pipes are open, rather than only one, twice as much water current will flow through. That is, with two equal pipes open, the net resistance to the flow of water will be reduced, by half. Note that if both pipes are closed, the dam offers infinite resistance to the flow of water. This corresponds in the electrical case to an open circuit—when no current flows—so the electrical resistance is infinite.

Notice that the forms of the equations for resistors, Eqs. 26–2 and 26–3, are just the reverse of their counterparts for capacitors, Chapter 24, Eqs. 24–3 and 24–4. That is, the formula for resistors in series has the same form as the formula for capacitors in parallel, and vice versa.

CONCEPTUAL EXAMPLE 26–2 **Series or Parallel?** (a) The lightbulbs in Fig. 26–6 are identical and have identical resistance R. Which configuration produces more light? (b) Which way do you think the headlights of a car are wired?

RESPONSE (a) The parallel combination has lower resistance ($= R/2$) than the series combination ($= 2R$). There will be more total current in configuration (2). The total power transformed, which is proportional to the light produced, is $P = IV$, so the greater current in (2) means more light.
(b) In parallel (2), because if one bulb goes out, the other bulb can stay lit. If they were in series (1), when one bulb burned out (the filament broke), the circuit would be open and no current would flow, so even the good bulb would not light.

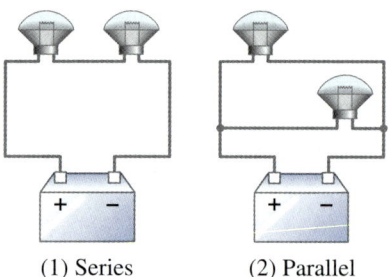

FIGURE 26–6 Example 26–2.

(1) Series   (2) Parallel

EXAMPLE 26–3 **Series and parallel resistors.** Two 100-Ω resistors are connected (a) in series, and (b) in parallel, to a 24.0-V battery. See Fig. 26–7. What is the current through each resistor and what is the equivalent resistance of each circuit?

SOLUTION (a) All the current that flows out of the battery passes first through R_1 and then R_2. So the current I is the same in both resistors; and the potential difference across the battery, V, equals the total change in potential across the two resistors:

$$V = V_1 + V_2 = IR_1 + IR_2.$$

Hence

$$I = \frac{V}{R_1 + R_2} = \frac{24.0\text{ V}}{100\text{ }\Omega + 100\text{ }\Omega} = 0.120\text{ A}.$$

The equivalent resistance, using Eq. 26–2, is $R_{eq} = R_1 + R_2 = 200\text{ }\Omega$. We could also get R_{eq} by thinking from the point of view of the battery: the total resistance R_{eq} must equal the battery voltage divided by the current it puts out: $R_{eq} = V/I = 24.0\text{ V}/0.120\text{ A} = 200\text{ }\Omega$.
[Note that the voltage across R_1 is $V_1 = IR_1 = (0.120\text{ A})(100\text{ }\Omega) = 12.0\text{ V}$, and that across R_2 is $V_2 = IR_2 = 12.0\text{ V}$, each being half of the battery voltage. A simple circuit like Fig. 26–7a is thus often called a simple "voltage divider."]
(b) Any given charge (or electron) can flow through only one or the other of the two resistors. Just as a river may break into two streams when going around an island, here too the total current I from the battery (Fig. 26–7b) equals the sum of the separate currents through the two resistors:

$$I = I_1 + I_2.$$

The potential difference across each resistor is the battery voltage $V = 24.0\text{ V}$. Hence

$$I = I_1 + I_2 = \frac{V}{R_1} + \frac{V}{R_2} = \frac{24.0\text{ V}}{100\text{ }\Omega} + \frac{24.0\text{ V}}{100\text{ }\Omega} = 0.24\text{ A} + 0.24\text{ A} = 0.48\text{ A}.$$

FIGURE 26–7 Example 26–3.

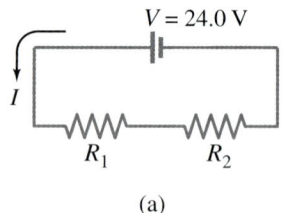

(a)

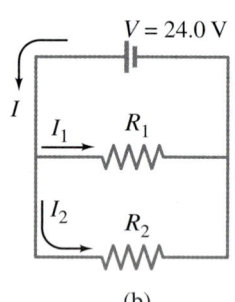

(b)

The equivalent resistance is

$$R_{eq} = \frac{V}{I} = \frac{24.0 \text{ V}}{0.48 \text{ A}} = 50 \text{ }\Omega.$$

We could also have obtained this result from Eq. 26-3:

$$\frac{1}{R_{eq}} = \frac{1}{100 \text{ }\Omega} + \frac{1}{100 \text{ }\Omega} = \frac{2}{100 \text{ }\Omega} = \frac{1}{50 \text{ }\Omega},$$

and so $R_{eq} = 50 \text{ }\Omega$.

EXAMPLE 26–4 Circuit with series and parallel. How much current flows from the battery shown in Fig. 26–8a?

SOLUTION The current I that flows out of the battery all passes through the 400-Ω resistor, but then it splits into I_1 and I_2 passing through the 500-Ω and 700-Ω resistors. The latter two are in parallel. We look for simplicity, something that we already know how to treat. So let's start by finding the equivalent resistance, R_P, of the parallel resistors, 500-Ω and 700-Ω:

$$\frac{1}{R_P} = \frac{1}{500 \text{ }\Omega} + \frac{1}{700 \text{ }\Omega} = 0.0020 \text{ }\Omega^{-1} + 0.0014 \text{ }\Omega^{-1} = 0.0034 \text{ }\Omega^{-1}.$$

This is $1/R_P$, so we take the reciprocal to find R_P. (It is a common mistake to forget to take this reciprocal. Notice that the units of reciprocal ohms, Ω^{-1}, are a reminder.) Thus

$$R_P = \frac{1}{0.0034 \text{ }\Omega^{-1}} = 290 \text{ }\Omega.$$

This 290 Ω is the equivalent resistance of the two parallel resistors, and is in series with the 400-Ω resistor as shown in the equivalent circuit of Fig. 26–8b. To find the total equivalent resistance R_{eq}, we add the 400-Ω and 290-Ω resistances together, since they are in series, and find

$$R_{eq} = 400 \text{ }\Omega + 290 \text{ }\Omega = 690 \text{ }\Omega.$$

The total current flowing from the battery is then

$$I = \frac{V}{R_{eq}} = \frac{12.0 \text{ V}}{690 \text{ }\Omega} = 0.017 \text{ A} = 17 \text{ mA}.$$

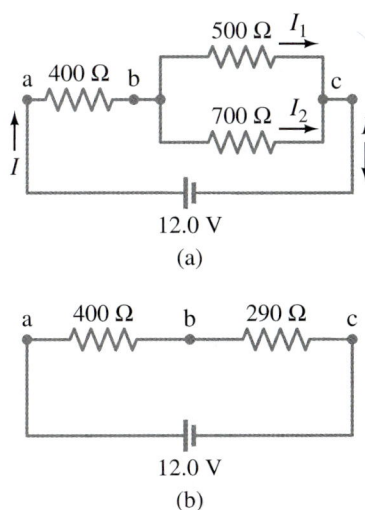

FIGURE 26–8 (a) Circuit for Examples 26–4 and 26–5. (b) Equivalent circuit, showing the equivalent resistance of 290 Ω for the two parallel resistors in (a).

EXAMPLE 26–5 Current in one branch. What is the current flowing through the 500-Ω resistor in Fig. 26–8a?

SOLUTION We need to find the voltage V_{bc} across the 500-Ω resistor, and then apply $V = IR$ to get the current. First we find the voltage across the 400-Ω resistor, V_{ab}, since we know that 17 mA passes through it; V_{ab} can be found using $V = IR$:

$$V_{ab} = (0.017 \text{ A})(400 \text{ }\Omega) = 6.8 \text{ V}.$$

Since the total voltage across the network of resistors is $V_{ac} = 12.0$ V, then V_{bc} must be $12.0 \text{ V} - 6.8 \text{ V} = 5.2 \text{ V}$. Then Ohm's law tells us that the current I_1 through the 500-Ω resistor is

$$I_1 = \frac{5.2 \text{ V}}{500 \text{ }\Omega} = 1.0 \times 10^{-2} \text{ A} = 10 \text{ mA}.$$

This is the answer we wanted. We can also calculate the current I_2 through the 700-Ω resistor since the voltage across it is also 5.2 V:

$$I_2 = \frac{5.2 \text{ V}}{700 \text{ }\Omega} = 7 \text{ mA}.$$

Notice that when I_1 combines with I_2 to form the total current I (at point c in Fig. 26–8a), their sum is $10 \text{ mA} + 7 \text{ mA} = 17 \text{ mA}$. This is, of course, the total current as calculated in Example 26–4.

(resistors in series)

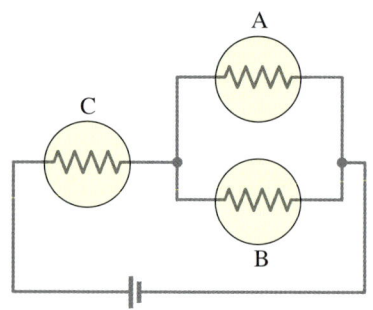

FIGURE 26–9 Example 26–6, three identical lightbulbs.

FIGURE 26–10 Example 26–7.

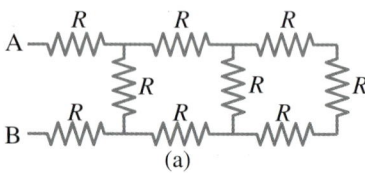
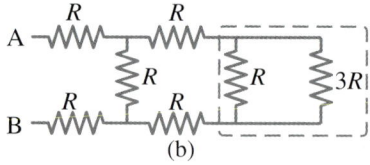
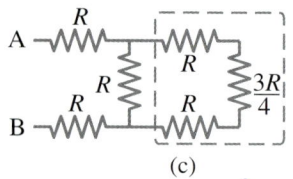
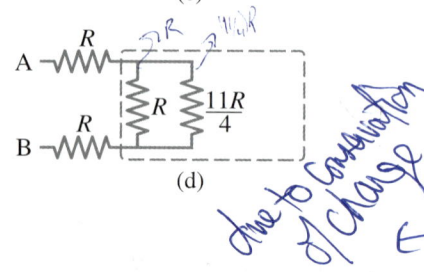

FIGURE 26–11 Currents can be calculated using Kirchhoff's rules.

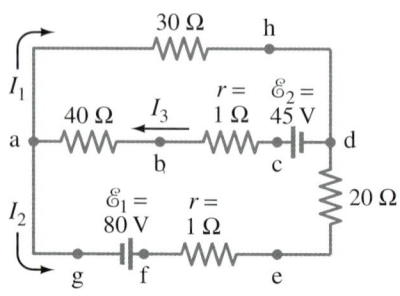

CONCEPTUAL EXAMPLE 26–6 **Bulb brightness in a circuit.** The circuit shown in Fig. 26–9 has three identical lightbulbs, each of resistance R. How will the brightness of bulbs A and B compare with that of bulb C?

RESPONSE The current that passes through bulb C splits into two equal parts when it reaches the junction leading to bulbs A and B, because the resistance of bulbs A and B are equal. Thus, bulbs A and B receive half the current, and will be less bright than bulb C.

EXAMPLE 26–7 **Resistor "ladder."** (a) Estimate the equivalent resistance of the "ladder" of equal 100-Ω resistors shown in Fig. 26–10a. In other words, what resistance would an ohmmeter read if connected between points A and B? (b) What is the current through each of the three resistors on the left if a 50.0 V battery is connected between points A and B?

SOLUTION It may seem that none of these resistors is in series or parallel. But there is a place to start: the three resistors on the far right are definitely in series with one another. Each has resistance $R(= 100\,\Omega)$, so those last three have a net resistance of $3R(= 300\,\Omega)$. Next we can see that this combination is in parallel with the next resistor to the left, as shown in the dashed box in Fig. 26–10b. The equivalent resistance of the resistors in the dashed box (b), call it R_{eq1}, is given by

$$\frac{1}{R_{eq1}} = \frac{1}{R} + \frac{1}{3R} = \frac{4}{3R}, \quad \text{so} \quad R_{eq1} = \frac{3R}{4}$$

which numerically is $300\,\Omega/4 = 75\,\Omega$. Next, this equivalent resistance of $3R/4$ is in series with the next two (Fig. 26–10c). The resistors in the dashed box (c) are in series, and are equivalent to $2R + 3R/4 = 11R/4$ (which equals $1100\,\Omega/4 = 275\,\Omega$). Now this $11R/4$ is in parallel with the next step on the ladder (Fig. 26–10d). The resistance in the dashed box (d) is equivalent to

$$\frac{1}{R_{eq2}} = \frac{1}{R} + \frac{4}{11R} = \frac{15}{11R}, \quad \text{so} \quad R_{eq2} = \frac{11R}{15}.$$

This in turn is in series with two more, yielding the final equivalent resistance of

$$R_{eq} = \frac{11R}{15} + R + R = \frac{41}{15}R.$$

We are given that $R = 100\,\Omega$, so $R_{eq} = 273\,\Omega$. Notice again the value of using algebra all along: we can have an answer that enables us to calculate the overall resistance regardless of the value of the individual resistor.

(b) If a 50.0 V battery is placed between points A and B, then the current issuing from the battery, and passing through the two resistors on the left, is $I = 50.0\,\text{V}/273\,\Omega = 0.183\,\text{A}$. The current through the first cross resistance is less than this since this R is in parallel with a net resistance of $2\frac{3}{4}R = \frac{11}{4}R$ (Fig. 26–10d). Let I_1 be the current through this R and I_2 the current in the $\frac{11}{4}R$, with $I_1 + I_2 = I$. The potential difference across R is the same as that across the $\frac{11}{4}R$, so $I_1 R = I_2(\frac{11}{4}R)$, and so $I_2 = \frac{4}{11}I_1$. Then

$$I = I_1 + I_2 = \left(1 + \frac{4}{11}\right)I_1 = \frac{15}{11}I_1.$$

Hence $I_1 = \frac{11}{15}I = \frac{11}{15}(0.183\,\text{A}) = 0.134\,\text{A}$.

26–3 Kirchhoff's Rules

In the last few Examples we have been able to find the currents flowing in circuits by combining resistances in series and parallel. This technique can be used for many circuits. However, some circuits are too complicated for that analysis. For example, we cannot find the currents flowing in each part of the circuit shown in Fig. 26–11 simply by combining resistances as we did before.

To deal with such complicated circuits, we use Kirchhoff's rules, devised by G. R. Kirchhoff (1824–1887) in the mid-nineteenth century. There are two of them, and they are simply convenient applications of the laws of conservation of charge and energy. **Kirchhoff's first** or **junction rule** is based on the conservation of charge, and we already used it in deriving the rule for parallel resistors. It states that

at any junction point, the sum of all currents entering the junction must equal the sum of all currents leaving the junction.

Junction rule
(conservation of charge)

That is, what goes in must come out. For example, at the junction point a in Fig. 26–11, I_3 is entering whereas I_1 and I_2 are leaving. Thus Kirchhoff's junction rule states that $I_3 = I_1 + I_2$. We already saw an instance of this at the end of Example 26–5.

Kirchhoff's second or **loop rule** is based on the conservation of energy. It states

the sum of the changes in potential around any closed path of a circuit must be zero.

Loop rule
(conservation of energy)

To see why this should hold, consider a rough analogy with the potential energy of a roller coaster on its track. When it starts from the station, it has a particular potential energy. As it climbs the first hill, its potential energy increases and reaches a peak at the top. As it descends the other side, its potential energy decreases and reaches a local minimum at the bottom of the hill. As the roller coaster continues on its path, its potential energy goes through more changes. But when it arrives back at the starting point, it has exactly as much potential energy as it had when it started at this point. Another way of saying this is that there was as much uphill as there was downhill.

Similar reasoning can be applied to an electric circuit. We will do the circuit of Fig. 26–11 shortly but first we consider the simpler circuit in Fig. 26–12. We have chosen it to be the same as the equivalent circuit of Fig. 26–8b already discussed. The current in this circuit is $I = (12.0 \text{ V})/(690 \, \Omega) = 0.017 \text{ A}$, as we calculated in Example 26–4. The positive side of the battery, point e in Fig. 26–12a, is at a high potential compared to point d at the negative side of the battery. That is, point e is like the top of a hill for a roller coaster. We can now follow the current around the circuit starting at any point we choose. Let us start at point e and follow a positive test charge completely around this circuit. As we go, we will note all changes in potential. When the test charge returns to point e, the potential there will be the same as when we started, so the total change in potential will be zero. It is useful to plot the changes in voltage around the circuit, and we do this in Fig. 26–12b; point d is arbitrarily taken as zero. As our positive test charge goes from point e to point a, there is no change in potential since there is no source of potential nor any resistance. However, as the charge passes through the 400-Ω resistor to get to point b, there is a decrease in potential of $V = IR = (0.017 \text{ A})(400 \, \Omega) = 6.8 \text{ V}$. In effect, the positive test charge is flowing "downhill" since it is heading toward the negative terminal of the battery. This is indicated in the graph of Fig. 26–12b. The decrease in potential between the two ends of a resistor ($= IR$) is called a **voltage drop**. Because this is a *decrease* in potential, we use a *negative* sign when applying Kirchhoff's loop rule; that is,

$$V_{ba} = V_b - V_a = -6.8 \text{ V}.$$

As the charge proceeds from b to c there is another voltage drop of $(0.017 \text{ A}) \times (290 \, \Omega) = 5.2 \text{ V}$, and since this is a decrease in potential, we write

$$V_{cb} = -5.2 \text{ V}.$$

There is no change in potential as our test charge moves from c to d. But when it moves from d, which is the negative or low potential side of the battery, to point e which is the positive terminal, the potential *increases* by 12.0 V. That is,

$$V_{ed} = +12.0 \text{ V}.$$

The sum of all the changes in potential in going around the circuit of Fig. 26–12 is

$$-6.8 \text{ V} - 5.2 \text{ V} + 12.0 \text{ V} = 0.$$

And this is exactly what Kirchhoff's loop rule said it would be.

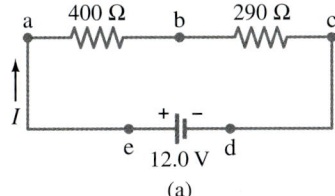
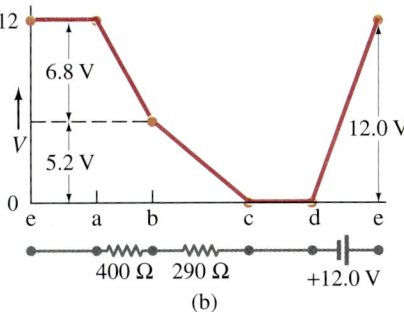

FIGURE 26–12 Changes in potential around the circuit in (a) are plotted in (b).

▶ **PROBLEM SOLVING**
Be consistent with signs

▶ **PROBLEM SOLVING**

Choose current directions arbitrarily

When using Kirchhoff's rules, we will designate the current in each separate branch of the given circuit by a different subscript, such as I_1, I_2, and I_3 in Fig. 26–13 (this is the same circuit as in Fig. 26–11). You do not have to know in advance in which direction these currents actually are moving. You make a guess and calculate the potentials around the circuit for that direction. If the current actually flows in the opposite direction, your answer will have a negative sign.

EXAMPLE 26–8 **Using Kirchhoff's rules.** Calculate the currents I_1, I_2, and I_3 in each of the branches of the circuit in Fig. 26–13.

SOLUTION Since (positive) current tends to move away from the positive terminal of a battery, we assume I_2 and I_3 to have the directions shown in Fig. 26–13. The direction of I_1 is not obvious in advance, so we arbitrarily chose the direction indicated. We have three unknowns and therefore we need three equations. We first apply Kirchhoff's junction rule to the currents at point a, where I_3 enters and I_2 and I_1 leave:

$$I_3 = I_1 + I_2. \tag{a}$$

This same equation holds at point d, so we get no new information there. We now apply Kirchhoff's loop rule to two different closed loops. First we apply it to the loop ahdcba. We start (and end) at point a. From a to h we have a voltage drop $V_{ha} = -(I_1)(30\,\Omega)$. From h to d there is no change, but from d to c the potential increases by 45 V: that is, $V_{cd} = +45$ V. From c to a the voltage drops through the two resistances by an amount $V_{ac} = -(I_3)(40\,\Omega + 1\,\Omega)$. Thus we have $V_{ha} + V_{cd} + V_{ac} = 0$, or

$$-30I_1 + 45 - (40 + 1)I_3 = 0 \tag{b}$$

where we have omitted the units. For our second loop, we take the complete circuit ahdefga. (We could have just as well taken abcdefg instead.) Again we start at point a and have $V_{ha} = -(I_1)(30\,\Omega)$, and $V_{dh} = 0$. But when we take our positive test charge from d to e, it actually is going uphill, against the flow of current—or at least against the *assumed* direction of the current, which is what counts in this calculation. Thus $V_{ed} = I_2(20\,\Omega)$ has a *positive* sign. Similarly, $V_{fe} = I_2(1\,\Omega)$. From f to g there is a decrease in potential of 80 V since we go from the high potential terminal of the battery to the low. Thus $V_{gf} = -80$ V. Finally, $V_{ag} = 0$, and the sum of the potential charges around this loop is then

$$-30I_1 + (20 + 1)I_2 - 80 = 0. \tag{c}$$

The physics is now done. The rest is algebra. We have three equations—labeled (*a*), (*b*), and (*c*)—in three unknowns. From Eq. (*c*) we have

$$I_2 = \frac{80 + 30I_1}{21} = 3.8 + 1.4I_1. \tag{d}$$

From Eq. (*b*) we have

$$I_3 = \frac{45 - 30I_1}{41} = 1.1 - 0.73I_1. \tag{e}$$

We substitute these into Eq. (*a*) and solve for I_1:

$$I_1 = I_3 - I_2 = 1.1 - 0.73I_1 - 3.8 - 1.4I_1$$
$$3.1I_1 = -2.7$$
$$I_1 = -0.87 \text{ A}.$$

The negative sign indicates that the direction of I_1 is actually opposite to that initially assumed and shown in Fig. 26–13. Note that the answer automatically comes out in amperes because all values were in volts and ohms. From Eq. (*d*) we have

$$I_2 = 3.8 + 1.4I_1 = 2.6 \text{ A},$$

and from Eq. (*e*)

$$I_3 = 1.1 - 0.73I_1 = 1.7 \text{ A}.$$

This completes the solution.

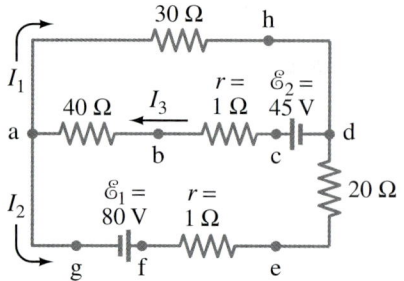

FIGURE 26–13 Currents can be calculated using Kirchhoff's rules. See Example 26–8.

| PROBLEM SOLVING | Kirchhoff's Rules |

1. Label + and − for each battery. The long side of a battery symbol is +.
2. Label the current in each branch of the circuit with a symbol and an arrow (as in Fig. 26–13): The direction of the arrow can be chosen arbitrarily. If the current is actually in the opposite direction, it will come out with a minus sign in the solution.
3. Apply Kirchhoff's junction rule at one or more junctions, and the loop rule for one or more loops. You will need as many independent equations as there are unknowns. You may write down more equations than this, but you will find that some of the equations will be redundant (that is, not be independent in the sense of providing new information). You may use $V = IR$ for each resistor, which sometimes will reduce the number of unknowns.
4. In applying the loop rule, follow each loop in one direction only. Pay careful attention to subscripts, and to signs:
   (a) For a resistor, the sign of the potential difference is negative if your chosen loop direction is the same as the chosen current direction through that resistor; the sign is positive if you are moving opposite to the chosen current direction.
   (b) For a battery, the sign of the potential difference is positive if your loop direction moves from the negative terminal toward the positive; the sign is negative if you are moving from the positive terminal toward the negative terminal.
5. Solve the equations algebraically for the unknowns. Be careful in manipulating equations not to err with signs. At the end, check your answers by plugging them into the original equations, or even by using any additional equations not used previously (either loop or junction rule equations).

EXAMPLE 26–9 **Wheatstone bridge.** A wheatstone bridge is a type of "bridge circuit" used to make measurements of resistance. The unknown resistance to be measured, R_x, is placed in the circuit with accurately known resistances R_1, R_2, and R_3. One of these, R_3, is a variable resistor which is adjusted so that when the switch is closed momentarily, the ammeter Ⓐ shows zero current flow. (We will see later in this Chapter how an ammeter works.) (a) Determine R_x in terms of R_1, R_2, and R_3. (b) If a Wheatstone bridge is "balanced" when $R_1 = 630\,\Omega$, $R_2 = 972\,\Omega$, and $R_3 = 42.6\,\Omega$, what is the value of the unknown resistance?

SOLUTION (a) We are told that R_3 has been adjusted until no current flows through the ammeter. Hence points B and D in Fig. 26–14 are at the same potential, so $V_{AB} = V_{AD}$ or

$$I_3 R_3 = I_1 R_1.$$

I_1 is the current that passes through R_1 and also through R_2 when the bridge is balanced; I_3 is the current through R_3 and R_x. When the bridge is balanced the voltage across R_x equals that across R_2, so

$$I_3 R_x = I_1 R_2.$$

We divide these two equations and find

$$R_x = \frac{R_2}{R_1} R_3.$$

In practice, the ammeter is very sensitive, and so when R_3 is being adjusted the switch is closed only momentarily to check if the current is zero or not.

(b) $$R_x = \frac{R_2}{R_1} R_3 = \left(\frac{972\,\Omega}{630\,\Omega}\right)(42.6\,\Omega) = 65.7\,\Omega.$$

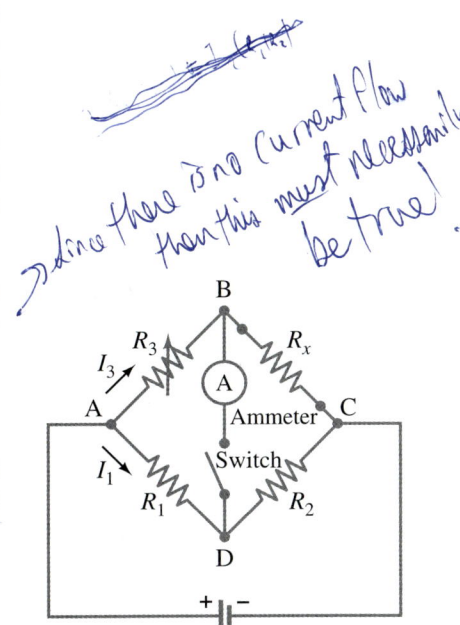

FIGURE 26–14 Example 26–9, Wheatstone bridge.

EMFs in Series and in Parallel; Charging a Battery

When two or more sources of emf, such as batteries, are arranged in series, as in Fig. 26–15a, the total voltage is the algebraic sum of their respective voltages. On the other hand, when a 20-V and a 12-V battery are connected oppositely, as shown in Fig. 26–15b, the net voltage V_{ca} is 8 V. That is, a positive test charge moved from a to b gains in potential by 20 V, but when it passes from b to c it drops by 12 V. So the net change is 20 V − 12 V = 8 V. You might think that connecting batteries in reverse like this would be wasteful. And for most purposes that would be true. But such a reverse arrangement is precisely how a battery charger works. In Fig. 26–15b, the 20-V source is charging up the 12-V battery. Because of its greater voltage, the 20-V source is forcing charge back into the 12-V battery: electrons are being forced into its negative terminal and removed from its positive terminal. An automobile alternator keeps the car battery charged in the same way. A voltmeter placed across the terminals of a (12-V) car battery with the engine running fairly fast can tell you whether or not the alternator is charging the battery. If it is, the voltmeter reads 13 or 14 V. If the battery is not being charged, the voltage will be 12 V, or less if the battery is discharging. Car batteries can be recharged, but other batteries may not be rechargeable, since the chemical reactions in many cannot be reversed. In such cases, the arrangement of Fig. 26–15b would simply waste energy.

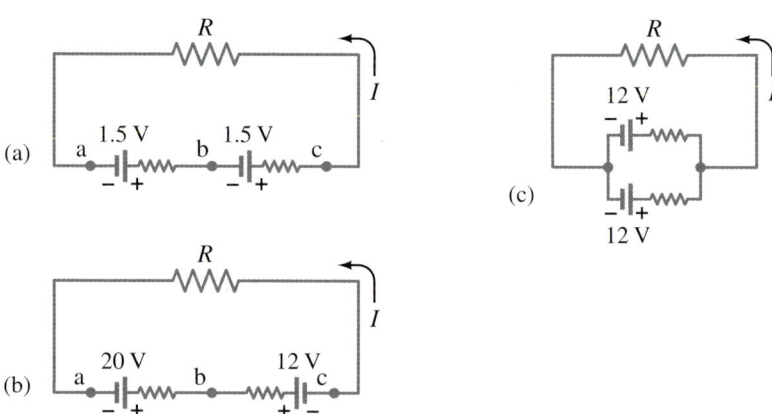

FIGURE 26–15 Batteries in series, (a) and (b), and in parallel, (c).

Sources of emf can also be arranged in parallel, Fig. 26–15c, which is useful normally only if the emfs are the same. A parallel arrangement is not used to increase voltage, but rather to provide more energy when large currents are needed. Each of the cells in parallel has to produce only a fraction of the total current, so the loss due to internal resistance is less than for a single cell; and the batteries will go dead less quickly.

EXAMPLE 26–10 Jump starting a car. A good car battery is being used to jump start a car with a weak battery. The good battery has an emf of 12.5 V and internal resistance 0.020 Ω. Suppose the weak battery has an emf of 10.1 V and internal resistance 0.10 Ω. Each copper jumper cable is 3.0 m long and 0.50 cm in diameter, and can be attached as shown in Fig. 26–16. Assume the starter motor can be represented as a resistor R_s = 0.15 Ω. Determine the current through the starter motor (a) if only the weak battery is connected to it, and (b) if the good battery is also connected, as shown in Fig. 26–16.

SOLUTION (a) The circuit is simple: an emf of 10.1 V connected to two resistances in series, 0.10 Ω + 0.15 Ω = 0.25 Ω. Hence the current is $I = V/R = (10.1 \text{ V})/(0.25 \text{ Ω}) = 40 \text{ A}$.

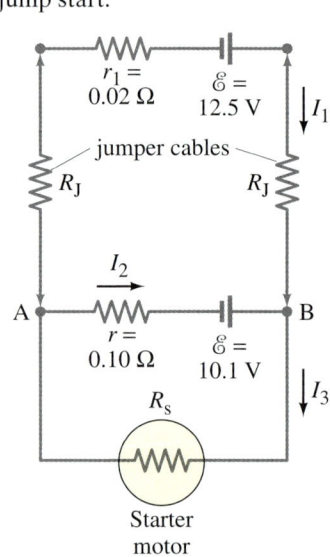

FIGURE 26–16 Example 26–10, a jump start.

(b) We need to find the resistance of the jumper cables that connect the good battery. From Eq. 25-3, each has resistance $R_J = \rho L/A = (1.68 \times 10^{-8}\,\Omega\cdot\text{m})(3.0\,\text{m})/(\pi)(0.25 \times 10^{-2}\,\text{m})^2 = 0.0026\,\Omega$. Kirchhoff's loop rule for the full outside loop gives

$$12.5\,\text{V} - I_1(2R_J + r) - I_3 R_S = 0$$
$$12.5\,\text{V} - I_1(0.025\,\Omega) - I_3(0.15\,\Omega) = 0. \tag{a}$$

The loop rule for the lower loop, including the weak battery and the starter, gives

$$10.1\,\text{V} - I_3(0.15\,\Omega) - I_2(0.10\,\Omega) = 0. \tag{b}$$

The junction rule at point B gives

$$I_1 + I_2 = I_3. \tag{c}$$

We have three equations in three unknowns. We combine equations (a) and (c), eliminating I_1, to obtain

$$12.5\,\text{V} - (I_3 - I_2)(0.025\,\Omega) - I_3(0.15\,\Omega) = 0$$
$$12.5\,\text{V} - I_3(0.175\,\Omega) + I_2(0.025\,\Omega) = 0.$$

Combining this last with (b) gives $I_3 = 71\,\text{A}$. Quite a bit better than in (a). The other currents are $I_2 = -6.2\,\text{A}$ and $I_1 = 77\,\text{A}$.

The circuit shown in Fig. 26–16, without the starter motor, is how a battery can be charged. The stronger battery pushes charge back into the weaker battery. Note in this Example that $I_2 = -6.2\,\text{V}$ and so is in the opposite direction from that assumed in Fig. 26–16. The terminal voltage of the weak 10.1 V battery is thus $V_{BA} = 10.1\,\text{V} + (6.2\,\text{A})(0.10\,\Omega) = 10.7\,\text{V}$

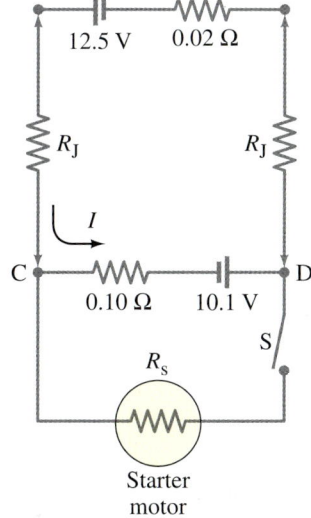

FIGURE 26–17 Example 26–11.

EXAMPLE 26–11 Jumper cables reversed. What would happen if the jumper cables of Example 26–10 were mistakenly connected in reverse, the positive terminal of each battery connected to the negative terminal of the other battery? Why could this be dangerous?

SOLUTION The circuit is shown in Fig. 26–17. Even before the starter motor is engaged (the switch S in Fig. 26–17 is open) there is trouble if the batteries are connected in this way. By Kirchhoff's loop rule in the single (upper) loop carrying current I, we have

$$12.5\,\text{V} - I(2R_J + 0.10\,\Omega + 0.02\,\Omega) + 10.1\,\text{V} = 0$$

where each $R_J = 0.0026\,\Omega$. We solve for I:

$$I = \frac{22.6\,\text{V}}{0.125\,\Omega} = 180\,\text{A}.$$

The extremely high current through the batteries could cause them to become very hot and explode. For example, the power dissipated in the weak battery would be $P = I^2 r = (180\,\text{A})^2(0.10\,\Omega) = 3200\,\text{W}!$

26–4 Circuits Containing Resistor and Capacitor (RC Circuits)

Our study of circuits in this chapter has, until now, dealt with steady currents that don't change in time. Now we examine circuits that contain both resistance and capacitance. Such a circuit is called an **RC circuit**. RC circuits are common in everyday life: they are used to control the speed of a car's windshield wiper, and the timing of the change of a traffic light from red to green. They are used in camera flashes and in heart pacemakers.

RC circuit

Charging the capacitor

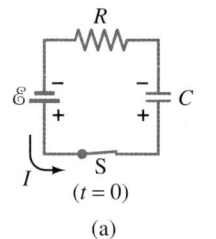

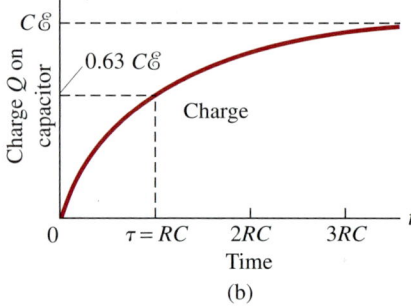

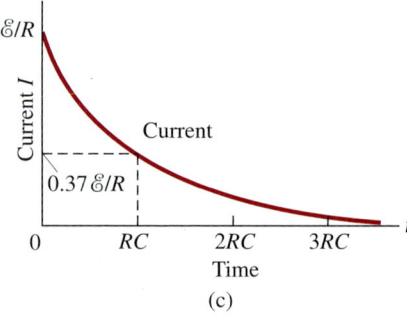

FIGURE 26–18 For the *RC* circuit shown in (a), the charge on the capacitor increases with time as shown in (b), and the current through the resistor decreases with time as shown in (c).

Charge on charging capacitor

Voltage across charging capacitor

Time constant = RC

Let us now examine the simple *RC* circuit shown in Fig. 26–18a. When the switch S is closed, current immediately begins to flow through the circuit. Electrons will flow out from the negative terminal of the battery, through the resistor *R*, and accumulate on the upper plate of the capacitor. And electrons will flow into the positive terminal of the battery, leaving a positive charge on the other plate of the capacitor. As charge accumulates on the capacitor, the potential difference across it increases; and the current is reduced until eventually the voltage across the capacitor equals the emf of the battery, $\mathcal{E}$. There is then no potential difference across the resistor, and no further current flows. The charge Q on the capacitor thus increases gradually as shown in Fig. 26–18b and reaches a maximum value equal to $C\mathcal{E}$ (Eq. 24–1, $Q_{\max} = CV_{ba} = C\mathcal{E}$). The mathematical form of this curve—that is, Q as a function of time—can be derived using conservation of energy (or Kirchhoff's loop rule). The emf $\mathcal{E}$ of the battery will equal the sum of the voltage drops across the resistor (IR) and the capacitor (Q/C):

$$\mathcal{E} = IR + \frac{Q}{C}. \quad (26\text{–}4)$$

The resistance R includes all resistance in the circuit, including the internal resistance of the battery; I is the current in the circuit at any instant, and Q is the charge on the capacitor at that same instant. Although $\mathcal{E}$, R, and C are constants, both Q and I are functions of time. The rate at which charge flows through the resistor ($I = dQ/dt$) is equal to the rate at which charge accumulates on the capacitor. Thus we can write

$$\mathcal{E} = R\frac{dQ}{dt} + \frac{1}{C}Q.$$

This equation can be solved by rearranging it:

$$\frac{dQ}{C\mathcal{E} - Q} = \frac{dt}{RC}.$$

We now integrate from $t = 0$, when $Q = 0$, to time t when a charge Q is on the capacitor:

$$\int_0^Q \frac{dQ}{C\mathcal{E} - Q} = \frac{1}{RC}\int_0^t dt$$

$$-\ln(C\mathcal{E} - Q) - (-\ln C\mathcal{E}) = \frac{t}{RC}$$

or

$$\ln(C\mathcal{E} - Q) - \ln(C\mathcal{E}) = -\frac{t}{RC}$$

so

$$\ln\left(1 - \frac{Q}{C\mathcal{E}}\right) = -\frac{t}{RC}.$$

We take the exponential of both sides

$$1 - \frac{Q}{C\mathcal{E}} = e^{-t/RC}$$

or

$$Q = C\mathcal{E}(1 - e^{-t/RC}). \quad (26\text{–}5\text{a})$$

The potential difference across the capacitor is $V_C = Q/C$, so

$$V_C = \mathcal{E}(1 - e^{-t/RC}). \quad (26\text{–}5\text{b})$$

From Eqs. 26–5 we see that the charge Q on the capacitor, and the voltage V_C across it, increase from zero at $t = 0$ to maximum values $Q_{\max} = C\mathcal{E}$ and $V_C = \mathcal{E}$ after a very long time. The quantity RC that appears in the exponent is called the **time constant** τ of the circuit:

$$\tau = RC.$$

(The units of RC are $\Omega \cdot F = (V/A)(C/V) = C/(C/s) = s$.) It represents the time required for the capacitor to reach $(1 - e^{-1}) = 0.63$ or 63 percent of its full charge. Thus the product RC is a measure of how quickly the capacitor gets charged. In a circuit, for example, where $R = 200\,\text{k}\Omega$ and $C = 3.0\,\mu\text{F}$, the time constant is $(2.0 \times 10^5\,\Omega)(3.0 \times 10^{-6}\,\text{F}) = 0.60\,\text{s}$. If the resistance is much lower, the time constant is much smaller. This makes sense, since a lower resistance will retard the flow of charge less. All circuits contain some resistance (if only in the connecting wires), so a capacitor never can be charged instantaneously when connected to a battery.

From Eqs. 26–5, it appears that Q and V_C never quite reach their maximum values within a finite time. However, they reach 86 percent of maximum in $2RC$, 95 percent in $3RC$, 98 percent in $4RC$, and so on. Q and V_C approach their maximum values asymptotically. For example, if $R = 20\,\text{k}\Omega$ and $C = 0.30\,\mu\text{F}$, the time constant is $(2.0 \times 10^4\,\Omega)(3.0 \times 10^{-7}\,\text{F}) = 6.0 \times 10^{-3}\,\text{s}$. So the capacitor is more than 98 percent charged in less than $\frac{1}{40}$ of a second.

The current I through the circuit of Fig. 26–18a at any time t can be obtained by differentiating Eq. 26–5a:

$$I = \frac{dQ}{dt} = \frac{\mathcal{E}}{R} e^{-t/RC}. \tag{26-6}$$

Current in resistor

Thus, at $t = 0$, the current is $I = \mathcal{E}/R$, as expected for a circuit containing only a resistor (there is not yet a potential difference across the capacitor). The current then drops exponentially in time with a time constant equal to RC. This is shown in Fig. 26–18c. The time constant RC represents the time required for the current to drop to $1/e \approx 0.37$ of its initial value.

EXAMPLE 26–12 *RC* **circuit, with emf.** The capacitance in the circuit of Fig. 26–18a is $C = 0.30\,\mu\text{F}$, the total resistance is $20\,\text{k}\Omega$, and the battery emf is 12 V. Determine (a) the time constant, (b) the maximum charge the capacitor could acquire, (c) the time it takes for the charge to reach 99 percent of this value, (d) the current I when the charge Q is half its maximum value, (e) the maximum current, and (f) the charge Q when, the current I is 0.20 its maximum value.

SOLUTION (a) The time constant is $RC = (2.0 \times 10^4\,\Omega)(3.0 \times 10^{-7}\,\text{F}) = 6.0 \times 10^{-3}\,\text{s}$.
(b) The maximum charge would be $Q = C\mathcal{E} = (3.0 \times 10^{-7}\,\text{F})(12\,\text{V}) = 3.6\,\mu\text{C}$.
(c) In Eq. 26–5a, we set $Q = 0.99C\mathcal{E}$:

$$0.99C\mathcal{E} = C\mathcal{E}(1 - e^{-t/RC}),$$

or

$$e^{-t/RC} = 1 - 0.99 = 0.01.$$

Then

$$\frac{t}{RC} = -\ln(0.01) = 4.6$$

so

$$t = 4.6RC = 28 \times 10^{-3}\,\text{s}$$

or 28 ms (less than $\frac{1}{30}\,\text{s}$).
(d) From part (b) the maximum charge is $3.6\,\mu\text{C}$. When the charge is half this value, $1.8\,\mu\text{C}$, the current I in the circuit can be found using the original differential equation, or Eq. 26–4:

$$I = \frac{1}{R}\left(\mathcal{E} - \frac{Q}{C}\right) = \frac{1}{2.0 \times 10^4\,\Omega}\left(12\,\text{V} - \frac{1.8 \times 10^{-6}\,\text{C}}{0.30 \times 10^{-6}\,\text{F}}\right) = 300\,\mu\text{A}.$$

(e) The current is a maximum when there is no charge on the capacitor ($Q = 0$):

$$I_{\text{max}} = \frac{\mathcal{E}}{R} = \frac{12\,\text{V}}{2.0 \times 10^4\,\Omega} = 600\,\mu\text{A}.$$

(f) Again using Eq. 26–4, with $I = 0.20 I_{\text{max}} = 120\,\mu\text{A}$, we have

$$Q = C(\mathcal{E} - IR) = (3.0 \times 10^{-7}\,\text{F})\left[12\,\text{V} - (1.2 \times 10^{-4}\,\text{A})(2.0 \times 10^4\,\Omega)\right] = 2.9\,\mu\text{C}.$$

FIGURE 26–19 For the *RC* circuit shown in (a), the charge *Q* on the capacitor decreases with time, as shown in (b), after the switch S is closed at $t = 0$. The voltage across the capacitor follows the same curve since $V \propto Q$.

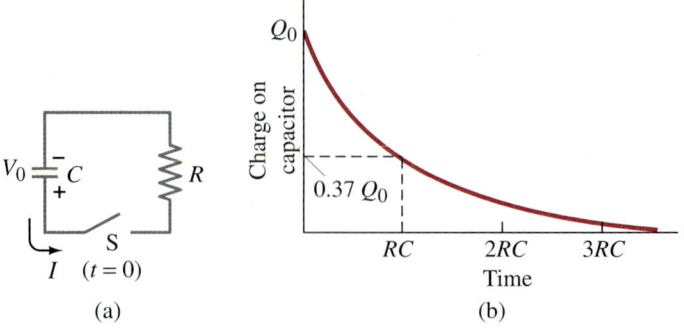

The circuit just discussed involved the *charging* of a capacitor by a battery through a resistance. Now let us look at another situation: when a capacitor is already charged (say to a voltage V_0), and it is allowed to *discharge* through a resistance *R* as shown in Fig. 26–19a. (In this case there is no battery.) When the switch S is closed, charge begins to flow through resistor *R* from one side of the capacitor toward the other side, until the capacitor is fully discharged. The voltage across the resistor at any instant equals that across the capacitor:

$$IR = \frac{Q}{C}.$$

The rate at which charge leaves the capacitor equals the negative of the current in the resistor, $I = -dQ/dt$, because the capacitor is discharging (*Q* is decreasing). So we write the above equation as

$$-\frac{dQ}{dt} R = \frac{Q}{C}.$$

We rearrange this to

$$\frac{dQ}{Q} = -\frac{dt}{RC}$$

and integrate it from $t = 0$ when the charge on the capacitor is Q_0, to some time *t* when the charge is *Q*:

$$\ln \frac{Q}{Q_0} = -\frac{t}{RC}$$

or

Charge on discharging capacitor

$$Q = Q_0 e^{-t/RC}. \qquad (26\text{--}7)$$

Thus the charge on the capacitor decreases exponentially in time with a time constant *RC*. This is shown in Fig. 26–19b. The current is

Current in resistor

$$I = -\frac{dQ}{dt} = \frac{Q_0}{RC} e^{-t/RC} = I_0 e^{-t/RC}, \qquad (26\text{--}8)$$

and it too is seen to decrease exponentially in time with the same time constant *RC*. Both the charge on the capacitor and the voltage across it ($V_C = Q/C$), as well as the current in the resistor, decrease to 37 percent of their original value in one time constant $t = \tau = RC$.

EXAMPLE 26–13 Discharging *RC* circuit. In the *RC* circuit shown in Fig. 26–20, the battery has fully charged the capacitor, so $Q_0 = C\mathscr{E}$. Then at $t = 0$ the switch is thrown from position a to b. The battery emf is 20.0 V, and the capacitance $C = 1.02 \, \mu\text{F}$. The current *I* is observed to decrease to 0.50 of its initial value in 40 μs. (*a*) What is the value of *R*? (*b*) What is the value of *Q*, the charge on the capacitor, at $t = 0$? (*c*) What is *Q* at $t = 60 \, \mu\text{s}$?

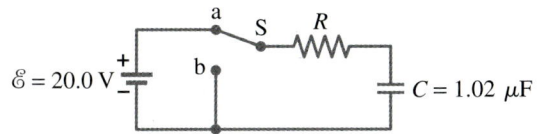

FIGURE 26–20 Example 26–13.

SOLUTION (a) At $t = 0$, the battery is removed from the circuit and the capacitor begins discharging through the resistor, as in Fig. 26–19. At any time t later (Eq. 26–7) we have

$$Q = Q_0 e^{-t/RC} = C\mathcal{E} e^{-t/RC},$$

and

$$I = -\frac{dQ}{dt} = \frac{\mathcal{E}}{R} e^{-t/RC} = I_0 e^{-t/RC}.$$

At $t = 40\,\mu\text{s}$, $I = 0.50 I_0$. Hence

$$0.50 I_0 = I_0 e^{-t/RC}$$

or, taking natural logs on both sides ($\ln 0.50 = -0.693$):

$$0.693 = \frac{t}{RC}$$

and

$$R = \frac{t}{(0.693)C} = \frac{(40 \times 10^{-6}\,\text{s})}{(0.693)(1.02 \times 10^{-6}\,\text{F})} = 57\,\Omega.$$

(b) At $t = 0$,

$$Q = Q_0 = C\mathcal{E} = (1.02 \times 10^{-6}\,\text{F})(20.0\,\text{V}) = 20.4\,\mu\text{C}.$$

(c) At $t = 60\,\mu\text{s}$,

$$Q = Q_0 e^{-\frac{60 \times 10^{-6}\,\text{s}}{(57\,\Omega)(1.02 \times 10^{-6}\,\text{F})}} = 7.3\,\mu\text{C}.$$

Applications of RC Circuits

The charging and discharging in an RC circuit can be used to produce voltage pulses at a regular frequency. The charge on the capacitor increases to a particular voltage, and then discharges. A simple way of initiating the discharge is by the use of a gas-filled tube that breaks down when the voltage across it reaches a certain value V_0. After the discharge is finished, the tube no longer conducts current and the recharging process repeats itself, starting at V_0'. Figure 26–21 shows a possible circuit, and the "sawtooth" voltage it produces.

An automobile turn signal indicator can be an application of a sawtooth oscillator circuit. Here the emf is supplied by the car battery ($\mathcal{E} = 12\,\text{V}$); the neon bulb, which flashes on at a rate of perhaps 2 cycles per second, is the turn signal indicator. The main component of the "flasher unit" is a moderately large capacitor.

The intermittent windshield wipers of a car can also use an RC circuit. The RC time constant, which can be changed using a multi-positioned switch for different values of R with fixed C, determines the rate at which the wipers come on.

Another interesting use of an RC circuit is the electronic heart pacemaker, which can make a stopped heart start beating again by applying an electric stimulus through electrodes attached to the chest. The stimulus can be repeated at the normal heartbeat rate if necessary. The heart itself contains *pacemaker* cells, which send out tiny electric pulses at a rate of 60 to 80 per minute. These signals induce the start of each heartbeat. In some forms of heart disease, the natural pacemaker fails to function properly, and the heart loses its beat. People suffering from this ailment now commonly make use of *electronic pacemakers* which produce a regular voltage pulse that starts and controls the frequency of the heartbeat. The electrodes are implanted in or near the heart and the circuit usually contains a capacitor and a resistor. The charge on the capacitor increases to a certain point and then discharges. Then it starts charging again. The pulsing rate depends on the values of R and C.

→ **PHYSICS APPLIED**

Sawtooth voltage; turn signals and wipers; pacemakers

FIGURE 26–21 (a) An RC circuit, coupled with a gas-filled tube as a switch, can produce a repeating "sawtooth" voltage, as shown in (b).

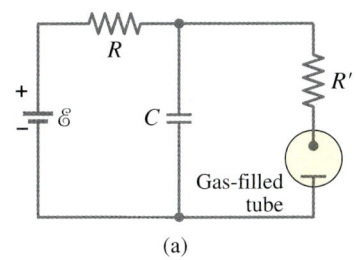

(a)

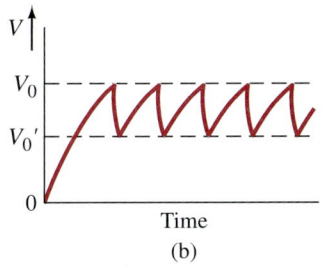

(b)

*26–5 DC Ammeters and Voltmeters

> **PHYSICS APPLIED**
> *Meters*

FIGURE 26–22 A multimeter used as a voltmeter.

Ammeter uses shunt resistor

An **ammeter** is used to measure current, and a **voltmeter** measures potential difference or voltage. The crucial part of an analog ammeter or voltmeter, in which the reading is by a pointer on a scale (Fig. 26–22), is a *galvanometer*. The galvanometer works on the principle of the force between a magnetic field and a current-carrying coil of wire, and will be discussed in Chapter 27. For now, we merely need to know that the deflection of the needle of a galvanometer is proportional to the current flowing through it. The *full-scale current sensitivity*, I_m, of a galvanometer is the current needed to make the needle deflect full scale.

A galvanometer can be used directly to measure small dc currents. For example, a galvanometer whose sensitivity I_m is 50 μA can measure currents from about 1 μA (currents smaller than this would be hard to read on the scale) up to 50 μA. To measure larger currents, a resistor is placed in parallel with the galvanometer. Thus an ammeter, represented by the symbol •—(A)—•, consists of a galvanometer (•—(G)—•) in parallel with a resistor called the **shunt resistor**, ("shunt" is a synonym for "in parallel"), as shown in Fig. 26–23. The shunt resistance is R_{sh}, and the resistance of the galvanometer coil, through which current passes, is r. The value of R_{sh} is chosen according to what full-scale deflection is desired and is normally very small, giving an ammeter a very small internal resistance.

FIGURE 26–23 An ammeter is a galvanometer in parallel with a (shunt) resistor with low resistance, R_{sh}.

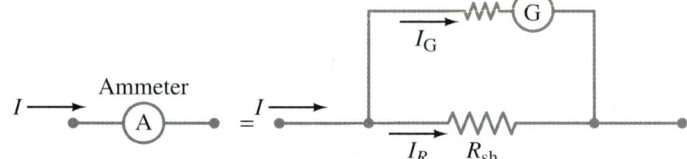

EXAMPLE 26–14 Ammeter design. Design an ammeter to read 1.0 A at full scale using a galvanometer with a full-scale sensitivity of 50 μA and a resistance $r = 30 \, \Omega$. Check if the scale is linear.

SOLUTION When the total current I entering the ammeter is 1.0 A, we want the current I_G through the galvanometer to be precisely 50 μA (to give full-scale deflection). See Fig. 26–23. Thus, when 1.0 A flows into the meter, we want 0.999950 A $(= I_R)$ to pass through the shunt resistor R_{sh}. Since the potential difference across the shunt is the same as across the galvanometer,

$$I_R R_{sh} = I_G r,$$

then

$$R_{sh} = \frac{I_G r}{I_R} = \frac{(5.0 \times 10^{-5} \, \text{A})(30 \, \Omega)}{(0.999950 \, \text{A})} = 1.5 \times 10^{-3} \, \Omega,$$

or 0.0015 Ω. The shunt resistor must thus have a very low resistance so that most of the current passes through it.

If the current I into the meter is 0.50 A, this will produce a current to the galvanometer equal to $I_G = I_R R_{sh}/r = (0.50 \, \text{A})(1.5 \times 10^{-3} \, \Omega)/30 \, \Omega = 25 \, \mu\text{A}$, which gives a deflection half of full scale, as required.

Voltmeter uses series resistor

A voltmeter (•—(V)—•) also consists of a galvanometer and a resistor. But the resistor R_{ser} is connected in series, Fig. 26–24, and it is usually large, giving a voltmeter a high internal resistance.

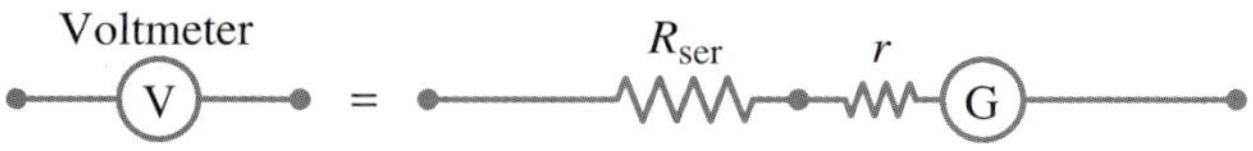

FIGURE 26–24 A voltmeter is a galvanometer in series with a resistor with high resistance, R_{ser}.

EXAMPLE 26–15 Voltmeter design. Using the same galvanometer with internal resistance $r = 30\,\Omega$ and full-scale current sensitivity of $50\,\mu\text{A}$, design a voltmeter that reads from 0 to 15 V. Is the scale linear?

SOLUTION When a potential difference of 15 V exists across the terminals of our voltmeter, we want $50\,\mu\text{A}$ to be passing through it so as to give a full-scale deflection. From Ohm's law, $V = IR$, we have (see Fig. 26–24)

$$15\,\text{V} = (50\,\mu\text{A})(r + R_{\text{ser}}),$$

so

$$R_{\text{ser}} = \frac{15\,\text{V}}{5.0 \times 10^{-5}\,\text{A}} - r = 300\,\text{k}\Omega - 30\,\Omega = 300\,\text{k}\Omega.$$

Notice that $r = 30\,\Omega$ is so small compared to the value of R_{ser} that it doesn't influence the calculation significantly.

The scale will again be linear: if the voltage to be measured is 6.0 V, the current passing through the voltmeter will be $(6.0\,\text{V})/(3.0 \times 10^5\,\Omega) = 2.0 \times 10^{-5}\,\text{A}$, or $20\,\mu\text{A}$. This will produce two-fifths of full-scale deflection, as required $(6.0\,\text{V}/15.0\,\text{V} = 2/5)$.

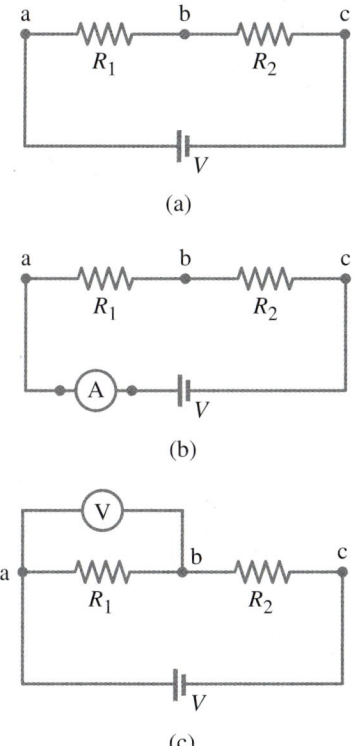

FIGURE 26–25 Measuring current and voltage.

The meters discussed above are for direct current. A dc meter can be modified to measure ac with the addition of diodes which allow current to flow in one direction only. An ac meter can be calibrated to read rms or peak values.

*Use of Voltmeters and Ammeters

Suppose you wish to determine the current I in the circuit shown in Fig. 26–25a, and the voltage V across the resistor R_1. How exactly are ammeters and voltmeters connected to the circuit being measured?

Because an ammeter is used to measure the current flowing in the circuit, it must be inserted directly into the circuit, in series with the other elements, as shown in Fig. 26–25b. The smaller its internal resistance, the less it will affect the circuit.

A voltmeter, on the other hand, is connected in parallel with the circuit element across which the voltage is to be measured. It is used to measure the potential difference between two points and its two wire leads (connecting wires) are connected to the two points, as shown in Fig. 26–25c where the voltage across R_1 is being measured. The larger its internal resistance $(R_{\text{ser}} + r$ in Fig. 26–24$)$ the less it affects the circuit being measured.

Voltmeters and ammeters can have several series or shunt resistors to offer a choice of range. **Multimeters** can measure voltage, current, and resistance. Sometimes they are called VOMs (Volt-Ohm-Meter). Meters with digital readout are called digital voltmeters (DVM) or digital multimeters (DMM), Fig. 26–26.

FIGURE 26–26 A digital multimeter being used to measure resistance.

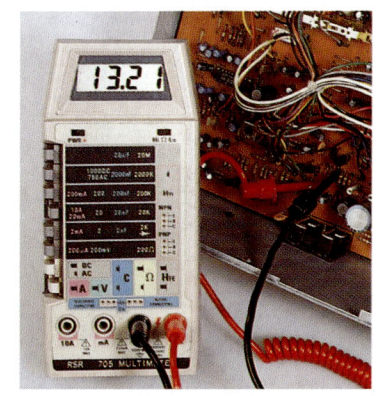

To measure resistance, the meter must contain a battery of known voltage connected in series to a resistor (R_{ser}) and to an ammeter. This makes an **ohmmeter**, Fig. 26–27. The resistor whose resistance is to be measured completes the circuit. The deflection is inversely proportional to the resistance. The calibration of the scale depends on the value of the series resistor. Since an ohmmeter sends a current through the device whose resistance is to be measured, it should not be used on very delicate devices that could be damaged by the current.

The **sensitivity** of a meter is generally specified on the face. It may be given as so many ohms per volt, which indicates how many ohms of resistance there are in the meter per volt of full-scale reading. For example, if the sensitivity is $30{,}000\,\Omega/\text{V}$, this means that on the 10-V scale the meter has a resistance of $300{,}000\,\Omega$. The full-scale current sensitivity, I_m, discussed earlier, is just the reciprocal of the sensitivity in Ω/V.

FIGURE 26–27 An ohmmeter.

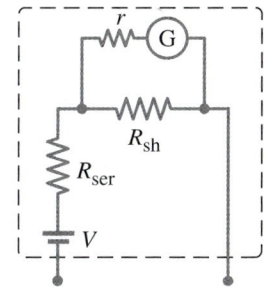

*Effects of Meter Resistance

It is important to know the sensitivity of a meter, for in many cases the resistance of the meter can seriously affect your results. Take the following Example.

EXAMPLE 26–16 Voltage reading versus true voltage. Suppose you are testing an electronic circuit which has two resistors, R_1 and R_2, each 15 kΩ, connected in series as shown in Fig. 26–28a. The battery maintains 8.0 V across them and has negligible internal resistance. A voltmeter whose sensitivity is 10,000 Ω/V is put on the 5.0-V scale. What voltage does the meter read when connected across R_1, and what error is caused by the finite resistance of the meter?

SOLUTION On the 5.0-V scale, the voltmeter has an internal resistance of $(5.0 \text{ V})(10,000 \text{ Ω/V}) = 50,000 \text{ Ω}$. When connected across R_1, as in Fig. 26–28b, we have this 50 kΩ in parallel with $R_1 = 15$ kΩ. The net resistance R_{eq} of these two is given by

$$\frac{1}{R_{eq}} = \frac{1}{50 \text{ kΩ}} + \frac{1}{15 \text{ kΩ}} = \frac{13}{150 \text{ kΩ}};$$

so $R_{eq} = 11.5$ kΩ. This $R_{eq} = 11.5$ kΩ is in series with $R_2 = 15$ kΩ, so the total resistance of the circuit is now 26.5 kΩ. Hence the current from the battery is

$$I = \frac{8.0 \text{ V}}{26.5 \text{ kΩ}} = 0.30 \text{ mA}.$$

Then the voltage drop across R_1, which is the same as that across the voltmeter, is $(3.0 \times 10^{-4} \text{ A})(11.5 \times 10^3 \text{ Ω}) = 3.5 \text{ V}$. [The voltage drop across R_2 is $(3.0 \times 10^{-4} \text{ A})(15 \times 10^3 \text{ Ω}) = 4.5 \text{ V}$, for a total of 8.0 V.] If we assume the meter is precise, it will read 3.5 V. In the normal circuit, without the meter, $R_1 = R_2$ so the voltage across R_1 is half that of the battery, or 4.0 V. Thus the voltmeter, because of its internal resistance, gives a low reading. In this case it is off by 0.5 V, or more than 10 percent.

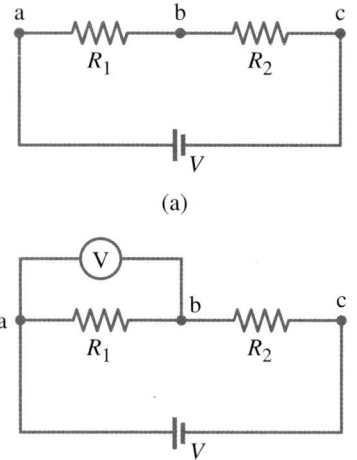

FIGURE 26–28 Example 26–16.

Example 26–16 illustrates how seriously a meter can affect a circuit and give a misleading reading. If the resistance of a voltmeter is much higher than the resistance of the circuit, however, it will have little effect and its readings can be trusted, at least to the manufactured precision of the meter, which for ordinary analog meters is typically 3 to 4 percent of full-scale deflection. An ammeter also can interfere with a circuit, but the effect is minimal if its resistance is much less than that of the circuit as a whole. For both voltmeters and ammeters, the more sensitive the galvanometer the less effect it will have. A 50,000-Ω/V meter is far better than a 1,000-Ω/V meter.

Electronic voltmeters using transistors, including digital meters, have very high input resistance (usually specified in ohms) in the range 10^6 to 10^8 Ω, and even higher. Hence they have very little effect on most circuits and their readings are reliable for most circuits. The precision of digital meters is typically one part in 10^4 (= 0.01 percent) or better. State-of-the-art instruments reach a precision of one part in 10^6.

*26–6 Transducers and the Thermocouple

A *transducer* is a device that converts one type of energy into another. A high fidelity loudspeaker is one kind of transducer—it transforms electric energy into sound energy (see Chapter 27). So is a microphone, which changes sound into an electrical signal.

FIGURE 26-29 Wire strain gauge.

Transducers are often used for measuring particular quantities. In this case they can be thought of as changing one kind of signal into another kind, usually electrical. The *resistance thermometer* and the *thermistor* mentioned in the previous chapter are examples; basically they translate a change in temperature into a change in an electrical property, resistance.

A *strain gauge*, Fig. 26-29, uses the elasticity of a wire that will stretch an amount proportional to the stress (or force) applied to it. When stretched, its resistance must increase since it is longer and its cross-sectional area is reduced (see Eq. 25-3). The fine wire of a strain gauge is usually bonded to a flexible backing material. When fixed tightly to a structure, the resistance of the strain gauge changes in direct proportion to any change in stress on the structure. Since the wire must not be strained beyond its elastic limit, the change in length is generally quite small. Hence the change in resistance is quite small (less than one part per thousand) and a very sensitive Wheatstone bridge is used to measure it. Strain gauges must be carefully calibrated and are used in many applications, such as by architects and engineers on models of proposed structures to determine the stress at critical points. A wire strain gauge can be attached to a membrane or diaphragm. A change in pressure against the membrane causes strain in the wire. Thus a strain gauge can be used to measure pressure, and is then called a *pressure transducer*.

Another kind of pressure transducer makes use of the *piezoelectric effect*. This effect occurs in certain crystals, such as quartz, that become polarized (Section 24-5) when a mechanical force is applied and produce an emf proportional to the force.

The *thermocouple* is a device that produces an electrical signal when subjected to differing temperatures. Unlike the resistance thermometer, in which it is the resistance that changes with temperature, a thermocouple produces an emf and is based on the "thermoelectric effect." When two dissimilar metals, say iron and copper, are joined at the ends, as shown in Fig. 26-30a, it is found that an emf is produced if the two junctions are at different temperatures.[†] The magnitude of this emf depends on the temperature difference. In operation, one junction of a thermocouple is kept at a known temperature. This "reference temperature" is often 0°C. The other junction, called the "test junction," is placed where the desired temperature is to be measured. The emf is measured by some precise means, Fig. 26-30b, whose terminals need to be made of the same metal, and kept at the same temperature, so that no additional emfs are produced. Often the thermocouple junctions are connected to lead-in wires, and these additional connections must also be kept at the same temperature.

A *microphone* is a transducer that changes a sound wave into an electrical signal. One type of microphone transducer is the capacitor (or condenser) microphone, Fig. 26-31. The changing air pressure in a sound wave causes one plate of the capacitor C to move back and forth. We saw in Chapter 24 that the capacitance is inversely proportional to the separation of the plates. Thus a sound wave causes the capacitance to change. This in turn causes the charge Q on the plates to change ($Q = CV$) so that an electric current is generated at the same frequencies as the incoming sound wave.

> **PHYSICS APPLIED**
>
> *Strain gauge, piezoelectric, thermocouple, microphones*

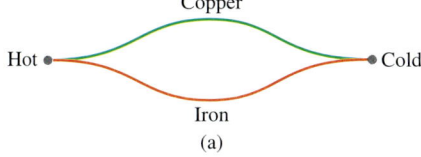
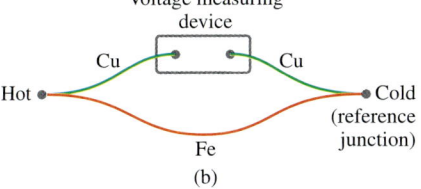

FIGURE 26-30 Thermocouple.

FIGURE 26-31 Diagram of a capacitor microphone.

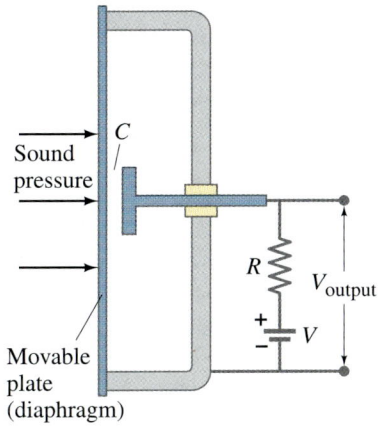

[†]Part of the theoretical explanation is that electrons in one metal occupy lower energy states than in the other, so some of them flow across the junction. This leaves one metal slightly more positive than the other, and so a *contact potential* exists between them. If the two junctions are at the same temperature, the same contact potential exists at each; they balance each other and no current flows. But if one of the junctions is at a higher temperature, the energy states are altered and the contact potentials will be different. In this case there will be a net emf and a current will flow.

Summary

A device that transforms one type of energy into electrical energy is called a **source** of **emf**. A battery behaves like a source of emf in series with an **internal resistance**. The emf is the potential difference determined by the chemical reactions in the battery and equals the terminal voltage when no current is drawn. When a current is drawn, the voltage at the battery's terminals is less than its emf by an amount equal to the Ir drop across the internal resistance.

When resistances are connected in **series** (end to end), the equivalent resistance is the sum of the individual resistances:

$$R_{eq} = R_1 + R_2 + \cdots.$$

When resistors are connected in **parallel**, the reciprocal of the total resistance equals the sum of the reciprocals of the individual resistances:

$$\frac{1}{R_{eq}} = \frac{1}{R_1} + \frac{1}{R_2} + \cdots.$$

In a parallel connection, the net resistance is less than any of the individual resistances.

Kirchhoff's rules are helpful in determining the currents and voltages in circuits. Kirchhoff's **junction rule** is based on conservation of electric charge and states that the sum of all currents entering any junction equals the sum of all currents leaving that junction. The second, or **loop rule**, is based on conservation of energy and states that the algebraic sum of the voltage changes around any closed path of the circuit must be zero.

When an **RC circuit** containing a resistor R in series with a capacitance C is connected to a dc source of emf, the voltage across the capacitor rises gradually in time characterized by an exponential of the form $\left(1 - e^{-t/RC}\right)$ where the **time constant**, $\tau = RC$, is the time it takes for the voltage to reach 63 percent of its maximum value. The current through the resistor decreases as $e^{-t/RC}$.

A capacitor discharging through a resistor is characterized by the same time constant: in a time $\tau = RC$, the voltage across the capacitor drops to 37 percent of its initial value. The charge on the capacitor, and voltage across it, decreases as $e^{-t/RC}$, as does the current.

Questions

1. Explain why birds can sit on power lines safely, while leaning a metal ladder up against one to fetch a stuck kite is extremely dangerous.
2. Discuss the advantages and disadvantages of Christmas tree lights connected in parallel versus those connected in series.
3. If all you have is a 120-V line, would it be possible to light several 6-V lamps without burning them out? How?
4. Two lightbulbs of resistance R_1 and R_2 ($>R_1$) are connected in series. Which is brighter? What if they are connected in parallel?
5. Describe carefully the difference between emf and potential difference.
6. Household outlets are often double outlets. Are these connected in series or parallel? How do you know?
7. With two identical lightbulbs and two identical batteries, how would you arrange the bulbs and batteries in a circuit in order to get the maximum possible total power out. (Assume that the batteries have negligible internal resistance.)
8. Explain why Kirchhoff's junction rule is based on conservation of electric charge.
9. Explain why Kirchhoff's loop rule is a result of the conservation of energy.
10. How does the overall resistance of your room's electric circuit change when instead of having a single 60-W lightbulb on, you turn on an additional 100-W bulb?
11. Given the circuit shown in Fig. 26–32, use the words "increases," "decreases," or "stays the same" to complete the following statements:
    (a) If R_7 increases, the potential difference between A and E (assume no resistance in Ⓐ and ℰ) _____.

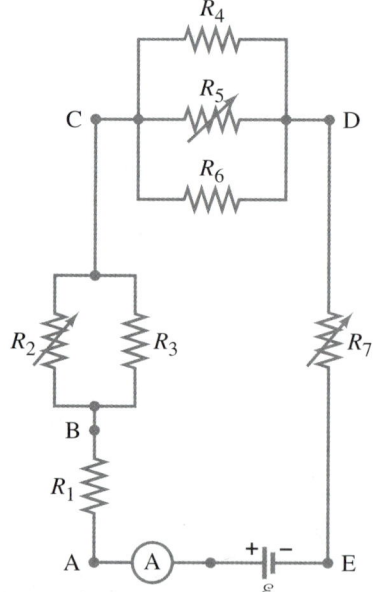

FIGURE 26–32 Question 11.

(b) If R_7 increases, the potential difference between A and E (assume Ⓐ and ℰ have resistance) _____.
(c) If R_7 increases, the voltage drop across R_4 _____.
(d) If R_2 decreases, the current through R_1 _____.
(e) If R_2 decreases, the current through R_6 _____.
(f) If R_2 decreases, the current through R_3 _____.
(g) If R_5 increases, the voltage drop across R_2 _____.
(h) If R_5 increases, the voltage drop across R_4 _____.
(i) If R_2, R_5, and R_7 increase, ℰ _____.

12. Why are batteries connected in series? Why in parallel? Does it matter if the batteries are nearly identical or not in either case?
13. Can the terminal voltage of a battery ever exceed its emf? Explain.
14. The 18-V source in Fig. 26–33 is "charging" the 12-V battery. Explain how it does this.

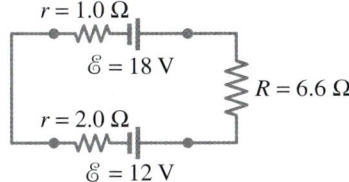

FIGURE 26–33 Questions 14 and 18; and Problem 24.

15. Explain in detail how you could measure the internal resistance of a battery.
16. Compare and discuss the formulas for the equivalent values for resistors and for capacitors when connected in series and in parallel.
17. Suppose that three identical capacitors are connected to a battery. Will they store more energy if connected in series or in parallel?
18. When applying Kirchhoff's loop rule (such as in Fig. 26–33), does the sign (or direction) of a battery's emf depend on the direction of current through the battery?
19. In an RC circuit, current flows from the battery until the capacitor is completely charged. Is the total energy supplied by the battery equal to the total energy stored by the capacitor? If not, where does the extra energy go?
20. Design a circuit in which two different switches of the type shown in Fig. 26–34 can be used to operate the same lightbulb from opposite sides of a room.

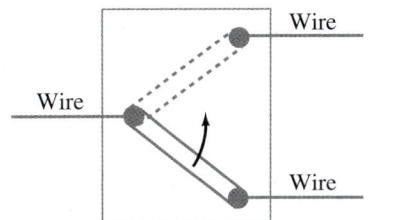

FIGURE 26–34 Question 20.

*21. What is the main difference between a voltmeter and an ammeter?
*22. What would happen if you mistakenly used an ammeter where you needed to use a voltmeter?
*23. Explain why an ideal ammeter would have zero resistance and an ideal voltmeter infinite resistance.

Problems

Section 26–1

1. (I) Calculate the terminal voltage for a battery with an internal resistance of $0.900\ \Omega$ and an emf of 8.50 V when the battery is connected in series with (a) a $68.0\text{-}\Omega$ resistor, and (b) a $680\text{-}\Omega$ resistor.
2. (I) Four 2.0-V cells are connected in series to a $12\text{-}\Omega$ lightbulb. If the resulting current flow is 0.62 A, what is the internal resistance of each cell, assuming they are identical and neglecting the wires?
3. (II) A 1.5-V dry cell can be tested by connecting it to a low-resistance ammeter. It should be able to supply at least 25 A. What is the internal resistance of the cell in this case?
4. (II) What is the internal resistance of a 12.0-V car battery whose terminal voltage drops to 9.8 V when the starter draws 60 A? What is the resistance of the starter?

Section 26–2

In the following Problems neglect the internal resistance of a battery unless the Problem refers to it.

5. (I) Four $90\text{-}\Omega$ lightbulbs are connected in series. What is the total resistance of the circuit? What is their resistance if they are connected in parallel?
6. (I) Three $40\text{-}\Omega$ lightbulbs and three $80\text{-}\Omega$ lightbulbs are connected in series. (a) What is the total resistance of the circuit? (b) What is their resistance if all six are wired in parallel?
7. (I) Given only one $25\text{-}\Omega$ and one $70\text{-}\Omega$ resistor, list all possible values of resistance that can be obtained.
8. (I) Suppose that you have a $500\text{-}\Omega$, a $900\text{-}\Omega$, and a $1.40\text{-k}\Omega$ resistor. What is (a) the maximum, and (b) the mininium resistance you can obtain by combining these?
9. (II) Suppose that you have a 6.0-V battery and you wish to apply a voltage of only 4.0 V. Given an unlimited supply of $1.0\text{-}\Omega$ resistors, how could you connect them so as to make a "voltage divider" that produced a 4.0-V output for a 6.0-V input?
10. (II) Three $1.20\text{-k}\Omega$ resistors can be connected together in four different ways, making combinations of series and/or parallel circuits. What are these four ways and what is the net resistance in each case?
11. (II) What is the net resistance of the circuit connected to the battery in Fig. 26–35? Each resistance has $R = 2.8\ \text{k}\Omega$.

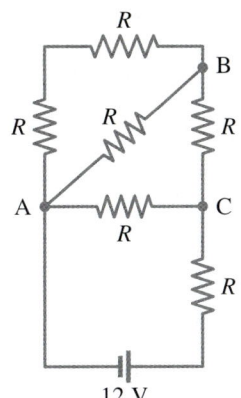

FIGURE 26–35 Problems 11 and 18.

12. (II) Eight lights are connected in series across a 110-V line. (a) What is the voltage across each bulb? (b) If the current is 0.60 A, what is the resistance of each bulb and the power dissipated in each?

13. (II) Eight lights are connected in parallel to a 110-V source by two long leads of total resistance 1.5 Ω. If 340 mA flows through each bulb, what is the resistance of each, and what fraction of the total power is wasted in the leads?

14. (II) Eight 7.0-W Christmas tree lights are connected in series to each other and to a 110-V source. What is the resistance of each bulb?

15. (II) A close inspection of an electric circuit reveals that a 480-Ω resistor was inadvertently soldered in the place where a 320-Ω resistor is needed. How can this be fixed without removing anything from the existing circuit?

16. (II) Two resistors when connected in series to a 110-V line use one fourth the power that is used when they are connected in parallel. If one resistor is 1.6 kΩ, what is the resistance of the other?

17. (II) A 75-W, 110-V bulb is connected in parallel with a 40-W, 110-V bulb. What is the net resistance?

18. (II) Calculate the current through each resistor in Fig. 26–35 if each resistance $R = 2.20 \text{ k}\Omega$. What is the potential difference between points A and B?

19. (II) Consider the network of resistors shown in Fig. 26–36. Answer qualitatively: (a) What happens to the voltage across each resistor when the switch S is closed? (b) What happens to the current through each when the switch is closed? (c) What happens to the power output of the battery when the switch is closed? (d) Let $R_1 = R_2 = R_3 = R_4 = 100 \text{ }\Omega$ and $V = 45.0 \text{ V}$. Determine the current through each resistor before and after closing the switch. Are your qualitative predictions confirmed?

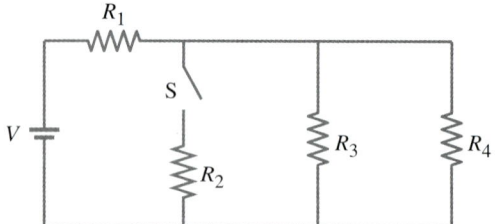

FIGURE 26–36 Problem 19.

20. (II) Three equal resistors (R) are connected to a battery, as shown in Fig. 26–37. Qualitatively, what happens to (a) the voltage drop across each of these resistors, (b) the current flow through each, and (c) the terminal voltage of the battery, when the switch S is opened, after having been closed for a long time? (d) If the emf of the battery is 12.0 V, what is its terminal voltage when the switch is closed if the internal resistance is 0.50 Ω and $R = 5.50 \text{ }\Omega$? (e) What is the terminal voltage when the switch S is open?

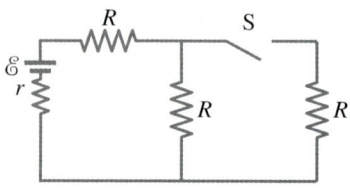

FIGURE 26–37 Problem 20.

21. (II) A battery with an emf of 12.0 V shows a terminal voltage of 11.8 V when operating in a circuit with two lightbulbs, each rated at 3.0 W (at 12.0 V) which are connected in parallel with it. What is the battery's internal resistance?

22. (III) A 3.8-kΩ and a 2.1-kΩ resistor are connected in parallel; this combination is connected in series with a 1.8-kΩ resistor. If each resistor is rated at $\frac{1}{2}$ W, what is the maximum voltage that can be applied across the whole network?

Section 26–3

23. (I) Calculate the current in the circuit of Fig. 26–38 and show that the sum of all the voltage changes around the circuit is zero.

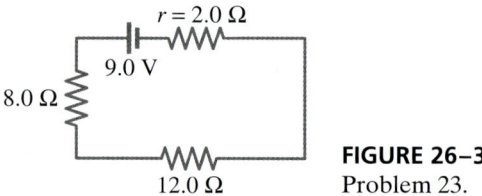

FIGURE 26–38 Problem 23.

24. (II) Determine the terminal voltage of each battery in Fig. 26–33.

25. (II) Determine the magnitudes and directions of the currents through R_1 and R_2 in Fig. 26–39.

26. (II) Repeat Problem 25, now assuming that each battery has an internal resistance $r = 1.2 \text{ }\Omega$.

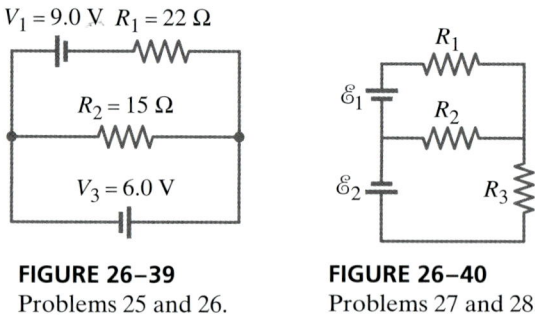

FIGURE 26–39 Problems 25 and 26.

FIGURE 26–40 Problems 27 and 28.

27. (II) Determine the magnitudes and directions of the currents through each resistor shown in Fig. 26–40. The batteries have emfs of $\mathcal{E}_1 = 9.0 \text{ V}$ and $\mathcal{E}_2 = 12.0 \text{ V}$, and the resistors have values of $R_1 = 15 \text{ }\Omega$, $R_2 = 20 \text{ }\Omega$, and $R_3 = 40 \text{ }\Omega$.

28. (II) Repeat Problem 27 assuming each battery has internal resistance $r = 1.0 \text{ }\Omega$.

29. (II) Determine the current through each of the resistors in Fig. 26–41.

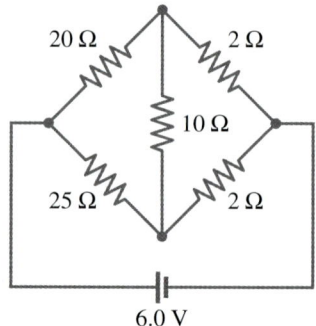

FIGURE 26–41 Problems 29 and 30.

680 CHAPTER 26 DC Circuits

30. (II) If the 20-Ω resistor in Fig. 26–41 is shorted out (resistance = 0), what then would be the current through the 10-Ω resistor?

31. (II) The current through the 4.0-kΩ resistor in Fig. 26–42 is 3.50 mA. What is the terminal voltage V_{ba} of the "unknown" battery? (There are two answers. Why?)

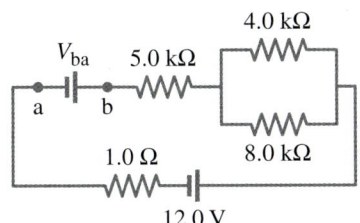

FIGURE 26–42 Problem 31.

32. (II) Suppose the 10-Ω resistor in Fig. 26–43 were replaced by an unknown resistance R. If the current through this unknown resistance is measured to be $I_2 = 0.90$ A to the right, what is the value of R?

33. (II) Suppose the 6.0-V battery in Fig. 26–43 is replaced by an unknown emf $\mathscr{E}$. If the current through the 10-Ω resistor is $I_2 = 0.30$ A to the left, what is $\mathscr{E}$? Assume $r = 1.0$ Ω.

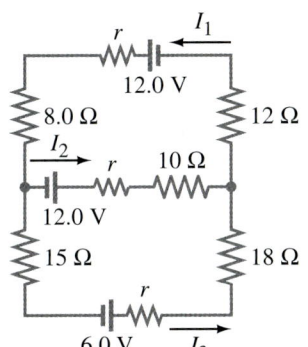

FIGURE 26–43 Problems 32, 33, 34, and 35.

34. (III) Determine the currents I_1, I_2, and I_3 in Fig. 26–43. Assume the internal resistance of each battery is $r = 1.0$ Ω. What is the terminal voltage of the 6.0-V battery?

35. (III) What would the current I_1 be in Fig. 26–43 ($r = 1.0$ Ω) if the 18-Ω resistor were shorted out?

36. (III) Determine the net resistance of the network shown in Fig. 26–44 (a) between points a and c, and (b) between points a and b. Assume $R' = R$. [Hint: Use symmetry.]

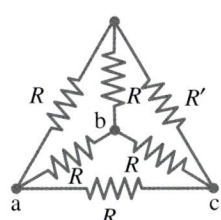

FIGURE 26–44 Problems 36 and 39.

37. (III) A voltage V is applied to n resistors connected in parallel. If the resistors are instead all connected in series with the applied voltage, show that the power transformed is decreased by a factor n^2.

38. (III) For the circuit shown in Fig. 26–45, determine (a) the current through the 14-V battery and (b) the potential difference between points a and b.

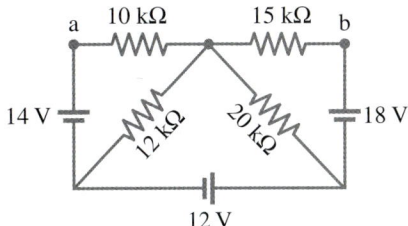

FIGURE 26–45 Problem 38.

39. (III) Determine the net resistance in Fig. 26–44 (a) between points a and c, and (b) between points a and b. Assume $R' \neq R$. [Hint: Apply an emf and determine currents; use symmetry at junctions.]

40. (III) Twelve resistors, each of resistance R, are connected as the edges of a cube as shown in Fig. 26–46. Determine the equivalent resistance (a) between points a and b, the ends of a side; (b) between points a and c, the ends of a face diagonal; (c) between points a and d, the ends of the volume diagonal. [Hint: Apply an emf and determine currents; use symmetry at junctions.]

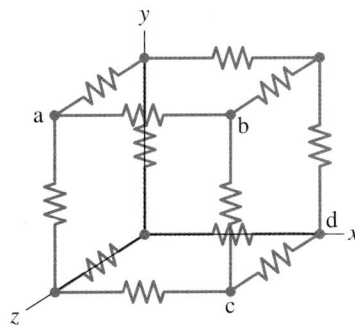

FIGURE 26–46 Problem 40.

Section 26–4

41. (II) In Fig. 26–18a, the total resistance is 15 kΩ, and the battery's emf is 24.0 V. If the time constant is measured to be 55 μs, calculate (a) the total capacitance of the circuit and (b) the time it takes for the voltage across the resistor to reach 16.0 V

42. (II) The RC circuit of Fig. 26–19a has $R = 6.7$ kΩ and $C = 6.0$ μF. The capacitor is at voltage V_0 at $t = 0$, when the switch is closed. How long does it take the capacitor to discharge to 1.0 percent of its initial voltage?

43. (II) How long does it take for the energy stored in a capacitor in a series RC circuit (Fig. 26–18a) to reach half its maximum value? Express answer in terms of the time constant $\tau = RC$.

44. (II) Two 6.0-μF capacitors, two 2.2-kΩ resistors, and a 12.0-V source are connected in series. Starting from the uncharged state, how long does it take for the current to drop from its initial value to 1.50 mA?

45. (III) Determine the time constant for charging the capacitor in the circuit of Fig. 26–47. [*Hint*: Use Kirchhoff's rules.] (b) What is the maximum charge on the capacitor?

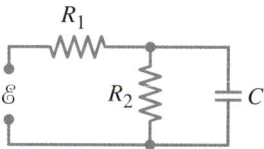

FIGURE 26–47
Problem 45.

46. (III) Two resistors and two uncharged capacitors are arranged as shown in Fig. 26–48. Then a potential difference of 24 V is applied across the combination as shown. (*a*) What is the potential at point a with S open? (Let $V = 0$ at the negative terminal of the source.) (*b*) What is the potential at point b with the switch open? (*c*) When the switch is closed, what is the final potential of point b? (*d*) How much charge flows through the switch S after it is closed?

47. (III) Suppose the switch S in Fig. 26–48 is closed. What is the time constant (or time constants) for charging the capacitors after the 24 V is applied.

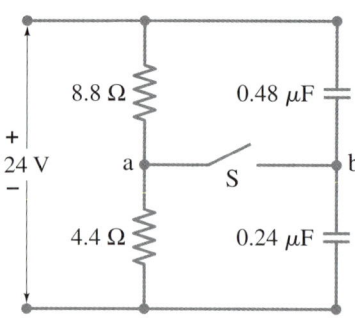

FIGURE 26–48
Problems 46 and 47.

* **Section 26–5**

* **48.** (I) What is the resistance of a voltmeter on the 250-V scale if the meter sensitivity is 50,000 Ω/V?

* **49.** (I) An ammeter has a sensitivity of 20,000 Ω/V. What current passing through the galvanometer produces full-scale deflection?

* **50.** (II) A galvanometer has an internal resistance of 30 Ω and deflects full scale for a 50-μA current. Describe how to use this galvanometer to make (*a*) an ammeter to read currents up to 30 A, and (*b*) a voltmeter to give a full-scale deflection of 1000 V.

* **51.** (II) A galvanometer has a sensitivity of 35 kΩ/V and internal resistance 20.0 Ω. How could you make this into (*a*) an ammeter that reads 2.0 A full scale, or (*b*) a voltmeter reading 1.00 V full scale?

* **52.** (II) A milliammeter reads 20 mA full scale. It consists of a 0.20-Ω resistor in parallel with a 30-Ω galvanometer. How can you change this ammeter to a voltmeter giving a full-scale reading of 10 V without taking the ammeter apart? What will be the sensitivity (Ω/V) of your voltmeter?

* **53.** (II) A 45-V battery of negligible internal resistance is connected to a 37-kΩ and a 28-kΩ resistor in series. What reading will a voltmeter, of internal resistance 100 kΩ, give when used to measure the voltage across each resistor? What is the percent inaccuracy due to meter resistance for each case?

* **54.** (II) An ammeter whose internal resistance is 60 Ω reads 4.25 mA when connected in a circuit containing a battery and two resistors in series whose values are 700 Ω and 400 Ω. What is the actual current when the ammeter is absent?

* **55.** (II) A battery with $\mathscr{E} = 12.0$ V and internal resistance $r = 1.0$ Ω is connected to two 9.0-kΩ resistors in series. An ammeter of internal resistance 0.50 Ω measures the current and at the same time a voltmeter with internal resistance 11.5 kΩ measures the voltage across one of the 9.0-kΩ resistors in the circuit. What do the ammeter and voltmeter read?

* **56.** (II) A 12.0-V battery (assume the internal resistance = 0) is connected to two resistors in series. A voltmeter whose internal resistance is 15.0 kΩ measures 5.5 V and 4.0 V, respectively, when connected across each of the resistors. What is the resistance of each resistor?

* **57.** (II) Two 8.4-kΩ resistors are placed in series and connected to a battery. A voltmeter of sensitivity 1000 Ω/V is on the 3.0-V scale and reads 2.0 V when placed across either of the resistors. What is the emf of the battery? (Ignore its internal resistance.)

* **58.** (III) The voltage across a 120-kΩ resistor in a circuit containing additional resistance (R_2) in series with a battery (V) is measured, by a 20,000-Ω/V meter on the 100-V scale, to be 25 V. On the 30-V scale, the reading is 23 V. What is the actual voltage in the absence of the voltmeter? What is R_2?

* **59.** (III) (*a*) A voltmeter and an ammeter can be connected as shown in Fig. 26–49a to measure a resistance R. The value of R will not quite be V/I where V is the voltmeter reading and I is the ammeter reading since some of the current actually goes through the voltmeter. Show that the actual value of R is given by

$$\frac{1}{R} = \frac{I}{V} - \frac{1}{R_v}$$

where R_v is the voltmeter resistance. Note that $R \approx V/I$ if $R_v \gg R$. (*b*) A voltmeter and an ammeter can also be connected as shown in Fig. 26–49b to measure a resistance R. Show in this case that

$$R = \frac{V}{I} - R_A$$

where V and I are the voltmeter and ammeter readings and R_A is the resistance of the ammeter. Note that $R \approx V/I$ if $R_A \ll R$.

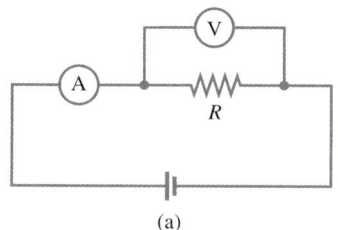

(a)

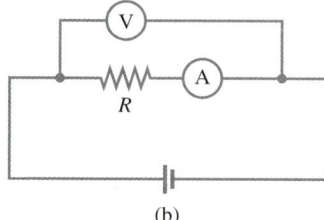

(b)

FIGURE 26–49
Problem 59.

* Section 26–6

*60. (I) A copper-constantan thermocouple produces emfs of about 40 μV/C°. If the reference temperature is 25°C, what must be the temperature of the test junction (assume a temperature increase) if the emf produced is 1.72 mV?

*61. (II) For iron-copper junctions near room temperature, the emf produced by a thermocouple is about 14 μV/C°. If emfs can be detected as low as 0.50 μV, to what accuracy can the temperature be read?

*62. (II) The *strain factor K* of a strain gauge is defined as the fractional change in resistance ($\Delta R/R$) divided by the fractional change in length:

$$K = \frac{\Delta R/R}{\Delta L/L}$$

and is relatively constant at a given temperature. In other words, the change in resistance ΔR is proportional to the change in length of the wire ΔL. (a) Show that this linear relationship is reasonable for small $\Delta L/L$. (b) A strain gauge with a strain factor of 1.8 is attached to a small muscle about 4.5 mm wide. The gauge is connected as the unknown arm in a Wheatstone bridge (Example 26–9). When the muscle is relaxed, the Wheatstone bridge is balanced when $R_2/R_1 = 1.4800$ and $R_3 = 40.700\,\Omega$. When the muscle contracts, balance occurs for $R_3 = 40.736\,\Omega$. How much has the muscle widened?

General Problems

63. Suppose that you wish to apply a 0.25-V potential difference between two points on the body. The resistance is about 2000 Ω, and you only have a 6.0-V battery. How can you connect up one or more resistors so that you can produce the desired voltage?

64. A three-way lightbulb can produce 50 W, 100 W, or 150 W, at 120 V. Such a bulb contains two filaments that can be connected to the 120 V individually or in parallel. Describe how the connections to the two filaments are made to give each of the three wattages. What must be the resistance of each filament?

65. Suppose you want to run some apparatus that is 115 m from an electric outlet. Each of the wires connecting your apparatus to the 120-V source has a resistance per unit length of 0.0065 Ω/m. If your apparatus draws 3.0 A, what will be the voltage drop across the connecting wires and what voltage will be applied to your apparatus?

66. Electricity can be a hazard in hospitals, particularly to patients who are connected to electrodes, such as an ECG. For example, suppose that the motor of a motorized bed shorts out to the bed frame, and the bed frame's connection to a ground has broken (or was not there in the first place). If a nurse touches the bed and the patient at the same time, she becomes a conductor and a complete circuit can be made through the patient to ground through the ECG apparatus. This is shown schematically in Fig. 26–50. Calculate the current through the patient.

67. A heart pacemaker is designed to operate at 72 beats/min using a 7.5-μF capacitor in a simple *RC* circuit. What value of resistance should be used if the pacemaker is to fire (capacitor discharge) when the voltage reaches 45 percent of maximum?

68. The internal resistance of a 1.35-V mercury cell is 0.030 Ω, whereas that of a 1.5-V dry cell is 0.35 Ω. Explain why three mercury cells can more effectively power a 2-W hearing aid that requires 4.0 V than can three dry cells.

69. Suppose that a person's body resistance is 1100 Ω. (a) What current passes through the body when the person accidentally is connected to 110 V? (b) If there is an alternative path to ground whose resistance is 40 Ω, what current passes through the person? (c) If the voltage source can produce at most 1.5 A, how much current passes through the person in case (b)?

70. An unknown length of platinum wire 0.920 mm in diameter is placed as the unknown resistance in a Wheatstone bridge (Example 26–9 and Fig. 26–51). Arms 1 and 2 have resistance of 38.0 Ω and 46.0 Ω, respectively. Balance is achieved when R_3 is 3.48 Ω. How long is the platinum wire?

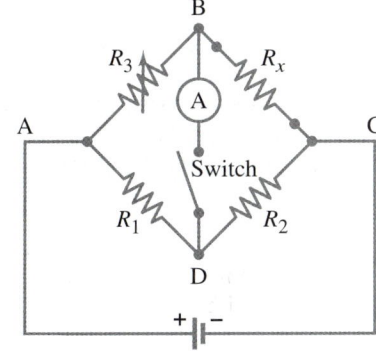

FIGURE 26–51 Wheatstone bridge. Problem 70.

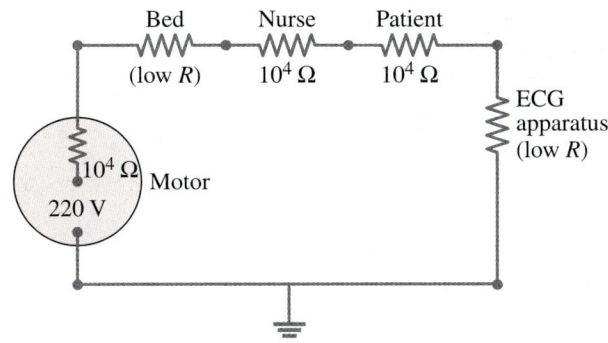

FIGURE 26–50 Problem 66.

71. Suppose two batteries, with unequal emfs of 2.0 V and 3.0 V, are connected as shown in Fig. 26–52. If each internal resistance is $r = 0.10\,\Omega$, and $R = 4.0\,\Omega$, what is the voltage across the resistor R?

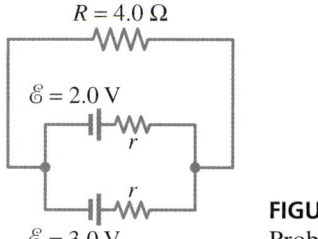

FIGURE 26–52 Problem 71.

72. Electrocardiographs are often connected as shown in Fig. 26–53. The leads are said to be capacitively coupled. A time constant of 3.0 s is typical and allows rapid changes in potential to be accurately recorded. If $C = 3.0\,\mu\text{F}$, what value must R have?

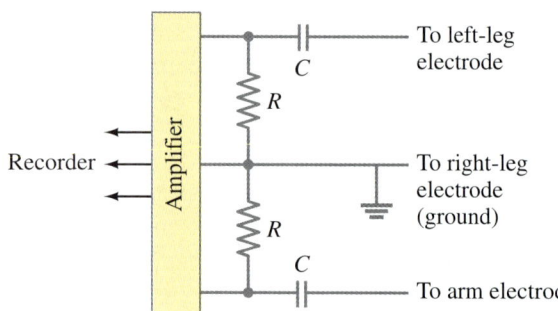

FIGURE 26–53   Problem 72.

73. A battery produces 40.8 V when 7.40 A are drawn from it and 44.5 V when 2.20 A are drawn. What is the emf and internal resistance of the battery?

74. How many $\frac{1}{2}$-W resistors, each of the same resistance, must be used to produce an equivalent 1.2-kΩ, 5-W resistor? What is the resistance of each, and how must they be connected?

75. Some light dimmer switches use a variable resistor as shown in Fig. 26–54. The slide moves from position $x = 0$ to $x = 1$, and the resistance up to slide position x is proportional to x (the total resistance is $R_{\text{pot}} = 100\,\Omega$). What is the power expended in the lightbulb if (a) $x = 1.00$, (b) $x = 0.50$, (c) $x = 0.25$?

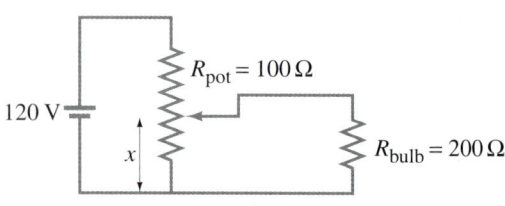

FIGURE 26–54   Problem 75.

76. A **potentiometer** is a device to precisely measure potential differences or emf, using a "null" technique. In the simple potentiometer circuit shown in Fig. 26–55, R' represents the total resistance of the resistor from A to B (which could be a long uniform "slide" wire), whereas R represents the resistance of only the part from A to the movable contact at C. When the unknown emf to be measured, $\mathcal{E}_x$, is placed into the circuit as shown, the movable contact C is moved until the galvanometer G gives a null reading (i.e., zero) when the switch S is closed. The resistance between A and C for this situation we call R_x. Next, a standard emf, $\mathcal{E}_s$, which is known precisely, is inserted into the circuit in place of $\mathcal{E}_x$ and again the contact C is moved until zero current flows through the galvanometer when the switch S is closed. The resistance between A and C now is called R_s. (a) Show that the unknown emf is given by

$$\mathcal{E}_x = \left(\frac{R_x}{R_s}\right)\mathcal{E}_s$$

where R_x, R_s, and $\mathcal{E}_s$ are all precisely known. The working battery is assumed to be fresh and give a constant voltage. (b) A slide-wire potentiometer is balanced against a 1.0182-V standard cell when the slide wire is set at 25.4 cm out of a total length of 100.0 cm. For an unknown source, the setting is 45.8 cm. What is the emf of the unknown? (c) The galvanometer of a potentiometer has an internal resistance of 30 Ω and can detect a current as small as 0.015 mA. What is the minimum uncertainty possible in measuring an unknown voltage? (d) Explain the advantage of using this "null" method of measuring emf.

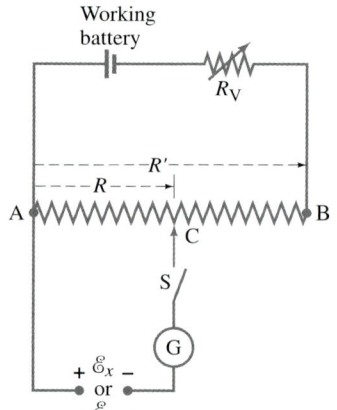

FIGURE 26–55
Potentiometer circuit.
Problem 76.

77. Electronic devices often use an RC circuit to protect against power outages as shown in Fig. 26–56. (a) If the device is supposed to keep the supply voltage at least 70 percent of nominal for as long as 0.20 s, how big a resistance is needed? The capacitor is 14 μF. (b) Between which two terminals should the device be connected, a and b, b and c, or a and c?

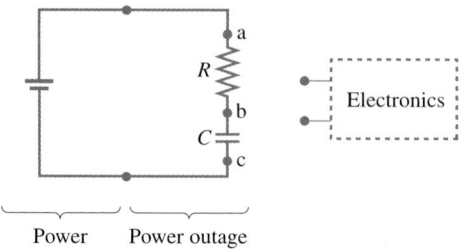

FIGURE 26–56
Problem 77.

78. For the circuit shown in Fig. 26–19a, show that the decrease in energy stored in the capacitor from $t = 0$ until one time constant has elapsed equals the energy dissipated as heat in the resistor.

79. A solar cell, 3.0 cm square, has an output of 350 mA at 0.80 V when exposed to full sunlight. A solar panel that delivers 1.0 A of current at an emf of 100 V to an external load is needed. How many cells will you need? How big a panel will you need, and how should you connect the cells to one another? How can you optimize the output of your solar panel?

80. Determine the current in each resistor of the circuit shown in Fig. 26–57.

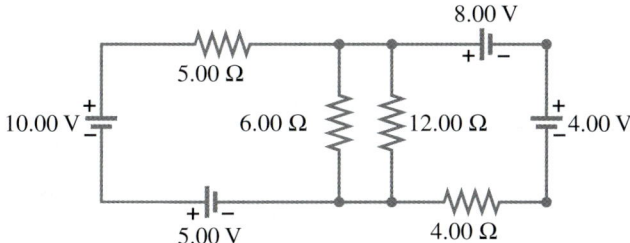

FIGURE 26–57 Problem 80.

81. In the circuit shown in Fig. 26–58, switch S is closed at time $t = 0$. (a) What is the current I_0 leaving the battery at $t = 0$, immediately after the switch is closed? (b) What is the current I a "long time" later? (c) What charge has accumulated on the capacitor after this long time? (d) If, finally, switch S is opened again, how long will it take after the switch is opened for the capacitor to lose 80 percent of its charge?

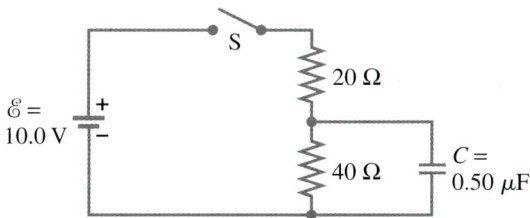

FIGURE 26–58 Problem 81.

82. A power supply has a fixed output voltage of 12.0 V, but you need $V_T = 3.0$ V for an experiment. (a) Using the voltage divider shown in Fig. 26–59, what should R_2 be if R_1 is 10.0 Ω? (b) What will the terminal voltage V_T be if you connect a load to the 3.0-volt terminal, which has a resistance of 7.0 Ω?

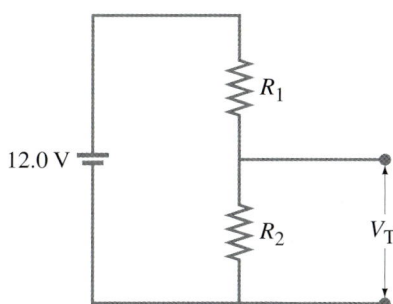

FIGURE 26–59 Problem 82.

83. In the circuit shown in Fig. 26–60, switch S is closed at time $t = 0$. (a) After the capacitor is fully charged, what is the voltage across it? How much charge is on it? (b) Switch S is now opened. How long does it now take for the capacitor to discharge until it has only 5.0 percent of its initial charge?

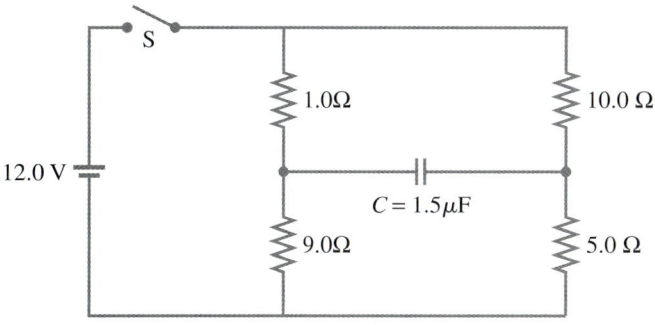

FIGURE 26–60 Problem 83.

84. Figure 26–61 shows the circuit for a simple *sawtooth oscillator*. At time $t = 0$, its switch S is closed. The neon bulb has initially infinite resistance until the voltage across it reaches 90.0 V, and then it begins to conduct with very little resistance (essentially zero). It stops conducting (its resistance becomes essentially infinite) when the voltage drops down to 70.0 V. (a) At what time t_1 does the neon bulb reach 90.0 V and start conducting? (b) At what time t_2 does the bulb reach 90.0 V for a second time and again become conducting? (c) Sketch the sawtooth waveform between $t = 0$ and $t = 0.50$ seconds.

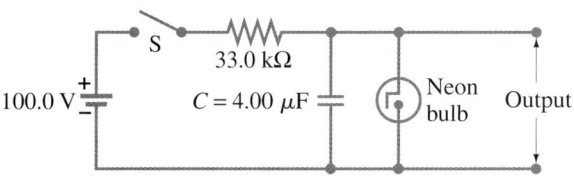

FIGURE 26–61 Problem 84.

General Problems 685

Magnets produce magnetic fields, but so do electric currents. An electric current flowing in this straight wire produces a magnetic field which causes the tiny pieces of iron (iron "filings") to align in the field. We will see in this chapter how magnetic field is defined, and that the magnetic field direction is along the iron filings. The magnetic field lines due to the electric field in this long wire form circles around the wire. We will also discuss how magnetic fields exert forces on electric currents and on charged particles, as well as useful applications of the interaction between magnetic fields and electric currents and moving electric charges.

CHAPTER 27

Magnetism

The history of magnetism begins thousands of years ago, with the ancient civilizations in Asia Minor. It was in a region of Asia Minor known as Magnesia that rocks were found that could attract each other. These rocks were called "magnets" after their place of discovery.

Not until the nineteenth century was it seen that magnetism and electricity are closely related. A crucial discovery was that electric currents produce magnetic effects (we will say "magnetic fields") like magnets do. All kinds of practical devices depend on magnetism, as we shall see in this and later chapters: from compasses to motors, loudspeakers, computer memory, and electric generators.

27–1 Magnets and Magnetic Fields

Poles of a magnet

We have all observed a magnet attract paper clips, nails, and other objects made of iron. Any magnet, whether it is in the shape of a bar or a horseshoe, has two ends or faces, called **poles**, which is where the magnetic effect is strongest. If a magnet is suspended from a fine thread, it is found that one pole of the magnet will always point toward the north. It is not known for sure when this fact was discovered, but it is known that the Chinese were making use of it as an aid to navigation by the eleventh century and perhaps earlier. This is, of course, the principle of a compass. A compass needle is simply a magnet which is supported at its center of gravity so that it can rotate freely. The pole of a freely suspended magnet that points toward geographic north is called the **north pole** of the magnet. The other pole points toward the south and is called the **south pole**.

It is a familiar fact that when two magnets are brought near one another, each exerts a force on the other. The force can be either attractive or repulsive and can be felt even when the magnets don't touch. If the north pole of one magnet is brought near the north pole of a second magnet, the force is repulsive. Similarly, if two south poles are brought close, the force is repulsive. But when a north pole is brought near a south pole, the force is attractive. These results are shown in Fig. 27–1, and are reminiscent of the force between electric charges; like poles repel, and unlike poles attract. *But do not confuse magnetic poles with electric charge.* They are not the same thing. One important difference is that a positive or negative electric charge can easily be isolated. But the isolation of a single magnetic pole seems impossible. If a bar magnet is cut in half, you do not obtain isolated north and south poles. Instead, two new magnets are produced, Fig. 27–2. If the cutting operation is repeated, more magnets are produced, each with a north and a south pole. Physicists have searched for isolated single magnetic poles (monopoles), but no magnetic monopole has ever been observed.

Only iron and a few other materials such as cobalt, nickel, gadolinium and certain alloys show strong magnetic effects. They are said to be **ferromagnetic** (from the Latin word *ferrum* for iron). All other materials show some slight magnetic effect, but it is extremely weak and can be detected only with delicate instruments. We will look in more detail at ferromagnetism in Sections 28–7 and 28–9.

We found it useful to speak of an electric field surrounding an electric charge. In the same way, we can imagine a **magnetic field** surrounding a magnet. The force one magnet exerts on another can then be described as the interaction between one magnet and the magnetic field of the other. Just as we drew electric field lines, we can also draw **magnetic field lines**. They can be drawn, as for electric field lines, so that (1) the direction of the magnetic field is tangent to a line at any point, and (2) the number of lines per unit area is proportional to the strength of the magnetic field.

The *direction* of the magnetic field at a given point can be defined as the direction that the north pole of a compass needle would point when placed at that point. (A more precise definition will be given in Section 27–3.) Figure 27–3a shows how one magnetic field line around a bar magnet is found using compass needles. The magnetic field determined in this way for the field outside a bar magnet is shown in Fig. 27–3b. Notice that because of our definition, the lines always point out from the north pole toward the south pole of a magnet (the north pole of a magnetic compass needle is attracted to the south pole of another magnet). Figure 27–4 shows how thin iron filings reveal the magnetic field lines by lining up like compass needles.

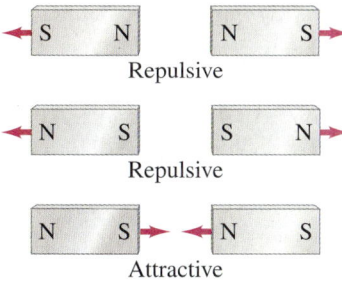

FIGURE 27–1 Like poles of a magnet repel; unlike poles attract.

FIGURE 27–2 If you break a magnet in half, you do not obtain isolated north and south poles; instead, two new magnets are produced, each with a north and a south pole.

Magnetic field lines

FIGURE 27–4 Thin iron filings indicate the magnetic field lines around a bar magnet.

FIGURE 27–3 (a) Plotting a magnetic field line of a bar magnet. (b) Magnetic field lines for a bar magnet.

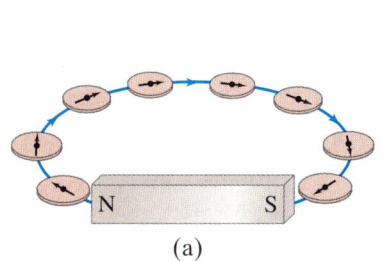
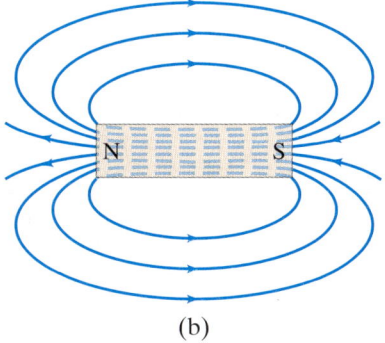

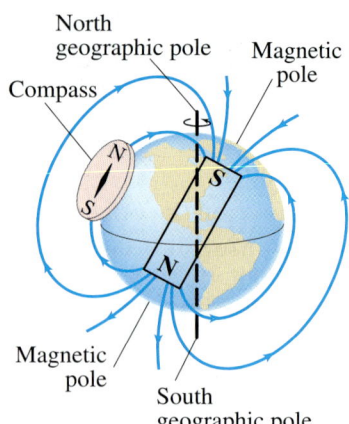

FIGURE 27–5 The Earth acts like a huge magnet; but its magnetic poles are not at the geographic poles, which are defined to be on the Earth's rotation axis.

FIGURE 27–6 Using a map and compass in the wilderness. First you align the compass case so the needle points away from true north (N) exactly the number of degrees of declination as stated on the map: 15° for the place shown on this topographic map of a part of California. Then align the map with true north, as shown, *not* with the compass needle.

→ **PHYSICS APPLIED**
Use of a compass.

Magnetic field lines continue inside a magnet, as indicated in Fig. 27–3b. Indeed, given the lack of single magnetic poles, magnetic field lines form closed loops, unlike electric field lines that begin on positive charges and end on negative charges.

The Earth's magnetic field is shown in Fig. 27–5. The pattern of field lines is as if there were an (imaginary) bar magnet inside the Earth. Since the north pole of a compass needle points north, the magnetic pole which is in the geographic north is actually a south pole magnetically, as indicated in Fig. 27–5 by the S on the schematic bar magnet inside the Earth. (Remember that the north pole of one magnet is attracted to the south pole of a second.) Nonetheless, this pole is still often called the "north magnetic pole," or "geomagnetic north," simply because it is in the north. Similarly, the Earth's southern magnetic pole, near the geographic south pole, is magnetically a north pole. The Earth's magnetic poles do not coincide with the *geographic* poles, which are on the Earth's axis of rotation. The north magnetic pole, for example, is in northern Canada, about 1300 km from the geographic north pole, or "true north." This must be taken into account when using a compass (Fig. 27–6). The angular difference between magnetic north, as indicated by a compass, and true (geographical) north, is called the **magnetic declination**. In the U.S. it varies from 0° to perhaps 25°, depending on location.

Notice in Fig. 27–5 that the Earth's magnetic field is not tangent to the Earth's surface at all points. The angle that the Earth's magnetic field makes with the horizontal at any point is referred to as the **angle of dip**.

The simplest magnetic field is one that is uniform—it doesn't change from one point to another. A perfectly uniform field over a large area is not easy to produce. But the field between two flat parallel pole pieces of a magnet is nearly uniform if the area of the pole faces is large compared to their separation, as shown in Fig. 27–7. At the edges, the field "fringes" out somewhat and is no longer uniform. The parallel evenly spaced field lines in the drawing indicate that the field is uniform at points not too near the edge, much like the electric field between two parallel plates (Fig. 23–1).

FIGURE 27–7 Magnetic field between two large poles of a magnet is nearly uniform except at the edges.

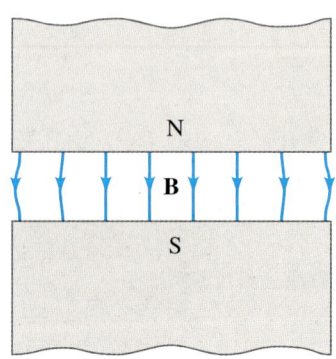

688 CHAPTER 27 Magnetism

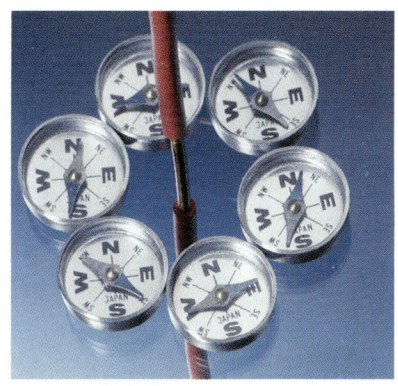

FIGURE 27–8 Deflection of compass needles near a current-carrying wire, showing the presence and direction of the magnetic field.

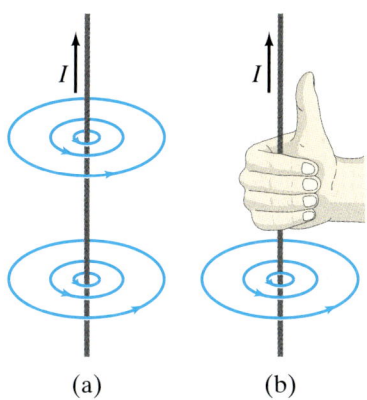

FIGURE 27–9 (a) Magnetic field lines around an electric current in a straight wire. (b) Right-hand rule for remembering the direction of the magnetic field: when the thumb points in the direction of the conventional current, the fingers wrapped around the wire point in the direction of the magnetic field.

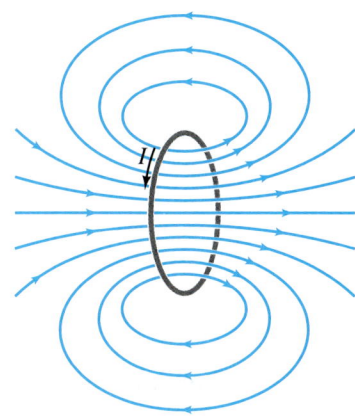

FIGURE 27–10 Magnetic field lines due to a circular loop of wire.

27–2 Electric Currents Produce Magnetism

During the eighteenth century, many scientists sought to find a connection between electricity and magnetism. A stationary electric charge and a magnet were shown not to have any influence on each other. But in 1820, Hans Christian Oersted (1777–1851) found that when a compass needle is placed near an electric wire, the needle deflects as soon as the wire is connected to a battery and a current flows. As we have seen, a compass needle can be deflected by a magnetic field. What Oersted found was that **an electric current produces a magnetic field**. He had found a connection between electricity and magnetism.

A compass needle placed near a straight section of current-carrying wire aligns itself so it is tangent to a circle drawn around the wire, Fig. 27–8. Thus, the magnetic field lines produced by a current in a straight wire are in the form of circles with the wire at their center, Fig. 27–9a. The direction of these lines is indicated by the north pole of the compass in Fig. 27–8. There is a simple way to remember the direction of the magnetic field lines in this case. It is called a **right-hand rule**: you grasp the wire with your right hand so that your thumb points in the direction of the conventional (positive) current; then your fingers will encircle the wire in the direction of the magnetic field, Fig. 27–9b. The magnetic field lines due to a circular loop of current-carrying wire can be determined in a similar way using a compass. The result is shown in Fig. 27–10. Again the right-hand rule can be used, as shown in Fig. 27–11.

Electric currents produce magnetic fields

Right-hand rule for magnetic field direction

FIGURE 27–11 Right-hand rule for determining the direction of the magnetic field relative to the current.

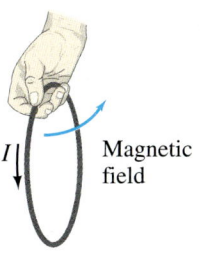

27–3 Force on an Electric Current in a Magnetic Field; Definition of B

In Section 27–2 we saw that an electric current exerts a force on a magnet, such as a compass needle. By Newton's third law, we might expect the reverse to be true as well: we should expect that *a magnet exerts a force on a current-carrying wire*. Experiments indeed confirm this effect, and it too was first observed by Oersted.

Magnet exerts a force on an electric current

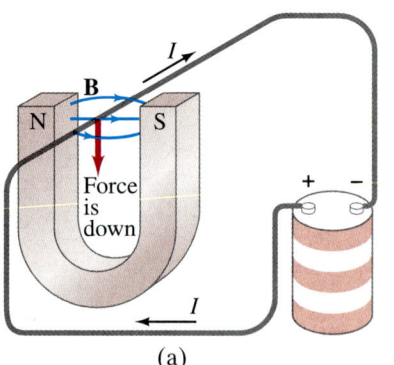
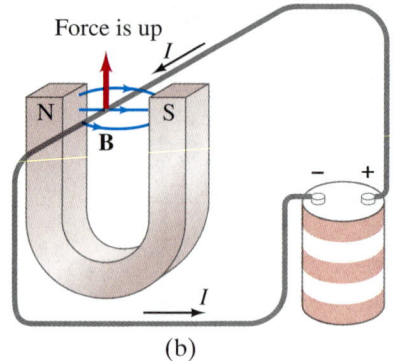
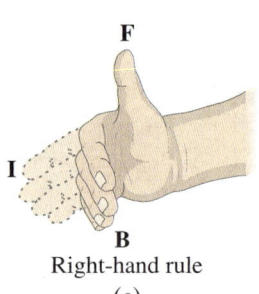

FIGURE 27–12 (a) Force on a current-carrying wire placed in a magnetic field **B**; (b) same, but current reversed; (c) right-hand rule for setup in (b).

Let us look at the force exerted on a current-carrying wire in detail. Suppose a straight wire is placed between the poles of a horseshoe magnet as shown in Fig. 27–12. When a current flows in the wire, a force is exerted on the wire. But this force is *not* toward one or the other pole of the magnet. Instead, the force is directed at right angles to the magnetic field direction, downward in Fig. 27–12a. If the current is reversed in direction, the force is in the opposite direction, Fig. 27–12b. It is found that *the direction of the force is always perpendicular to the direction of the current and also perpendicular to the direction of the magnetic field*, **B**. This statement does not completely describe the direction, however: the force could be either up or down in Fig. 27–12b and still be perpendicular to both the current and to **B**. Experimentally, the direction of the force is given by another **right-hand rule**, as illustrated in Fig. 27–12c. Orient your right hand so that outstretched fingers can point in the direction of the current, and when you bend your fingers they point in the direction of the magnetic field lines. Then your thumb points in the direction of the force on the wire.

*Right-hand rule for force on current due to **B***

This describes the direction of the force. What about its magnitude? It is found experimentally that the magnitude of the force is directly proportional to the current I in the wire, and to the length l of wire in the magnetic field (assumed uniform). Furthermore, if the magnetic field is made stronger, the force is proportionally greater. The force also depends on the angle θ between the current direction and the magnetic field (Fig. 27–13). When the current is perpendicular to the field lines, the force is strongest. When the wire is parallel to the magnetic field lines, there is no force at all. At other angles, the force is proportional to $\sin\theta$ (Fig. 27–13). Thus for a current I in a wire, with length l in a uniform magnetic field B, we have

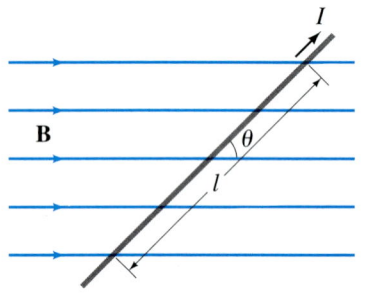

FIGURE 27–13 Current-carrying wire in a magnetic field. Force on the wire is directed into the page.

$$F \propto IlB \sin\theta.$$

Up to now we have not defined the magnetic field strength precisely. In fact, the magnetic field B can be conveniently defined in terms of the above proportion so that the proportionality constant is precisely 1. Thus we have

Force on electric current in a uniform magnetic field

$$F = IlB \sin\theta. \qquad (27\text{–}1)$$

If the direction of the current is perpendicular to the field ($\theta = 90°$), then the force is

$$F_{\text{max}} = IlB. \qquad [\mathbf{I} \perp \mathbf{B}] \qquad (27\text{–}2)$$

Definition of magnetic field

If the current is parallel to the field ($\theta = 0°$), the force is zero. The magnitude of **B** can then be defined as $B = F_{\text{max}}/Il$ where F_{max} is the magnitude of the force on a straight length l of wire carrying a current I when the wire is perpendicular to **B**.

The relation between the force **F** on a wire carrying current I, and the magnetic field **B** that causes the force, can be written as a vector equation. To do so, we recall that the direction of **F** is given by the right-hand rule (Fig. 27–13c), and the magnitude by Eq. 27–1. This is consistent with the definition of the vector cross product (see Section 11–1), so we can write

$$\mathbf{F} = I\mathbf{l} \times \mathbf{B}; \tag{27-3}$$

Force on electric current in a uniform magnetic field

here, $\mathbf{l}$ is a vector whose magnitude is the length of the wire and its direction is along the wire (assumed straight) in the direction of the conventional (positive) current.

The above discussion applies if the magnetic field is uniform and the wire is straight. If **B** is not uniform, or if the wire does not everywhere make the same angle θ with **B**, then Eq. 27–3 can be written

$$d\mathbf{F} = I\,d\mathbf{l} \times \mathbf{B}, \tag{27-4}$$

Force on electric current segment in any magnetic field

where $d\mathbf{F}$ is the infinitesimal force acting on a differential length $d\mathbf{l}$ of the wire. The total force on the wire is then found by integrating.

Equation 27–4 can serve (just as well as Eq. 27–2 or 27–3) as a practical definition of **B**. An equivalent way to define **B**, in terms of the force on a moving electric charge, is discussed in the next Section.

The SI unit for magnetic field B is the **tesla** (T). From Eqs. 27–1, 2, 3, or 4, it is clear that $1\,\text{T} = 1\,\text{N}/\text{A}\cdot\text{m}$. An older name for the tesla is the "weber per meter squared" $(1\,\text{Wb}/\text{m}^2 = 1\,\text{T})$. Another unit commonly used to specify magnetic field is a cgs unit, the **gauss** (G): $1\,\text{G} = 10^{-4}\,\text{T}$. A field given in gauss should always be changed to teslas before using with other SI units. To get a "feel" for these units, we note that the magnetic field of the Earth at its surface is about $\frac{1}{2}$ G or $0.5 \times 10^{-4}\,\text{T}$. On the other hand, strong electromagnets can produce fields on the order of several teslas and superconducting magnets over 10 T.

The tesla and the gauss (units)

On a diagram, when we want to represent a magnetic field that is pointing out of the page (toward us) or into the page, we use $\odot$ or $\times$. The $\odot$ is meant to resemble the tip of an arrow pointing directly toward the reader, whereas the $\times$ or $\otimes$ resembles the tail of an arrow going away. (See Figs. 27–14 and 27–15.)

EXAMPLE 27–1 Measuring a magnetic field. A rectangular loop of wire hangs vertically as shown in Fig. 27–14. A magnetic field **B** is directed horizontally, perpendicular to the wire, and points out of the page at all points as represented by the symbol $\odot$. The magnetic field **B** is very nearly uniform along the horizontal portion of wire ab (length $l = 10.0\,\text{cm}$) which is near the center of a large magnet producing the field. The top portion of the wire loop is free of the field. The loop hangs from a balance which measures a downward force (in addition to the gravitational force) of $F = 3.48 \times 10^{-2}\,\text{N}$ when the wire carries a current $I = 0.245\,\text{A}$. What is the magnitude of the magnetic field B at the center of the magnet?

SOLUTION The magnetic forces on the two vertical sections of the wire loop point to the left and right, respectively. They are equal and in opposite directions and so add up to zero. Hence, the net magnetic force on the loop is that on the horizontal section ab whose length is $l = 0.100\,\text{m}$ (and $\theta = 90°$ so $\sin\theta = 1$); thus

$$B = \frac{F}{Il} = \frac{3.48 \times 10^{-2}\,\text{N}}{(0.245\,\text{A})(0.100\,\text{m})} = 1.42\,\text{T}.$$

This technique is a highly precise means of determining magnetic fields.

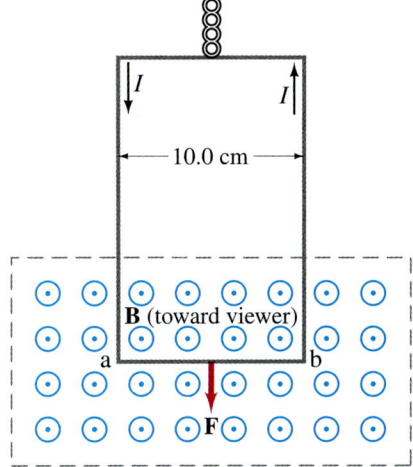

FIGURE 27–14 Measuring a magnetic field **B**. Example 27–1.

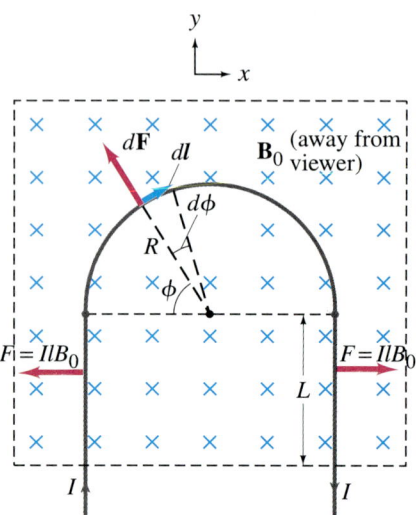

FIGURE 27–15 Example 27–2.

EXAMPLE 27–2 **Magnetic force on a semicircular wire.** A rigid wire, carrying a current I, consists of a semicircle of radius R and two straight portions as shown in Fig. 27–15. The wire lies in a plane perpendicular to a uniform magnetic field $\mathbf{B}_0$. The straight portions each have length l within the field. Determine the net force on the wire due to the magnetic field $\mathbf{B}_0$.

SOLUTION The forces on the two straight sections are equal ($= IlB_0$) and in opposite directions, so they cancel. Hence the net force is that on the semicircular portion. We divide the semicircle into short lengths $dl = R\, d\phi$ as indicated and make use of Eq. 27–4, $d\mathbf{F} = I\, d\mathbf{l} \times \mathbf{B}$, to find

$$dF = IB_0 R\, d\phi,$$

where dF is the force on the length $dl = R\, d\phi$, and the angle between $d\mathbf{l}$ and $\mathbf{B}_0$ is 90° (so $\sin\theta = 1$ in the cross product). The x component of the force $d\mathbf{F}$ on the segment $d\mathbf{l}$ shown, and the x component of $d\mathbf{F}$ for a symmetrically located $d\mathbf{l}$ on the other side of the semicircle, will cancel each other. Thus for the entire semicircle there will be no x component of force. Hence we need be concerned only with the y components, each equal to $dF \sin\phi$, and the total force will have magnitude

$$F = \int_0^\pi dF \sin\phi = IB_0 R \int_0^\pi \sin\phi\, d\phi = -IB_0 R \cos\phi \Big|_0^\pi = 2IB_0 R,$$

with direction vertically upward along the y axis in Fig. 27–15.

27–4 Force on an Electric Charge Moving in a Magnetic Field

We have seen that a current-carrying wire experiences a force when placed in a magnetic field. Since a current in a wire consists of moving electric charges, we might expect that freely moving charged particles (not in a wire) would also experience a force when passing through a magnetic field. Indeed, this is the case.

From what we already know we can predict the force on a single moving electric charge. If N such particles of charge q pass by a given point in time t, they constitute a current $I = Nq/t$. We let t be the time for a charge q to travel a distance L in a magnetic field $\mathbf{B}$; then $\mathbf{l} = \mathbf{v}t$ where $\mathbf{v}$ is the velocity of the particle. Thus, the force on these N particles is, by Eq. 27–3, $\mathbf{F} = I\mathbf{l} \times \mathbf{B} = (Nq/t)(\mathbf{v}t) \times \mathbf{B} = Nq\mathbf{v} \times \mathbf{B}$. The force on *one* of the N particles is then

$$\mathbf{F} = q\mathbf{v} \times \mathbf{B}. \quad (27\text{–}5\mathrm{a})$$

This basic and important result can be considered as an alternative way of defining the magnetic field $\mathbf{B}$, in place of Eq. 27–4 or 27–3. The magnitude of the force in Eq. 27–5a is

$$F = qvB \sin\theta. \quad (27\text{–}5\mathrm{b})$$

This gives the magnitude of the force on a particle of charge q moving with velocity $\mathbf{v}$ at a point where the magnetic field has magnitude B. The angle between $\mathbf{v}$ and $\mathbf{B}$ is θ. The force is greatest when the particle moves perpendicular to $\mathbf{B}$ ($\theta = 90°$):

$$F_{\max} = qvB. \quad [\mathbf{v} \perp \mathbf{B}]$$

The force is *zero* if the particle moves *parallel* to the field lines ($\theta = 0°$). The *direction* of the force is perpendicular to the magnetic field $\mathbf{B}$ and to the velocity $\mathbf{v}$ of the particle. It is given again by a right-hand rule, as for any cross product (for $q > 0$): orient your right hand so that your outstretched fingers point along the direction of motion of the particle ($\mathbf{v}$) and when you bend your fingers they point along the direction of $\mathbf{B}$. Then your thumb will point in the direction of the force. This is true only for *positively* charged particles, and will be "down" for the situation shown in Fig. 27–16. For negatively charged particles, the force is in exactly the opposite direction, "up" in Fig. 27–16.

Force on moving charge in magnetic field

FIGURE 27–16 Force on charged particles due to a magnetic field is perpendicular to the magnetic field direction.

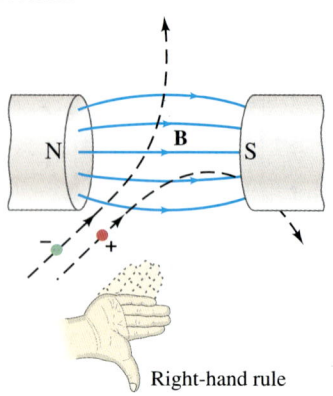

Right-hand rule

EXAMPLE 27-3 Magnetic force on a proton. A proton having a speed of 5.0×10^6 m/s in a magnetic field feels a force of 8.0×10^{-14} N toward the west when it moves vertically upward. When moving horizontally in a northerly direction, it feels zero force. What is the magnitude and direction of the magnetic field in this region? (The charge on a proton is $q = +e = 1.6 \times 10^{-19}$ C.)

SOLUTION Since the proton feels no force when moving north, the field must be in a north–south direction. In order to produce a force to the west when the proton moves upward, the right-hand rule tells us that **B** must point toward the north. (Your thumb points west and the outstretched fingers of your right hand point upward only when your bent fingers point north.) The magnitude of **B**, from Eq. 27–5 with $\theta = 90°$, is

$$B = \frac{F}{qv} = \frac{8.0 \times 10^{-14} \text{ N}}{(1.6 \times 10^{-19} \text{ C})(5.0 \times 10^6 \text{ m/s})} = 0.10 \text{ T}.$$

The path of a charged particle moving in a plane perpendicular to a uniform magnetic field is a circle. See Fig. 27–17, where the magnetic field is directed *into* the paper, as represented by ×'s. An electron at point P is moving to the right, and the force on it at this point is downward as shown (use the right-hand rule and reverse the direction for negative charge). The electron is thus deflected downward. A moment later, say when it reaches point Q, the force is still perpendicular to the velocity and is in the direction shown. Since the force is always perpendicular to **v**, the magnitude of **v** does not change—it moves at constant speed. But the particle changes direction, and moves in a circular path with constant centripetal acceleration (see Example 27–4) due to the magnetic force directed toward the center of this circle at all points. The electron moves clockwise in Fig. 27–17. A positive particle would feel a force in the opposite direction and would thus move counterclockwise.

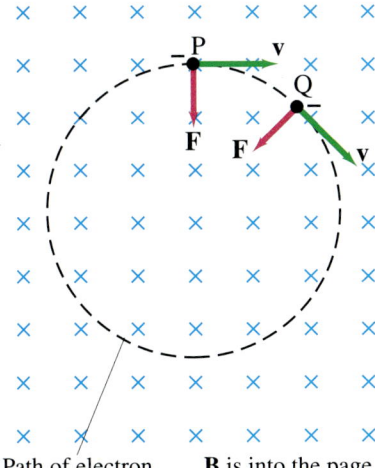

Path of electron   **B** is into the page

FIGURE 27-17 Force exerted by a uniform magnetic field on a moving charged particle (in this case, an electron) produces a circular path.

EXAMPLE 27-4 Electron's path in a uniform magnetic field. An electron travels at 2.0×10^7 m/s in a plane perpendicular to a 0.010-T magnetic field. Describe its path.

SOLUTION The electron moves at speed v in a curved path whose radius of curvature is found using Newton's second law, $F = ma$. We have a centripetal acceleration $a = v^2/r$ (Eq. 3–14). The force is given by Eq. 27–5 with $\sin \theta = 1$, $F = qvB$, so we have

$$F = ma$$

$$qvB = \frac{mv^2}{r}.$$

We solve for r and find

$$r = \frac{mv}{qB}.$$

Since **F** is perpendicular to **v**, the magnitude of **v** doesn't change. From this equation we see that if **B** = constant, then r = constant, and the curve must be a circle as we claimed above. To get r we put in the numbers:

$$r = \frac{(9.1 \times 10^{-31} \text{ kg})(2.0 \times 10^7 \text{ m/s})}{(1.6 \times 10^{-19} \text{ C})(0.010 \text{ T})} = 1.1 \times 10^{-2} \text{ m} = 1.1 \text{ cm}.$$

The time T required for a particle of charge q moving with constant speed v to make one circular revolution in a uniform magnetic field **B** ($\perp$ **v**) is $T = 2\pi r/v$, where $2\pi r$ is the circumference of its circular path. From Example 27–4, $r = mv/qB$, so

$$T = \frac{2\pi m}{qB}.$$

SECTION 27-4   Force on an Electric Charge Moving in a Magnetic Field

Since T is the period of rotation, the frequency of rotation is

$$f = \frac{1}{T} = \frac{qB}{2\pi m}. \qquad (27-6)$$

This is often called the **cyclotron frequency** of a particle in a field because this is the rotation frequency of particles in a cyclotron (see Problem 59). Note that f does not depend on the speed v; if v is large, r is large ($r = mv/qB$) for a given B, but the frequency is independent of v and r, as long as v is not near the speed of light (see Section 7–5 or Chapter 37).

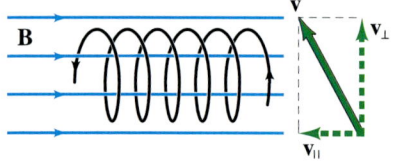

FIGURE 27–18 Conceptual Example 27–5.

CONCEPTUAL EXAMPLE 27–5 **A spiral path.** What is the path of a charged particle in a uniform magnetic field if its velocity is *not* perpendicular to the magnetic field?

RESPONSE The velocity vector can be broken down into components parallel and perpendicular to the field. The velocity component parallel to the field lines experiences no force, and so this component remains constant. The velocity component perpendicular to the field results in circular motion about the field lines. Putting these two motions together produces a helical (spiral) motion around the field lines as shown in Fig. 27–18.

➡ **PHYSICS APPLIED**

The aurora borealis

CONCEPTUAL EXAMPLE 27–6 **Aurora borealis.** Charged ions approach the Earth from the Sun (the "solar wind") and enter the atmosphere mainly near the poles, sometimes causing a phenomenon called the *aurora borealis* or "northern lights" in northern latitudes. Why toward the poles?

RESPONSE A glance at Fig. 27–19 (see also Fig. 27–18) provides the answer. Imagine a stream of charged particles approaching the Earth as shown. The velocity component perpendicular to the field for each particle becomes a circular orbit around the field lines, whereas the velocity component parallel to the field carries the particle along the field lines toward the poles. The high concentration of charged particles ionizes the air, and the recombining of electrons with atoms emits light (Chapter 38), which is the aurora, especially during periods of high sunspot activity when the solar wind is greater.

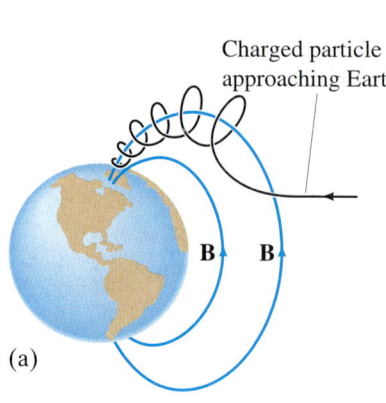

FIGURE 27–19 (a) Diagram showing a charged particle approaching the Earth which is "captured" by the magnetic field of the Earth. Such particles follow the field lines toward the poles as shown. (b) Photo of aurora borealis.

If a particle of charge q moves with velocity **v** in the presence of both a magnetic field **B** and an electric field **E**, it will feel a force

Lorentz equation
$$\mathbf{F} = q(\mathbf{E} + \mathbf{v} \times \mathbf{B}) \qquad (27-7)$$

where we have made use of Eqs. 21–3 and 27–5. Equation 27–7 is often called the **Lorentz equation** and is considered one of the basic equations in physics.

EXAMPLE 27–7 **Velocity selector, or filter: Crossed E and B fields.** Some electronic devices and experiments need a beam of charged particles all moving at nearly the same velocity. This can be achieved using both a uniform electric field and a uniform magnetic field, arranged so they are at right angles to each other. As shown in Fig. 27–20a, particles of charge q pass through slit S_1 and enter the region where **B** points into the page and **E** points from the positive plate toward the negative plate. If the particles enter with different velocities, show how this device "selects" a particular velocity, and determine what this velocity is.

SOLUTION After passing through slit S_1, each particle is subject to two forces as shown in Fig. 27–20b. If q is positive, the magnetic force is upwards and the electric force downwards. (Vice versa if q is negative.) The exit slit, S_2, is assumed to be directly in line with S_1 and the particles' velocity **v**. Depending on the magnitude of **v**, some particles will be bent upwards and some downwards. The only ones to make it through the slit S_2 will be those for which the net force is zero: $\sum F = qvB - qE = 0$. Hence this device selects particles whose velocity is

$$v = \frac{E}{B}. \tag{27-8}$$

This result doesn't depend on the sign of the charge q.

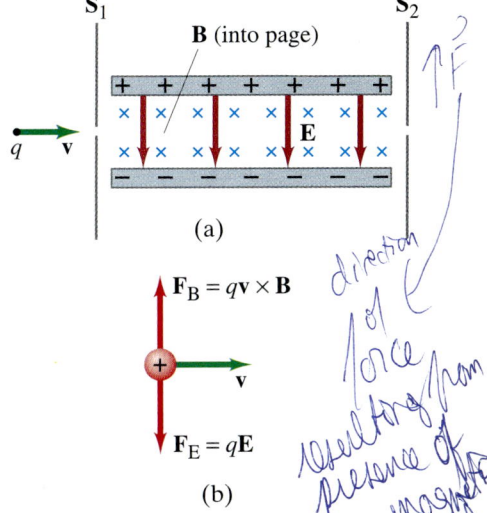

FIGURE 27–20 A velocity selector: if $v = E/B$, the particles make it through S_1 and S_2.

27–5 Torque on a Current Loop; Magnetic Dipole Moment

When an electric current flows in a closed loop of wire placed in a magnetic field, as shown in Fig. 27–21, the magnetic force on the current can produce a torque. This is the basic principle behind a number of important practical devices, including voltmeters, ammeters, and motors. (We discuss these applications in the next Section.) The interaction between a current and a magnetic field is important in other areas as well, including atomic physics.

When current flows through the loop in Fig. 27–21a, whose face we assume is parallel to **B** and is rectangular, the magnetic field exerts a force on both vertical sections of wire as shown, $\mathbf{F}_1$ and $\mathbf{F}_2$ (see also top view, Fig. 27–21b). Notice that, by the right-hand rule, the direction of the force $\mathbf{F}_1$ on the upward current on the left is in the opposite direction from the equal magnitude force $\mathbf{F}_2$ on the descending current on the right. These forces give rise to a net torque that tends to rotate the coil about its vertical axis.

Let us calculate the magnitude of this torque. From Eq. 27–2, the force has magnitude $F = IaB$, where a is the length of the vertical arm of the coil. The lever arm for each force is $b/2$, where b is the width of the coil and the "axis" is at the midpoint. The total torque is the sum of the torques due to each of the forces, so

$$\tau = IaB\frac{b}{2} + IaB\frac{b}{2} = IabB = IAB,$$

where $A = ab$ is the area of the coil. If the coil consists of N loops of wire, the torque on N wires becomes

$$\tau = NIAB.$$

If the coil makes an angle θ with the magnetic field, as shown in Fig. 27–21c, the forces are unchanged, but each lever arm is reduced from $\frac{1}{2}b$ to $\frac{1}{2}b\sin\theta$. Note that the angle θ is chosen to be the angle between **B** and the perpendicular to the face of the coil, Fig. 27–21c. So the torque becomes

$$\tau = NIAB\sin\theta. \tag{27-9}$$

This formula, derived here for a rectangular coil, is valid for any shape of flat coil.

FIGURE 27–21 Calculating the torque on a current loop in a magnetic field **B**. (a) Loop face parallel to **B** field lines; (b) top view; (c) loop makes an angle to **B**, reducing the torque since the lever arm is reduced.

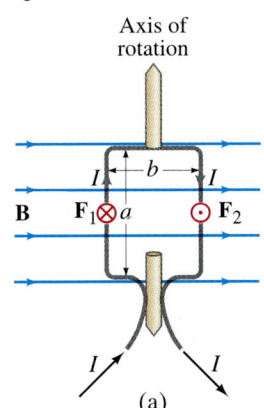

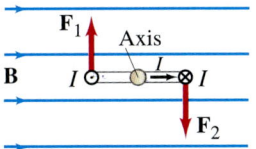

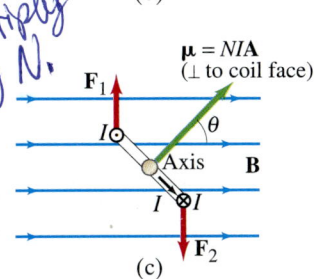

The quantity NIA is called the **magnetic dipole moment** of the coil and is considered a vector:

Magnetic dipole moment

$$\boldsymbol{\mu} = NI\mathbf{A}, \quad (27\text{-}10)$$

where the direction of $\mathbf{A}$ (and therefore of $\boldsymbol{\mu}$) is *perpendicular* to the plane of the coil (the green arrow in Fig. 27–21c) consistent with the right-hand rule (cup your right hand so your fingers wrap around the loop in the direction of current flow, then your thumb points in the direction of $\boldsymbol{\mu}$ and $\mathbf{A}$). With this definition of $\boldsymbol{\mu}$, we can rewrite Eq. 27–9 in vector form:

$$\boldsymbol{\tau} = NI\mathbf{A} \times \mathbf{B}$$

or

$$\boldsymbol{\tau} = \boldsymbol{\mu} \times \mathbf{B}, \quad (27\text{-}11)$$

which gives the correct magnitude and direction for $\boldsymbol{\tau}$.

Equation 27–11 has the same form as Eq. 21–9b for an electric dipole (with electric dipole moment $\mathbf{p}$) in an electric field $\mathbf{E}$, which is $\boldsymbol{\tau} = \mathbf{p} \times \mathbf{E}$. And just as an electric dipole has potential energy given by $U = -\mathbf{p} \cdot \mathbf{E}$ when in an electric field, we expect a similar form for a magnetic dipole in a magnetic field. In order to rotate a current loop (Fig. 27–21) so as to increase θ, we must do work against the force due to the magnetic field. Hence the potential energy depends on angle (see Eq. 10–22, the work-energy principle for rotational motion) as

$$U = \int \tau \, d\theta = \int NIAB \sin\theta \, d\theta = -\mu B \cos\theta + C.$$

If we choose $U = 0$ at $\theta = \pi/2$, then the arbitrary constant C is zero and the potential energy is

$$U = -\mu B \cos\theta = -\boldsymbol{\mu} \cdot \mathbf{B}, \quad (27\text{-}12)$$

as expected. Bar magnets and compass needles, as well as current loops, can be considered as magnetic dipoles. Note the striking similarities of the fields produced by a bar magnet and a current loop, Figs. 27–3b and 27–10.

EXAMPLE 27–8 Torque on a coil. A circular coil of wire has a diameter of 20.0 cm and contains 10 loops. The current in each loop is 3.00 A, and the coil is placed in a 2.00-T magnetic field. Determine the maximum and minimum torque exerted on the coil by the field.

SOLUTION Equation 27–9 is valid for any shape of coil, including circular, where the area is

$$A = \pi r^2 = \pi(0.100 \text{ m})^2 = 3.14 \times 10^{-2} \text{ m}^2.$$

The maximum torque occurs when the coil's face is parallel to the magnetic field, so $\theta = 90°$ in Fig. 27–21c, and $\sin\theta = 1$ in Eq. 27–9:

$$\tau = NIAB \sin\theta = (10)(3.00 \text{ A})(3.14 \times 10^{-2} \text{ m}^2)(2.00 \text{ T})(1) = 1.88 \text{ m}\cdot\text{N}.$$

The minimum torque occurs if $\sin\theta = 0$, for which $\theta = 0°$, and then $\tau = 0$ from Eq. 27–9.

EXAMPLE 27–9 **Magnetic moment of a hydrogen atom.** Determine the magnetic dipole moment of the electron orbiting the proton of a hydrogen atom, assuming (in the Bohr model) it is in its ground state with a circular orbit of radius 0.529×10^{-10} m. [Note: This is a very rough picture of atomic structure, but nonetheless gives an accurate result.]

SOLUTION From Newton's second law, $F = ma$, we have, since the electron is held in its orbit by the coulomb force,

$$\frac{e^2}{4\pi\epsilon_0 r^2} = \frac{mv^2}{r};$$

so

$$v = \sqrt{\frac{e^2}{4\pi\epsilon_0 mr}} = \sqrt{\frac{(8.99 \times 10^9 \, \text{N} \cdot \text{m}^2/\text{C}^2)(1.60 \times 10^{-19} \, \text{C})^2}{(9.11 \times 10^{-31} \, \text{kg})(0.529 \times 10^{-10} \, \text{m})}} = 2.19 \times 10^6 \, \text{m/s}.$$

Since current is the electric charge that passes a given point per unit time, the revolving electron is equivalent to a current

$$I = \frac{e}{T} = \frac{ev}{2\pi r},$$

where $T = 2\pi r/v$ is the time required for one orbit. Since the area of the orbit is $A = \pi r^2$, the magnetic dipole moment is

$$\mu = IA = \frac{ev}{2\pi r}(\pi r^2) = \tfrac{1}{2}evr$$
$$= \tfrac{1}{2}(1.60 \times 10^{-19} \, \text{C})(2.19 \times 10^6 \, \text{m/s})(0.529 \times 10^{-10} \, \text{m}) = 9.27 \times 10^{-24} \, \text{A} \cdot \text{m}^2,$$

or 9.27×10^{-24} J/T.

*27–6 Applications: Galvanometers, Motors, Loudspeakers

The basic component of analog meters (those with pointer and dial), including analog ammeters, voltmeters, and ohmmeters, is a galvanometer. We have already seen how these meters are designed (Section 26–5), and now we can examine how the crucial element, a galvanometer, works. As shown in Fig. 27–22, a **galvanometer** consists of a coil of wire (with attached pointer) suspended in the magnetic field of a permanent magnet. When current flows through the loop of wire, the magnetic field exerts a torque on the loop, as given by Eq. 27–9, $\tau = NIAB \sin\theta$. This torque is opposed by a spring which exerts a torque τ_s approximately proportional to the angle ϕ through which it is turned (Hooke's law). That is, $\tau = k\phi$, where k is the stiffness constant of the spring. Thus the coil and the attached pointer will rotate only to the point where the spring torque balances the torque due to the magnetic field. From Eq. 27–9 we then have $k\phi = NIAB \sin\theta$ or

$$\phi = \frac{NIAB \sin\theta}{k}.$$

Thus the deflection of the pointer, ϕ, is directly proportional to the current I flowing in the coil. But it also depends on the angle θ the coil makes with **B**. For a useful meter we need ϕ to depend only on the current I, independent of θ. To solve this problem, curved pole pieces are used and the galvanometer coil is wrapped around a cylindrical iron core as shown in Fig. 27–23. The iron tends to concentrate the magnetic field lines so that **B** always points parallel to the face of the coil at the wire outside the core. The force is then always perpendicular to the face of the coil and the torque will not vary with angle. Thus ϕ will be proportional to I, as required.

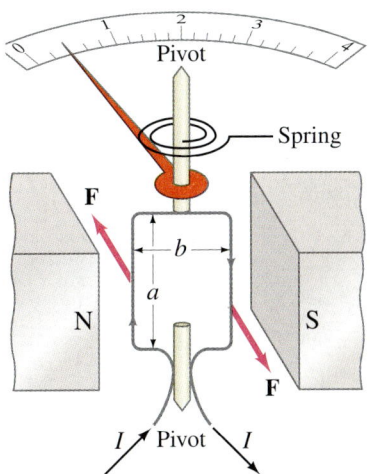

FIGURE 27–22 Galvanometer.

FIGURE 27–23 Galvanometer coil wrapped on an iron core.

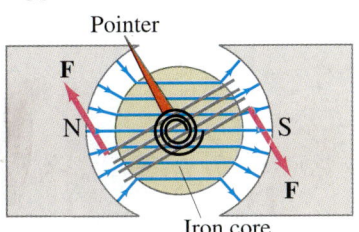

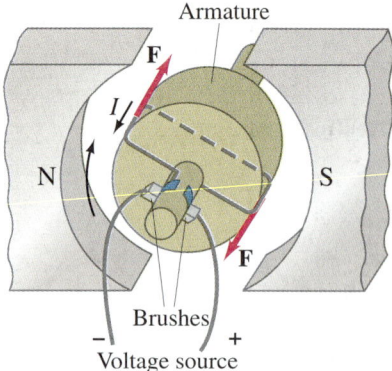

FIGURE 27–24 Diagram of a simple dc motor.

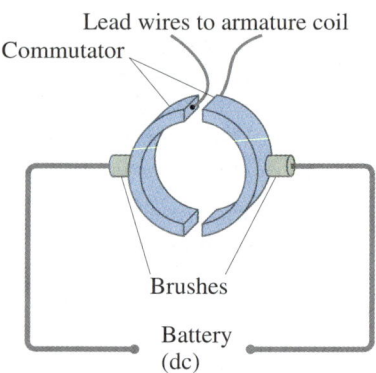

FIGURE 27–25 The commutator-brush arrangement in a dc motor assures alternation of the current in the armature to keep rotation continuous. The commutators are attached to the motor shaft and turn with it, whereas the brushes remain stationary.

> **PHYSICS APPLIED**
> *Motor*

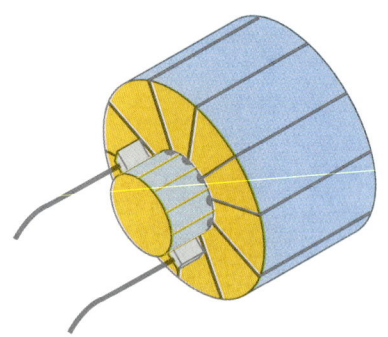

FIGURE 27–26 Motor with many windings.

FIGURE 27–27 Loudspeaker.

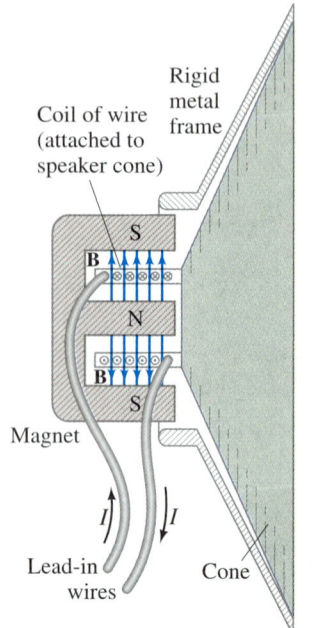

An **electric motor** changes electric energy into (rotational) mechanical energy. A motor works on the same principle as a galvanometer, except that there is no spring so the coil can rotate continuously in one direction. The coil is larger and is mounted on a large cylinder called the **rotor** or **armature**, Fig. 27–24. Actually, there are several coils, although only one is indicated in the Figure. The armature is mounted on a shaft or axle. At the moment shown in Fig. 27–24, the magnetic field exerts forces on the current in the loop as shown. However, when the coil, which is rotating clockwise in Fig. 27–24, passes beyond the vertical position the forces would then act to return the coil back to vertical if the current remained the same. But if the current could somehow be reversed at that critical moment, the forces would reverse, and the coil would continue rotating in the same direction. Thus, alternation of the current is necessary if a motor is to turn continuously in one direction. This can be achieved in a **dc motor** with the use of **commutators** and **brushes**: as shown in Fig. 27–25, the brushes are stationary contacts that rub against the conducting commutators mounted on the motor shaft. At every half revolution, each commutator changes its connection to the other brush. Thus the current in the coil reverses every half revolution as required for continuous rotation. Most motors contain several coils, called "windings," each located in a different place on the armature, Fig. 27–26. Current flows through each coil only during a small part of a revolution, at the time when its orientation results in the maximum torque. In this way, a motor produces a much steadier torque than can be obtained from a single coil. An **ac motor**, with ac current as input, can work without commutators since the current itself alternates. Many motors use wire coils to produce the magnetic field (electromagnets) instead of a permanent magnet. Indeed the design of most motors is more complex than described here, but the general principles remain the same.

A **loudspeaker** also works on the principle that a magnet exerts a force on a current-carrying wire. The electrical output of a stereo or TV set is connected to the wire leads of the speaker. The speaker leads are connected internally to a coil of wire, which is itself attached to the speaker cone, Fig. 27–27. The speaker cone is usually made of stiffened cardboard and is mounted so that it can move back and forth freely. A permanent magnet is mounted directly in line with the coil of wire. When the alternating current of an audio signal flows through the wire coil, which is free to move within the magnet, the coil experiences a force due to the magnetic field of the magnet. As the current alternates at the frequency of the audio signal, the coil and attached speaker cone move back and forth at the same frequency, causing alternate compressions and rarefactions of the adjacent air, and sound waves are produced. A speaker thus changes electrical energy into sound energy, and the frequencies and intensities of the emitted sound waves can be an accurate reproduction of the electrical input.

27–7 Discovery and Properties of the Electron

The electron plays a basic role in our understanding of electricity and magnetism today. But its existence was not suggested until the 1890s. We discuss it here because magnetic fields were crucial for measuring its properties.

Toward the end of the nineteenth century, studies were being done on the discharge of electricity through rarefied gases. One apparatus, diagrammed in Fig. 27–28, was a glass tube fitted with electrodes and evacuated so only a small amount of gas remained inside. When a very high voltage was applied to the electrodes, a dark space seemed to extend outward from the cathode (negative electrode) toward the opposite end of the tube; and that far end of the tube would glow. If one or more screens containing a small hole were inserted as shown, the glow was restricted to a tiny spot on the end of the tube. It seemed as though something being emitted by the cathode traveled to the opposite end of the tube. These "somethings" were named **cathode rays**.

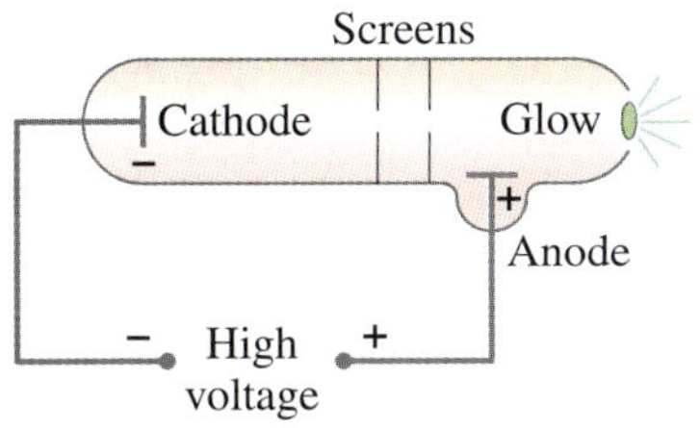

There was much discussion at the time about what these rays might be. Some scientists thought they might resemble light. But the observation that the bright spot at the end of the tube could be deflected to one side by an electric or magnetic field suggested that cathode rays could be charged particles; and the direction of the deflection was consistent with a negative charge. Furthermore, if the tube contained certain types of rarefied gas, the path of the cathode rays was made visible by a slight glow.

Estimates of the charge e of the (assumed) cathode-ray particles, as well as of their charge-to-mass ratio e/m, had been made by 1897. But in that year, J. J. Thomson (1856–1940) was able to measure e/m directly, using the apparatus shown in Fig. 27–29. Cathode rays are accelerated by a high voltage and then pass between a pair of parallel plates built into the tube. The voltage applied to the plates produces an electric field, and a pair of coils produces a magnetic field. When only the electric field is present, say with the upper plate positive, the cathode rays are deflected upward as in path a in Fig. 27–29. If only a magnetic field exists, say inward, the rays are deflected downward along path c. These observations are just what is expected for a negatively charged particle. The force on the rays due to the magnetic field is $F = evB$, where e is the charge and v is the velocity of the cathode rays. In the absence of an electric field, the rays are bent into a curved path, so we have, from $F = ma$,

$$evB = m\frac{v^2}{r},$$

and thus

$$\frac{e}{m} = \frac{v}{Br}.$$

The radius of curvature r can be measured and so can B. The velocity v can be found by applying an electric field in addition to the magnetic field. The electric field E is adjusted so that the cathode rays are undeflected and follow path b in Fig. 27–29. This

FIGURE 27–28 Discharge tube. In some models, one of the screens is the anode (positive plate).

FIGURE 27–29 Cathode rays deflected by electric and magnetic fields.

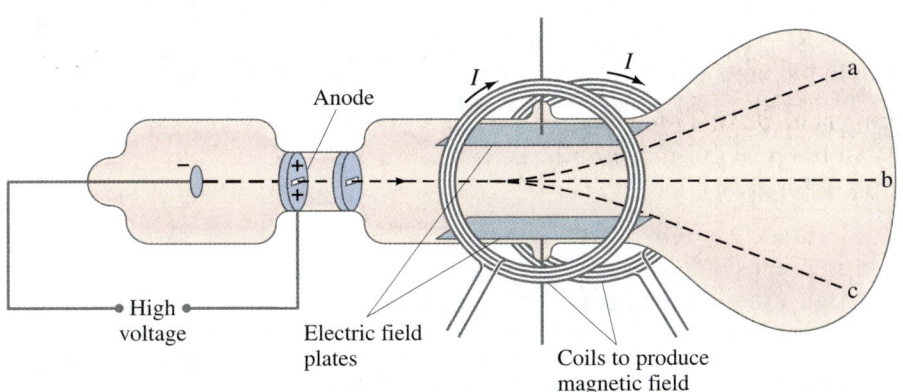

(a)

is just like the velocity selector of Example 27–7 where the force due to the electric field, $F = eE$, is balanced by the force due to the magnetic field, $F = evB$. Thus $eE = evB$ and $v = E/B$. Combining this with the above equation we have

e/m measured

$$\frac{e}{m} = \frac{E}{B^2 r}. \qquad (27\text{–}13)$$

The quantities on the right side can all be measured so that although e and m could not be determined separately, the ratio e/m could be determined. The accepted value today is $e/m = 1.76 \times 10^{11}$ C/kg. Cathode rays soon came to be called **electrons**.

"Discovery" of the electron

It is worth noting that the "discovery" of the electron, like many others in science, is not quite so obvious as discovering gold or oil. Should the discovery of the electron be credited to the person who first saw a glow in the tube? Or to the person who first called them cathode rays? Perhaps neither one, for they had no conception of the electron as we know it today. In fact, the credit for the discovery is generally given to Thomson, but not because he was the first to see the glow in the tube. Rather it is because he believed that this phenomenon was due to tiny negatively charged particles and made careful measurements on them. Furthermore he argued that these particles were constituents of atoms, and not ions or atoms themselves as many thought, and he developed an electron theory of matter. His view is close to what we accept today, and this is why Thomson is credited with the "discovery." Note, however, that neither he nor anyone else ever actually saw an electron itself. We discuss this briefly, for it illustrates the fact that discovery in science is not always a clear-cut matter. In fact some philosophers of science think the word "discovery" is often not appropriate, such as in this case.

Thomson believed that an electron was not an atom, but rather a constituent, or part, of an atom. Convincing evidence for this came soon with the determination of the charge and the mass of the cathode rays. Thomson's student J. S. Townsend made the first direct (but rough) measurements of e in 1897. But it

Millikan oil-drop experiment

was the more refined **oil-drop experiment** of Robert A. Millikan (1868–1953) that yielded a precise value for the charge on the electron and showed that charge comes in discrete amounts. In this experiment, tiny droplets of mineral oil carrying an electric charge were allowed to fall under gravity between two parallel plates, Fig. 27–30. The electric field E between the plates was adjusted until the drop was suspended in midair. The downward pull of gravity, mg, was then just balanced by the upward force due to the electric field. Thus $qE = mg$, so the charge $q = mg/E$. The mass of the droplet was determined by measuring its terminal velocity in the absence of the electric field. Sometimes the drop was charged negatively, and sometimes positively, suggesting that the drop had acquired or lost electrons (by friction, leaving the atomizer). Millikan's painstaking observations and analysis presented convincing evidence that any charge was an integral multiple of a smallest charge, e, that was ascribed to the electron, and that the value of e was 1.6×10^{-19} C. (Today's value of e, as mentioned in Chapter 21, is $e = 1.602 \times 10^{-19}$ C.) This value of e, combined with the measurement of e/m, gives the mass of the electron to be $(1.6 \times 10^{-19} \text{ C})/(1.76 \times 10^{11} \text{ C/kg}) = 9.1 \times 10^{-31}$ kg. This mass is less than a thousandth the mass of the smallest atom, and thus confirmed the idea that the electron is only a part of an atom. The accepted value today for the mass of the electron is $m_e = 9.11 \times 10^{-31}$ kg.

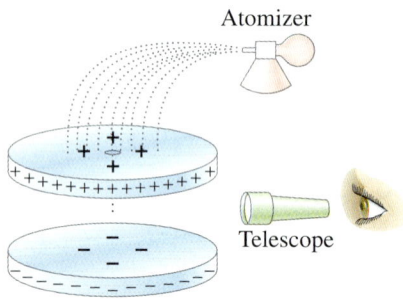

FIGURE 27–30 Millikan's oil-drop experiment.

CRT, Revisited

The cathode ray tube (CRT), which is the picture tube of TV sets, oscilloscopes, and computer monitors, was discussed in Chapter 23. There, in Fig. 23–19, we saw a design using electric deflection plates to maneuver the electron beam. Many CRTs, however, make use of the magnetic field produced by coils to maneuver the electron beam. They operate much like the coils shown in Fig. 27–29. Both types—electrostatic deflection and magnetic deflection—are in use today.

*27–8 The Hall Effect

When a current-carrying conductor is held firmly in a magnetic field, the field exerts a sideways force on the charges moving in the conductor. For example, if electrons move to the right in the rectangular conductor shown in Fig. 27–31a, the inward magnetic field will exert a downward force on the electrons $\mathbf{F_B} = -e\mathbf{v_d} \times \mathbf{B}$, where $\mathbf{v_d}$ is the drift velocity of the electrons (Section 25–8). So the electrons will tend to move nearer face D than face C. There will thus be a potential difference between faces C and D of the conductor. This potential difference builds up until the electric field $\mathbf{E_H}$ it produces exerts a force, $e\mathbf{E_H}$, on the moving charges that is equal and opposite to the magnetic force. This effect is called the **Hall effect** after E. H. Hall, who discovered it in 1879. The difference of potential produced is called the **Hall emf**.

The electric field due to the separation of charge is called the *Hall field*, $\mathbf{E_H}$, and points downward in Fig. 27–31a, as shown. In equilibrium, the force due to this electric field is balanced by the magnetic force $ev_d B$, so

$$eE_H = ev_d B.$$

Hence $E_H = v_d B$. The Hall emf is then (assuming the conductor is long and thin so E_H is uniform)

$$\mathscr{E}_H = E_H l = v_d B l, \quad (27\text{–}14)$$

where l is the width of the conductor.

A current of negative charges moving to the right is equivalent to positive charges moving to the left, at least for most purposes. But the Hall effect can distinguish these two. As can be seen in Fig. 27–31b, positive particles moving to the left are deflected downward, so that the bottom surface is positive relative to the top surface. This is the reverse of part (a). Indeed, the direction of the emf in the Hall effect first revealed that it is negative particles that move in metal conductors.

The magnitude of the Hall emf is proportional to the strength of the magnetic field. The Hall effect can thus be used to measure magnetic field strengths. First the conductor, called a *Hall probe*, is calibrated with known magnetic fields. Then, for the same current, its emf output will be a measure of B. Hall probes can be made very small and are convenient and accurate to use.

The Hall effect can also be used to measure the drift velocity of charge carriers when the external magnetic field B is known. Such a measurement also allows us to determine the density of charge carriers in the material.

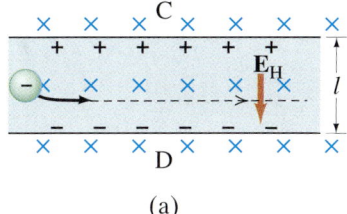
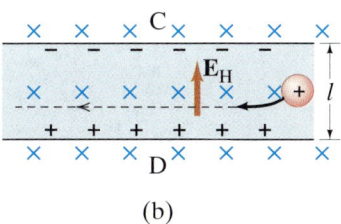

FIGURE 27–31 The Hall effect. (a) Negative charges moving to the right as the current. (b) Positive charges moving to the left as the current.

EXAMPLE 27–10 Drift velocity using the Hall effect. A long copper strip 1.8 cm wide and 1.0 mm thick is placed in a 1.2-T magnetic field as in Fig. 27–31a. When a steady current of 15 A passes through it, the Hall emf is measured to be 1.02 μV. Determine the drift velocity of the electrons and the density of free (conducting) electrons (number per unit volume) in the copper.

SOLUTION The drift velocity (Eq. 27–14) is

$$v_d = \frac{\mathscr{E}_H}{Bl} = \frac{1.02 \times 10^{-6}\,\text{V}}{(1.2\,\text{T})(1.8 \times 10^{-2}\,\text{m})} = 4.7 \times 10^{-5}\,\text{m/s}.$$

The density of charge carriers n is obtained from Eq. 25–13, $I = nev_d A$, where A is the cross-sectional area through which the current I flows. Then

$$n = \frac{I}{ev_d A} = \frac{15\,\text{A}}{(1.6 \times 10^{-19}\,\text{C})(4.7 \times 10^{-5}\,\text{m/s})(1.8 \times 10^{-2}\,\text{m})(1.0 \times 10^{-3}\,\text{m})}$$

$$= 11 \times 10^{28}\,\text{m}^{-3}.$$

This value for the density of free electrons in copper, $n = 11 \times 10^{28}$ per m³, is the experimentally measured value. It represents *more* than one free electron per atom, which as we saw in Example 25–12 is $8.4 \times 10^{28}\,\text{m}^{-3}$.

*27–9 Mass Spectrometer

PHYSICS APPLIED
The mass spectrometer

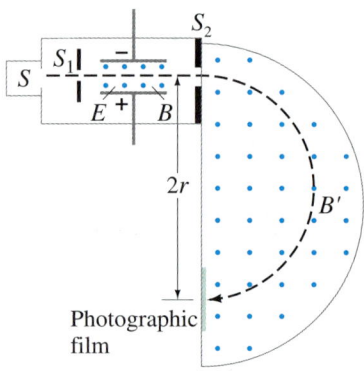

FIGURE 27–32 Bainbridge mass spectrometer. The magnetic fields B and B' point out of the paper (indicated by the dots).

Various methods were developed in the early part of this century to measure the masses of atoms. One of the most accurate was the **mass spectrometer** of Fig. 27–32. Ions are produced by heating, or by an electric current, in the source S. Often the particles are accelerated and then pass through slit S_1 into a velocity selector of crossed electric and magnetic fields (as in Example 27–7). Only those ions whose speed is $v = E/B$ will pass through undeflected and emerge through slit S_2. In the second region, after S_2, there is only a magnetic field B' so the ions follow a circular path. The radius of their path can be measured because the ions darken the photographic plate where they strike. Since $qvB' = mv^2/r$ and $v = E/B$, then we have

$$m = \frac{qB'r}{v} = \frac{qBB'r}{E}.$$

By measuring the quantities on the right, m can be determined. Note that for ions of the same charge, the mass of each is proportional to the radius of its path.

The masses of many atoms were measured in this way. When a pure substance was used, it was sometimes found that two or more closely spaced marks would appear on the film. For example, neon produced two marks whose radii corresponded to atoms of mass 20 and 22 atomic mass units (u). Impurities were ruled out and it was concluded that there must be two types of neon with different masses. These different forms were called **isotopes**. It was soon found that most elements are mixtures of isotopes, and that the difference in mass is due to different numbers of neutrons. Mass spectrometers can be used to separate not only different elements and isotopes, but different molecules as well.

EXAMPLE 27–11 Mass spectrometry. Carbon atoms of atomic mass 12.0 u are found to be mixed with another, unknown, element. In a mass spectrometer, the carbon traverses a path of radius 22.4 cm and the unknown's path has a 26.2 cm radius. What is the unknown element? Assume they have the same charge.

SOLUTION Since mass is proportional to the radius, we have

$$\frac{m_x}{m_C} = \frac{26.2 \text{ cm}}{22.4 \text{ cm}} = 1.17.$$

Thus $m_x = 1.17 \times 12.0 \text{ u} = 14.0 \text{ u}$. The other element is probably nitrogen (see the Periodic Table, inside the back cover). However, it could also be an isotope of carbon or oxygen. Further physical or chemical analysis would be needed.

Summary

A magnet has two **poles**, north and south. The north pole is that end which points toward geographic north when the magnet is freely suspended. Unlike poles of two magnets attract each other, whereas like poles repel.

We can imagine that a **magnetic field** surrounds every magnet. The SI unit for magnetic field is the **tesla** (T).

Electric currents produce magnetic fields. For example, the lines of magnetic field due to a current in a straight wire form circles around the wire and the field exerts a force on magnets placed near it.

A magnetic field exerts a force on an electric current. The force on an infinitesimal length of wire $d\mathbf{l}$ carrying a current I in a magnetic field $\mathbf{B}$ is

$$d\mathbf{F} = I\, d\mathbf{l} \times \mathbf{B}.$$

If the field $\mathbf{B}$ is uniform over a straight length l of wire, then the force is

$$\mathbf{F} = I\mathbf{l} \times \mathbf{B}$$

which has magnitude

$$F = IlB \sin\theta$$

where θ is the angle between magnetic field $\mathbf{B}$ and the wire. The direction of the force is perpendicular to the wire and to the magnetic field, and is given by the right-hand rule. This relation serves as the definition of magnetic field $\mathbf{B}$.

Similarly, a magnetic field **B** exerts a force on a charge q moving with velocity **v** given by

$$\mathbf{F} = q\mathbf{v} \times \mathbf{B}.$$

The magnitude of the force is

$$F = qvB \sin\theta,$$

where θ is the angle between **v** and **B**. The path of a charged particle moving perpendicular to a uniform magnetic field is a circle.

If both electric and magnetic fields are present, the force is

$$\mathbf{F} = q\mathbf{E} + q\mathbf{v} \times \mathbf{B}.$$

The torque on a current loop in a magnetic field **B** is

$$\boldsymbol{\tau} = \boldsymbol{\mu} \times \mathbf{B},$$

where $\boldsymbol{\mu}$ is the **magnetic dipole moment** of the loop:

$$\boldsymbol{\mu} = NI\mathbf{A}.$$

Here N is the number of coils carrying current I in the loop and **A** is a vector perpendicular to the plane of the loop (use right-hand rule) and has magnitude equal to the area of the loop.

The measurement of the charge-to-mass ratio (e/m) of the electron was done using magnetic and electric fields. The charge e on the electron was first measured in the Millikan oil-drop experiment and then its mass was obtained from the measured value of the e/m ratio.

Questions

1. A compass needle is not always balanced parallel to the Earth's surface, but one end may dip downward. Explain.
2. Draw the magnetic field lines around a straight section of wire carrying a current horizontally to the left.
3. In what direction are the magnetic field lines surrounding a straight wire carrying a current that is moving directly toward you?
4. A horseshoe magnet is held vertically with the north pole on the left and south pole on the right. A wire passing between the poles, equidistant from them, carries a current directly away from you. In what direction is the force on the wire?
5. In the relation, $\mathbf{F} = I\boldsymbol{l} \times \mathbf{B}$, which pairs of the vectors (**F**, ***l***, **B**) are always at 90°? Which can be at other angles?
6. The magnetic field due to current in wires in your home can affect a compass. Discuss the effect in terms of currents, including if they are ac or dc.
7. If a negatively charged particle enters a region of uniform magnetic field which is perpendicular to the particle's velocity, will the kinetic energy of the particle increase, decrease, or stay the same. Explain your answer. (Neglect gravity.)
8. In Fig. 27–33, charged particles move in the vicinity of a current-carrying wire. For each charged particle the arrow indicates the direction of motion of the particle and the + or − indicates the sign of the charge. For each of the particles, indicate the direction of the magnetic force due to the magnetic field produced by the wire.

10. Note that the pattern of magnetic field lines surrounding a bar magnet is similar to that of the electric field around an electric dipole. From this fact, predict how the magnetic field will change with distance (a) when near one pole of a very long bar magnet, and (b) when far from a magnet as a whole.
11. Explain why a strong magnet held near a television screen causes the picture to become distorted. Also, explain why the picture sometimes goes completely black where the field is the strongest. [But don't risk damage to your TV by trying this.]
12. Describe the trajectory of a negatively charged particle in the velocity filter of Fig. 27–20 if its speed exceeds E/B. What is its trajectory if $v < E/B$? Would it make any difference if the particle were positively charged?
13. Can you set a resting electron into motion with a magnetic field? With an electric field?
14. A charged particle is moving in a circle under the influence of a uniform magnetic field. If an electric field that points in the same direction as the magnetic field is turned on, describe the path the charged particle will take.
15. The force on a particle in a magnetic field is the idea behind *electromagnetic pumping*. It is used to pump metallic fluids (such as sodium) and more recently to pump blood in artificial heart machines. The basic design is shown in Fig. 27–35. An electric field is applied perpendicular to a blood vessel and to a magnetic field. Explain how ions are caused to move. Do positive and negative ions feel a force in the same direction?

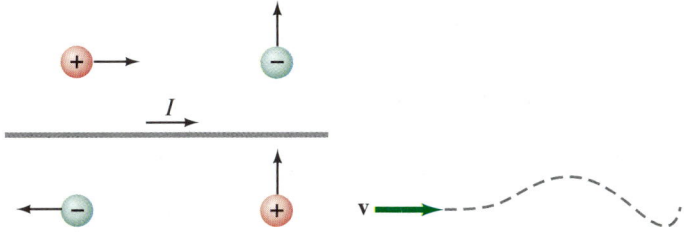

FIGURE 27–33 Question 8. **FIGURE 27–34** Question 9.

9. A positively charged particle in a nonuniform magnetic field follows the trajectory shown in Fig. 27–34. Indicate the direction of the magnetic field everywhere in space, assuming the path is always in the plane of the page, and indicate the relative magnitudes of the field in each region.

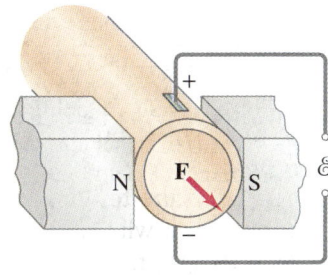

FIGURE 27–35 Electromagnetic pumping in a blood vessel. Question 15.

16. A beam of electrons is directed toward a horizontal wire carrying a current from left to right (Fig. 27–36). In what direction is the beam deflected?

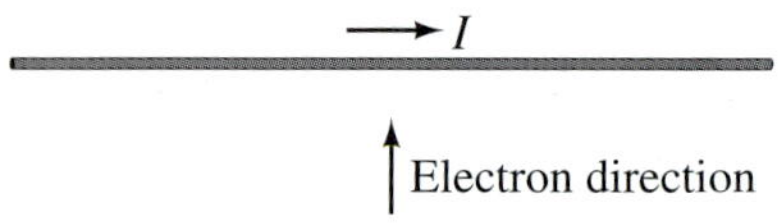

FIGURE 27–36 Question 16.

17. What kind of field or fields surround a moving electric charge?
18. Could we have defined the direction of the magnetic field **B** to be in the direction of the force on a moving charged particle? Explain.
19. A charged particle moves in a straight line through a particular region of space. Could there be a nonzero magnetic field in this region? If so, give two possible situations.
20. If a moving charged particle is deflected sideways in some region of space, can we conclude, for certain, that $\mathbf{B} \neq 0$ in that region? Explain.
21. In a particular region of space there is a uniform magnetic field **B**. Outside this region, $B = 0$. Can you inject an electron from outside into the field perpendicularly so it will move in a closed circular path in the field? What if the electron is injected near the center?
22. How could you tell whether moving electrons in a certain region of space are being deflected by an electric field or by a magnetic field (or by both)?
23. How can you make a compass without using iron or other ferromagnetic material?
24. Describe how you could determine the dipole moment of a bar magnet or compass needle.
25. In what positions (if any) will a current loop placed in a uniform magnetic field be in (a) stable equilibrium, and (b) unstable equilibrium?
*26. A rectangular piece of semiconductor is inserted in a magnetic field and a battery is connected to its ends as shown in Fig. 27–37. When a sensitive voltmeter is connected between points a and b, it is found that point a is at a higher potential than b. What is the sign of the charge carriers in this semiconductor material?

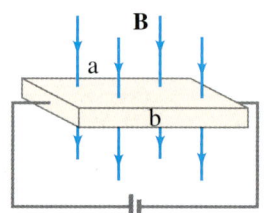

FIGURE 27–37 Question 26.

*27. Two ions have the same mass, but one is singly ionized and the other is doubly ionized. How will their positions on the film of the mass spectrometer of Fig. 27–32 differ?

Problems

Section 27–3

1. (I) (a) What is the force per meter of length on a straight wire carrying a 7.40-A current when perpendicular to a 0.90-T uniform magnetic field? (b) What if the angle between the wire and field is 45.0°?
2. (I) Calculate the magnetic force on a 240-m length of wire stretched between two towers carrying a 150-A current. The Earth's magnetic field of 5.0×10^{-5} T makes an angle of 60° with the wire.
3. (I) How much current is flowing in a wire 4.20 m long if the maximum force on it is 0.900 N when placed in a uniform 0.0800-T field?
4. (I) A 1.5-m length of wire carrying 4.5 A of current is oriented horizontally. At that point on the Earth's surface, the dip angle of the Earth's magnetic field makes an angle of 40° to the wire. Estimate the magnetic force on the wire due to the Earth's magnetic field of 5.5×10^{-5} T at this point.
5. (I) The force on a wire carrying 8.75 A is a maximum of 1.18 N when placed between the pole faces of a magnet. If the pole faces are 55.5 cm in diameter, what is the approximate strength of the magnetic field?
6. (II) The magnetic force per meter on a wire is measured to be only 45 percent of its maximum possible value. Sketch the relationship of the wire and the field if the force were a maximum, and sketch the relationship as it actually is, calculating the angle between the wire and the magnetic field.
7. (II) The force on a wire is a maximum of 5.30 N when placed between the pole faces of a magnet. The current flows horizontally to the right and the magnetic field is vertical. The wire is observed to "jump" toward the observer when the current is turned on. (a) What type of magnetic pole is the top pole face? (b) If the pole faces have a diameter of 10.0 cm, estimate the current in the wire if the field is 0.15 T. (c) If the wire is tipped so that it now makes an angle of 10° with the horizontal, what force will it now feel?
8. (II) Suppose the straight wires connected to the conductor bent into a semicircle in Fig. 27–15 were bent outwardly, still in the plane of the page, so they were horizontal at the base of the semicircle. If a length L of each remained in the field **B**, what would be the total force on the conductor as a whole?
9. (II) A straight 2.0-mm-diameter copper wire can just "float" horizontally in air because of the force of the Earth's magnetic field **B**, which is horizontal, perpendicular to the wire, and of magnitude 5.0×10^{-5} T. What current does the wire carry?
10. (II) A long wire stretches along the x axis and carries a 3.0-A current to the right ($+x$). The wire passes through a uniform magnetic field $\mathbf{B} = (0.20\mathbf{i} - 0.30\mathbf{j} + 0.25\mathbf{k})$T. Determine the components of the force on the wire per cm of length.

11. (III) A curved wire, connecting two points a and b, lies in a plane perpendicular to a uniform magnetic field **B** and carries a current I. Show that the resultant magnetic force on the wire, no matter what its shape, is the same as that on a straight wire connecting the two points carrying the same current I. See Fig. 27–38.

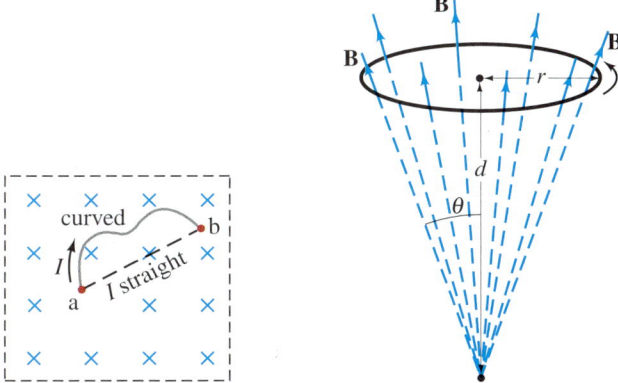

FIGURE 27–38 Problem 11.   **FIGURE 27–39** Problem 12.

12. (III) A circular loop of wire, of radius r, carries current I. It is placed in a magnetic field whose straight lines seem to diverge from a point a distance d below the ring on its axis. (That is, the field makes an angle θ with the loop at all points, Fig. 27–39, where $\tan\theta = r/d$.) Determine the force on the loop.

Section 27–4

13. (I) Determine the magnitude and direction of the force on an electron traveling 7.75×10^5 m/s horizontally to the east in a vertically upward magnetic field of strength 0.85 T.

14. (I) Find the direction of the force on a negative charge for each diagram shown in Fig. 27–40, where **v** is the velocity of the charge and **B** is the direction of the magnetic field. ($\otimes$ means the vector points inward. $\odot$ means it points outward, toward the viewer.)

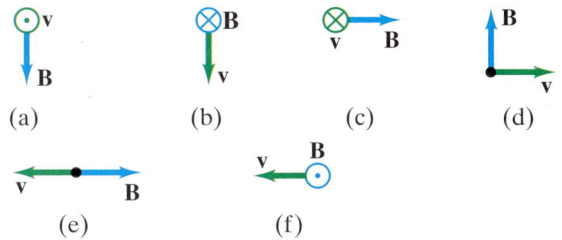

FIGURE 27–40 Problem 14.

15. (I) Determine the direction of **B** for each case in Fig. 27–41, where **F** represents the force on a positively charged particle moving with velocity **v**.

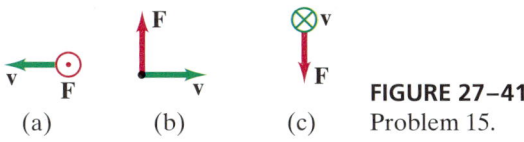

FIGURE 27–41 Problem 15.

16. (I) An electron is projected vertically upward with a speed of 1.80×10^6 m/s into a uniform magnetic field of 0.250 T that is directed horizontally away from the observer. Describe the electron's path in this field.

17. (I) A particle of charge q moves in a circular path of radius r in a uniform magnetic field **B**. Show that its momentum is $p = qBr$.

18. (II) What is the velocity of a beam of electrons that go undeflected when passing through crossed electric and magnetic fields of magnitude 8.8×10^3 V/m and 3.5×10^{-3} T, respectively? What is the radius of the electron orbit if the electric field is turned off?

19. (II) For a particle of mass m and charge q moving in a circular path in a magnetic field B, (a) show that its kinetic energy is proportional to r^2, the square of the radius of curvature of its path, and (b) show that its angular momentum is $L = qBr^2$, about the center of the circle.

20. (II) An electron moves with velocity $\mathbf{v} = (4.0\mathbf{i} - 6.0\mathbf{j}) \times 10^4$ m/s in a magnetic field $\mathbf{B} = (-0.80\mathbf{i} + 0.60\mathbf{j})$ T. Determine the magnitude and direction of the force on the electron.

21. (II) A 5.0-MeV (kinetic energy) proton enters a 0.20-T field, in a plane perpendicular to the field. What is the radius of its path?

22. (II) An electron experiences the greatest force as it travels 2.9×10^6 m/s in a magnetic field when it is moving northward. The force is upward and of magnitude 7.2×10^{-13} N. What is the magnitude and direction of the magnetic field?

23. (II) A doubly charged helium atom whose mass is 6.6×10^{-27} kg is accelerated by a voltage of 2100 V. (a) What will be its radius of curvature if it moves in a plane perpendicular to a uniform 0.340-T field? (b) What is its period of revolution?

24. (II) A 3.40-g bullet moves with a speed of 160 m/s perpendicular to the Earth's magnetic field of 5.00×10^{-5} T. If the bullet possesses a net charge of 13.5×10^{-9} C, by what distance will it be deflected from its path due to the Earth's magnetic field after it has traveled 1.00 km?

25. (II) Suppose the Earth's magnetic field at the equator has magnitude 0.40×10^{-4} T and a northerly direction at all points. How fast must a singly ionized uranium ion ($m = 238$ u, $q = e$) move so as to circle the Earth 5.0 km above the equator? Can you ignore gravity?

26. (II) A proton (mass m_p), a deuteron ($m = 2m_p$, $Q = e$), and an alpha particle ($m = 4m_p$, $Q = 2e$) are accelerated by the same potential difference V and then enter a uniform magnetic field **B** where they move in circular paths perpendicular to **B**. Determine the radius of the paths for the deuteron and alpha particle in terms of that for the proton.

27. (II) A proton moves through a region of space where there is a magnetic field, $\mathbf{B} = (0.45\mathbf{i} + 0.20\mathbf{j})$ T and an electric field $\mathbf{E} = (3.0\mathbf{i} - 4.2\mathbf{j}) \times 10^3$ V/m. At a given instant, the proton's velocity is $\mathbf{v} = (6.0\mathbf{i} + 3.0\mathbf{j} - 5.0\mathbf{k}) \times 10^3$ m/s. Determine the components of the total force on the proton.

28. (II) An electron experiences a force $\mathbf{F} = (3.8\mathbf{i} - 2.7\mathbf{j}) \times 10^{-13}$ N when passing through a magnetic field $\mathbf{B} = (0.35\text{ T})\mathbf{k}$. Determine the electron's velocity.

29. (II) An electron enters a uniform magnetic field $B = 0.23$ T at a 45° angle to **B**. Determine the radius r and pitch p (distance between loops) of the electron's helical path assuming its speed is 3.0×10^6 m/s. See Fig. 27–42.

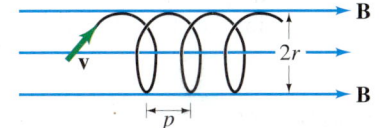

FIGURE 27–42 Problem 29.

30. (II) The path of protons emerging from an accelerator must be bent by 90° by a "bending magnet" so as not to strike a barrier in their path a distance l from their exit hole in the accelerator. Show that the field **B** in the bending magnet, which we assume is uniform and can extend over an area $l \times l$, must have magnitude of at least $B \geq (2mK/e^2l^2)^{\frac{1}{2}}$, where m is the mass of a proton and K is its kinetic energy.

31. (II) A proton moving with speed $v = 2.0 \times 10^5$ m/s in a field-free region abruptly enters an essentially uniform magnetic field $B = 0.850$ T as shown in Fig. 27–43 ($\mathbf{B} \perp \mathbf{v}$). If the proton enters the magnetic field region at a 45° angle as shown, (a) at what angle does it leave and (b) at what distance x does it exit from the field?

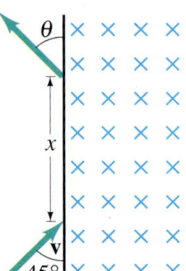

FIGURE 27–43 Problem 31.

Section 27–5

32. (I) A 13.0-cm-diameter circular loop of wire is placed with its face parallel to the uniform magnetic field between the pole pieces of a large magnet. When 7.10 A flows in the coil, the torque on it is 0.185 m·N. What is the magnetic field strength?

33. (I) How much work is required to rotate the current loop (Fig. 27–21) in a uniform magnetic field **B** from (a) $\theta = 0°$ ($\boldsymbol{\mu} \parallel \mathbf{B}$) to $\theta = 180°$, (b) $\theta = 90°$ to $\theta = -90°$?

34. (II) Show that the magnetic dipole moment μ of an electron orbiting the proton nucleus of a hydrogen atom is related to the orbital momentum L of the electron by

$$\mu = \frac{e}{2m}L.$$

35. (II) A circular coil 17.0 cm in diameter and containing twelve loops lies flat on the ground. The Earth's magnetic field at this location has magnitude 5.50×10^{-5} T and points into the Earth at an angle of 66.0° below a line pointing due north. If a 7.10-A clockwise current passes through the coil, (a) determine the torque on the coil, and (b) which edge of the coil rises up, north, east, south, or west.

36. (II) A 20-loop circular coil 20 cm in diameter lies in the xy plane. The current in each loop of the coil is 7.6 A clockwise, and an external magnetic field $\mathbf{B} = (0.80\mathbf{i} + 0.60\mathbf{j} - 0.65\mathbf{k})$ T passes through the coil. Determine: (a) the magnetic moment of the coil, $\boldsymbol{\mu}$; (b) the torque on the coil due to the external magnetic field; (c) the potential energy U of the coil in the field (take the same zero for U as we did in our discussion of Fig. 27–21).

37. (III) Suppose a nonconducting rod of length l carries a uniformly distributed charge Q. It is rotated with angular velocity ω about an axis perpendicular to the rod at one end. Show that the magnetic dipole moment of this rod is $\frac{1}{6}Q\omega l^2$. [*Hint*: Consider the motion of each infinitesimal length of the rod.]

*Section 27–6

*38. (I) A galvanometer needle deflects full scale for a 63.0-μA current. What current will give full-scale deflection if the magnetic field weakens to 0.860 of its original value?

*39. (I) If the restoring spring of a galvanometer weakens by 20 percent over the years, what current will give full-scale deflection if it originally required 36 μA?

*40. (I) If the current to a motor drops by 18 percent, by what factor does the output torque change?

Section 27–7

41. (I) What is the value of q/m for a particle that moves in a circle of radius 8.0 mm in a 0.46-T magnetic field if a crossed 200-V/m electric field will make the path straight?

42. (II) An oil drop whose mass is determined to be 3.3×10^{-15} kg is held at rest between two large plates separated by 1.0 cm as in Fig. 27–30. If the potential difference between the plates is 340 V, how many excess electrons does this drop have?

*Section 27–8

*43. (II) A rectangular sample of a metal is 3.0 cm wide and 500 μm thick. When it carries a 42-A current and is placed in a 0.80-T magnetic field it produces a 6.5-μV Hall emf. Determine: (a) the Hall field in the conductor; (b) the drift speed of the conduction electrons; (c) the density of free electrons in the metal.

*44. (II) In a probe that uses the Hall effect to measure magnetic fields, a 12.0-A current passes through a 1.50-cm-wide 1.30-mm-thick strip of sodium metal. If the Hall emf is 2.42 μV, what is the magnitude of the magnetic field (take it perpendicular to the flat face of the strip)? Assume one free electron per atom of Na, and take its specific gravity to be 0.971.

*45. (II) The Hall effect can be used to measure blood flow rate because the blood contains ions that constitute an electric current. (a) Does the sign of the ions influence the emf? (b) Determine the flow velocity in an artery 3.3 mm in diameter if the measured emf is 0.10 mV and B is 0.070 T. (In actual practice, an alternating magnetic field is used.)

*Section 27–9

*46. (I) In a mass spectrometer, germanium atoms have radii of curvature equal to 21.0, 21.6, 21.9, 22.2, and 22.8 cm. The largest radius corresponds to an atomic mass of 76 u. What are the atomic masses of the other isotopes?

*47. (II) Suppose the electric field between the electric plates in the mass spectrometer of Fig. 27–32 is 2.48×10^4 V/m and the magnetic fields $B = B' = 0.58$ T. The source contains carbon isotopes of mass numbers 12, 13, and 14 from a long-dead piece of a tree. (To estimate atomic masses, multiply by 1.66×10^{-27} kg.) How far apart are the lines formed by the singly charged ions of each type on the photographic film? What if the ions were doubly charged?

*48. (II) A mass spectrometer is being used to monitor air pollutants. It is difficult, however, to separate molecules with nearly equal mass such as CO (28.0106 u) and N_2 (28.0134 u). How large a radius of curvature must a spectrometer have if these two molecules are to be separated on the film by 0.50 mm?

*49. (II) One form of mass spectrometer accelerates ions by a voltage V before they enter a magnetic field B. The ions are assumed to start from rest. Show that the mass of an ion is $m = qB^2R^2/2V$, where R is the radius of the ions' path in the magnetic field and q is their charge.

General Problems

50. Protons move in a circle of radius 5.10 cm in a 0.725-T magnetic field. What value of electric field could make their paths straight? In what direction must the electric field point?

51. Protons with momentum 4.8×10^{-16} kg·m/s are magnetically steered clockwise in a circular path 2.0 km in diameter at Fermi National Accelerator Laboratory in Illinois. What is the magnitude and direction of the field in the magnets surrounding the beam pipe?

52. A proton and an electron have the same kinetic energy upon entering a region of constant magnetic field. What is the ratio of the radii of their circular paths?

53. Near the equator, the Earth's magnetic field points almost horizontally to the north and has magnitude $B = 0.50 \times 10^{-4}$ T. What should be the magnitude and direction for the velocity of an electron if its weight is to be exactly balanced by the magnetic force?

54. Calculate the force on an airplane which has acquired a net charge of 1550 μC and moves with a speed of 120 m/s perpendicular to the Earth's magnetic field of 5.0×10^{-5} T.

55. The power cable for an electric trolley carries a horizontal dc current of 330 A toward the east. The Earth's magnetic field has a strength 5.0×10^{-5} T and makes an angle of dip of 22° at this location. Calculate the magnitude and direction of the magnetic force on a 10-m length of this cable.

56. Two stiff parallel wires a distance l apart in a horizontal plane act as rails to support a light metal rod of mass m (perpendicular to each rail), Fig. 27–44. A magnetic field **B**, directed vertically upward (outward in diagram), acts throughout. At $t = 0$, wires connected to the rails are connected to a constant current source and a current I begins to flow through the system. Determine the speed of the rod, which starts from rest at $t = 0$, as a function of time (a) assuming no friction between the rod and the rails, and (b) if the coefficient of friction is μ_k. (c) In which direction does the rod move, east or west, if the current through it heads north?

57. Suppose the rod in Fig. 27–44 (Problem 56) has mass $m = 0.40$ kg and length 22 cm and the current through it is $I = 40$ A. If the coefficient of static friction is $\mu_s = 0.50$, determine the minimum magnetic field **B** (not necessarily vertical) that will just cause the rod to slide. Give the magnitude of **B** and its direction relative to the vertical.

58. Estimate the approximate maximum deflection of the electron beam near the center of a TV screen due to the Earth's 5.0×10^{-5} T field. Assume the screen is 20 cm from the electron gun where the electrons are accelerated (a) by 2.0 kV, or (b) by 30 kV. Note that in color TV sets, the beam must be directed accurately to within less than 1 mm in order to strike the correct phosphor. Because the Earth's field is significant here, mu-metal shields are used to reduce the Earth's field in the CRT. (See Section 23–9.)

59. The **cyclotron** (Fig. 27–45) is a device used to accelerate elementary particles such as protons to high speeds. Particles starting at point A with some initial velocity travel in circular orbits in the magnetic field B. The particles are accelerated to higher speeds each time they pass in the gap between the metal "dees," where there is an electric field E. (There is no electric field within the cavity of the metal dees.) The electric field changes direction each half-cycle, owing to an ac voltage $V = V_0 \sin 2\pi ft$, so that the particles are increased in speed at each passage through the gap. (a) Show that the frequency f of the voltage must be $f = Bq/2\pi m$, where q is the charge on the particles and m their mass. (b) Show that the kinetic energy of the particles increases by $2qV_0$ each revolution, assuming that the gap is small. (c) If the radius of the cyclotron is 2.0 m and the magnetic field strength is 0.50 T, what will be the maximum kinetic energy of accelerated protons in MeV?

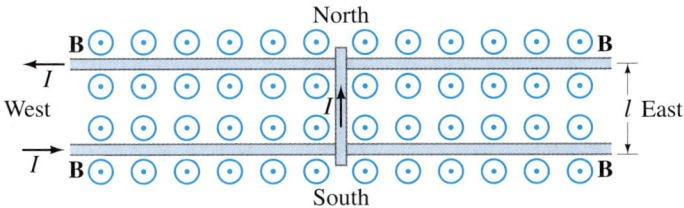

FIGURE 27–44 Looking down on a rod sliding on rails. Problems 56 and 57.

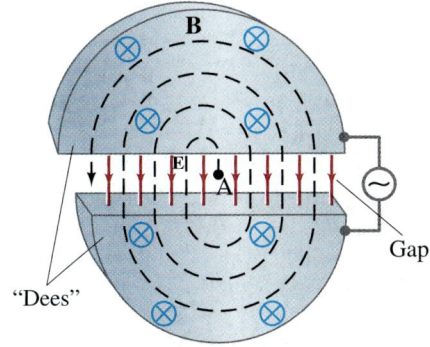

FIGURE 27–45 A cyclotron. Problem 59.

60. The rectangular loop of wire shown in Fig. 27–21 has mass m and carries current I. Show that if the loop is oriented at an angle $\theta \ll 1$ (in radians), then when it is released it will execute simple harmonic motion about $\theta = 0$. Calculate the period of the motion.

61. Magnetic fields are very useful in particle accelerators for "beam steering"; that is, the magnetic fields can be used to change the beam's direction without altering its speed (Fig. 27–46). Discuss how this could work with a beam of protons. What happens to protons that are not moving with the speed that the magnetic field is designed for? If the field extends over a region 5.0 cm wide and has a magnitude of 0.33 T, by approximately what angle will a beam of protons traveling at 0.75×10^7 m/s be bent?

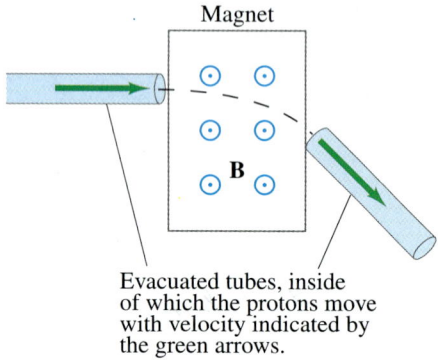

FIGURE 27–46 Problem 61.

62. A square loop of aluminum wire is 20.0 cm on a side. It is to carry 25.0 A and rotate in a uniform 1.65-T magnetic field as shown in Fig. 27–47. (a) Determine the minimum diameter of the wire so that it will not fracture from tension or shear. Assume a safety factor of 10. (See Table 12–2.) (b) What is the resistance of a single loop of this wire?

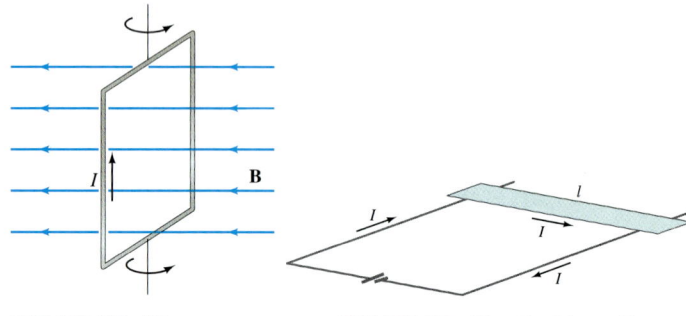

FIGURE 27–47 Problem 62.

FIGURE 27–48 Problem 63.

63. A sort of "projectile launcher" is shown in Fig. 27–48. A large current moves in a closed loop composed of fixed rails, a power supply, and a very light, almost frictionless bar touching the rails. A magnetic field is perpendicular to the plane of the circuit. If the bar has a length of 20 cm, a mass of 1.5 g, and is placed in a field of 1.7 T, what constant current flow is needed in order for it to accelerate to 30 m/s in a distance of 1.0 m? In what direction must the field point?

64. (a) What value of magnetic field would make a beam of electrons, traveling to the right at a speed of 4.8×10^6 m/s, go undeflected through a region where there is a uniform electric field of 10,000 V/m pointing vertically up? (b) What is the direction of the magnetic field if it is known to be perpendicular to the electric field? (c) What is the frequency of the circular orbit of the electrons if the electric field is turned off?

65. In a certain cathode ray tube, electrons are accelerated horizontally by 25 kV. They then pass through a uniform magnetic field B for a distance of 3.5 cm, which deflects them upward so they reach the top of the screen 22 cm away, 11 cm above the center. Estimate the value of B.

66. **Zeeman effect.** In the Bohr model of the hydrogen atom, the electron is held in its circular orbit of radius r about its proton nucleus by electrostatic attraction. If the atoms are placed in a weak magnetic field **B**, the rotation frequency of electrons rotating in a plane perpendicular to **B** is changed by an amount

$$\Delta f = \pm \frac{eB}{4\pi m}$$

where e and m are the charge and mass of an electron. (a) Derive this result, assuming the force due to **B** is much less than that due to electrostatic attraction of the nucleus. (b) What does the $\pm$ sign indicate?

67. A proton follows a spiral path through a gas in a magnetic field of 0.010 T, perpendicular to the plane of the spiral, as shown in Fig. 27–49. In two successive loops, P and Q, the radii are 10.0 mm and 8.5 mm, respectively. Calculate the change in the kinetic energy of the proton as it travels from P to Q.

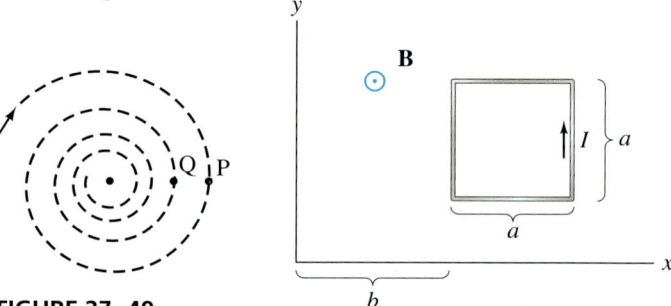

FIGURE 27–49 Problem 67.

FIGURE 27–50 Problem 68.

68. The net force on a current loop in a uniform magnetic field is zero, since contributions to the net force from opposite sides of the loop cancel. However, if the field varies in magnitude from one side of the loop to the other, then a net force can be generated on the loop. Consider a square loop with sides whose length is a, located with one side at $x = b$ in the xy plane (Fig. 27–50). A magnetic field is directed along z, with a magnitude that varies with x according to

$$B = B_0\left(1 - \frac{x}{d}\right).$$

If the current in the loop circulates counter-clockwise (that is, the magnetic dipole moment of the loop is along the z axis), find an expression for the net force on the loop.

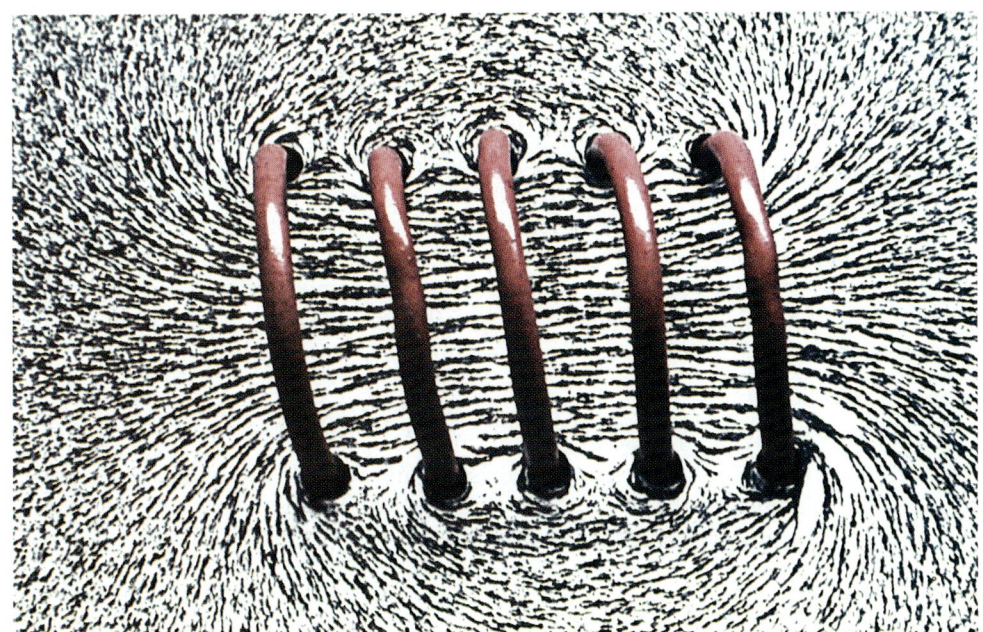

A long coil of wire with many closely spaced loops is called a solenoid. When a solenoid carries an electric current, a nearly uniform magnetic field is produced within the loops as suggested by the alignment of the iron filings in this photo. The magnitude of the field inside a solenoid is readily found using Ampère's law, one of the great general laws of electromagnetism, relating magnetic fields and electric currents. We examine these connections in detail in this chapter, as well as other means for producing magnetic fields.

CHAPTER 28

Sources of Magnetic Field

In the previous chapter, we discussed primarily the effects (forces and torques) that a magnetic field has on electric currents and on moving electric charges. We did see, however, that magnetic fields are produced not only by magnets but also by electric currents (Oersted's great discovery). It is this aspect of magnetism, the production of magnetic fields, that we discuss in this chapter. We will now see how magnetic field strengths are determined for some simple situations, and discuss some general relations between magnetic fields and their sources. We begin with the simplest case, the magnetic field created by a long straight wire carrying a steady electric current. We then look at how such a field, created by one wire, exerts a force on a second current-carrying wire. Interestingly enough, this interaction is used for the precise definitions of both the units of electric current and electric charge, the ampere and the coulomb.

Then we develop an elegant general approach to finding the connection between current and magnetic field known as Ampère's law, one of the fundamental equations of physics. We also examine a second technique for determining the magnetic field due to a current, known as the Biot-Savart law; though harder to visualize intuitively, the Biot-Savart law does allow us to solve problems more readily in many cases than Ampère's law.

Finally, we end this chapter with a discussion of what we understand about iron and other magnetic materials and how they produce magnetic fields.

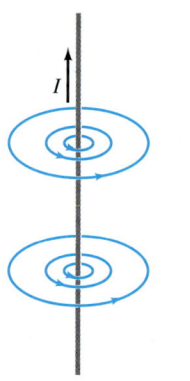

FIGURE 28–1 Same as Fig. 27–9a, magnetic field lines around a long straight wire carrying an electric current I.

Magnetic field due to current in a long straight wire

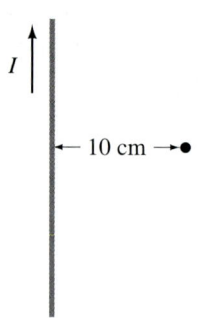

FIGURE 28–2 Example 28–1.

→ **PHYSICS APPLIED**

A compass, near a current, may not point North

FIGURE 28–3 (a) Two parallel conductors carrying currents I_1 and I_2. (b) Magnetic field produced by I_1. (Field produced by I_2 is not shown.)

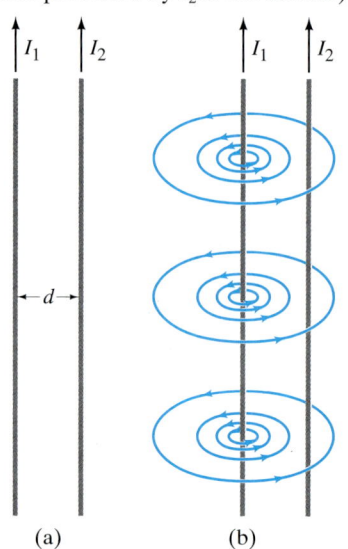

28–1 Magnetic Field Due to a Straight Wire

We saw in Section 27–2, Fig. 27–9, that the magnetic field due to the electric current in a long straight wire is such that the field lines are circles with the wire at the center (Fig. 28–1). You might expect that the field strength at a given point would be greater if the current flowing in the wire were greater; and that the field would be less at points farther from the wire. This is indeed the case. Careful experiments show that the magnetic field B at a point near a long straight wire is directly proportional to the current I in the wire and inversely proportional to the distance r from the wire:

$$B \propto \frac{I}{r}.$$

This relation is valid as long as r, the perpendicular distance to the wire, is much less than the distance to the ends of the wire (i.e., the wire is long).

The proportionality constant is written[†] as $\mu_0/2\pi$; thus,

$$B = \frac{\mu_0}{2\pi} \frac{I}{r}. \qquad \text{[outside a long straight wire]} \quad (28\text{–}1)$$

The value of the constant μ_0, which is called the **permeability of free space**, is $\mu_0 = 4\pi \times 10^{-7}\,\text{T·m/A}$ (see Section 28–3).

EXAMPLE 28–1 Calculation of B near a wire. A vertical electric wire in the wall of a building carries a dc current of 25 A upward. What is the magnetic field at a point 10 cm due north of this wire (Fig. 28–2)?

SOLUTION According to Eq. 28–1:

$$B = \frac{\mu_0 I}{2\pi r} = \frac{(4\pi \times 10^{-7}\,\text{T·m/A})(25\,\text{A})}{(2\pi)(0.10\,\text{m})} = 5.0 \times 10^{-5}\,\text{T},$$

or 0.50 G. By the right-hand rule (Fig. 27–9b), the field points to the west (into the page in Fig. 28–2) at this point. Since this field has about the same magnitude as Earth's, a compass placed at this location would not point north, but in a northwesterly direction.

28–2 Force Between Two Parallel Wires

We have seen that a wire carrying a current produces a magnetic field (magnitude given by Eq. 28–1 for a long straight wire), and furthermore that such a wire feels a force when placed in a magnetic field (Section 27–3, Eq. 27–1). Thus, we expect that two current-carrying wires would exert a force on each other.

Consider two long parallel conductors separated by a distance d, as in Fig. 28–3a. They carry currents I_1 and I_2, respectively. Each current produces a magnetic field that is "felt" by the other so that each must exert a force on the other, as Ampère first pointed out. For example, the magnetic field B_1 produced by I_1 is given by Eq. 28–1. At the location of the second conductor, the magnitude of the magnetic field produced by I_1 is

$$B_1 = \frac{\mu_0}{2\pi} \frac{I_1}{d}.$$

See Fig. 28–3b where the field due *only* to I_1 is shown. According to Eq. 27–2, the

---

[†]The constant is chosen in this complicated way so that Ampère's law (Section 28–4), which is considered more fundamental, will have a simple and elegant form.

force F per unit length l on the conductor carrying current I_2, when the field and current are perpendicular, is

$$\frac{F}{l} = I_2 B_1.$$

Note that the force on I_2 is due only to the field produced by I_1. Of course I_2 also produces a field, but it does not exert a force on itself. We substitute in the above formula for B_1 and find that the force per unit length on I_2 is

$$\frac{F}{l} = \frac{\mu_0}{2\pi} \frac{I_1 I_2}{d}. \qquad (28\text{-}2)$$

If we use the right-hand rule of Fig. 27–9b, we see that the lines of B_1 are as shown in Fig. 28–3b. Then using the right-hand rule of Fig. 27–12c, we see that the force exerted on I_2 will be to the left in Fig. 28–3b. That is, I_1 exerts an attractive force on I_2 (Fig. 28–4a). This is true as long as the currents are in the same direction. If I_2 is in the opposite direction, the right-hand rule indicates that the force is in the opposite direction. That is, I_1 exerts a repulsive force on I_2 (Fig. 28–4b).

Reasoning similar to that above shows that the magnetic field produced by I_2 exerts an equal but opposite force on I_1. We expect this to be true also, of course, from Newton's third law. Thus, as shown in Fig. 28–4, parallel currents in the same directions attract each other, whereas parallel currents in opposite directions repel.

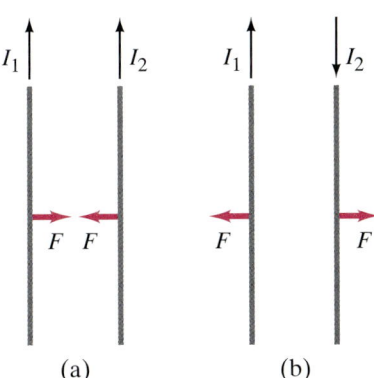

FIGURE 28–4 (a) Parallel currents in the same direction exert an attractive force on each other. (b) Antiparallel currents (in opposite directions) exert a repulsive force on each other.

EXAMPLE 28–2 Force between two current carrying wires. The two wires of a 2.0-m-long appliance cord are 3.0 mm apart and carry a current of 8.0 A dc. Calculate the force between these wires.

SOLUTION Equation 28–2 gives us

$$F = \frac{(2.0 \times 10^{-7} \,\text{T·m/A})(8.0 \,\text{A})^2(2.0 \,\text{m})}{(3.0 \times 10^{-3} \,\text{m})} = 8.5 \times 10^{-3} \,\text{N},$$

where we have written $\mu_0/2\pi = 2.0 \times 10^{-7} \,\text{T·m/A}$. Since the currents are in opposite directions, the force would tend to spread them apart.

EXAMPLE 28–3 Suspending a current with a current. A horizontal wire carries a current $I_1 = 80$ A dc. A second parallel wire 20 cm below it (Fig. 28–5) must carry how much current I_2 so that it doesn't fall due to gravity? The lower wire has a mass of 0.12 g per meter of length.

SOLUTION The force of gravity on wire 2 is downward, and per each meter of length has magnitude

$$\frac{F}{l} = \frac{mg}{l} = \frac{(0.12 \times 10^{-3} \,\text{kg})(9.8 \,\text{m/s}^2)}{1.0 \,\text{m}} = 1.18 \times 10^{-3} \,\text{N/m}.$$

The magnetic force on wire 2 must be upward (hence I_2 must have the same direction as I_1) and, with $d = 0.20$ m and $I_1 = 80$ A, has magnitude

$$\frac{F}{l} = \frac{\mu_0}{2\pi} \frac{I_1 I_2}{d}.$$

We solve for I_2 and find

$$I_2 = \frac{2\pi d}{\mu_0 I_1}\left(\frac{F}{l}\right) = \frac{2\pi(0.20 \,\text{m})}{(4\pi \times 10^{-7} \,\text{T·m/A})(80 \,\text{A})}(1.18 \times 10^{-3} \,\text{N/m}) = 15 \,\text{A}.$$

FIGURE 28–5 Example 28–3.

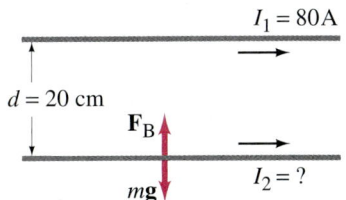

28-3 Operational Definitions of the Ampere and the Coulomb

You may have wondered how the constant μ_0 in Eq. 28-1 could be exactly $4\pi \times 10^{-7}\,\text{T·m/A}$. Here is how it happened. With an older definition of the ampere, μ_0 was measured experimentally to be very close to this value. Today, however, μ_0 is *defined* to be exactly $4\pi \times 10^{-7}\,\text{T·m/A}$. This, of course, could not be done if the ampere were defined independently. The ampere, the unit of current, is now defined in terms of the magnetic field B it produces using the defined value of μ_0.

In particular, we use the force between two parallel current-carrying wires, Eq. 28-2, to define the ampere precisely. If $I_1 = I_2 = 1\,\text{A}$ exactly, and the two wires are exactly 1 m apart, then

$$\frac{F}{l} = \frac{\mu_0}{2\pi}\frac{I_1 I_2}{d} = \frac{(4\pi \times 10^{-7}\,\text{T·m/A})}{(2\pi)}\frac{(1\,\text{A})(1\,\text{A})}{(1\,\text{m})} = 2 \times 10^{-7}\,\text{N/m}.$$

Definitions of ampere and coloumb

Thus, *one **ampere** is defined as that current flowing in each of two long parallel conductors 1 m apart, which results in a force of exactly $2 \times 10^{-7}\,\text{N/m}$ of length of each conductor.*

This is the precise definition of the ampere. The **coulomb** is then defined as being *exactly* one ampere-second: $1\,\text{C} = 1\,\text{A·s}$. The value of k or ϵ_0 in Coulomb's law (Section 21-5) is obtained from experiment.

This may seem a rather roundabout way of defining quantities. The reason behind it is the desire for **operational definitions** of quantities—that is, definitions of quantities that can actually be measured given a definite set of operations to carry out. For example, the unit of charge, the coulomb, could be defined in terms of the force between two equal charges after defining a value for ϵ_0 or k in Eqs. 21-1 or 21-2. However, to carry out an actual experiment to measure the force between two charges is very difficult. For one thing, any desired amount of charge is not easily obtained precisely; and charge tends to leak from objects into the air. The amount of current in a wire, on the other hand, can be varied accurately and continuously (by putting a variable resistor in a circuit). Thus the force between two current-carrying conductors is far easier to measure precisely. And this is why the ampere is defined first and then the coulomb in terms of the ampere. At the National Institute of Standards and Technology in Maryland, precise measurement of current is made using circular coils of wire rather than straight lengths because it is more convenient and accurate.

Electric and magnetic field strengths are also defined operationally: the electric field in terms of the measurable force on a charge, via Eq. 21-3; and the magnetic field in terms of the force per unit length on a current-carrying wire, via Eq. 27-2.

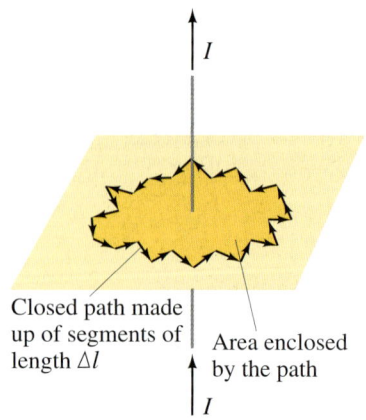

FIGURE 28-6 Arbitrary path enclosing a current, for Ampère's law. The path is broken down into segments of equal length Δl.

28-4 Ampère's Law

In Section 28-1 we saw that Eq. 28-1 gives the relation between the current in a long straight wire and the magnetic field it produces. This equation is valid only for a long straight wire. The following question arises: is there a general relation between a current in a wire of any shape and the magnetic field around it? The answer is yes: the French scientist André Marie Ampère (1775–1836), proposed such a relation shortly after Oersted's discovery. Consider an arbitrary closed path around a current as shown in Fig. 28-6, and imagine this path as being made up of short segments each of length Δl. First, we take the product of the length of each segment times the component of **B** parallel to that segment (call this component $B_\parallel$). If we now sum all these terms, according to Ampère, the result will be equal to μ_0 times the net current I_encl that passes through the surface enclosed by the path:

$$\sum B_\parallel \Delta l = \mu_0 I_\text{encl}.$$

The lengths Δl are chosen so that $B_\parallel$ is essentially constant along each length. The sum must be made over a *closed path*; and I_{encl} is the net current passing through the surface bounded by this closed path. In the limit $\Delta l \to 0$, this relation becomes

$$\oint \mathbf{B} \cdot d\mathbf{l} = \mu_0 I_{encl}, \quad \text{[B parallel to the surface]} \quad (28\text{–}3)$$

where $d\mathbf{l}$ is an infinitesimal length vector and the vector dot product assures that the parallel component of $\mathbf{B}$ is taken. Equation 28–3 is known as **Ampère's law**. The integrand in Eq. 28–3 is taken around a closed path, and I_{encl} is the current passing through the area enclosed by the chosen path.

To understand Ampère's law better, let us apply it to the simple case of a long straight wire carrying a current I which we've already examined, and which served as an inspiration for Ampère himself. Suppose we want to find the magnitude of $\mathbf{B}$ at some point A which is a distance r from the wire (Fig. 28–7). We know the magnetic field lines are circles with the wire at their center. So to apply Eq. 28–3 we choose as our path of integration a circle of radius r. The choice of path is ours, so we choose one that will be convenient: at any point on this circular path, $\mathbf{B}$ will be tangent to the circle. Furthermore, since all points on the path are the same distance from the wire, by symmetry we expect B to have the same magnitude at each point. Thus for any short segment of the circle (Fig. 28–7), $\mathbf{B}$ will be parallel to that segment, and hence

$$\mu_0 I = \oint \mathbf{B} \cdot d\mathbf{l} = \oint B\, dl = B \oint dl = B(2\pi r),$$

where $\oint dl = 2\pi r$, the circumference of the circle, and $I_{encl} = I$. We solve for B and obtain

$$B = \frac{\mu_0 I}{2\pi r}.$$

This is just Eq. 28–1 for the field near a long straight wire as discussed earlier.

Ampère's law thus works for this simple case. A great many experiments indicate that Ampère's law is valid in general. However, as with Gauss's law for the electric field, its practical value as a means to calculate the magnetic field is limited mainly to simple or symmetric situations. Its importance is that it relates the magnetic field to the current in a direct and mathematically elegant way. Ampère's law is thus considered one of the basic laws of electricity and magnetism. It is valid for any situation where the currents and fields are steady and not changing in time, and no magnetic materials are present.

We now can see why the constant in Eq. 28–1 is written $\mu_0/2\pi$. This is done so that only μ_0 appears in Eq. 28–3, rather than, say, $2\pi k$ if we had used k in Eq. 28–1. In this way, the more fundamental equation, Ampère's law, has the simpler form.

It should be noted, that the $\mathbf{B}$ in Ampère's law is not necessarily due only to the current I_{encl}. Ampère's law, like Gauss's law for the electric field, is valid in general. $\mathbf{B}$ is the field at each point in space along the chosen path due to all sources—including the current I enclosed by the path, but also due to any other sources. For example, the field surrounding two parallel current-carrying wires is the vector sum of the fields produced by each, and the field lines are shown in Fig. 28–8. If the path chosen for the integral (Eq. 28–3) is a circle centered on one of the wires with radius less than the distance between the wires (the dashed line in Fig. 28–8), only the current (I_1) in the encircled wire is included on the right side of Eq. 28–3. $\mathbf{B}$ on the left side of the equation must be the total $\mathbf{B}$ at each point due to both wires. Note also that $\oint \mathbf{B} \cdot d\mathbf{l}$ for the path shown in Fig. 28–8 is the same whether the second wire is present or not (in both cases, it equals $\mu_0 I_1$). How can this be? It can be so because although the fields due to the two wires tend to cancel one another at points between them, such as point N in the diagram ($\mathbf{B} = 0$ at a point midway between the wires if $I_1 = I_2$), at a point such as M in the Figure, the fields add together to produce a larger field. In the *sum*, $\oint \mathbf{B} \cdot d\mathbf{l}$, these effects just balance so that $\oint \mathbf{B} \cdot d\mathbf{l} = \mu_0 I_1$, whether the second wire is there or not. The integral $\oint \mathbf{B} \cdot d\mathbf{l}$ will be the same in each case, even though $\mathbf{B}$ will not be the same at every point for each of the two cases.

AMPÈRE'S LAW

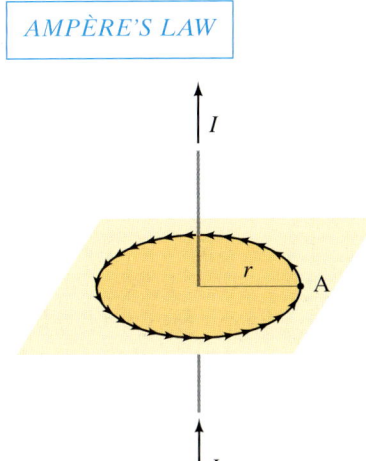

FIGURE 28–7 Circular path of radius r.

B for straight wire using Ampère's law

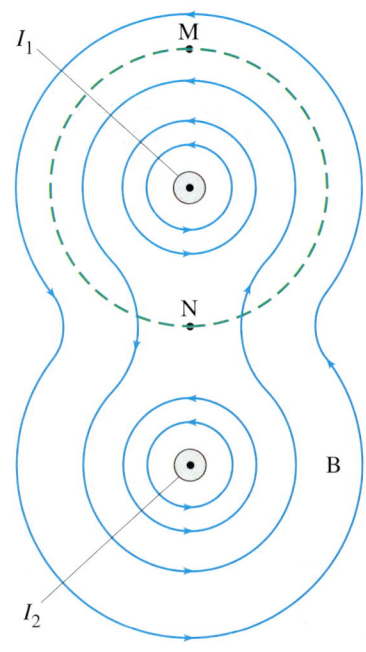

FIGURE 28–8 Magnetic field lines around two long parallel wires whose equal currents, I_1 and I_2, are coming out of the paper toward the viewer.

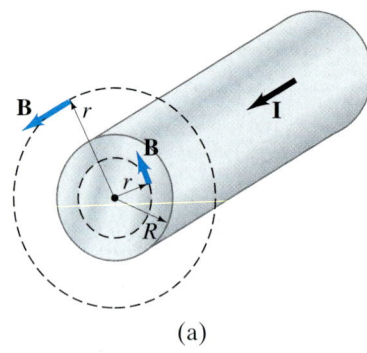

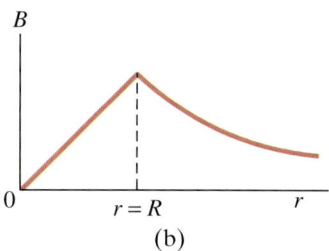

FIGURE 28–9 Magnetic field inside and outside a cylindrical conductor (Example 28–4).

EXAMPLE 28–4 **Field inside and outside a wire.** A long straight cylindrical wire conductor of radius R carries a current I of uniform current density in the conductor. Determine the magnetic field at (a) points outside the conductor ($r > R$), and (b) points inside the conductor ($r < R$). See Fig. 28–9. Assume that r, the radial distance from the axis, is much less than the length of the wire. (c) If $R = 2.0$ mm and $I = 60$ A, what is B at $r = 1.0$ mm, $r = 2.0$ mm, and $r = 3.0$ mm?

SOLUTION (a) Because the wire is long, straight, and cylindrical, we expect from the symmetry of the situation that the magnetic field must be the same at all points that are the same distance from the center of the conductor—there is no reason why any such point should have preference over the others at the same distance from the wire (they are physically equivalent), and so B must have the same value at all points the same distance from the center. We also expect **B** to be tangent to circles around the wire (Fig. 28–1), so let us choose a circular path of integration outside the wire ($r > R$), but concentric with it, as we did in Fig. 28–7. Then $I_{encl} = I$, so

$$\oint \mathbf{B} \cdot d\mathbf{l} = B(2\pi r) = \mu_0 I_{encl}$$

or

$$B = \frac{\mu_0 I}{2\pi r}. \qquad [r > R]$$

which is the same result as for a thin wire.

(b) Inside the wire ($r < R$), we again choose a circular path concentric with the cylinder; we expect **B** to be tangential to this path, and again, because of the symmetry, it will have the same magnitude at all points on the circle. The current enclosed in this case is less than I by a factor of the ratio of the areas:

$$I_{encl} = I \frac{\pi r^2}{\pi R^2}.$$

So Ampère's law gives

$$\oint \mathbf{B} \cdot d\mathbf{l} = \mu_0 I_{encl}$$

$$B(2\pi r) = \mu_0 I \left(\frac{\pi r^2}{\pi R^2} \right)$$

so

$$B = \frac{\mu_0 I r}{2\pi R^2}. \qquad [r < R]$$

The field is zero at the center of the conductor and increases linearly with r until $r = R$; beyond $r = R$, B decreases as $1/r$. This is shown in Fig. 28–9b. Note that these results are valid only for points close to the center of the conductor as compared to its length. For a current to flow, there must be connecting wires (to a battery, say), and the field due to these conducting wires, if not very far away, will destroy the assumed symmetry.

(c) At $r = 2.0$ mm, the surface of the wire, $r = R$, so

$$B = \frac{\mu_0 I}{2\pi R} = \frac{(4\pi \times 10^{-7}\,\mathrm{T \cdot m/A})(60\,\mathrm{A})}{(2\pi)(2.0 \times 10^{-3}\,\mathrm{m})} = 6.0 \times 10^{-3}\,\mathrm{T}.$$

We saw in (b) that inside the wire B is linear in r. So at $r = 1.0$ mm, B will be half what it is at $r = 2.0$ mm or 3.0×10^{-3} T. Outside the wire, B falls off as $1/r$, so at $r = 3.0$ mm it will be two-thirds as great as at $r = 2.0$ mm, or $B = 4.0 \times 10^{-3}$ T. To check, we use our result in (a), $B = \mu_0 I/2\pi r$, which gives the same result.

CONCEPTUAL EXAMPLE 28–5 **Coaxial cable.** A *coaxial cable* is a single wire surrounded by a cylindrical metallic braid, as shown in Fig. 28–10. The two conductors are separated by an insulator. The central wire carries current to the other end of the cable, and the outer braid carries the return current and is usually considered ground. Describe the magnetic field (*a*) in the space between the conductors, and (*b*) outside the cable.

RESPONSE (*a*) In the space between the conductors, we can apply Ampère's law for a circular path around the center wire, just as we did for the case shown in Fig. 28–7, and the magnitude is as given by Eq. 28–1. The current in the outer conductor has no bearing on this result. (Ampère's law uses only the current enclosed *inside* the path; as long as the currents outside the path don't affect the symmetry of the field, they do not contribute to the field along the path at all).
(*b*) Outside the cable, we can draw a similar circular path, for we expect the field to have the same circular symmetry. Now, however, there are two currents enclosed by the path, and they add up to zero. The field outside the cable is zero.

The nice feature of coaxial cables is that they are self-shielding: no stray magnetic fields escape outside the cable. The outer cylindrical conductor also shields external electric fields from coming in (see also Example 21–13). This makes them ideal for carrying signals near sensitive equipment. Audiophiles use coaxial cables between stereo equipment components and even to the loudspeakers.

➡ **PHYSICS APPLIED**
Coaxial cable (shielding)

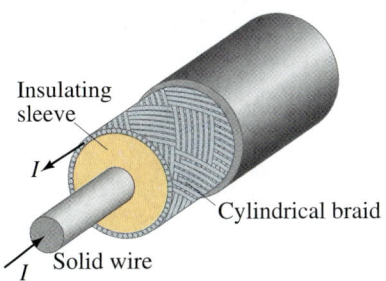

FIGURE 28–10 Coaxial cable. Conceptual Example 28–5.

EXAMPLE 28–6 **A nice use for Ampère's law.** Use Ampère's law to show that in any region of space where there are no currents the magnetic field cannot be both unidirectional and nonuniform as shown in Fig. 28–11a.

SOLUTION The wider spacing of lines near the top of Fig. 28–11a indicates the field has a smaller magnitude at the top than it does lower down. We now apply Ampère's law to the rectangular path abcd shown dashed in the diagram. Since no current is enclosed by this path,

$$\oint \mathbf{B} \cdot d\mathbf{l} = 0.$$

The integral along sections ab and cd is zero, since $\mathbf{B} \perp d\mathbf{l}$. Thus

$$\oint \mathbf{B} \cdot d\mathbf{l} = B_{bc}l - B_{da}l = (B_{bc} - B_{da})l,$$

which is not zero since the field B_{bc} along the path bc is less than the field B_{da} along path da. Hence we have a contradiction: $\oint \mathbf{B} \cdot d\mathbf{l}$ cannot be both zero (since $I = 0$) and nonzero. Thus we have shown that a nonuniform unidirectional field is not consistent with Ampère's law. A nonuniform field whose direction also changes, as in Fig. 28–11b, is consistent with Ampère's law (convince yourself this is so), and possible. The fringing of a permanent magnet's field (Fig. 27–7) has this shape.

FIGURE 28–11 Example 28–6.

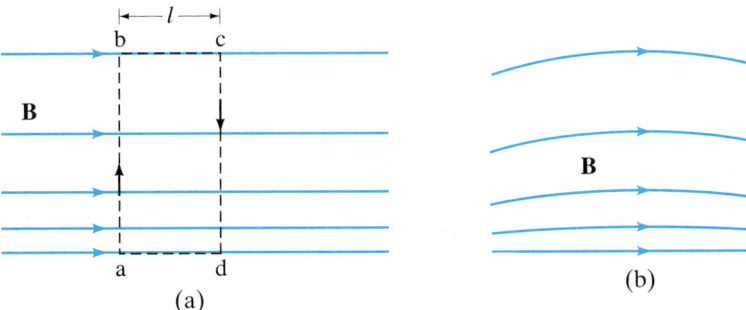

PROBLEM SOLVING: Ampère's Law

1. Ampère's Law, like Gauss' law, is always a valid statement. But as a calculation tool it is limited primarily to systems with a high degree of symmetry. The first step in applying Ampère's Law is to identify any useful symmetry.

2. Choose an integration path that reflects the symmetry (see the Examples to get a feel for this). In particular, search for paths where B has constant magnitude along the entire path or along segments of the path. Make sure your integration path passes through the point where you wish to evaluate the magnetic field.

3. Use symmetry to determine the direction of **B** along the integration path. With a smart choice of path, **B** will be either parallel or perpendicular to the path.

4. Evaluate the right-hand side of Ampère's Law by determining the enclosed current. Be careful with signs. Let the fingers of your right hand curl along the direction of **B** so that your thumb shows the direction of positive current. If the problem involves a solid conductor and your integration path does not enclose the full current, calculate the enclosed current using the current density (current per unit area) multiplied by the enclosed area (as in Example 28–4).

28–5 Magnetic Field of a Solenoid and a Toroid

A long coil of wire consisting of many loops is called a **solenoid**. Each loop produces a magnetic field as was shown in Fig. 27–10. In Fig. 28–12a, we see the field due to a solenoid when the coils are far apart. Near each wire, the field lines are very nearly circles as for a straight wire (that is, at distances that are small compared to the curvature of the wire). Between any two wires, the fields due to each loop tend to cancel. Toward the center of the solenoid, the fields add up to give a field that can be fairly large and fairly uniform. For a long solenoid, with closely packed coils, the field is nearly uniform and parallel to the solenoid axes within the entire cross section, as shown in Fig. 28–12b. The field outside the solenoid is very small compared to the field inside, except near the ends. Note that the same number of field lines that are concentrated inside the solenoid, spread out into the vast open space outside.

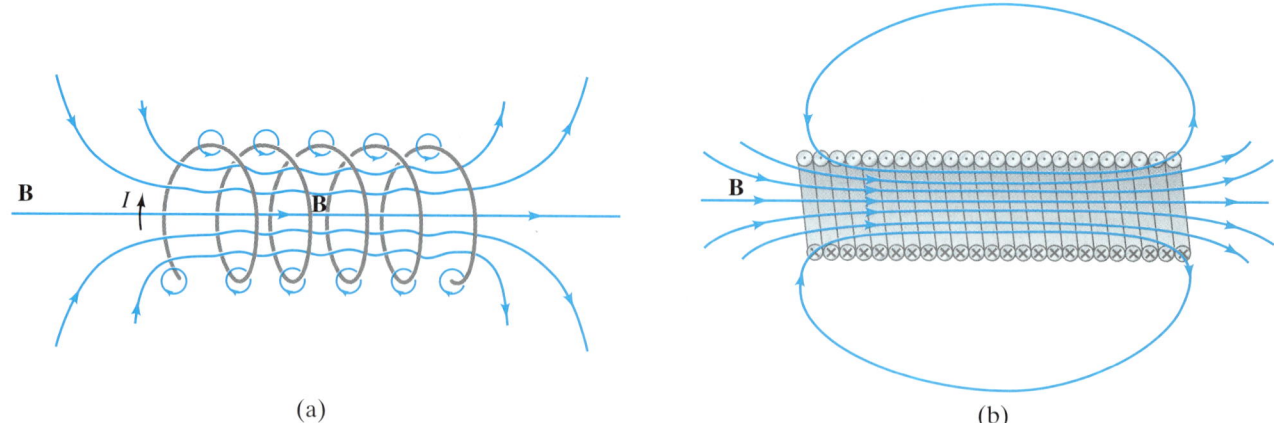

FIGURE 28–12 Magnetic field due to a solenoid: (a) loosely spaced turns, (b) closely spaced turns.

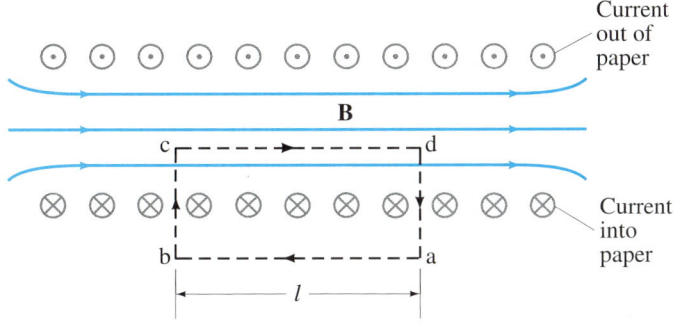

FIGURE 28–13 Magnetic field inside a long solenoid is uniform. Dashed lines indicate the path chosen for use in Ampère's law.

We now use Ampère's law to determine the magnetic field inside a very long (ideally, infinitely long) closely packed solenoid. We choose the path abcd shown in Fig. 28–13, far from either end, for applying Ampère's law. We will consider this path as made up of four segments, the sides of the rectangle: ab, bc, cd, da. Then the left side of Eq. 28–3, Ampère's law, becomes

$$\oint \mathbf{B} \cdot d\mathbf{l} = \int_a^b \mathbf{B} \cdot d\mathbf{l} + \int_b^c \mathbf{B} \cdot d\mathbf{l} + \int_c^d \mathbf{B} \cdot d\mathbf{l} + \int_d^a \mathbf{B} \cdot d\mathbf{l}.$$

The field outside the solenoid is so small as to be negligible compared to the field inside. Thus the first term in this sum will be zero. Furthermore, **B** is perpendicular to the segments bc and da inside the solenoid, and is nearly zero between and outside the coils, so these terms too are zero. Therefore we have reduced the integral to the segment cd where **B** is the nearly uniform field inside the solenoid, and is parallel to $d\mathbf{l}$, so

$$\oint \mathbf{B} \cdot d\mathbf{l} = \int_c^d \mathbf{B} \cdot d\mathbf{l} = Bl,$$

where l is the length cd. Now we determine the current enclosed by this loop for the right side of Ampère's law, Eq. 28–3. If a current I flows in the wire of the solenoid, the total current enclosed by our path abcd is NI where N is the number of loops our path encircles (five in Fig. 28–13). Thus Ampère's law gives us

$$Bl = \mu_0 NI.$$

If we let $n = N/l$ be the *number of loops per unit length*, then

$$B = \mu_0 nI. \qquad \text{[solenoid]} \qquad (28\text{–}4)$$

Magnetic field inside a solenoid

This is the magnitude of the magnetic field within a solenoid. Note that B depends only on the number of loops per unit length, n, and the current I. The field does not depend on position within the solenoid, so B is uniform. This is strictly true only for an infinite solenoid, but it is a good approximation for real ones for points not close to the ends.

EXAMPLE 28–7 Field inside a solenoid. A thin 10-cm-long solenoid used for fast electromechanical switching has a total of 400 turns of wire and carries a current of 2.0 A. Calculate the field inside near the center.

SOLUTION The number of turns per unit length is $n = 400/0.10\,\text{m} = 4.0 \times 10^3\,\text{m}^{-1}$. Thus

$$B = \mu_0 nI = (12.57 \times 10^{-7}\,\text{T·m/A})(4.0 \times 10^3\,\text{m}^{-1})(2.0\,\text{A})$$
$$= 1.0 \times 10^{-2}\,\text{T}.$$

A close look at Fig. 28–12 shows that the field outside of a solenoid is much like that of a bar magnet (Fig. 27–3). Indeed, a solenoid acts like a magnet, with one end acting as a north pole and the other as south pole, depending on the direction of the current in the loops. Since magnetic field lines leave the north pole of a magnet, the north poles of the solenoids in Fig. 28–12 are on the right.

Solenoids have many practical applications, and we discuss some of them later in the chapter, in Section 28–8.

EXAMPLE 28–8 Toroid. Use Ampère's law to determine the magnetic field (a) inside and (b) outside a toroid, which is like a solenoid bent into the shape of a circle as shown in Fig. 28–14a.

SOLUTION The magnetic field lines inside the toroid will be circles concentric with the toroid. (If you think of the toroid as a solenoid bent into a circle, the field lines bend along with the solenoid.) We choose as our path of integration one of these field lines of radius r inside the toroid as shown by the dashed line labeled "path 1" in Fig. 28–14a. We make this choice to use the symmetry of the situation, so B must be the same at all points along the path (although it is not necessarily the same across the whole cross-section of the toroid); so Ampère's law

$$\oint \mathbf{B} \cdot d\mathbf{l} = \mu_0 I_{\text{encl}}$$

becomes

$$B(2\pi r) = \mu_0 NI,$$

where N is the total number of coils and I is the current in each of the coils. Thus

$$B = \frac{\mu_0 NI}{2\pi r}.$$

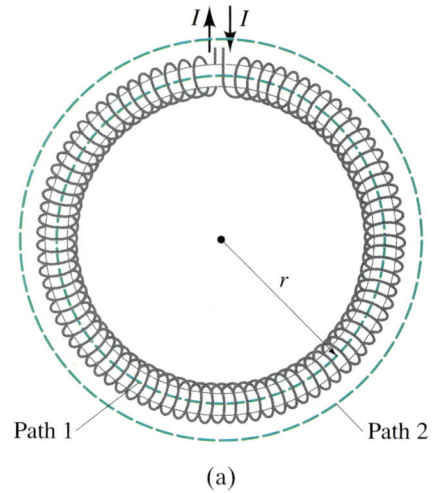

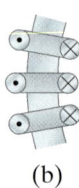

FIGURE 28–14 (a) A toroid. (b) A section of the toroid showing direction of the current for three loops: ⊙ means current toward viewer, ⊗ means current away from viewer.

The magnetic field B is not uniform within the toroid: it is largest along the inner edge (where r is smallest) and smallest at the outer edge. However, if the toroid is large, but thin (so that the difference between the inner and outer radii is small compared to the average radius) the field will be essentially uniform within the toroid. In this case, the formula for B reduces to that for a straight solenoid $B = \mu_0 nI$ where $n = N/(2\pi r)$ is the number of coils per unit length. (b) Outside the toroid, we choose as our path of integration a circle concentric with the toroid, "path 2" in Fig. 28–14a. This path encloses N loops carrying current I in one direction and N loops carrying the same current in the opposite direction. (Figure 28–14b shows the directions of the current for the parts of the loop on the inside and outside of the toroid.) Thus the net current enclosed by path 2 is zero. For a very tightly packed toroid, all points on path 2 are equidistant from the toroid and equivalent, so we expect B to be the same at all points along the path. Hence, Ampère's law gives

$$\oint \mathbf{B} \cdot d\mathbf{l} = \mu_0 I_{\text{encl}}$$

$$B(2\pi r) = 0$$

or

$$B = 0.$$

The same is true for a path taken at a radius smaller than that of the toroid. So there is no field exterior to a very tightly wound toroid. It is all inside the loops.

28-6 Biot-Savart Law

The usefulness of Ampère's law for determining the magnetic field **B** due to particular electric currents is restricted to situations where the symmetry of the given currents allows us to evaluate $\oint \mathbf{B} \cdot d\mathbf{l}$ readily. This does not, of course, invalidate Ampère's law nor does it reduce its fundamental importance. Recall the electric case, where Gauss's law is considered fundamental but is limited in its use for actually calculating **E**. We must often determine the electric field **E** by another method summing over contributions due to infinitesimal charge elements dq via Coulomb's law: $dE = (1/4\pi\epsilon_0)(dq/r^2)$. A magnetic equivalent to this infinitesimal form of Coulomb's law would be helpful for currents that do not have great symmetry. Such a law was developed by Jean Baptiste Biot (1774–1862) and Felix Savart (1791–1841) shortly after Oersted's discovery in 1820 that a current produces a magnetic field.

According to Biot and Savart, a current I flowing in any path can be considered as many tiny (infinitesimal) current elements, such as in the wire of Fig. 28–15. If $d\mathbf{l}$ represents any infinitesimal length along which the current is flowing, then the magnetic field, $d\mathbf{B}$, at any point P in space, due to this element of current, is given by

$$d\mathbf{B} = \frac{\mu_0 I}{4\pi} \frac{d\mathbf{l} \times \hat{\mathbf{r}}}{r^2}, \quad (28\text{-}5) \quad \textit{Biot-Savart law}$$

where **r** is the displacement vector from the element $d\mathbf{l}$ to the point P, and $\hat{\mathbf{r}} = \mathbf{r}/r$ is the unit vector in the direction of **r** (see Fig. 28–15). Equation 28–5 is known as the **Biot-Savart law**. The magnitude of $d\mathbf{B}$ is

$$dB = \frac{\mu_0 I \, dl \sin\theta}{4\pi r^2}, \quad (28\text{-}6)$$

where θ is the angle between $d\mathbf{l}$ and **r** (Fig. 28–15). The total magnetic field at point P is then found by summing (integrating) over all current elements:

$$\mathbf{B} = \int d\mathbf{B}.$$

Note that this is a *vector* sum. The Biot-Savart law is the magnetic equivalent of Coulomb's law in its infinitesimal form. It is even an inverse square law, like Coulomb's law.

An important difference between the Biot-Savart law and Ampère's law (Eq. 28–3) is that in Ampère's law $[\oint \mathbf{B} \cdot d\mathbf{l} = \mu_0 I_{\text{encl}}]$, **B** is not necessarily due only to the current enclosed by the path of integration. But in the Biot-Savart law the field $d\mathbf{B}$ in Eq. 28–5 is due only, and entirely, to the current element $I \, d\mathbf{l}$. To find the total **B** at any point in space, it is necessary to include *all* currents.

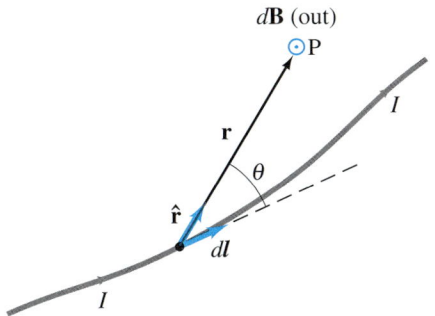

FIGURE 28–15 Biot-Savart law: the field at P due to current element $I \, d\mathbf{l}$ is $d\mathbf{B} = (\mu_0 I/4\pi)(d\mathbf{l} \times \hat{\mathbf{r}}/r^2)$.

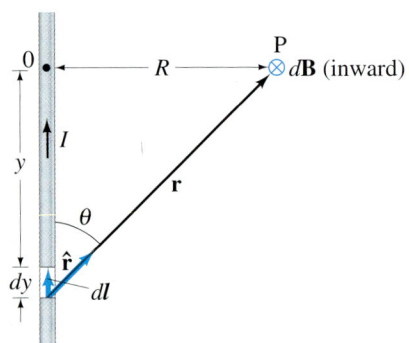

FIGURE 28–16 Determining **B** due to a long straight wire using the Biot-Savart law.

EXAMPLE 28–9 *B due to current I in straight wire.* For the field near a long straight wire carrying a current I, show that the Biot-Savart law gives the same result as Eq. 28–1, $B = \mu_0 I / 2\pi r$.

SOLUTION We calculate the magnetic field in Fig. 28–16 at point P, which is a perpendicular distance R from an infinitely long wire. The current is moving upwards, and both $d\mathbf{l}$ and $\hat{\mathbf{r}}$, which appear in the cross product of Eq. 28–5, are in the plane of the page. Hence the direction of the field $d\mathbf{B}$ due to each element of current must be directed into the plane of the page as shown (right-hand rule for the cross product $d\mathbf{l} \times \hat{\mathbf{r}}$). Thus all the $d\mathbf{B}$ have the same direction at point P, and add up to give **B** the same direction consistent with our previous results (Figs. 28–1, 28–7, and 28–9). The magnitude of **B** will be

$$B = \frac{\mu_0 I}{4\pi} \int_{y=-\infty}^{+\infty} \frac{dy \sin\theta}{r^2},$$

where $dy = dl$ and $r^2 = R^2 + y^2$. Note that we are integrating over y (the length of the wire) so R is considered constant. Both y and θ are variables, but they are not independent. In fact, $y = -R/\tan\theta$. Note that we measure y as positive upward from point 0, so for the current element we are considering $y < 0$. Then

$$dy = +R \csc^2\theta \, d\theta = \frac{R \, d\theta}{\sin^2\theta} = \frac{R \, d\theta}{(R/r)^2} = \frac{r^2 \, d\theta}{R}.$$

So our integral becomes

$$B = \frac{\mu_0 I}{4\pi} \frac{1}{R} \int_{\theta=0}^{\pi} \sin\theta \, d\theta = -\frac{\mu_0 I}{4\pi R} \cos\theta \Big|_0^\pi = \frac{\mu_0 I}{2\pi R}.$$

This is just Eq. 28–1 for the field near a long wire, where R has been used instead of r.

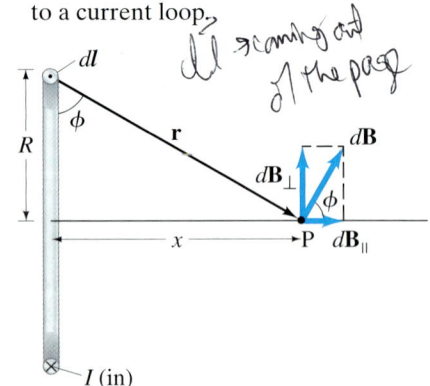

FIGURE 28–17 Determining **B** due to a current loop.

EXAMPLE 28–10 *Current loop.* Determine **B** for points on the axis of a circular loop of wire of radius R carrying a current I, Fig. 28–17.

SOLUTION For an element of current at the top of the loop, the magnetic field $d\mathbf{B}$ at point P on the axis has the direction shown, and magnitude (Eq. 28–5)

$$dB = \frac{\mu_0 I \, dl}{4\pi r^2}$$

since $d\mathbf{l}$ is perpendicular to $\mathbf{r}$ so $|d\mathbf{l} \times \hat{\mathbf{r}}| = dl$. We can break $d\mathbf{B}$ down into components $dB_\parallel$ and $dB_\perp$, which are parallel and perpendicular to the axis as shown. When we sum over all the elements of the loop symmetry tells us that the perpendicular components will cancel on opposite sides, so $B_\perp = 0$. Hence, the total **B** will point along the axis, and will have magnitude

$$B = B_\parallel = \int dB \cos\phi = \int dB \frac{R}{r} = \int dB \frac{R}{(R^2 + x^2)^{\frac{1}{2}}},$$

where x is the distance of P from the center of the ring, and $r^2 = R^2 + x^2$. Now we put in dB from the equation above and integrate around the current loop, noting that all segments $d\mathbf{l}$ of current are the same distance, $(R^2 + x^2)^{\frac{1}{2}}$, from point P:

$$B = \frac{\mu_0 I}{4\pi} \frac{R}{(R^2 + x^2)^{\frac{3}{2}}} \int dl = \frac{\mu_0 I R^2}{2(R^2 + x^2)^{\frac{3}{2}}}$$

since $\int dl = 2\pi R$, the circumference of the loop. At the very center of the loop (where $x = 0$) the field has its maximum value

$$B = \frac{\mu_0 I}{2R}. \qquad \text{[at center of loop]}$$

Recall from Section 27–5 that a current loop, such as that just discussed (Fig. 28–17) is considered a *magnetic dipole*. We saw there that a magnetic dipole has a dipole moment

$$\mu = NIA,$$

where A is the area of the loop and N is the number of coils in the loop, each carrying current I. We also saw in Chapter 27 that a magnetic dipole placed in an external magnetic field experiences a torque and possesses potential energy, just like an electric dipole. In Example 28–10, we have looked at another aspect of a magnetic dipole: the magnetic field *produced by* a magnetic dipole has magnitude, along the dipole axis, of

$$B = \frac{\mu_0 I R^2}{2(R^2 + x^2)^{\frac{3}{2}}}.$$

We can write this in terms of the magnetic dipole moment $\mu = IA = I\pi R^2$ (for a single loop $N = 1$):

$$B = \frac{\mu_0}{2\pi} \frac{\mu}{(R^2 + x^2)^{\frac{3}{2}}}. \qquad \text{[magnetic dipole]} \quad \textbf{(28–7a)}$$

(Be careful to distinguish μ for dipole moment from μ_0, the magnetic permeability constant.) For distances far from the loop, $x \gg R$, this becomes

$$B \approx \frac{\mu_0}{2\pi} \frac{\mu}{x^3}. \qquad \begin{bmatrix}\text{on axis,} \\ \text{magnetic dipole, } x \gg R\end{bmatrix} \quad \textbf{(28–7b)}$$

The magnetic field on the axis of a magnetic dipole decreases with the cube of the distance, just as for an electric dipole. B decreases as the cube of the distance also for points not on the axis, although the multiplying factor is not the same. The magnetic field due to a current loop can be determined at various points using the Biot-Savart law and the results are in accord with experiment. The field lines around a current loop are shown in Fig. 28–18.

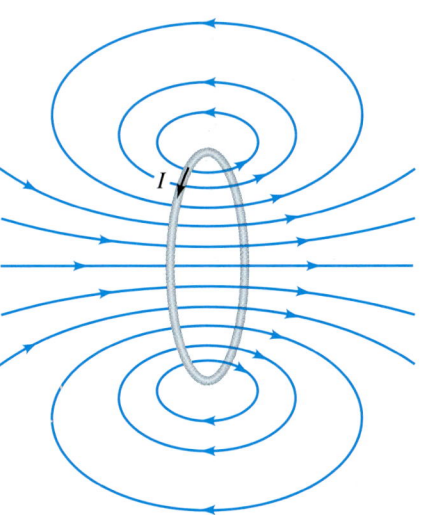

FIGURE 28–18 Magnetic field due to a circular loop of wire. (Same as Fig. 27–10.)

EXAMPLE 28–11 B due to a wire segment. One quarter of a circular loop of wire carries a current I as shown in Fig. 28–19. The current I enters and leaves on straight segments of wire, as shown; the straight wires are along the radial direction from the center C of the circular portion. Find the magnetic field at point C.

FIGURE 28–19 Example 28–11.

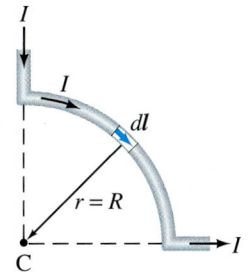

SOLUTION The current in the straight sections produces no magnetic field at point C because $d\mathbf{l}$ and $\hat{\mathbf{r}}$ in the Biot-Savart law (Eq. 28–5) are parallel and therefore $d\mathbf{l} \times \hat{\mathbf{r}} = 0$. (Said another way, the magnitude of $d\mathbf{l} \times \mathbf{r}$ is $(dl)(r) \sin \theta$ where $\theta = 0$.) Each piece $d\mathbf{l}$ of the curved section of wire produces a field $d\mathbf{B}$ that points into the paper at C (right-hand rule). The magnitude of each $d\mathbf{B}$ is (Eq. 28–6)

$$dB = \frac{\mu_0 I \, dl}{4\pi R^2}$$

where $r = R$ is the radius of the curved section, and $\sin \theta$ in Eq. 28–6 is $\sin 90° = 1$. Because $r = R$ for all pieces $d\mathbf{l}$, then

$$B = \int dB = \frac{\mu_0 I}{4\pi R^2} \int dl = \frac{\mu_0 I}{4\pi R^2} \left(\frac{1}{4} 2\pi R\right) = \frac{\mu_0 I}{8R}.$$

*28–7 Magnetic Materials—Ferromagnetism

Magnetic fields can be produced (*a*) by magnetic materials (magnets) and (*b*) by electric currents. We have studied the latter extensively in this chapter. Now we take a brief look at magnetic materials, which are so present in everyday life: ordinary magnets, iron cores in motors and electromagnets, magnetic recording tape and computer data storage disks, even the magnetic stripe on credit cards. We saw in Section 27–1 that iron (and a few other materials) can be made into strong magnets. These materials are said to be **ferromagnetic**. We now look more deeply into the sources of ferromagnetism.

A bar magnet, with its two opposite poles near either end, resembles an electric dipole (equal-magnitude positive and negative charges separated by a distance). Indeed, a bar magnet is sometimes referred to as a "magnetic dipole." There are opposite "poles" separated by a distance. And the magnetic field lines of a bar magnet form a pattern much like that for the electric field of an electric dipole: compare Fig. 21–33a with Fig. 27–3b.

Microscopic examination reveals that a magnet is actually made up of tiny regions known as **domains**, which are at most about 1 mm in length or width. Each domain behaves like a tiny magnet with a north and a south pole. In an unmagnetized piece of iron, these domains are arranged randomly, as shown in Fig. 28–20a. The magnetic effects of the domains cancel each other out, so this piece of iron is not a magnet. In a magnet, the domains are preferentially aligned in one direction as shown in Fig. 28–20b (downward in this case). A magnet can be made from an unmagnetized piece of iron by placing it in a strong magnetic field. (You can make a needle magnetic, for example, by stroking it with one pole of a strong magnet.) Careful observations show in this case that the magnetization of domains may actually rotate slightly so as to be more nearly parallel to the external field. Or, more commonly, the borders of domains move so that those domains whose magnetic orientation is parallel to the external field grow in size at the expense of other domains, as can be seen by comparing Figs. 28–20a and b. This explains how a magnet can pick up unmagnetized pieces of iron like paper clips or bobby pins. The magnet's field causes a slight alignment of the domains in the unmagnetized object so that the object becomes a temporary magnet with its north pole facing the south pole of the permanent magnet, and vice versa; thus, attraction results. In the same way, elongated iron filings will arrange themselves in a magnetic field just as a compass needle does, and will reveal the shape of the magnetic field, Fig. 28–21.

An iron magnet can remain magnetized for a long time, and thus it is referred to as a "permanent magnet." However, if you drop a magnet on the floor or strike it with a hammer, you may jar the domains into randomness. The magnet can thus lose some or all of its magnetism. Heating a magnet too can cause a loss of magnetism, for raising the temperature increases the random thermal motion of the atoms which tends to randomize the domains. Above a certain temperature known as the **Curie temperature** (1043 K for iron), a magnet cannot be made at all.[†]

The striking similarity between the fields produced by a bar magnet (Fig. 27–3b) and by a loop of electric current (Fig. 28–18) suggests that the magnetic field produced by a current may have something to do with ferromagnetism, an idea proposed by Ampère in the nineteenth century. According to modern atomic theory, the atoms that make up any material can be roughly visualized as containing electrons that orbit around a central nucleus. Since the electrons are charged, they constitute an electric current and therefore produce a magnetic

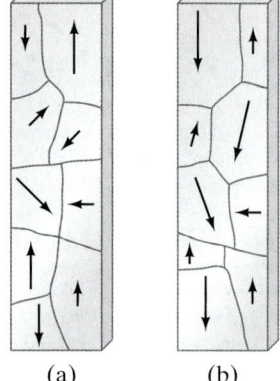

FIGURE 28–20 (a) An unmagnetized piece of iron is made up of domains that are randomly arranged. Each domain is like a tiny magnet; the arrows represent the magnetization direction, with the arrowhead being the N pole. (b) In a magnet, the domains are preferentially aligned in one direction, and may be altered in size by the magnetization process.

FIGURE 28–21 Iron filings line up along magnetic field lines.

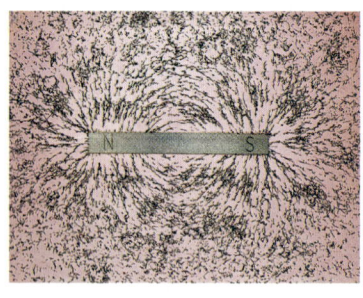

[†] Iron, nickel, cobalt, gadolinium, and certain alloys are ferromagnetic at room temperature; several other elements and alloys have low Curie temperature and thus are ferromagnetic only at low temperatures.

field. But if there is no external field, the electron orbits in different atoms are arranged randomly, so the magnetic effects due to the many orbits in all the atoms in a material cancel out. However, electrons by themselves produce an additional intrinsic magnetic field—they have an intrinsic magnetic moment referred to as their "spin" magnetic moment.[†] It is the magnetic field due to electron spin that is now believed to produce ferromagnetism. In most materials, the magnetic fields due to electron spin cancel out because they are oriented at random. But in iron and other ferromagnetic materials, a complicated cooperative mechanism, known as "exchange coupling," operates. The result is that the spin of the electrons contributing to the ferromagnetism in a domain point in the same direction. Thus the tiny magnetic fields due to each of these electrons add up to give the magnetic field of a domain. And when the domains are aligned, as we have seen, a strong magnet results.

*28–8 Electromagnets and Solenoids

A long coil of wire consisting of many loops of wire, as discussed in Section 28–5, is called a solenoid. The magnetic field within a solenoid can be fairly large since it will be the sum of the fields due to the current in each loop (see Fig. 28–22). The solenoid acts like a magnet; one end can be considered the north pole and the other the south pole, depending on the direction of the current in the loops (use the right-hand rule). Since the magnetic field lines leave the north pole of a magnet, the north pole of the solenoid in Fig. 28–22 is on the right.

If a piece of iron is placed inside a solenoid, the magnetic field is increased greatly because the domains of the iron are aligned by the magnetic field produced by the current. The resulting magnetic field is the sum of that due to the current and that due to the iron, and can be hundreds or thousands of times that due to the current alone (see Section 28–9). This arrangement is called an **electromagnet**. The iron used in electromagnets acquires and loses its magnetism quite readily when the current is turned on or off, and so is referred to as "soft iron." (It is "soft" only in a magnetic sense.) Iron that holds its magnetism even when there is no externally applied field is called "hard iron." Hard iron is used in permanent magnets. Soft iron is usually used in electromagnets so that the field can be turned on and off readily. Whether iron is hard or soft depends on heat treatment and other factors.

Electromagnets find use in many practical applications, from use in motors and generators to producing large magnetic fields for research. For some applications an iron core is not present—the magnetic field comes only from the current in the wire coils. When the current flows continuously in a normal electromagnet, a great deal of waste heat (I^2R power) can be produced. Cooling coils, which are tubes carrying water, must be used to absorb the heat in bigger installations. For some applications, superconducting magnets are used. The current-carrying wires are made of superconducting material (Section 25–9) kept below the transition temperature. Very high fields can be produced in the absence of an iron core, higher than what could be reached with an iron core (due to saturation of the iron—see the next Section). No electric power is needed to maintain large current in the superconducting coils, which means large savings of electricity, nor must huge amounts of heat be dissipated. Of course, energy is needed to keep the superconducting coils at the necessary low temperature.

Another useful device consists of a solenoid into which a rod of iron is partially inserted. This combination is also referred to as a solenoid. One simple use is

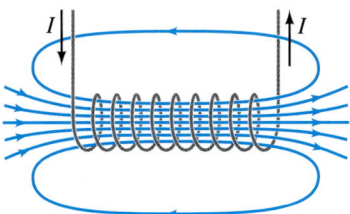

FIGURE 28–22 Magnetic field of a solenoid. The north pole of this solenoid, thought of as a magnet, is on the right, and the south pole is on the left.

▶ **PHYSICS APPLIED**
Electromagnets and solenoids

[†]The name "spin" comes from an early suggestion that this intrinsic magnetic moment arises from the electron "spinning" on its axis (as well as "orbiting" the nucleus) to produce the extra field. However this view of a spinning electron is oversimplified and not valid.

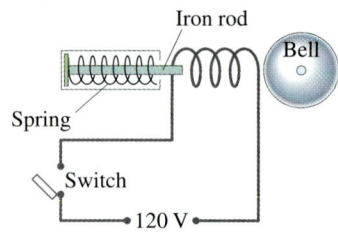

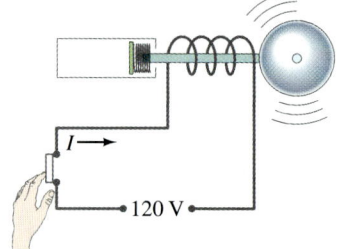

FIGURE 28–23 Solenoid used as a doorbell.

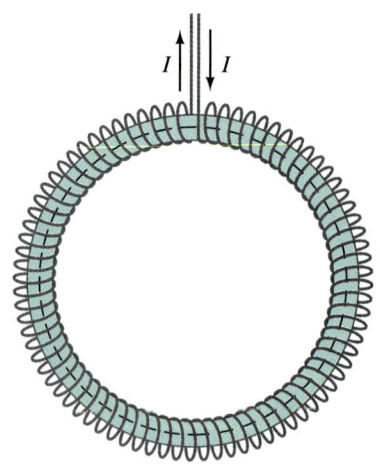

FIGURE 28–24 Iron-core toroid.

FIGURE 28–25 Total magnetic field B in an iron-core toroid as a function of the external field B_0 (B_0 is caused by the current I in the coil).

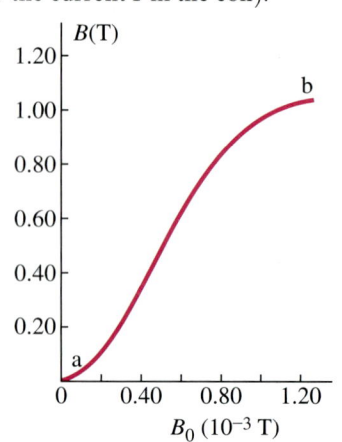

as a doorbell (Fig. 28–23). When the circuit is closed by pushing the button, the coil effectively becomes a magnet and exerts a force on the iron rod. The rod is pulled into the coil and strikes the bell. A larger solenoid is used in the starters of cars; when you engage the starter, you are closing a circuit that not only turns the starter motor, but activates a solenoid that first moves the starter into direct contact with the gears on the engine's flywheel. Solenoids are used as switches in many other devices, such as tape recorders. They have the advantage of moving mechanical parts quickly and accurately.

*28–9 Magnetic Fields in Magnetic Materials; Hysteresis

The field of a long solenoid is directly proportional to the current. Indeed, Eq. 28–4 tells us that the field B_0 inside a solenoid is given by

$$B_0 = \mu_0 nI.$$

This is valid if there is only air inside the coil. If we put a piece of iron or other ferromagnetic material inside the solenoid, the field will be greatly increased, often by hundreds or thousands of times. This occurs because the domains in the iron become preferentially aligned by the external field. The resulting magnetic field is the sum of that due to the current and that due to the iron. It is sometimes convenient to write the total field in this case as a sum of two terms:

$$\mathbf{B} = \mathbf{B}_0 + \mathbf{B}_M. \qquad (28\text{–}8)$$

Here, $\mathbf{B}_0$ refers to the field due only to the current in the wire (the "external field"). It is the field that would be present in the absence of a ferromagnetic material. Then $\mathbf{B}_M$ represents the additional field due to the ferromagnetic material itself; often $\mathbf{B}_M \gg \mathbf{B}_0$.

The total field inside a solenoid in such a case can also be written by replacing the constant μ_0 in Eq. 28–4 by another constant, μ, characteristic of the material inside the coil:

$$B = \mu nI; \qquad (28\text{–}9)$$

μ is called the **magnetic permeability** of the material (do not confuse it with **μ** for magnetic moment). For ferromagnetic materials, μ is much greater than μ_0. For all other materials, its value is very close to μ_0 (Section 28–10). The value of μ, however, is not constant for ferromagnetic materials; it depends on the value of the external field B_0, as the following experiment shows.

Measurements on magnetic materials are generally done using a toroid, which is essentially a long solenoid bent into the shape of a circle (Fig. 28–24), so that practically all the lines of **B** remain within the toroid. Suppose the toroid has an iron core that is initially unmagnetized and there is no current in the windings of the toroid. Then the current I is slowly increased, and B_0 increases linearly with I. The total field B also increases, but follows the curved line shown in the graph of Fig. 28–25. (Note the different scales: $B \gg B_0$.) Initially, point a, the domains (Section 28–7) are randomly oriented. As B_0 increases, the domains become more and more aligned until at point b, nearly all are aligned. The iron is said to be approaching **saturation**. Point b is typically 70 percent of full saturation. (If B_0 is increased further, the curve continues to rise very slowly, and reaches 98 percent saturation only when B_0 reaches a value about a thousandfold above that at point b; the last few domains are very difficult to align.) Next, suppose the external field B_0 is reduced by decreasing the current in the coils. As the current is reduced to zero, shown as point c in Fig. 28–26, the domains do not become completely random. Some permanent magnetism remains. If the current is then reversed in direction, enough domains can be turned around so $B = 0$ (point d). As the reverse current is increased further, the iron approaches saturation in the opposite direction (point e). Finally, if the current is again reduced to zero and then increased in the original direction, the total field follows the path efgb, again approaching saturation at point b.

Notice that the field did not pass through the origin (point a) in this cycle. The fact that the curves do not retrace themselves on the same path is called **hysteresis**. The curve bcdefgb is called a **hysteresis loop**. In such a cycle, much energy is transformed to thermal energy (friction) due to realigning of the domains. It can be shown that the energy dissipated in this way is proportional to the area of the hysteresis loop.

At points c and f, the iron core is magnetized even though there is no current in the coils. These points correspond to a permanent magnet. For a permanent magnet, it is desired that ac and af be as large as possible. Materials for which this is true are said to have high **retentivity**.

Materials with a broad hysteresis curve as in Fig. 28–26 are said to be magnetically "hard." On the other hand, a hysteresis curve such as that in Fig. 28–27 occurs for "soft" iron. This is preferred for *electromagnets* and transformers (Section 29–6) since the field can be more readily switched off, and the field can be reversed with less loss of energy.

A ferromagnetic material can be demagnetized—that is, made unmagnetized. This can be done by reversing the magnetizing current repeatedly while decreasing its magnitude. This results in the curve of Fig. 28–28. The heads of a tape recorder are demagnetized in this way. The alternating magnetic field acting at the heads due to a demagnetizer is strong when the demagnetizer is placed near the heads and decreases as it is moved slowly away. Video and audio tapes themselves can be erased and ruined by a magnetic field, as can computer disks.

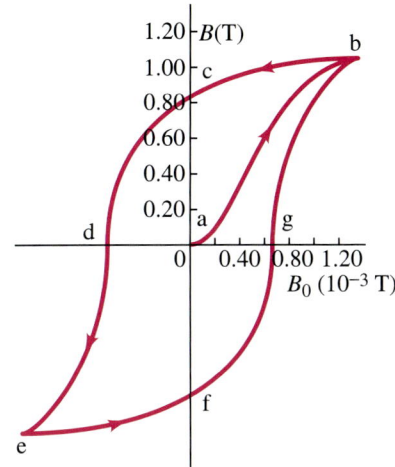

FIGURE 28–26 Hysteresis curve.

FIGURE 28–27 Hysteresis curve for soft iron.

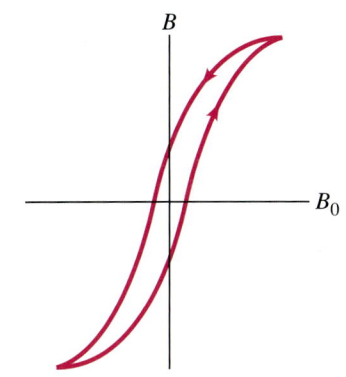

*28–10 Paramagnetism and Diamagnetism

All materials are magnetic to some extent. Nonferromagnetic materials fall into two principal classes: *paramagnetic*, in which the magnetic permeability μ is slightly greater than μ_0; and *diamagnetic*, in which, μ is slightly less than μ_0. The ratio of μ to μ_0 for any material is called the **relative permeability** K_m:

$$K_m = \frac{\mu}{\mu_0}.$$

Another useful parameter is the **magnetic susceptibility** χ_m defined as

$$\chi_m = K_m - 1.$$

Paramagnetic substances have $K_m > 1$ and $\chi_m > 0$, whereas diamagnetic substances have $K_m < 1$ and $\chi_m < 0$. See Table 28–1.

FIGURE 28–28 Successive hysteresis loops during demagnetization.

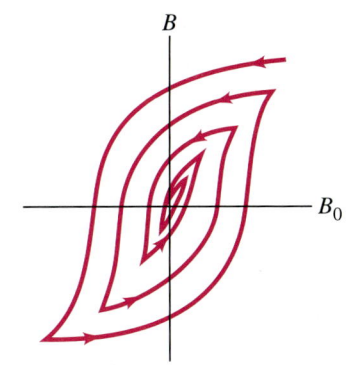

TABLE 28–1 Paramagnetism and Diamagnetism: Magnetic Susceptibilities

Paramagnetic substance	χ_m	Diamagnetic substance	χ_m
Aluminum	2.3×10^{-5}	Copper	-9.8×10^{-6}
Calcium	1.9×10^{-5}	Diamond	-2.2×10^{-5}
Magnesium	1.2×10^{-5}	Gold	-3.6×10^{-5}
Oxygen (STP)	2.1×10^{-6}	Lead	-1.7×10^{-5}
Platinum	2.9×10^{-4}	Nitrogen (STP)	-5.0×10^{-9}
Tungsten	6.8×10^{-5}	Silicon	-4.2×10^{-6}

Paramagnetism

The difference between paramagnetic and diamagnetic materials can be understood theoretically at the molecular level on the basis of whether or not the molecules have a permanent magnetic dipole moment. **Paramagnetism** occurs in materials whose molecules (or ions) have a permanent magnetic dipole moment.[†] In the absence of an external field, the molecules are randomly oriented and no magnetic effects are observed. However, when an external magnetic field is applied, say, by putting the material in a solenoid, the applied field exerts a torque on the magnetic dipoles (Section 27–5), tending to align them parallel to the field. The total magnetic field (external plus that due to aligned magnetic dipoles) will be slightly greater than B_0. The thermal motion of the molecules reduces the alignment, however. A useful quantity is the **magnetization vector**, **M**, defined as the magnetic dipole moment per unit volume,

$$\mathbf{M} = \frac{\boldsymbol{\mu}}{V},$$

where $\boldsymbol{\mu}$ is the magnetic dipole moment of the sample and V its volume. It is found experimentally that M is directly proportional to the external magnetic field (tending to align the dipoles) and inversely proportional to the Kelvin temperature T (tending to randomize dipole directions). This is called *Curie's law*, after Pierre Curie (1859–1906), who first noted it:

$$M = C\frac{B}{T},$$

where C is a constant. If the ratio B/T is very large (B very large or T very small) Curie's law is no longer accurate; as B is increased (or T decreased), the magnetization approaches some maximum value, M_{max}. This makes sense, of course, since M_{max} corresponds to complete alignment of all the permanent magnetic dipoles. However, even for very large magnetic fields, ≈ 2.0 T, deviations from Curie's law are normally noted only at very low temperatures, on the order of a few kelvins.

Ferromagnetic materials, as mentioned in Section 28–7, are no longer ferromagnetic above a characteristic temperature called the Curie temperature (1043 K for iron). Above this Curie temperature, they generally are paramagnetic.

Diamagnetism

Diamagnetic materials (for which μ_m is slightly less than μ_0) are made up of molecules that have no permanent magnetic dipole moment. When an external magnetic field is applied, magnetic dipoles are induced, but the induced magnetic dipole moment is in the direction opposite to that of the field. Hence the total field will be slightly less than the external field. The effect of the external field—in the crude model of electrons orbiting nuclei—is to increase the "orbital" speed of electrons revolving in one direction, and to decrease the speed of electrons revolving in the other direction; the net result is a net dipole moment opposing the external field. Diamagnetism is present in all materials, but is weaker even than paramagnetism and so is overwhelmed by paramagnetic and ferromagnetic effects in materials that display these other forms of magnetism.

[†]Other types of paramagnetism also occur whose origin is different from that described here, such as in metals where free electrons can contribute.

Summary

Ampère's law states that the line integral of the magnetic field **B** around any closed loop is equal to μ_0 times the total net current I_{encl} enclosed by the loop:

$$\oint \mathbf{B} \cdot d\mathbf{l} = \mu_0 I_{encl}.$$

The magnetic field B at a distance r from a long straight wire is directly proportional to the current I in the wire and inversely proportional to r. The magnetic field lines are circles centered at the wire.

The magnetic field inside a long tightly wound solenoid is $B = \mu_0 n I$ where n is the number of coils per unit length and I is the current in each coil.

The force that one long current-carrying wire exerts on a second parallel current-carrying wire a distance l away serves as the definition of the ampere unit, and ultimately of the coulomb as well.

The **Biot-Savart law** is useful for determining the magnetic field due to a known arrangement of currents. It states that

$$d\mathbf{B} = \frac{\mu_0 I}{4\pi} \frac{d\mathbf{l} \times \hat{\mathbf{r}}}{r^2},$$

where $d\mathbf{B}$ is the contribution to the total field at some point P due to a current I along an infinitesimal length $d\mathbf{l}$ of its path, and $\hat{\mathbf{r}}$ is the unit vector along the direction of the displacement vector $\mathbf{r}$ from $d\mathbf{l}$ to P. The total field **B** will be the integral over all $d\mathbf{B}$.

Iron and a few other materials can be made into strong permanent magnets. They are said to be **ferromagnetic**. Ferromagnetic materials are made up of tiny **domains**—each a tiny magnet—which are preferentially aligned in a permanent magnet, but randomly aligned in a nonmagnetized sample.

When a ferromagnetic material is placed in a magnetic field B_0 due to a current, say inside a solenoid or toroid, the material becomes magnetized. When the current is turned off, however, the material remains magnetized, and when the current is increased in the opposite direction (and then again reversed), a graph of the total field B versus B_0 is a **hysteresis loop**, and the fact that the curves do not retrace themselves is called **hysteresis**.

Questions

1. The magnetic field due to current in wires in your home can affect a compass. Discuss the problem in terms of currents, depending on whether they are ac or dc, and their distance away.

2. Compare and contrast the magnetic field due to a long straight current and the electric field due to a long straight line of electric charge at rest (Section 21–7).

3. Two long wires carrying equal currents I are at right angles to each other, but don't quite touch. Describe the magnetic force one exerts on the other.

4. A horizontal wire carries a large current. A second wire carrying a current in the same direction is suspended below. Can the current in the upper wire hold the lower wire in suspension against gravity? Under what conditions will it be in equilibrium?

5. A horizontal current-carrying wire, free to move in Earth's gravitational field, is suspended directly above a second, parallel, current-carrying wire. (a) In what direction is the current in the lower wire? (b) Can the upper wire be held in stable equilibrium due to the magnetic force of the lower wire? Explain.

6. (a) Write Ampère's law for a path that surrounds both conductors in Fig. 28–8. (b) Repeat, assuming the lower current I_2, is in the opposite direction ($I_2 = -I_1$).

7. Suppose the cylindrical conductor of Fig. 28–9a has a concentric cylindrical hollow cavity inside it (so it looks like a pipe). What can you say about **B** in the cavity?

8. Explain why a field such as that shown in Fig. 28–11b is consistent with Ampère's law. Could the lines curve upward instead of downward?

9. What would be the effect on B inside a long solenoid if (a) the diameter of all the loops was doubled, or (b) the spacing between loops was doubled, or (c) the solenoid's length was doubled along with a doubling in the total number of loops.

10. Use the Biot-Savart law to convince yourself that the field of the current loop in Fig. 28–18 is correct as shown for points off the axis.

11. Do you think **B** will be the same for all points in the plane of the current loop of Fig. 28–18?

12. Why does twisting the lead-in wires to electrical devices reduce the magnetic effects of the leads?

13. Compare the Biot-Savart law with Coulomb's law: What are the similarities and differences?

14. A type of magnetic switch similar to a solenoid is a **relay**. A relay is an electromagnet (the iron rod inside the coil doesn't move) which, when activated, attracts a piece of soft iron on a pivot. Design a relay (a) to make a doorbell and (b) to close an electrical switch. A relay is used in the latter case when you need to switch on a circuit carrying a very large current but you do not want that large current flowing through the main switch. For example the starter switch of a car is connected to a relay so that the large current needed for the starter doesn't pass to the dashboard switch.

15. How might you measure the magnetic dipole moment of the Earth?
16. How might you define or determine the magnetic pole strength (the magnetic equivalent of a single electric charge) for (a) a bar magnet, (b) a current loop?
17. A heavy magnet attracts, from rest, a heavy block of iron. Before striking the magnet the block has acquired considerable kinetic energy. (a) What is the source of this kinetic energy? (b) When the block strikes the magnet, some of the latter's domains are jarred into randomness; describe the energy transformations.
18. Will a magnet attract any metallic object, or only those made of iron? (Try it and see.) Why is this so?
19. An unmagnetized nail will not attract an unmagnetized paper clip. However, if one end of the nail is in contact with a magnet, the other end *will* attract a paper clip. Explain.
20. How do you suppose the first magnets found in Magnesia were formed?
21. Why will either pole of a magnet attract an unmagnetized piece of iron?
22. Suppose you have three iron rods, two of which are magnetized but the third is not. How would you determine which two are the magnets without using any additional objects?
23. Two iron bars attract each other no matter which ends are placed close together. Are both magnets? Explain.
*24. Describe the magnetization curve for (a) paramagnetic substance and (b) a diamagnetic substance, and compare to that for a ferromagnetic substance (Fig. 28–26).
*25. Can all materials be considered (a) diamagnetic, (b) paramagnetic, (c) ferromagnetic?

Problems

Sections 28–1 and 28–2

1. (I) Jumper cables used to start a stalled vehicle often carry a 65-A current. How strong is the magnetic field 7.5 cm away? Compare to the Earth's magnetic field.
2. (I) If an electric wire is allowed to produce a magnetic field no larger than that of the Earth $(0.55 \times 10^{-4} \text{ T})$ at a distance of 25 cm, what is the maximum current the wire can carry?
3. (I) What is the magnitude and direction of the force between two parallel wires 45 m long and 6.0 cm apart, each carrying 35 A in the same direction?
4. (I) A vertical straight wire carrying an upward 22-A current exerts an attractive force per unit length of $8.8 \times 10^{-4} \text{ N/m}$ on a second parallel wire 7.0 cm away. What current (magnitude and direction) flows in the second wire?
5. (I) In Fig. 28–29, a long straight wire carries current I out of the page toward the viewer. Indicate, with appropriate arrows, the direction of **B** at each of the points C, D, and E in the plane of the page.

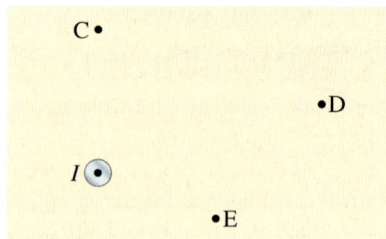

FIGURE 28–29 Problem 5.

6. (II) An experiment on the Earth's magnetic field is being carried out 1.00 m from an electric cable. What is the maximum allowable current in the cable if the experiment is to be accurate to ±1.0 percent?

7. (II) Two long thin parallel wires 15.0 cm apart carry 25-A currents in the same direction. Determine the magnetic field strength at a point 12.0 cm from one wire and 5.0 cm from the other (Fig. 28–30).

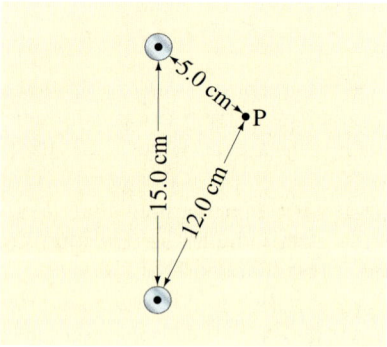

FIGURE 28–30
Problem 7.

8. (II) A horizontal compass is placed 20 cm due south from a straight vertical wire carrying a 40-A current downward. In what direction does the compass needle point at this location? Assume the horizontal component of the Earth's field at this point is $0.45 \times 10^{-4} \text{ T}$ and the magnetic declination is 0°.
9. (II) A long horizontal wire carries 22.0 A of current due north. What is the net magnetic field 20.0 cm due west of the wire if the Earth's field there points downward, 40° below the horizontal, and has magnitude $5.0 \times 10^{-5} \text{ T}$?
10. (II) A straight stream of protons passes a given point in space at a rate of 1.5×10^9 protons/s. What magnetic field do they produce 2.0 m from the beam?
11. (II) Determine the magnetic field midway between two long straight wires 2.0 cm apart in terms of the current I in one when the other carries 15 A. Assume these currents are (a) in the same direction, and (b) in opposite directions.

728 CHAPTER 28 Sources of Magnetic Field

12. (II) A long pair of wires serves to conduct 25.0 A of dc current to and from an instrument. If the wires are of negligible diameter but are 2.8 mm apart, what is the magnetic field 10.0 cm from their midpoint, in their plane (Fig. 28–31)? Compare to the magnetic field of the Earth.

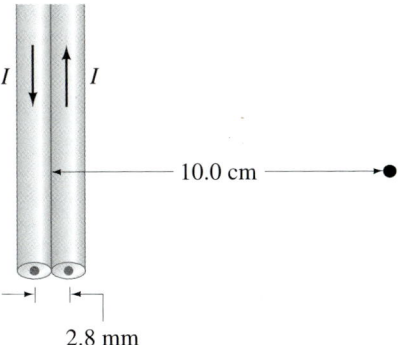

FIGURE 28–31 Problem 12.

13. (II) A compass needle points 20° E of N outdoors. However, when it is placed 12.0 cm to the east of a vertical wire inside a building, it points 55° E of N. What is the magnitude and direction of the current in the wire? The Earth's field there is 0.50×10^{-4} T and is horizontal.

14. (II) A rectangular loop of wire is placed next to a straight wire, as shown in Fig. 28–32. There is a current of 2.5 A in both wires. What is the magnitude and direction of the net force on the loop?

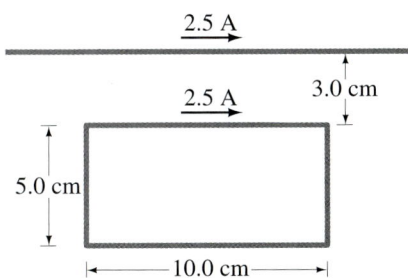

FIGURE 28–32 Problem 14.

15. (II) Let two long parallel wires, a distance d apart, carry equal currents I in the same direction. One wire is at $x = 0$, the other at $x = d$, Fig. 28–33. Determine **B** between the wires as a function of x.

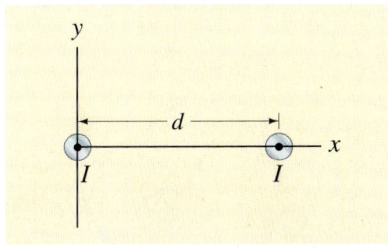

FIGURE 28–33 Problems 15 and 16.

16. (II) Repeat Problem 15 if the wire at $x = 0$ carries twice the current ($2I$) as the other wire, and in the opposite direction.

17. (II) Two long wires are oriented so that they are perpendicular to each other, and at their closest, they are 20.0 cm apart (Fig. 28–34). What is the magnitude of the magnetic field at a point midway between them if the top one carries a current of 20.0 A and the bottom one carries 5.0 A?

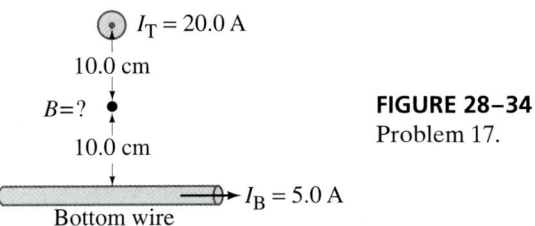

FIGURE 28–34 Problem 17.

18. (II) Two long parallel wires 7.00 cm apart carry 16.5-A currents in the same direction. Determine the magnetic field strength at a point P 12.0 cm from one wire and 13.0 cm from the other. See Fig. 28–35.

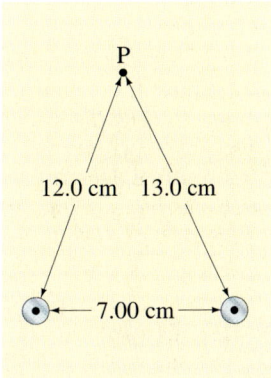

FIGURE 28–35 Problem 18.

19. (III) A very long flat conducting strip of width L and negligible thickness lies in a horizontal plane and carries a uniform current I across its cross section. (a) Show that at points a distance y directly above its center, the field is given by

$$B = \frac{\mu_0 I}{\pi L} \tan^{-1} \frac{L}{2y},$$

assuming the strip is infinitely long. [*Hint*: Divide the strip into many thin "wires," and sum (integrate) over these.] (b) What value does B approach for $y \gg L$? Does this make sense? Explain.

20. (III) An electron is moving in a plane that also contains a long straight current-carrying wire. The electron is heading at a 45° angle toward the wire, with a speed of 3.4×10^6 m/s, when it is 50 cm away. The electron reaches only as close as 1.0 cm, before being repelled away, always moving in the same plane. What is the current in the wire?

Sections 28–4 and 28–5

21. (I) A 40.0-cm long solenoid 1.35 cm in diameter is to produce a field of 0.385 mT at its center. How much current should the solenoid carry if it has 1000 turns of wire?

22. (I) A 32-cm-long solenoid, 1.8 cm in diameter, is to produce a 0.30-T magnetic field at its center. If the maximum current is 5.7 A, how many turns must the solenoid have?

23. (I) A 2.5-mm-diameter copper wire carries a 40-A current. Determine the magnetic field: (a) at the surface of the wire; (b) inside the wire, 0.50 mm below the surface; (c) outside the wire 2.5 mm from the surface.

24. (II) A toroid (Fig. 28–14) has a 50.0-cm inner diameter and a 54.0-cm outer diameter. It carries a 25.0 A current in its 500 coils. Determine the range of values for B inside the toroid.

25. (II) A 20.0 m long copper wire, 2.00 mm in diameter including insulation, is tightly wrapped in a single layer with adjacent coils touching, to form a solenoid of diameter 2.50 cm. What is (a) the length of the solenoid and (b) the field at the center when the current in the wire is 20.0 A?

26. (II) (a) Use Eq. 28–1, and the vector nature of **B**, to show that the magnetic field lines around two long parallel wires carrying equal currents $I_1 = I_2$ are as shown in Fig. 28–8. (b) Draw the equipotential lines around two stationary positive electric charges. (c) Are these two diagrams similar? Identical? Why or why not?

27. (II) A coaxial cable consists of a solid inner conductor of radius R_1, surrounded by a concentric cylindrical tube or inner radius R_2 and outer radius R_3 (Fig. 28–36). The conductors carry equal and opposite currents I_0 distributed uniformly across their cross-sections. Determine the magnetic field at a distance R from the axis for: (a) $R < R_1$; (b) $R_1 < R < R_2$; (c) $R_2 < R < R_3$; (d) $R > R_3$.

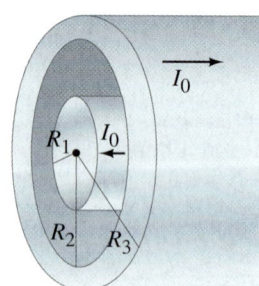

FIGURE 28–36 Problems 27 and 28.

28. (III) Suppose the current in the coaxial cable of Problem 27, Fig. 28–36, is not uniformly distributed, but instead the current density j varies linearly with distance from the center: $j_1 = C_1 R$ for the inner conductor and $j_2 = C_2 R$ for the outer conductor. Each conductor still carries the same total current I_0, in opposite directions. Determine the magnetic field in terms of I_0 in the same four regions of space as in Problem 27.

Section 28–6

29. (I) The Earth's magnetic field is essentially that of a magnetic dipole. If the field near the North Pole is about 1.0×10^{-4} T, what will it be (approximately) 13,000 km above the surface at the North Pole?

30. (II) A wire, in a plane, has the shape shown in Fig. 28–37, two arcs of a circle connected by radial lengths of wire. Determine **B** at point C in terms of R_1, R_2, θ, and the current I.

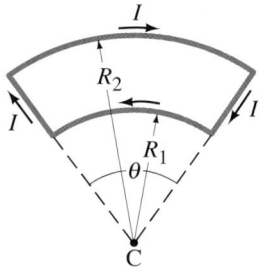

FIGURE 28–37 Problem 30.

31. (II) A circular conducting ring of radius R is connected to two exterior straight wires ending at two ends of a diameter (Fig. 28–38). The current I splits into unequal portions while passing through the ring as shown. What is **B** at the center of the ring?

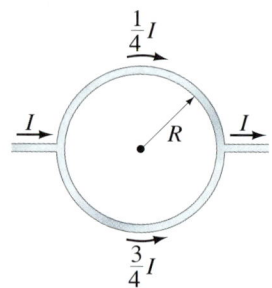

FIGURE 28–38 Problem 31.

32. (II) A small loop of wire of radius 1.8 cm is placed at the center of a 25.0-cm wire loop. The planes of the loops are perpendicular to each other, and a 7.0-A current flows in each. Estimate the torque the large loop exerts on the smaller one. What simplifying assumption did you make?

33. (II) A wire is formed into the shape of two half circles connected by equal-length straight sections as shown in Fig. 28–39. A current I flows in the circuit clockwise as shown. Determine (a) the magnitude and direction of the magnetic field at the center, C, and (b) the magnetic dipole moment of the circuit.

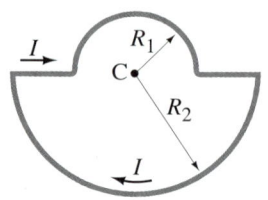

FIGURE 28–39 Problem 33.

34. (II) Use the Biot-Savart law to show that the magnetic field **B**, due to a single point charge q moving with velocity **v**, at a point P whose position vector relative to the charge is **r** (Fig. 28–40) is given by

$$\mathbf{B} = \frac{\mu_0}{4\pi} \frac{q\mathbf{v} \times \mathbf{r}}{r^3}.$$

(Assume v is much less than the speed of light.)

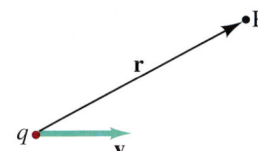

FIGURE 28–40 Problem 34.

35. (II) A nonconducting circular disk, of radius R, carries a uniformly distributed electric charge Q. The plate is set spinning with angular velocity ω about an axis perpendicular to the plate through its center (Fig. 28–41). Determine (a) its magnetic dipole moment and (b) the magnetic field at points on its axis a distance x from its center; (c) does Eq. 28–7b apply in this case for $x \gg R$?

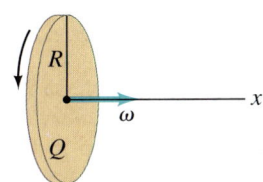

FIGURE 28–41 Problem 35.

36. (II) Consider a straight section of wire of length l, as in Fig. 28–42, which carries a current I. (a) Show that the magnetic field at a point P a distance R from the wire along its perpendicular bisector is

$$B = \frac{\mu_0 I}{2\pi R} \frac{l}{(l^2 + 4R^2)^{\frac{1}{2}}}.$$

(b) Show that this is consistent with Example 28–9 for an infinite wire.

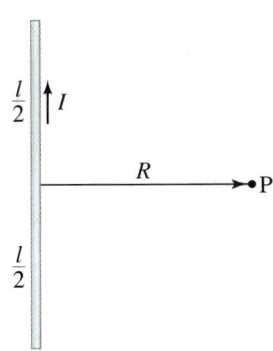

FIGURE 28–42 Problem 36.

37. (II) A segment of wire of length l carries a current I as shown in Fig. 28–43. (a) Show that for points along the positive x axis (the axis of the wire), such as point Q, the magnetic field **B** is zero. (b) Determine a formula for the field at points along the y axis, such as point P.

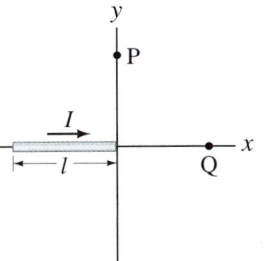

FIGURE 28–43 Problem 37.

38. (II) Use the result of Problem 37 to find the magnetic field at point P in Fig. 28–44 due to the current in the square loop.

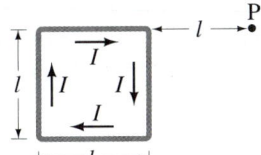

FIGURE 28–44 Problem 38.

39. (II) A wire is bent into the shape of a regular polygon with n sides whose vertices are a distance R from the center. (See Fig. 28–45, which shows the special case of $n = 6$.) If the wire carries a current I_0, (a) determine the magnetic field at the center; (b) if n is allowed to become very large ($n \to \infty$), show that the formula in part (a) reduces to that for a circular loop (Example 28–10).

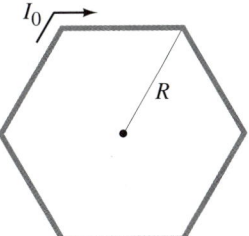

FIGURE 28–45 Problem 39.

40. (III) Start with the result of Example 28–10 for the magnetic field along the axis of a single loop to obtain the field inside a very long solenoid (Eq. 28–4).

41. (III) A single rectangular loop of wire, with sides a and b, carries a current I. An xy coordinate system has its origin at the lower left corner of the rectangle with the x axis parallel to side b (Fig. 28–46). Determine the magnetic field B at all points (x, y) within the loop.

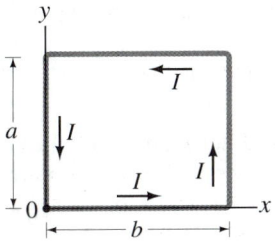

FIGURE 28–46 Problem 41.

42. (III) A square loop of wire, of side l, carries a current I. (a) Determine the magnetic field B at points on a line perpendicular to the plane of the square which passes through the center of the square (Fig. 28–47). Express B as a function of x, the distance along the line from the center of the square. (b) For $x \gg l$, does the square appear to be a magnetic dipole? If so, what is its dipole moment?

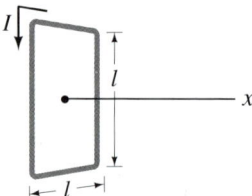

FIGURE 28–47 Problem 42.

*Section 28–7

* **43.** (II) An iron atom has a magnetic dipole moment of about $1.8 \times 10^{-23}\,\text{A}\cdot\text{m}^2$. (a) Determine the dipole moment of an iron bar 12 cm long, 1.2 cm wide, and 1.2 cm thick, if it is 100 percent saturated. (b) What torque would be exerted on this bar when placed in a 1.2-T field acting at right angles to the bar?

*Section 28–9

* **44.** (I) The following are some values of B and B_0 for a piece of annealed iron as it is being magnetized:

$B_0(10^{-4}\,\text{T})$	0	0.13	0.25	0.50	0.63	0.78	1.0	1.3
$B(\text{T})$	0	0.0042	0.010	0.028	0.043	0.095	0.45	0.67
$B_0(10^{-4}\,\text{T})$		1.9	2.5	6.3	13.0	130	1,300	10,000
$B(\text{T})$		1.01	1.18	1.44	1.58	1.72	2.26	3.15

Determine the magnetic permeability μ for each value and plot a graph of μ versus B_0.

* **45.** (I) A large thin toroid has 400 loops of wire per meter, and a 20-A current flows through the wire. If the relative permeability of the iron is 3000, what is the total field B inside the toroid?

* **46.** (II) An iron-core solenoid is 38 cm long, 1.8 cm in diameter, and has 600 turns of wire. A magnetic field of 2.2 T is produced when 48 A flows in the wire. What is the permeability μ at this high field strength?

General Problems

47. Three long parallel wires are 38.0 cm from one another. (Looking along them, they are at three corners of an equilateral triangle.) The current in each wire is 8.00 A, but that in wire M is opposite to that in wires N and P (Fig. 28–48). Determine the magnetic force per unit length on each wire due to the other two.

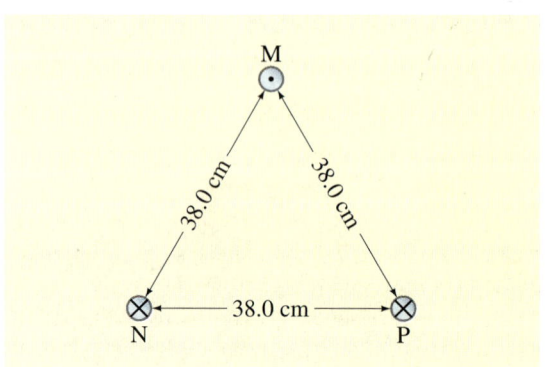

FIGURE 28–48 Problems 47 and 48.

48. In Fig. 28–48 the top wire is 2.00-mm-diameter copper wire and is suspended in air due to the two magnetic forces from the bottom two wires. The current is 20.0 A in each of the two bottom wires. Calculate the required current flow in the suspended wire.

49. An electron enters a large solenoid at a 7.0° angle to the axis. If the field is a uniform $3.3 \times 10^{-2}\,\text{T}$, determine the radius and pitch (distance between loops) of the electron's helical path if its speed is $1.3 \times 10^7\,\text{m/s}$.

50. A rectangular loop of wire carries a 2.0-A current and lies in a plane which also contains a very long straight wire carrying a 10.0-A current as shown in Fig. 28–49. Determine (a) the net force and (b) the net torque on the loop due to the straight wire.

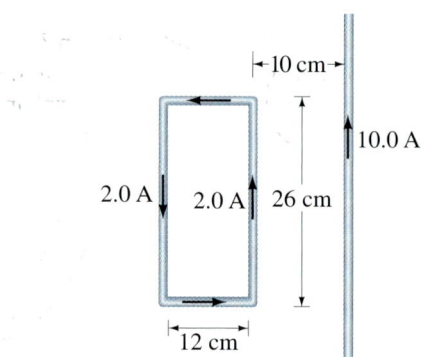

FIGURE 28–49 Problem 50.

51. A very large flat conducting sheet of thickness t carries a uniform current density $\mathbf{j}$ throughout (Fig. 28–50). Determine the magnetic field (magnitude and direction) at a distance y above the plane. (Assume the plane is infinitely long and wide.)

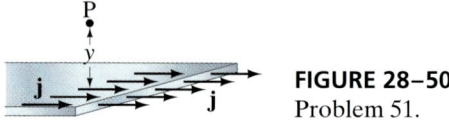

FIGURE 28–50 Problem 51.

52. A long horizontal wire carries a current of 48 A. A second wire, made of 2.5-mm-diameter copper wire and parallel to the first but 15 cm below it, is held in suspension magnetically (Fig. 28–51). (*a*) What is the magnitude and direction of the current in the lower wire? (*b*) Is the lower wire in stable equilibrium? (*c*) Repeat parts (*a*) and (*b*) if the second wire is suspended 15 cm *above* the first due to the first's magnetic field.

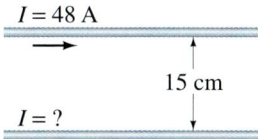

FIGURE 28–51 Problem 52.

53. You have 1.0 kg of copper and want to make a practical solenoid that produces the greatest possible magnetic field. Consider variables such as solenoid diameter, length, and so on, to determine if you should make your copper wire long and thin, short and fat, or something else.

54. For two long parallel wires separated by a distance L, carrying currents I_1 and I_2 as in Fig. 28–8, show that for the circular path of radius r $(r < L)$ centered on I_1, that

$$\oint \mathbf{B} \cdot d\mathbf{l} = \mu_0 I_1$$

in accord with Ampère's law. (But do not use Ampère's law.)

55. Near the Earth's poles the magnetic field is about 1 G $(1 \times 10^{-4}\,\text{T})$. Imagine a simple model in which the Earth's field is produced by a single current loop around the equator. What current would this loop carry?

56. A square loop of wire, of side L, carries a current I. Show that the magnetic field at the center of the square is

$$B = \frac{2\sqrt{2}\,\mu_0 I}{\pi L}.$$

[*Hint*: Determine $\mathbf{B}$ for each segment of length L.]

57. In Problem 56, if you reshaped the square wire into a circle, would B increase or decrease at the center? Explain.

58. Two large coils of wire, each with N turns carrying a current I and separated by a distance equal to the radius R of the coils, are called *Helmholtz coils* (see Fig. 28–52). (*a*) Determine B at points x along the line joining their centers. Let $x = 0$ at the center of one coil, $x = R$ at the center of the other. (*b*) Plot B versus x from $x = 0$ to $x = R$. (*c*) Determine B on the axis midway between the coils $(x = R/2)$ if $R = 20.0\,\text{cm}$, $I = 35\,\text{A}$, and each coil contains $N = 350$ turns. (*d*) Show that the field midway between the coils is particularly uniform by showing that when the separation of the coils is R, then $dB/dx = 0$ and $d^2B/dx^2 = 0$ at the midpoint.

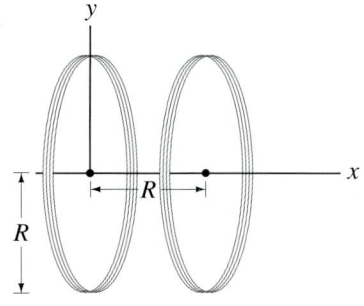

FIGURE 28–52 Problem 58.

59. A 175-g model airplane charged to 18.0 mC and traveling at 2.8 m/s passes within 8.6 cm of a wire, nearly parallel to its path, carrying a 30-A current. What acceleration (in g's) does this interaction give the airplane?

60. Suppose that an electromagnet uses a coil 2.0 m in diameter made from square copper wire 2.0 mm on a side; the power supply produces 50 V at a maximum power output of 1.0 kW. (*a*) How many turns are needed to run the power supply at maximum power? (*b*) What is the magnetic field strength at the center of the coil? (*c*) If you use a greater number of turns and this same power supply, will a greater magnetic field strength result? Explain.

61. Four long straight parallel wires located at the corners of a square of side l carry equal currents I_0 perpendicular to the page as shown in Fig. 28–53. Determine the magnitude and direction of **B** at the center of the square.

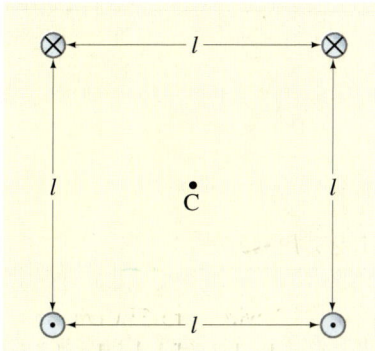

FIGURE 28–53 Problem 61.

62. Determine the magnetic field at the point P due to the long wire with a square bend shown in Fig. 28–54. The point P is halfway between the two corners. [*Hint*: You can use the results of Problems 36 and 37.]

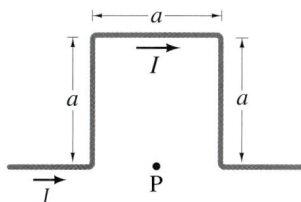

FIGURE 28–54 Problem 62.

63. You want to get an idea of the magnitude of magnetic fields produced by overhead power lines. You estimate that the two wires run about 30 m above the ground about 3 m apart. A call to the local power company provides the information that the lines operate at 10 kilovolts and provide a maximum of 40 MW to the local area. Estimate the maximum magnetic field you might experience walking under these power lines, and compare to the Earth's field. [Note that if the current is ac, the magnetic field will be changing too.]

One of the great laws of physics is Faraday's law of induction, which says that a changing magnetic field produces an induced emf. This photo shows a bar magnet moving inside a coil, and the galvanometer registers an induced current. This phenomenon of electromagnetic induction is the basis for many practical devices, from generators to alternators to transformers, tape recording and reading computer memory.

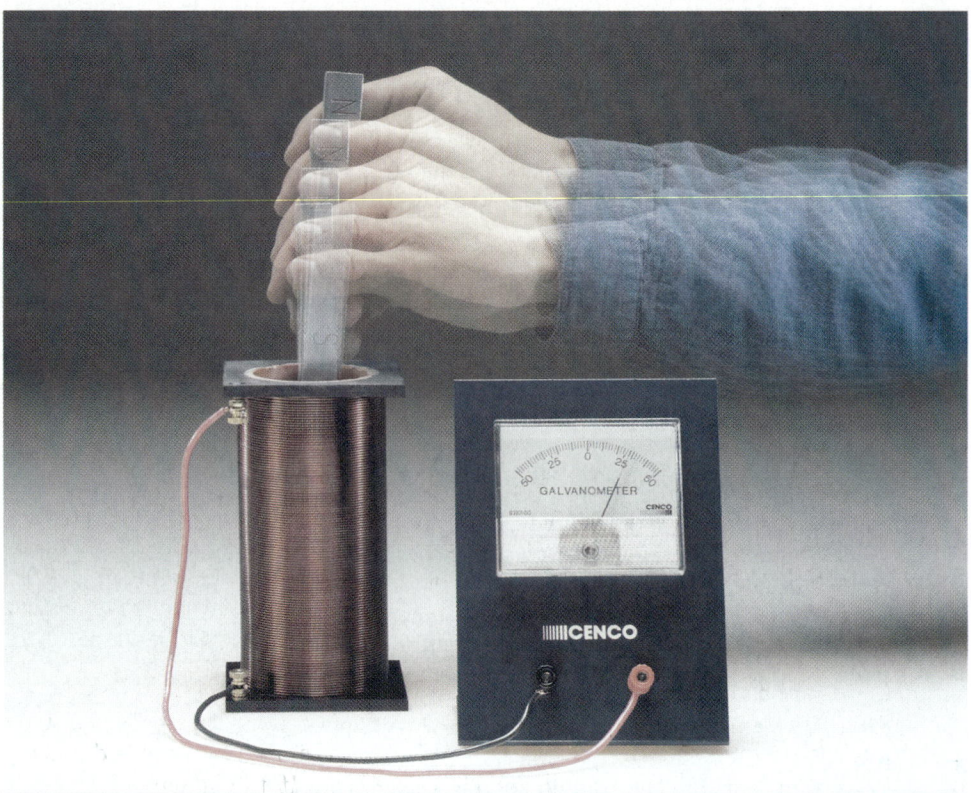

CHAPTER 29

Electromagnetic Induction and Faraday's Law

In Chapter 27, we discussed two ways in which electricity and magnetism are related: (1) an electric current produces a magnetic field; and (2) a magnetic field exerts a force on an electric current or moving electric charge. These discoveries were made in 1820–1821. Scientists then began to wonder: if electric currents produce a magnetic field, is it possible that a magnetic field can produce an electric current? Ten years later the American Joseph Henry (1797–1878) and the Englishman Michael Faraday (1791–1867) independently found that it was possible. Henry actually made the discovery first. But Faraday published his results earlier and investigated the subject in more detail. We now discuss this phenomenon and some of its world-changing applications such as the electric generator.

29–1 Induced EMF

In his attempt to produce an electric current from a magnetic field, Faraday used an apparatus like that shown in Fig. 29–1. A coil of wire, X, was connected to a battery. The current that flowed through X produced a magnetic field that was intensified by the iron core. Faraday hoped that by using a strong enough battery, a steady current in X would produce a great enough magnetic field to produce a current in a second coil Y. This second circuit, Y, contained a galvanometer to detect any current but contained no battery. He met no success with steady currents.

*Constant **B** induces no emf*

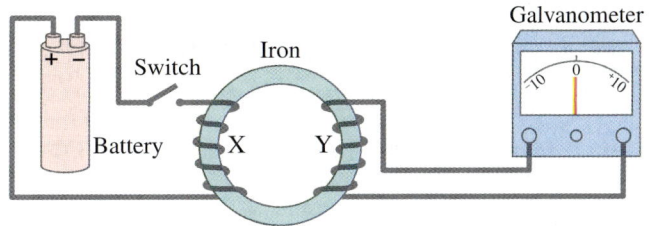

FIGURE 29–1 Faraday's experiment to induce an emf.

But the long-sought effect was finally observed when Faraday saw the galvanometer in circuit Y deflect strongly at the moment he closed the switch in circuit X. And the galvanometer deflected strongly in the opposite direction when he opened the switch. A *steady* current in X had produced *no* current in Y. Only when the current in X was starting or stopping was a current produced in Y.

Faraday concluded that although a steady magnetic field produces no current, a *changing* magnetic field can produce an electric current! Such a current is called an **induced current**. When the magnetic field through coil Y changes, a current flows as if there were a source of emf in the circuit. We therefore say that

an induced emf is produced by a changing magnetic field.

Changing **B** *induces an emf*

Faraday did further experiments on **electromagnetic induction**, as this phenomenon is called. For example, Fig. 29–2 shows that if a magnet is moved quickly into a coil of wire, a current is induced in the wire. If the magnet is quickly removed, a current is induced in the opposite direction. Furthermore, if the magnet is held steady and the coil of wire is moved toward or away from the magnet, or the coil is rotated, again an emf is induced and a current flows. Motion or change is required to induce an emf. It doesn't matter whether the magnet or the coil moves. It is the relative motion that counts.

FIGURE 29–2 (a) A current is induced when a magnet is moved toward a coil. (b) The induced current is opposite when the magnet is moved away from the coil. Note that the galvanometer zero is at the center of the scale and the needle deflects left or right, depending on the direction of the current. In (c) no current is induced if the magnet does not move relative to the coil. It is the relative motion that counts: the magnet can be held steady and the coil moved, which also induces an emf.

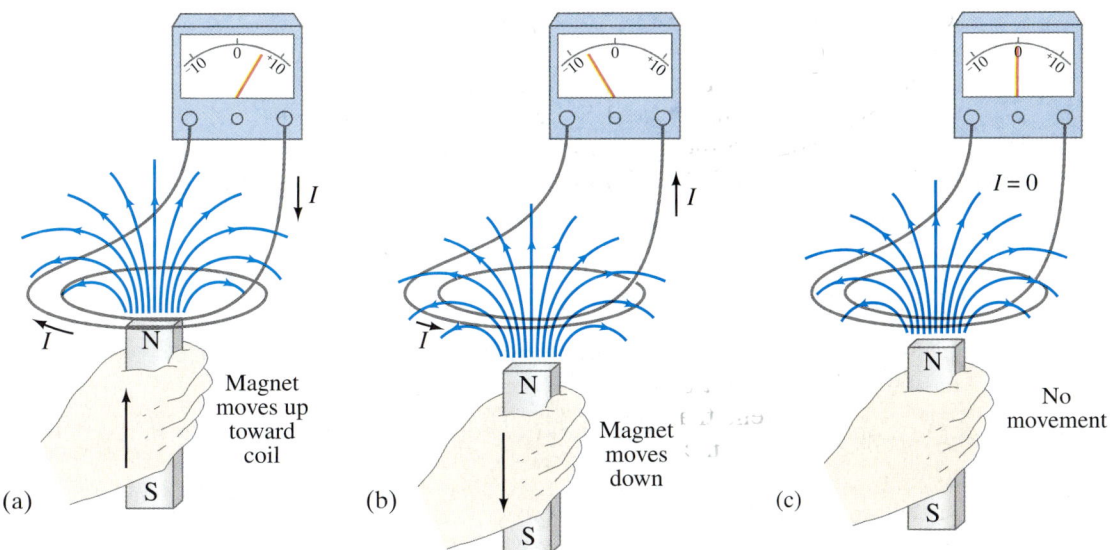

SECTION 29–1 Induced EMF 735

29–2 Faraday's Law of Induction; Lenz's Law

Faraday investigated quantitatively what factors influence the magnitude of the emf induced. He found first of all that the more rapidly the magnetic field changes, the greater the induced emf. But the emf is not simply proportional to the rate of change of the magnetic field, **B**. Rather the emf is proportional to the rate of change of the **magnetic flux**, Φ_B, passing through the circuit or loop of area A. Magnetic flux for a uniform magnetic field is defined as

$$\Phi_B = B_\perp A = BA\cos\theta = \mathbf{B}\cdot\mathbf{A}. \qquad [\text{B uniform}] \quad (29\text{–}1\text{a})$$

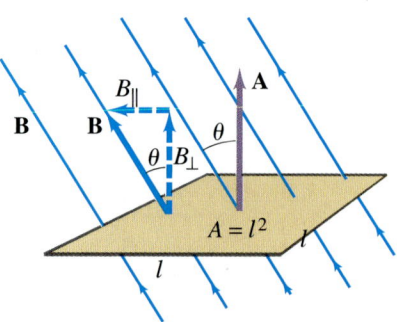

FIGURE 29–3 Determining the flux through a flat loop of wire. This loop is square, of side l and area $A = l^2$.

Here $B_\perp$ is the component of the magnetic field **B** perpendicular to the face of the loop, and θ is the angle between **B** and the vector **A** (representing the area) whose direction is perpendicular to the face of the loop. These quantities are shown in Fig. 29–3 for a square loop of side l whose area is $A = l^2$. If the area is of some other shape, or **B** is not uniform, the magnetic flux can be written[†]

$$\Phi_B = \int \mathbf{B}\cdot d\mathbf{A}. \qquad (29\text{–}1\text{b})$$

FIGURE 29–4 Magnetic flux Φ_B is proportional to the number of lines of **B** that pass through the loop.

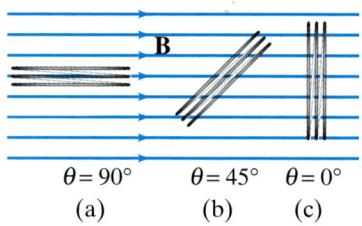

As we saw earlier, the lines of **B** (like lines of **E**) can be drawn such that the number of lines per unit area is proportional to the field strength. Then the flux Φ_B can be thought of as being proportional to the *total number of lines passing through the loop*. This is illustrated in Fig. 29–4, where the loop is viewed from the side (on edge). For $\theta = 90°$, no lines pass through the loop and $\Phi_B = 0$, whereas Φ_B is a maximum when $\theta = 0°$. The unit of magnetic flux is the tesla-meter²; this is called a **weber**: $1\,\text{Wb} = 1\,\text{T}\cdot\text{m}^2$.

With this definition of the flux, we can now write down the results of Faraday's investigations—namely, that the emf induced in a circuit is equal to the rate of change of magnetic flux through the circuit:

$$\mathcal{E} = -\frac{d\Phi_B}{dt}. \qquad (29\text{–}2\text{a})$$

FARADAY'S LAW OF INDUCTION

This fundamental result is known as **Faraday's law of induction**, and is one of the basic laws of electromagnetism.

If the circuit contains N closely wrapped loops, the emfs induced in each add together, so

$$\mathcal{E} = -N\frac{d\Phi_B}{dt}. \qquad (29\text{–}2\text{b})$$

The minus sign in Eq. 29–2 is placed there to remind us in which direction the induced emf acts. Experiments show that:

Lenz's law

An induced emf gives rise to a current whose magnetic field opposes the original change in flux.

This is known as **Lenz's law**. Said another way, valid even if no current can flow (as when a circuit is not complete), is:

An induced emf is always in a direction that opposes the original change in flux that caused it.

Let us apply Lenz's law to the case of relative motion between a magnet and a

---

[†]The integral is taken over an open surface—that is, one bounded by a closed curve such as a circle or square. In the present discussion, the area is that enclosed by the loop under discussion. The area is not an enclosed surface as we used in Gauss's law, Chapter 22.

coil, Fig. 29–2. The changing flux through the coil induces an emf, which produces a current in the coil. And this induced current produces its own magnetic field. In Fig. 29–2a the distance between the coil and the magnet decreases. The magnetic field (and number of field lines), and therefore the flux, through the coil increases. The magnetic field of the magnet points upward. To oppose this upward increase, the field inside the coil produced by the induced current points *downward*. Thus, Lenz's law tells us that the current moves as shown (use the right-hand rule). In Fig. 29–2b, the flux *decreases* (because the magnet is moved away), so the induced current produces an *upward* magnetic field inside the coil that is "trying" to maintain the status quo. Thus the current is as shown.

Let us consider what would happen if Lenz's law were not true, but were just the reverse. The induced current in this imaginary situation would produce a flux in the same direction as the original change. This greater change in flux would produce an even larger current followed by a still greater change in flux, and so on. The current would continue to grow indefinitely, producing power $(= I^2R)$ even after the original stimulus ended. This would violate the conservation of energy. Such "perpetual motion" devices do not exist. Thus, Lenz's law as stated above (and not its opposite) is consistent with the law of conservation of energy.

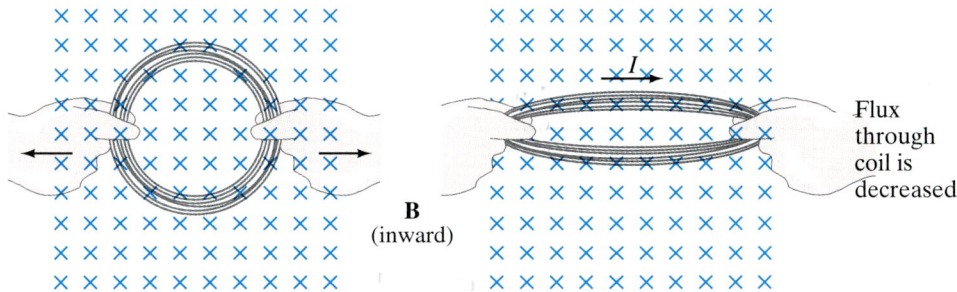

FIGURE 29–5 A current can be induced by changing the area of the coil. In both this case and that of Fig. 29–6, the flux through the coil is reduced. Here the brief induced current acts in the direction shown so as to try to maintain the original flux ($\Phi = BA$) by producing its own magnetic field into the page. That is, as the area A decreases, the current acts to increase B in the original (inward) direction.

It is important to note that an emf is induced whenever there is a change in flux. Since magnetic flux $\Phi_B = \int \mathbf{B} \cdot d\mathbf{A} = \int B \cos\theta \, dA$, we see that an emf can be induced in three ways: (1) by a changing magnetic field B; (2) by changing the area A of the loop in the field; or (3) by changing the loop's orientation θ with respect to the field. Figures 29–1 and 29–2 illustrated case 1. Examples of cases 2 and 3 are illustrated in Figs. 29–5 and 29–6, respectively.

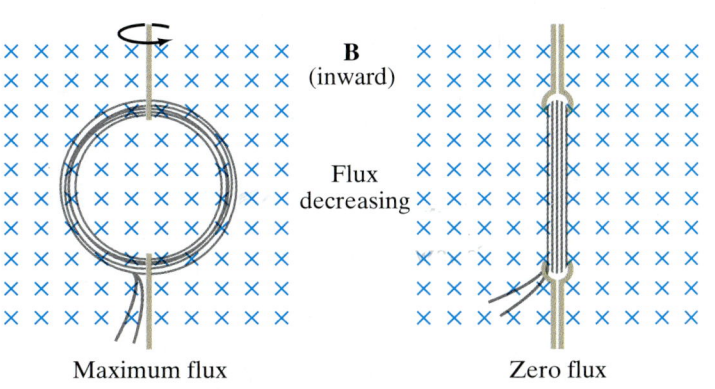

FIGURE 29–6 A current can be induced by rotating a coil in a magnetic field.

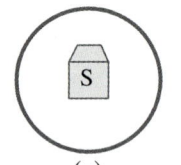 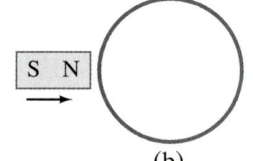 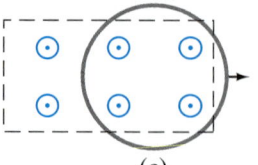 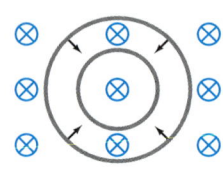 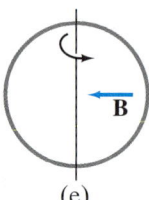

(a) N magnetic pole moving toward loop into the page

(b) N magnetic pole moving toward the loop in the plane of the page

(c) Pulling the loop to the right out of a magnetic field that points out of the page

(d) Shrinking a loop in a magnetic field pointing into the page

(e) Rotating the loop about the vertical diameter by pulling the left side toward the reader and pushing the right side away from the reader in a magnetic field that points from right to left in the plane of the page

FIGURE 29–7 Example 29–1.

CONCEPTUAL EXAMPLE 29–1 Practice with Lenz's law. In which direction is the current induced in the loop for each situation in Fig. 29–7?

RESPONSE (a) Magnetic field lines point out from the N pole of a magnet, so as the magnet moves toward the loop, the field points into the page and is getting stronger. The current will be induced in the counterclockwise direction to produce a field **B** *out* of the page so that its own flux counteracts the externally imposed change.
(b) The field is in the plane of the page, so the flux through the loop is zero throughout the process; hence there is no change in magnetic flux with time, and there will be no induced emf or current in the loop.
(c) Initially, the magnetic flux pointing out of the page passes through the loop. If you pull the loop out of the field, the induced current will be in a direction to make up the deficiency: the current flow will be counterclockwise to produce an outward (toward the reader) magnetic field.
(d) The flux is into the page and the coil area shrinks so the flux will decrease; hence the induced current will be clockwise to try to produce its own flux into the page to make up for the flux decrease.
(e) Initially there is no flux through the loop (why?). When you start to rotate the loop, the flux begins passing through the loop increasing to the left. To counteract this, the loop will have current induced in a counterclockwise direction so as to produce its own flux to the right.

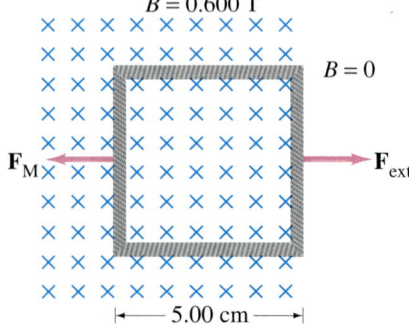

FIGURE 29–8 Example 29–2. The square coil in a magnetic field $B = 0.600\,\text{T}$ is pulled abruptly to the right to a region where $B = 0$.

EXAMPLE 29–2 Pulling a coil from a magnetic field. A square coil of wire with side 5.00 cm contains 100 loops and is positioned perpendicular to a uniform 0.600-T magnetic field, as shown in Fig. 29–8. It is quickly and uniformly pulled from the field (moving perpendicular to **B**) to a region where B drops abruptly to zero. At $t = 0$, the right edge of the coil is at the edge of the field. It takes 0.100 s for the whole coil to reach the field-free region. Find (a) the rate of change in flux through the coil, (b) the emf and current induced, and (c) how much energy is dissipated in the coil if its resistance is 100 Ω. (d) What was the average force required?

SOLUTION (a) First we find how the magnetic flux changes during the time interval $\Delta t = 0.100\,\text{s}$. The area of the coil is $A = (5.00 \times 10^{-2}\,\text{m})^2 = 2.50 \times 10^{-3}\,\text{m}^2$. The flux is initially $\Phi_B = BA = (0.600\,\text{T})(2.50 \times 10^{-3}\,\text{m}^2) = 1.50 \times 10^{-3}\,\text{Wb}$. After 0.100 s, the flux is zero. The rate of change in flux is constant (because the coil is square), equal to

$$\frac{\Delta \Phi_B}{\Delta t} = \frac{0 - 1.50 \times 10^{-3}\,\text{Wb}}{0.100\,\text{s}} = -1.50 \times 10^{-2}\,\text{Wb/s}.$$

(b) The emf induced (Eq. 29–2) during this period is

$$\mathscr{E} = -N\frac{d\Phi_B}{dt} = -(100)(-1.50 \times 10^{-2}\,\text{Wb/s}) = 1.50\,\text{V}.$$

The current is

$$I = \frac{\mathcal{E}}{R} = \frac{1.50 \text{ V}}{100\,\Omega} = 15.0 \text{ mA},$$

and, by Lenz's law, must be clockwise to oppose the decreasing flux into the page.
(c) The total energy dissipated is

$$E = Pt = I^2 Rt = (1.50 \times 10^{-2}\text{ A})^2 (100\,\Omega)(0.100\text{ s}) = 2.25 \times 10^{-3}\text{ J}.$$

(d) We can calculate the force directly using $\mathbf{F} = I\mathbf{l} \times \mathbf{B}$, Eq. 27–3, for constant $\mathbf{B}$. The force the magnetic field exerts on the top and bottom sections of the square loop of Fig. 29–8 are in opposite directions and cancel each other. The magnetic force $\mathbf{F}_M$ exerted on the left vertical section of the square loop acts to the left as shown because the current is up (clockwise). The right side of the loop is in the region where $\mathbf{B} = 0$. Hence the needed external force, to the right, has magnitude

$$F_{\text{ext}} = NIlB = (100)(0.0150\text{ A})(0.0500\text{ m})(0.600\text{ T}) = 0.0450\text{ N},$$

since there are $N = 100$ current-carrying loops.
We can also calculate the average force using the result of part (c): From conservation of energy, the energy dissipated E is equal to the work W needed to pull the coil out of the field. Because $W = \bar{F}d$ where $d = 5.00$ cm, then

$$\bar{F} = \frac{W}{d} = \frac{2.25 \times 10^{-3}\text{ J}}{5.00 \times 10^{-2}\text{ m}} = 0.0450\text{ N},$$

which is the same answer, as we expect.

29–3 EMF Induced in a Moving Conductor

Another way to induce an emf is shown in Fig. 29–9, and this situation helps illuminate the nature of the induced emf. Assume that a uniform magnetic field $\mathbf{B}$ is perpendicular to the area bounded by the U-shaped conductor and the movable rod resting on it. If the rod is made to move at a speed v, it travels a distance $dx = v\,dt$ in a time dt. Therefore, the area of the loop increases by an amount $dA = l\,dx = lv\,dt$ in a time dt. By Faraday's law, there is an induced emf $\mathcal{E}$ whose magnitude is given by

$$\mathcal{E} = \frac{d\Phi_B}{dt} = \frac{B\,dA}{dt} = \frac{Blv\,dt}{dt} = Blv. \quad (29\text{–}3)$$

This equation is valid as long as B, l, and v are mutually perpendicular. (If they are not, we use only the components of each that are mutually perpendicular.) An emf induced on a conductor moving in a magnetic field is sometimes called a **motional emf**.

We can also obtain Eq. 29–3 without using Faraday's law. We saw in Chapter 27 that a charged particle moving perpendicular to a magnetic field B with speed v experiences a force $\mathbf{F} = q\mathbf{v} \times \mathbf{B}$. When the rod of Fig. 29–9a moves to the right with speed v, the electrons in the rod move with this same speed. Therefore, since $\mathbf{v} \perp \mathbf{B}$, each electron feels a force $F = qvB$, which acts upward as shown in Fig. 29–9b. If the rod were not in contact with the U-shaped conductor, free electrons would collect at the upper end of the rod, leaving the lower end positive. There must thus be an induced emf. If the rod does slide on the U-shaped conductor, the electrons will flow into it. There will then be a clockwise (conventional) current flowing in the loop. To calculate the emf, we determine the work W needed to move a charge q from one end of the rod to the other against this potential difference: $W = \text{force} \times \text{distance} = (qvB)(l)$. The emf equals the work done per unit charge, so $\mathcal{E} = W/q = qvBl/q = Blv$, just as above.[†]

FIGURE 29–9 (a) A conducting rod is moved to the right on a U-shaped conductor in a uniform magnetic field $\mathbf{B}$ that points out of the page. (b) Rod only, showing force on one electron.

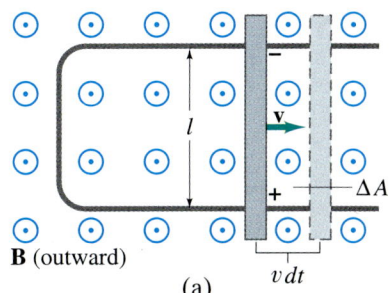

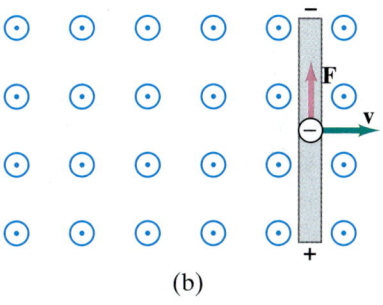

[†]This argument, which is basically the same as for the Hall effect (Section 27–8), explains this one way of inducing an emf. It does not explain the general case of electromagnetic induction.

This emf produces an electric field E in the rod which moves the electrons along (as in Section 25–8). Assuming a uniform E in the rod, then $E = \mathcal{E}/l = Bv$.

FIGURE 29–10 Example 29–3.

EXAMPLE 29–3 Does a moving airplane develop a large emf? An airplane travels 1000 km/h in a region where the Earth's magnetic field is 5.0×10^{-5} T and is nearly vertical (Fig. 29–10). What is the potential difference induced between the wing tips that are 70 m apart?

SOLUTION Since $v = 1000$ km/h $= 280$ m/s, and $\mathbf{v} \perp \mathbf{B}$, we have
$$\mathcal{E} = Blv = (5.0 \times 10^{-5}\,\text{T})(70\,\text{m})(280\,\text{m/s}) = 1.0\,\text{V}.$$
Not much to worry about.

EXAMPLE 29–4 Force on the rod. To make the rod of Fig. 29–9a move to the right at speed v, you need to apply a force. (a) Explain and determine the magnitude of the required force. (b) What external power is needed to move the rod?

SOLUTION (a) When the rod moves to the right, a current flows downward in the rod, as already discussed. We can see this also from Lenz's law: the upward magnetic flux through the loop is increasing, so the induced current must oppose the increase. Thus the current is clockwise so as to produce a magnetic field into the page (right-hand rule). The force on the moving rod is $\mathbf{F} = I\mathbf{l} \times \mathbf{B}$ for a constant $\mathbf{B}$ (Eq. 27–3). The right-hand rule tells us this force is to the left, and is thus a "drag force" opposing our effort to move the rod to the right.
The magnitude of the external force, to the right, needs to be $F = IlB$, where the current $I = \mathcal{E}/R = Blv/R$. The resistance R is that of the whole circuit, the rod and the U-shaped conductor. The force F required to move the rod is thus
$$F = IlB = \frac{B^2l^2}{R}v.$$
If B, l, and R are constant, then a constant speed v is produced by a constant force. Constant R implies that all the resistance is in the rod and none in the U-shaped conductor.
(b) The external power needed to move the rod for constant R is
$$P = Fv = \frac{B^2l^2v^2}{R}.$$
The power dissipated in the resistance is $P = I^2R$. With $I = \mathcal{E}/R = Blv/R$,
$$P = I^2R = \frac{B^2l^2v^2}{R}$$
so the power input equals that dissipated in the resistance at any moment.

29–4 Electric Generators

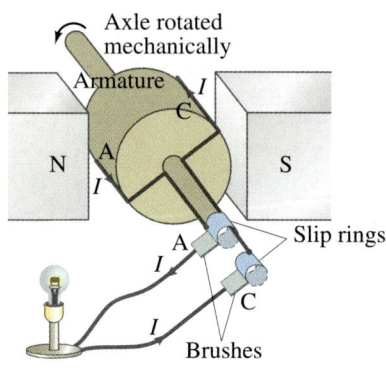

FIGURE 29–11 An ac generator.

Probably the most important practical result of Faraday's great discovery was the development of the **electric generator** or **dynamo**. A generator transforms mechanical energy into electric energy. This is just the opposite of what a motor does. Indeed, a generator is basically the inverse of a motor. A simplified diagram of an **ac generator** is shown in Fig. 29–11. A generator consists of many coils of wire (only one is shown) wound on an armature that can rotate in a magnetic field. The axle is turned by some mechanical means (falling water, car motor belt), and an emf is induced in the rotating coil. An electric current is thus the *output* of a generator. In Fig. 29–11, the equation $\mathbf{F} = q\mathbf{v} \times \mathbf{B}$ tells us that, with the armature rotating counterclockwise, the (conventional) current in the wire labeled A on the armature is outward; therefore it is outward at brush A, as shown. (Each brush presses against a continuous slip ring.) After one-half revolution, wire A will be where wire C is now in the drawing, and the current then at brush A will be inward. Thus the current produced is alternating. Let us look at this in more detail.

Let us assume the loop is being made to rotate in a uniform magnetic field **B** with constant angular velocity ω. From Faraday's law (Eq. 29–2a), the induced emf is

$$\mathcal{E} = -\frac{d\Phi_B}{dt} = -\frac{d}{dt}\int \mathbf{B}\cdot d\mathbf{A} = -\frac{d}{dt}[BA\cos\theta]$$

where A is the area of the loop and θ is the angle between **B** and **A**. Since $\omega = d\theta/dt$, then $\theta = \theta_0 + \omega t$. We arbitrarily take $\theta_0 = 0$, so

$$\mathcal{E} = -BA\frac{d}{dt}(\cos\omega t) = BA\omega\sin\omega t.$$

If the rotating coil contains N loops,

$$\mathcal{E} = NBA\omega\sin\omega t$$
$$= \mathcal{E}_0 \sin\omega t. \tag{29-4}$$

Thus the output emf is sinusoidal (Fig. 29–12) with amplitude $\mathcal{E}_0 = NBA\omega$. Such a rotating coil in a magnetic field is the basic operating principle of an ac generator.

Over 99 percent of the electricity used in the United States is produced from generators (Fig. 29–13). The frequency $f = \omega/2\pi$ is 60 Hz for general use in the United States and Canada, although 50 Hz is used in many countries. In electric power generating plants, the armature is mounted on a heavy axle connected to a turbine, which is the modern equivalent of a waterwheel. Water pressure at a dam can turn the turbine at a hydroelectric plant. Most of the power generated at present in the United States, however, is done at steam plants, where the burning of fossil fuels (coal, oil, natural gas) boils water to produce high-pressure steam that turns the turbines. Likewise, at nuclear power plants, the nuclear energy released is used to produce steam to turn turbines. Thus, a heat engine (Chapter 20) connected to a generator is the principal means of generating electric power.

The frequency of 60 Hz is maintained very precisely by power companies, and in doing Problems, we will assume it is at least as precise as other numbers given.

EXAMPLE 29–5 An ac generator. The armature of a 60-Hz ac generator rotates in a 0.15-T magnetic field. If the area of the coil is $2.0 \times 10^{-2}\,\text{m}^2$, how many loops must the coil contain if the peak output is to be $\mathcal{E}_0 = 170\,\text{V}$?

SOLUTION From Eq. 29–4 we see that the maximum emf is $\mathcal{E}_0 = NBA\omega$. Since $\omega = 2\pi f = (6.28)(60\,\text{s}^{-1}) = 377\,\text{s}^{-1}$, we have

$$N = \frac{\mathcal{E}_0}{BA\omega} = \frac{170\,\text{V}}{(0.15\,\text{T})(2.0\times 10^{-2}\,\text{m}^2)(377\,\text{s}^{-1})} = 150\,\text{turns}.$$

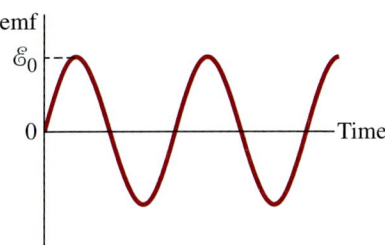

FIGURE 29–12 An ac generator produces an alternating current. The output emf $\mathcal{E} = \mathcal{E}_0 \sin\omega t$, where $\mathcal{E}_0 = NAB\omega$ (Eq. 29–4).

➡ **PHYSICS APPLIED**
Power plants

FIGURE 29–13 Water-driven generators at the base of Boulder Dam, Nevada.

A **dc generator** is much like an ac generator, except the slip rings are replaced by split-ring commutators, Fig. 29–14a, just as in a dc motor. The output of such a generator is as shown and can be smoothed out by placing a capacitor in parallel with the output (Section 26–4). More common is the use of many armature windings, as in Fig. 29–14b, which produces a smoother output.

➡ **PHYSICS APPLIED**
DC generator

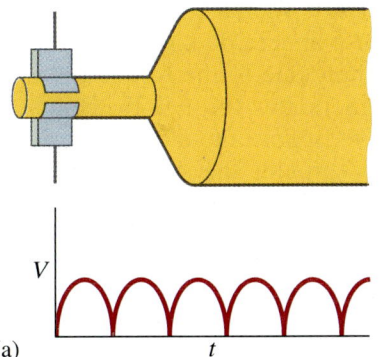
(a)

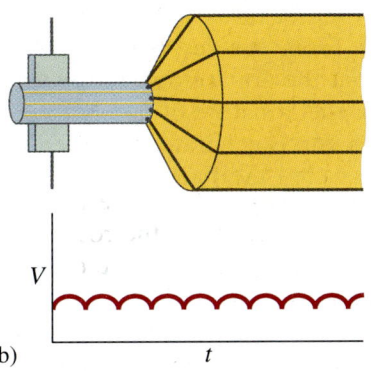
(b)

FIGURE 29–14 (a) A dc generator with one set of commutators, and (b) a dc generator with many sets of commutators and windings.

> **PHYSICS APPLIED**
> *Alternators*

> *Output emf of alternator*

In the past, automobiles used dc generators. More common now, however, are ac generators or **alternators**, which avoid the problems of wear and electrical arcing (sparks) across the split-ring commutators of dc generators. Alternators differ from the generators discussed above in the following way. In an alternator, current from the battery produces a magnetic field in an electromagnet, called the *rotor*, which is made to rotate by a belt from the engine. Surrounding the rotating rotor are a set of stationary coils called the *stator*, Fig. 29–15. The magnetic field of the rotor passes through the stator coils and, since the rotor is rotating, the field through the fixed stator coils is changing. Hence an alternating current is induced in the stator coils, which is the output. This ac output is changed to dc for charging the battery by the use of semiconductor diodes, which allow current flow in one direction only.

FIGURE 29–15 (a) Simplified schematic diagram of an alternator. The input electromagnet current to the rotor is connected through continuous slip rings. Sometimes the rotor electromagnet is replaced by a permanent magnet. (b) Actual shape of an alternator. The rotor is made to turn by a belt from the engine. The current in the wire coil of the rotor produces a magnetic field inside it on its axis that points horizontally from left to right, thus making north and south poles of the plates attached at either end. These end plates are made with triangular fingers that are bent over the coil—hence there are alternating N and S poles quite close to one another, with magnetic field lines between them as shown by the blue lines. As the rotor turns, these field lines pass through the fixed stator coils (shown on the right for clarity, but in operation the rotor rotates within the stator), inducing a current in them, which is the output.

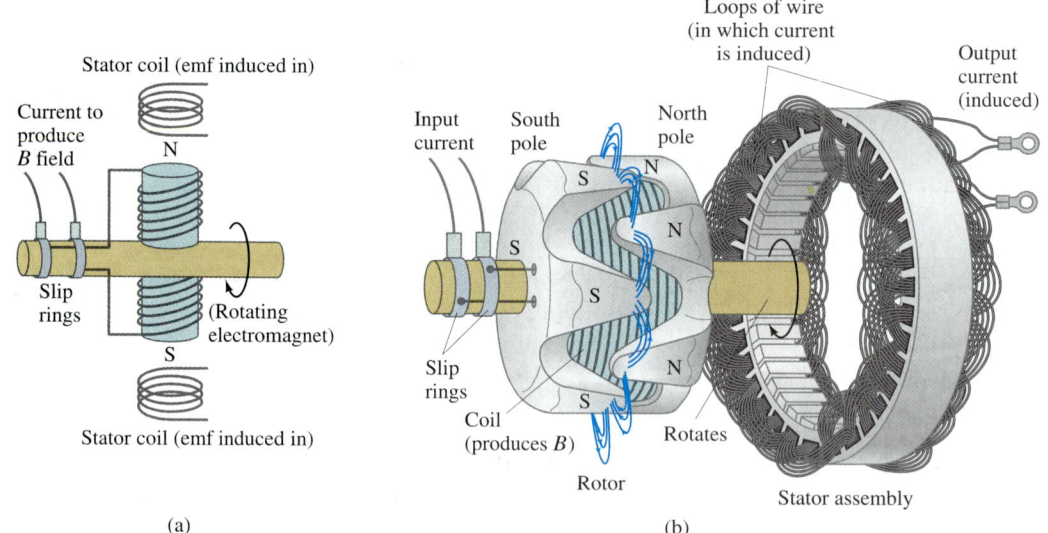

*29–5 Counter EMF and Torque; Eddy Currents

*Counter EMF

A motor turns and produces mechanical energy when a current is made to flow in it. From our description in Section 27–6 of a simple dc motor, you might expect that the armature would accelerate indefinitely due to the torque on it. However, as the armature of the motor turns, the magnetic flux through the coil changes and an emf is generated. This induced emf acts to oppose the motion (Lenz's law) and is called the **back emf** or **counter emf**. The greater the speed of the motor, the greater the counter emf. A motor normally turns and does work on something, but if there were no load, the motor's speed would increase until the counter emf equaled the input voltage. When there is a mechanical load, the speed of the motor may be limited also by the load. The counter emf will then be less than the external applied voltage. The greater the mechanical load, the slower the motor rotates and the lower is the counter emf ($\mathcal{E} \propto \omega$, Eq. 29–4).

> *Back emf*

EXAMPLE 29-6 **Counter emf in a motor.** The armature windings of a dc motor have a resistance of 5.0 Ω. The motor is connected to a 120-V line, and when the motor reaches full speed against its normal load, the counter emf is 108 V. Calculate (*a*) the current into the motor when it is just starting up, and (*b*) the current when it reaches full speed.

SOLUTION (*a*) Initially, the motor is not turning (or turning very slowly), so there is no induced counter emf. Hence, from Ohm's law, the current is

$$I = \frac{V}{R} = \frac{120 \text{ V}}{5.0 \text{ Ω}} = 24 \text{ A}.$$

(*b*) At full speed, the counter emf is a source of emf that opposes the exterior source. We represent this counter emf as a battery in the equivalent circuit shown in Fig. 29–16. In this case, Ohm's law (or Kirchhoff's rule) gives

$$120 \text{ V} - 108 \text{ V} = I(5.0 \text{ Ω}).$$

Therefore

$$I = \frac{12 \text{ V}}{5.0 \text{ Ω}} = 2.4 \text{ A}.$$

This result shows that the current can be very high when a motor first starts up. This is why the lights in your house may dim when the motor of the refrigerator (or other large motor) starts up. The large initial current causes the voltage at the outlets to drop, since the house wiring has resistance and there is some voltage drop across it when large currents are drawn.

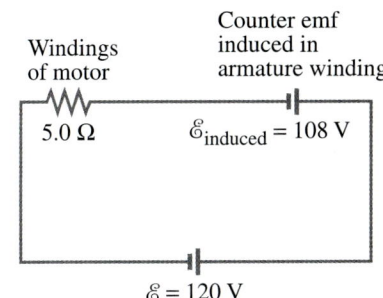

FIGURE 29–16 Circuit of a motor showing induced counter emf.

Effect of back emf on current

CONCEPTUAL EXAMPLE 29-7 **Motor overload.** When using an appliance such as a blender, electric drill, or sewing machine, if the appliance is overloaded or jammed so that the motor slows appreciably or stops while the power is still connected, the device can burn out and be ruined. Explain why this happens.

RESPONSE The motors are designed to run at a certain speed for a given applied voltage and the designer must take the expected counter emf into account. If the rotation speed is reduced, the counter emf will not be as high as expected ($\mathcal{E} \propto \omega$, Eq. 29-4), and the current will increase, and may become large enough that the windings of the motor heat up to the point of ruining the motor.

➡ **PHYSICS APPLIED**
Burning out a motor

*Counter Torque

In a generator, the situation is the reverse of that for a motor. As we saw, the mechanical turning of the armature induces an emf in the loops, which is the output. If the generator is not connected to an external circuit, the emf exists at the terminals but no current flows. In this case, it takes little effort to turn the armature. But if the generator *is* connected to a device that draws current, then a current flows in the coils of the armature. Because this current-carrying coil is in a magnetic field, there will be a torque exerted on it (as in a motor), and this torque opposes the motion (use the right-hand rule for the force on a wire, in Fig. 29–11). This is called a **counter torque**. The greater the electrical load—that is, the more current that is drawn—the greater will be the counter torque. Hence the external applied torque will have to be greater to keep the generator turning. This of course makes sense from the conservation-of-energy principle. More mechanical-energy input is needed to produce more electrical-energy output.

Counter torque

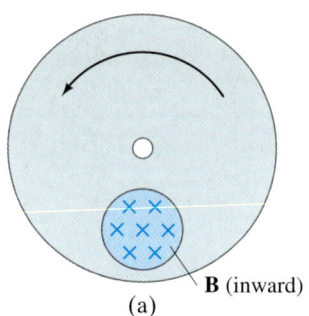
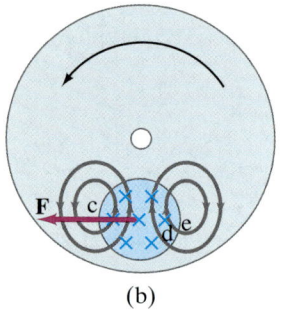

FIGURE 29-17 Production of eddy currents in a rotating wheel.

FIGURE 29-18 Repairing a step-down transformer on a utility pole.

FIGURE 29-19 Step-up transformer ($N_P = 4, N_S = 12$).

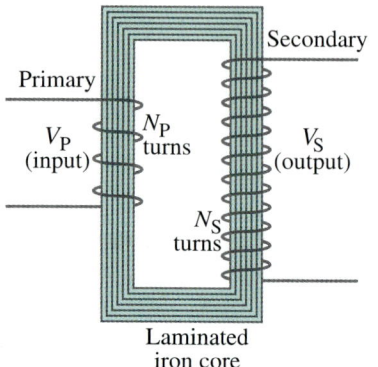

*Eddy Currents

Induced currents are not always confined to well-defined paths such as in wires. Consider, for example, the rotating metal wheel in Fig. 29–17a. A magnetic field is applied to a limited area as shown and points into the paper. The section of wheel in the magnetic field has an emf induced in it because the conductor is moving, carrying electrons with it. The flow of (conventional) current is upward in the region of the magnetic field (Fig. 29–17b), and the current follows a downward return path outside that region. Why? According to Lenz's law, the induced currents oppose the change that causes them. Consider the part of the wheel labeled c in Fig. 29–17b, where the magnetic field is zero but is just about to enter a region where **B** points into the page. To oppose this change, the induced current is counterclockwise to produce a field pointing out of the page (right-hand rule). Similarly region d is about to move to e, where **B** is zero; hence the current is clockwise to produce an inward field opposed to this change. These currents are referred to as **eddy currents** and can be present in any conductor that is moving across a magnetic field or through which the magnetic flux is changing.

In Fig. 29–17, the magnetic field exerts a force **F** on the induced currents it has created, and that force opposes the rotational motion. Eddy currents can be used in this way as a smooth braking device on, say, a rapid-transit car. In order to stop the car, an electromagnet can be turned on that applies its field either to the wheels or to the moving steel rail below. Eddy currents can also be used to dampen (reduce) the oscillation of a vibrating system. Eddy currents, however, can be a problem. For example, eddy currents induced in the armature of a motor or generator produce heat ($P = I\mathscr{E}$) and waste energy. To reduce the eddy currents, the armatures are *laminated*; that is, they are made of very thin sheets of iron that are well insulated from one another. (See Fig. 29–19 in the next Section.) Thus the total path length of the eddy currents is confined to each slab, which increases the total resistance; hence the current is less and there is less wasted energy.

29-6 Transformers and Transmission of Power

A transformer is a device for increasing or decreasing an ac voltage. Transformers are found everywhere: in TV sets to give the high voltage needed for the picture tube, in converters for plugging in a portable "Walkman," on utility poles (Fig. 29–18) to reduce the high voltage from the electric company to that usable in houses (110 V or 220 V), and in many other applications. A **transformer** consists of two coils of wire known as the **primary** and **secondary** coils. The two coils can be interwoven (with insulated wire); or they can be linked by a soft iron core which is laminated to prevent eddy-current losses (Section 29–5), as shown in Fig. 29–19. Transformers are designed so that (nearly) all the magnetic flux produced by the current in the primary also passes through the secondary coil, and we assume this is true in what follows. We also assume that energy losses in the resistance of the coils and hysteresis in the iron can be ignored—a good approximation for real transformers, which are often better than 99 percent efficient.

When an ac voltage is applied to the primary, the changing magnetic field it produces will induce an ac voltage of the same frequency in the secondary. However, the voltage will be different according to the number of loops in each coil. From Faraday's law, the voltage or emf induced in the secondary is

$$V_S = N_S \frac{d\Phi_B}{dt},$$

where N_S is the number of turns in the secondary coil, and $d\Phi_B/dt$ is the rate at which the magnetic flux changes.

The input primary voltage, V_P, is also related to the rate at which the flux changes

$$V_P = N_P \frac{d\Phi_B}{dt},$$

where N_P is the number of turns in the primary coil. This follows because the changing flux produces a counter emf, $N_P d\Phi_B/dt$, in the primary that exactly balances the applied voltage V_P if the resistance of the primary can be ignored (Kirchhoff's rules). We divide these two equations, assuming little or no flux is lost, to find

$$\frac{V_S}{V_P} = \frac{N_S}{N_P}. \qquad (29\text{-}5) \quad \textit{Transformer equation}$$

This *transformer equation* tells how the secondary (output) voltage is related to the primary (input) voltage; V_S and V_P in Eq. 29–5 can be the rms values for both, or peak values for both. (DC voltages don't work in a transformer because there would be no changing magnetic flux.)

If N_S is greater than N_P, we have a **step-up transformer**. The secondary voltage is greater than the primary voltage. For example, if the secondary has twice as many turns as the primary, then the secondary voltage will be twice that of the primary. If N_S is less than N_P, we have a **step-down transformer**.

Although ac voltage can be increased (or decreased) with a transformer, we don't get something for nothing. Energy conservation tells us that the power output can be no greater than the power input. A well-designed transformer can be greater than 99 percent efficient, so little energy is lost to heat. The power output thus essentially equals the power input. Since power $P = VI$ (Eq. 25–6), we have

$$V_P I_P = V_S I_S,$$

or

$$\frac{I_S}{I_P} = \frac{N_P}{N_S}. \qquad (29\text{-}6) \quad \textit{Transformer equation II}$$

EXAMPLE 29–8 Portable radio transformer. A transformer for home use of a portable radio reduces 120-V ac to 9.0-V ac. (Such a device also contains diodes to change the 9.0-V ac to dc.) The secondary contains 30 turns and the radio draws 400 mA. Calculate: (*a*) the number of turns in the primary; (*b*) the current in the primary; and (*c*) the power transformed.

SOLUTION (*a*) This is a step-down transformer, and from Eq. 29–5 we have

$$N_P = N_S \frac{V_P}{V_S} = \frac{(30)(120 \text{ V})}{(9.0 \text{ V})} = 400 \text{ turns}.$$

(*b*) From Eq. 29–6,

$$I_P = I_S \frac{N_S}{N_P} = (0.40 \text{ A})\left(\frac{30}{400}\right) = 0.030 \text{ A}.$$

(*c*) The power transformed is

$$P = I_S V_S = (9.0 \text{ V})(0.40 \text{ A}) = 3.6 \text{ W},$$

which is, assuming 100 percent efficiency, the same as the power in the primary, $P = (120 \text{ V})(0.030 \text{ A}) = 3.6 \text{ W}.$

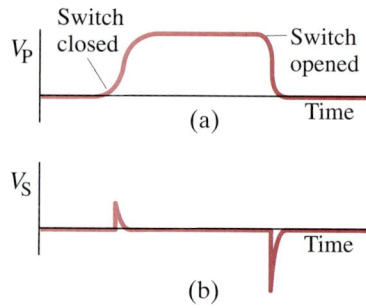

FIGURE 29–20 A dc voltage turned on and off as shown in (a) produces voltage pulses in the secondary (b). Voltage scales in (a) and (b) are not necessarily the same.

A transformer operates only on ac. A dc current in the primary does not produce a changing flux and therefore induces no emf in the secondary. However, if a dc voltage is applied to the primary through a switch, at the instant the switch is opened or closed there will be an induced current in the secondary. For example, if the dc is turned on and off as shown in Fig. 29–20a, the voltage induced in the secondary is as shown in Fig. 29–20b. Notice that the secondary voltage drops to zero when the dc voltage is steady. This is basically how, in the ignition system of an automobile, the high voltage is created to produce the spark across the gap of a spark plug that ignites the gas-air mixture. The transformer is referred to simply as the "ignition coil," and transforms the 12 V of the battery (when switched off) into a spike of as much as 25 kV.

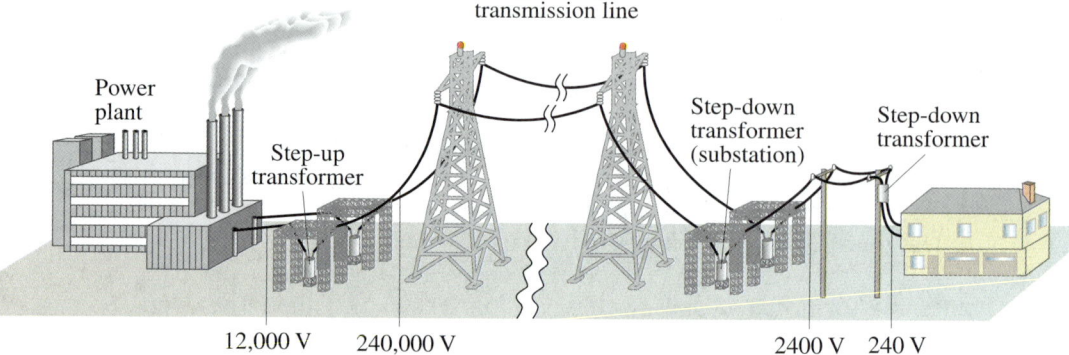

FIGURE 29–21 The transmission of electric power from power plants to homes makes use of transformers at various stages.

➡ **PHYSICS APPLIED**

Transformers help power transmission

Transformers play an important role in the transmission of electricity. Power plants are often situated some distance from metropolitan areas, so electricity must often be transmitted over long distances (Fig. 29–21). There is always some power loss in the transmission lines, and this loss can be minimized if the power is transmitted at high voltage, using transformers, as the following Example shows.

EXAMPLE 29–9 Transmission lines. An average of 120 kW of electric power is sent to a small town from a power plant 10 km away. The transmission lines have a total resistance of 0.40 Ω. Calculate the power loss if the power is transmitted at (a) 240 V and (b) 24,000 V.

SOLUTION We cannot use $P = V^2/R$ because if R is the resistance of the transmission lines, we don't know the voltage drop along them; the given voltages are applied across the lines plus the load (the town). But for each case we can determine the current I in the lines, and then find the power loss from $P = I^2R$.
(a) If 120 kW is sent at 240 V, the total current will be

$$I = \frac{P}{V} = \frac{1.2 \times 10^5 \text{ W}}{2.4 \times 10^2 \text{ V}} = 500 \text{ A}.$$

The power loss in the lines, P_L, is then

$$P_L = I^2R = (500 \text{ A})^2(0.40 \text{ Ω}) = 100 \text{ kW}.$$

Thus, over 80 percent of all the power would be wasted as heat in the power lines!

(b) If 120 kW is sent at 24,000 V, the total current will be

$$I = \frac{P}{V} = \frac{1.2 \times 10^5 \text{ W}}{2.4 \times 10^4 \text{ V}} = 5.0 \text{ A}.$$

The power loss in the lines is then

$$P_L = I^2 R = (5.0 \text{ A})^2 (0.40 \text{ }\Omega) = 10 \text{ W},$$

which is less than $\frac{1}{100}$ of 1 percent. We see that the greater the voltage, the less the current and thus the less power is wasted in the transmission lines. It is for this reason that power is usually transmitted at very high voltages, as high as 700 kV.

The great advantage of ac, and a major reason it is in nearly universal use, is that the voltage can easily be stepped up or down by a transformer. The output voltage of an electric generating plant is stepped up prior to transmission. Upon arrival in a city, it is stepped down in stages at electric substations prior to distribution. The voltage in lines along city streets is typically 2400 V and is stepped down to 240 V or 120 V for home use by transformers (Figs. 29–18 and 29–21).

29–7 A Changing Magnetic Flux Produces an Electric Field

We have seen in earlier chapters (especially Chapter 25, Section 25–8) that when an electric current flows in a wire, there is an electric field in the wire that does the work of moving the electrons in the wire. In this chapter we have seen that a changing magnetic flux induces a current in the wire, which implies that there is an electric field in the wire induced by the changing magnetic flux. Thus we come to the important conclusion that

a changing magnetic flux produces an electric field.

This result applies not only to wires and other conductors, but is actually a general result that applies to any region in space. Indeed, an electric field will be produced at any point in space where there is a changing magnetic field.

Faraday's Law—General Form

We can put these ideas into mathematical form by generalizing our relation between an electric field and the potential difference between two points a and b: $V_{ab} = \int_a^b \mathbf{E} \cdot d\mathbf{l}$ (Eq. 23–3) where $d\mathbf{l}$ is an element of displacement along the path of integration. The emf $\mathcal{E}$ induced in a circuit is equal to the work done per unit charge by the electric field, which equals the integral of $\mathbf{E} \cdot d\mathbf{l}$ around the closed path:

$$\mathcal{E} = \oint \mathbf{E} \cdot d\mathbf{l}. \tag{29-7}$$

We combine this with Eq. 29–2a, to obtain a more elegant and general form of Faraday's law

$$\oint \mathbf{E} \cdot d\mathbf{l} = -\frac{d\Phi_B}{dt} \tag{29-8}$$

FARADAY'S LAW (GENERAL FORM)

which relates the changing magnetic flux to the electric field it produces. The integral on the left is taken around a path enclosing the area through which the magnetic flux Φ_B is changing. This more elegant statement of Faraday's law (Eq. 29–8) is valid not only in conductors, but in any region of space. To illustrate this, let us take an Example.

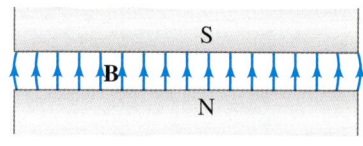

(a)

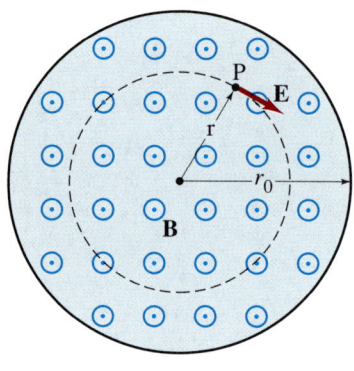

(b)

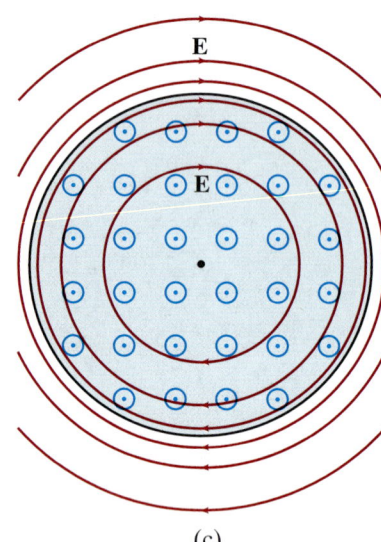

(c)

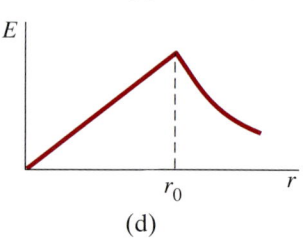

(d)

FIGURE 29–22 Example 29–10. (a) Side view of nearly constant **B**. (b) Top view, for determining the electric field **E** at point P. (c) Lines of **E** produced by increasing **B** (pointing outward). (d) Graph of E vs. r.

EXAMPLE 29–10 **E produced by changing B.** A magnetic field **B** between the pole faces of an electromagnet is nearly uniform at any instant over a circular area of radius r_0 as shown in Figs. 29–22a and b. The current in the windings of the electromagnet is increasing in time so that **B** changes in time at a constant rate dB/dt at each point. Beyond the circular region ($r > r_0$), we assume **B** = 0 at all times. Determine the electric field **E** at any point P a distance r from the center of the circular area.

SOLUTION The changing magnetic flux through a circle of radius r, shown dashed in Fig. 29–22b, will produce an emf around this circle. Because all points on the dashed circle are equivalent physically, the electric field too will show this symmetry and will be in the plane perpendicular to **B**. Thus we can expect **E** to be perpendicular to **B** and to be tangent to the circle of radius r; the direction of **E** will be as shown in Fig. 29–22b and c, since by Lenz's law the induced **E** needs to be capable of producing a current that generates a magnetic field opposing the original change in **B**. By symmetry, we also expect **E** to have the same magnitude at all points on the circle of radius r. We therefore take this circle as our path of integration in Eq. 29–8 (ignoring the minus sign so we can concentrate on magnitude since we got the direction of **E** from Lenz's law) and obtain

$$E(2\pi r) = (\pi r^2)\frac{dB}{dt}, \qquad [r < r_0]$$

since $\Phi_B = BA = B(\pi r^2)$ at any instant. We solve for E and obtain

$$E = \frac{r}{2}\frac{dB}{dt}. \qquad [r < r_0]$$

This expression is valid up to the edge of the circle ($r \leq r_0$), beyond which **B** = 0. If we now consider a point where $r > r_0$, the flux through a circle of radius r is $\Phi_B = \pi r_0^2 B$. Then Eq. 29–8 gives

$$E(2\pi r) = \pi r_0^2 \frac{dB}{dt} \qquad [r > r_0]$$

or

$$E = \frac{r_0^2}{2r}\frac{dB}{dt}. \qquad [r > r_0]$$

Thus the magnitude of the electric field increases linearly from zero at the center of the magnet to $E = (dB/dt)(r_0/2)$ at the edge, and then decreases inversely with distance in the region beyond the edge of the magnetic field. The electric field lines are circles as shown in Fig. 29–22c. A graph of E vs. r is shown in Fig. 29–22d.

*Forces Due to Changing B are Nonconservative

Example 29–10 illustrates an important difference between electric fields produced by changing magnetic fields and electric fields produced by electric charges at rest (electrostatic fields). Electric field lines produced in the electrostatic case (Chapters 21 to 24) start and stop on electric charges. But the electric field lines produced by a changing magnetic field are continuous; they form closed loops. This distinction goes even further and is an important one. In the electrostatic case, the potential difference between two points is given by

$$V_{ab} = V_a - V_b = \int_a^b \mathbf{E} \cdot d\mathbf{l}.$$

If the integral is around a closed loop, so points a and b are the same, then $V_{ab} = 0$. Hence the integral of $\mathbf{E} \cdot d\mathbf{l}$ around a closed path is zero:

$$\oint \mathbf{E} \cdot d\mathbf{l} = 0. \qquad \text{[electrostatic field]}$$

This followed from the fact that the electrostatic force (Coulomb's law) is a conservative force, and so a potential energy function could be defined. Indeed, the relation above, $\oint \mathbf{E} \cdot d\mathbf{l} = 0$, tells us that the work done per unit charge around any closed path is zero (or the work done between any two points is independent of path—see Chapter 8), which is a property only of a conservative force. But in the nonelectrostatic case, when the electric field is produced by a changing magnetic field, the integral around a closed path is *not* zero, but is given by Eq. 29–8:

$$\oint \mathbf{E} \cdot d\mathbf{l} = -\frac{d\Phi_B}{dt}.$$

We thus come to the conclusion that the forces due to changing magnetic fields are *nonconservative*. We are not able therefore to define a potential energy, or potential function, at a given point in space for the nonelectrostatic case. Although static electric fields are *conservative fields*, the electric field produced by a changing magnetic field is a **nonconservative field**.

*29–8 Applications of Induction: Sound Systems, Computer Memory, the Seismograph

There are various types of *microphones*, and many operate on the principle of induction. In one form, a microphone is just the inverse of a loudspeaker (Section 27–6). A small coil connected to a membrane is suspended close to a small permanent magnet, as shown in Fig. 29–23. The coil moves in the magnetic field when sound waves strike the membrane. The frequency of the induced emf will be just that of the impinging sound waves, and this emf is the "signal" that can be amplified and sent to loudspeakers, or sent to a tape recorder to be recorded on tape. In a "ribbon" microphone, a thin metal ribbon is suspended between the poles of a permanent magnet. The ribbon vibrates in response to sound waves, and the emf induced in the ribbon is proportional to its velocity.

Tape recording and tape playback is done by *tape heads* inside a tape recorder (or cassette deck). Recording tape for use in audio and video tape recorders contains a thin layer of magnetic oxide on a thin plastic tape. During recording, the audio or video signal voltage is sent to the recording head, which acts as a tiny electromagnet (Fig. 29–24) that magnetizes the tiny section of tape passing over the narrow gap in the head at each instant. In playback, the changing magnetism of the moving tape at the gap causes corresponding changes in the magnetic field within the soft-iron head, which in turn induces an emf in the coil (Faraday's law). This induced emf is the output signal that can be amplified and sent to a loudspeaker (or, in the case of a video signal, to the picture tube). In audio and video recorders, the signals may be *analog*—they vary continuously in amplitude over time. The variation in degree of magnetization of the tape at any point reflects the variation in amplitude of the audio or video signal.

Digital information, such as used on computer disks (floppy disks or hard disks) or on magnetic computer tape and some types of digital tape recorders, is read and written using heads that are basically the same as just described (Fig. 29–24). The essential difference is in the signals, which are not analog, but are digital, and in particular binary, meaning that only two values are possible for each of the extremely high number of predetermined spaces on the tape or disk. The two possible values are usually referred to as 1 and 0. The signal voltage does not vary continuously but rather takes on only two values, say +5 volts and 0 volts, corresponding to the 1 or 0. Thus, information is carried as a series of "bits," each of which can have only one of two values, 1 or 0.

In another field, geophysics, an important device based on electromagnetic induction is one type of *seismograph* or *geophone*. A seismograph is placed in direct contact with the Earth and converts the motion of the Earth—whether due to an earthquake or to an explosion (such as for mineral prospecting or for detecting a bomb test)—into an electrical signal. A seismograph contains a magnet and a

→ PHYSICS APPLIED
Microphones

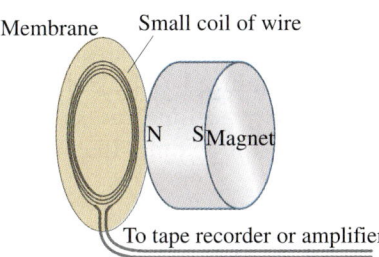

FIGURE 29–23 Diagram of a microphone that works by induction.

FIGURE 29–24 Recording and/or playback head for tape or disk. In recording (or "writing"), the electric input signal to the head, which acts as an electromagnet, magnetizes the passing tape or disk. In playback (or "reading"), the changing magnetic field of the passing tape or disk induces a changing magnetic field in the head, which in turn induces in the coil an emf that is the output signal.

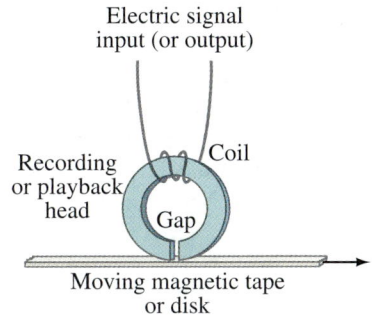

coil of wire, one of which is fixed rigidly to the case which moves as the Earth does where it is planted. The other element is inertial and is suspended from the case by a spring. In the type shown in Fig. 29–25, the coil moves with the Earth, and the relative motion of the magnet and coil produces an induced emf in the coil, which is the output of the device. In many geophones, the coil is inertial and the magnet moves with the Earth.

FIGURE 29–25 (a) A seismograph or geophone. (b) A seismograph reading (Northridge, California earthquake, January 17, 1994).

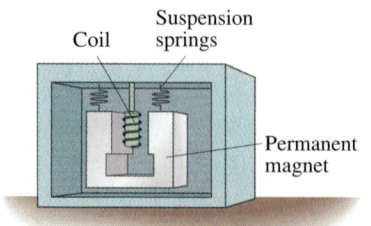
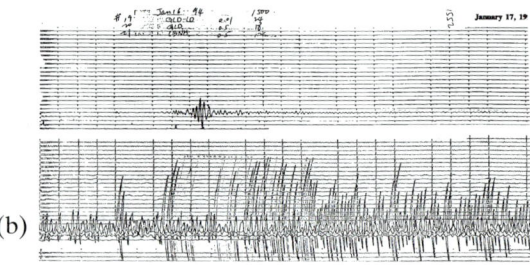

(a) (b)

Summary

The **magnetic flux** passing through a loop is equal to the product of the area of the loop times the perpendicular component of the (uniform) magnetic field: $\Phi_B = B_\perp A = BA \cos\theta$. If **B** is not uniform, then

$$\Phi_B = \int \mathbf{B} \cdot d\mathbf{A}.$$

If the magnetic flux through a coil of wire changes in time, an emf is induced in the coil. The magnitude of the induced emf equals the time rate of change of the magnetic flux through the loop times the number N of loops in the coil:

$$\mathcal{E} = -N \frac{d\Phi_B}{dt}.$$

This is **Faraday's law of induction**.

The induced emf, and the current it produces, are in a direction that opposes the change in flux which caused them (**Lenz's law**).

We can also see from Faraday's law that a straight wire of length l moving with speed v perpendicular to a magnetic field of strength B has an emf induced between its ends equal to:

$$\mathcal{E} = Blv.$$

Faraday's law also tells us that *a changing magnetic field produces an electric field*. The mathematical relation is

$$\oint \mathbf{E} \cdot d\mathbf{l} = -\frac{d\Phi_B}{dt}$$

and is the general form of Faraday's law. The integral on the left is taken around the loop through which the magnetic flux Φ_B is changing.

An electric **generator** changes mechanical energy into electrical energy. Its operation is based on Faraday's law: a coil of wire is made to rotate uniformly by mechanical means in a magnetic field, and the changing flux through the coil induces a sinusoidal current, which is the output of the generator.

A motor, which operates in the reverse of a generator, acts like a generator in that a **counter emf** is induced in its rotating coil; since this counter emf opposes the input voltage, it can act to limit the current in a motor coil.

Similarly, a generator acts somewhat like a motor in that a **counter torque** acts on its rotating coil.

A **transformer**, which is a device to change the magnitude of an ac voltage, consists of a primary and a secondary coil. The changing flux due to an ac voltage in the primary induces an ac voltage in the secondary. In a 100 percent efficient transformer, the ratio of output to input voltages (V_S/V_P) equals the ratio of the number of turns N_S in the secondary to the number N_P in the primary:

$$\frac{V_S}{V_P} = \frac{N_S}{N_P}.$$

The ratio of secondary to primary current is in the inverse ratio of turns:

$$\frac{I_S}{I_P} = \frac{N_P}{N_S}.$$

Questions

1. What would be the advantage, in Faraday's experiments (Fig. 29–1), of using coils with many turns?

2. Suppose you are holding a circular loop of wire and suddenly thrust a magnet, south pole first, toward the center of the circle. Is a current induced in the wire? Is a current induced when the magnet is held steady within the loop? Is a current induced when you withdraw the magnet? In each case, if your answer is yes, specify the direction.

3. Suppose you are looking along a line through the centers of two circular (but separate) wire loops, one behind the other. A battery is suddenly connected to the front loop, establishing a clockwise current. (a) Will a current be induced in the second loop? (b) If so, when does this current start? (c) When does it stop? (d) In what direction is this current? (e) Is there a force between the two loops? (f) If so, in what direction?

4. The battery discussed in Question 3 is disconnected. Will a current be induced in the second loop? If so, when does it start and stop? In what direction is this current?

5. Two loops of wire are moving in the vicinity of a very long straight wire carrying a steady current as shown in Fig. 29–26. Find the direction of the induced current in each loop.

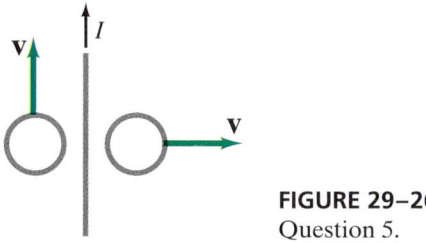

FIGURE 29–26 Question 5.

6. Is there a force between the two loops discussed in Question 5? If so, in what direction?

7. In what direction will the current flow in Fig. 29–9 if the rod moves to the left, which decreases the area of the loop to the left?

8. Some modern stove burners are based on induction. That is, an ac current passes around a coil that is the "burner," a burner that never gets hot. Explain why it will heat a metal pan but not a glass container.

9. A region where no magnetic field is desired is surrounded by a sheet of low-resistivity metal. (*a*) Will this sheet shield the interior from a rapidly changing magnetic field outside? Explain. (*b*) Will it act as a shield to a static magnetic field? (*c*) What if the sheet is superconducting (resistivity = 0)?

10. Show, using Lenz's law, that the emf induced in the moving rod in Fig. 29–9 is positive at the bottom and negative at the top so that the current flows clockwise in the circuit loop on the left.

11. What is the advantage of placing the two insulated electric wires carrying ac close together or even twisted about each other?

12. Explain why, exactly, the lights may dim briefly when a refrigerator motor starts up. When an electric heater is turned on, the lights may stay dimmed as long as it is on. Explain the difference.

* 13. Use Fig. 29–11 and the right-hand rules to show why the counter torque in a generator *opposes* the motion.

* 14. Will an eddy-current brake (Fig. 29–17) work on a copper or aluminum wheel, or must it be ferromagnetic?

* 15. It has been proposed that eddy currents be used to help sort solid waste for recycling. The waste is first ground into tiny pieces and iron removed with a dc magnet. The waste then is allowed to slide down an incline over permanent magnets. How will this aid in the separation of nonferrous metals (Al, Cu, Pb, brass) from nonmetallic materials?

* 16. The pivoted metal bar with slots in Fig. 29–27 falls much more quickly through a magnetic field than does a solid bar. Explain in detail.

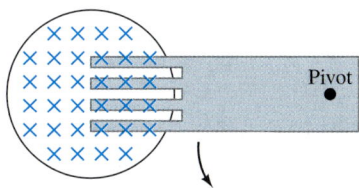

FIGURE 29–27 Question 16.

* 17. If an aluminum sheet is held between the poles of a large bar magnet, it requires some force to pull it out of the magnetic field even though the sheet is not ferromagnetic and does not touch the pole faces. Explain.

* 18. A metal bar, pivoted at its upper end, oscillates freely in the absence of a magnetic field; but in a magnetic field, its oscillations are quickly damped out. Explain. (This *magnetic damping* is used in a number of practical devices.)

19. An enclosed transformer has four wire leads coming from it. How could you determine the ratio of turns on the two coils without taking the transformer apart? How would you know which wires paired with which?

20. A transformer designed for a 120-V ac input will often "burn out" if connected to a 120-V dc source. Explain. [*Hint*: The resistance of the primary coil is usually very low.]

* 21. Since a magnetic microphone is basically like a loudspeaker, could a loudspeaker (Section 27–6) actually serve as a microphone? That is, could you speak into a loudspeaker and obtain an output signal that could be amplified? Explain. Discuss, in light of your response, how a microphone and loudspeaker differ in construction.

Problems

Sections 29–1 and 29–2

1. (I) The magnetic flux through a coil of wire containing two loops changes uniformly from -80 Wb to $+58\text{ Wb}$ in 0.72 s. What is the emf induced in the coil?

2. (I) A 26-cm-diameter circular loop of wire lies in a plane perpendicular to a 0.90-T magnetic field. It is removed from the field in 0.15 s. What is the average induced emf?

3. (I) The rectangular loop shown in Fig. 29–28 is pushed into the magnetic field which points inward as shown. In what direction is the induced current? Explain.

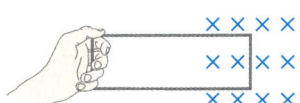

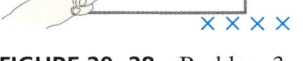

FIGURE 29–28 Problem 3.

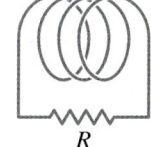

FIGURE 29–29 Problem 4.

4. (I) The north pole of the magnet in Fig. 29–29 is being inserted into the coil. In which direction is the induced current flowing through the resistor R? Explain.

5. (I) A 7.2-cm-diameter loop of wire is initially oriented perpendicular to a 1.3-T magnetic field. It is rotated so that its plane is parallel to the field direction in 0.20 s. What is the average induced emf in the loop?

6. (I) A 9.2-cm-diameter wire coil is initially oriented so that its plane is perpendicular to a magnetic field of 0.63 T pointing up. During the course of 0.15 s, the field is changed to one of 0.25 T pointing down. What is the average induced emf in the coil?

7. (II) What is the direction of the induced current in the circular loop due to the current shown in each part of Fig. 29–30? Explain.

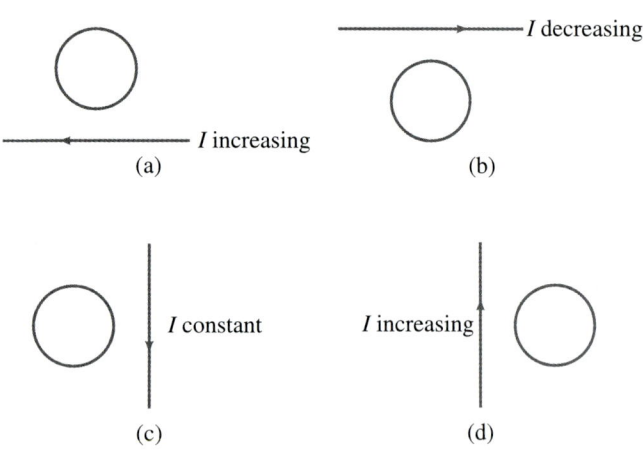

FIGURE 29–30 Problem 7.

8. (II) (a) If the resistance of the resistor in Fig. 29–31 is slowly increased, what is the direction of the current induced in the small circular loop inside the larger loop? (b) What would it be if the small loop were placed outside and to the left of the larger one? Explain.

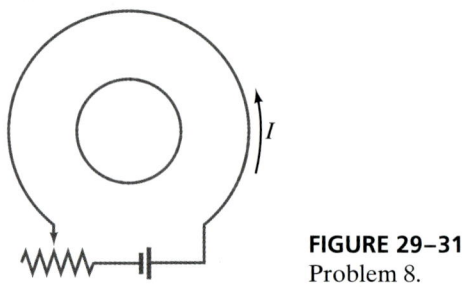

FIGURE 29–31
Problem 8.

9. (II) If the solenoid in Fig. 29–32 is being pulled away from the loop shown, in what direction is the induced current in the loop? Explain.

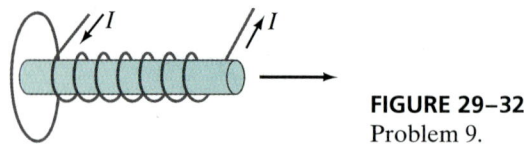

FIGURE 29–32
Problem 9.

10. (II) The magnetic field perpendicular to a circular loop of wire 20 cm in diameter is changed from +0.52 T to −0.45 T in 180 ms, where + means the field points away from an observer and − toward the observer. (a) Calculate the induced emf. (b) In what direction does the induced current flow?

11. (II) An elastic circular loop in the plane of the paper lies in a 0.75 T magnetic field pointing into the paper. If the loop's diameter changes from 20.0 cm to 6.0 cm in 0.50 s, (a) what is the direction of the induced current, (b) what is the magnitude of the average induced emf, and (c) if the loop's resistance is 2.5 Ω, what is the average induced current during the 0.50 s?

12. (II) A square loop of wire 15 cm on a side is rotated uniformly about an axis through its center and parallel to two sides. It rotates 360° in a magnetic field **B** perpendicular to the axis in 45 ms. If the rms induced emf is 70 mV, what is the average value of B?

13. (II) A 20-cm-diameter circular loop of wire has a resistance of 150 Ω. It is initially in a 0.40-T magnetic field, with its plane perpendicular to **B**, but is removed from the field in 100 ms. Calculate the electric energy dissipated in this process.

14. (II) A single rectangular loop of wire with the dimensions shown in Fig. 29–33 is situated so that part is inside a region of uniform magnetic field of 0.450 T and part is outside the field. The total resistance of the loop is 0.230 Ω. Calculate the force required to pull the loop from the field (to the right) at a constant velocity of 3.40 m/s. Neglect gravity.

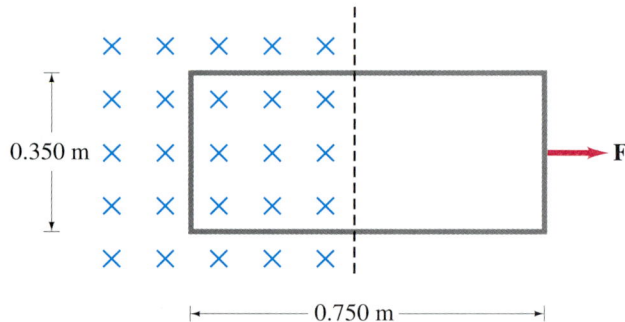

FIGURE 29–33 Problem 14.

15. (II) The magnetic field perpendicular to a single 15.6-cm-diameter circular loop of copper wire decreases uniformly from 0.550 T to zero. If the wire is 2.05 mm in diameter, how much charge moves through the coil during this operation?

16. (II) The magnetic flux through each loop of a 60-loop coil is given by $(8.8t - 0.51t^3) \times 10^{-2}$ T·m², where the time t is in seconds. (a) Determine the emf $\mathscr{E}$ as a function of time. (b) What is $\mathscr{E}$ at $t = 1.0$ s and $t = 5.0$ s?

17. (II) A 35.0-cm-diameter coil consists of 20 turns of circular copper wire 2.0 mm in diameter. A uniform magnetic field, perpendicular to the plane of the coil, changes at a rate of 3.20×10^{-3} T/s. Determine (a) the current in the loop, and (b) the rate at which thermal energy is produced.

18. (II) A single circular loop of wire is placed inside a long solenoid with its plane perpendicular to the axis of the solenoid. The area of the loop is A_1 and that of the solenoid, which has n turns per unit length, is A_2. A current $I = I_0 \cos \omega t$ flows in the solenoid turns. What is the induced emf in the small loop?

19. (II) The area of an elastic circular loop decreases at a constant rate, $dA/dt = -3.50 \times 10^{-2}$ m²/s. The loop is in a magnetic field $B = 0.48$ T whose direction is perpendicular to the plane of the loop. At $t = 0$, the loop has area $A = 0.285$ m². Determine the induced emf at $t = 0$, and at $t = 2.00$ s.

20. (II) Suppose the radius of the elastic loop in Problem 19 increases at a constant rate, $dr/dt = 7.00$ cm/s. Determine the emf induced in the loop at $t = 0$ and at $t = 1.00$ s.

21. (III) Determine the magnetic flux through a square loop of side a (Fig. 29–34) if one side is parallel to, and a distance a from, a straight wire that carries a current I.

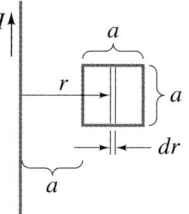

FIGURE 29–34 Problem 21.

Section 29–3

22. (I) The moving rod in Fig. 29–9 is 19.0 cm long and moves with a speed of 25.0 cm/s. If the magnetic field is 0.750 T, calculate the emf developed.

23. (II) In Fig. 29–9 the rod moves with a speed of 1.8 m/s, is 24.0 cm long, and has 2.2 Ω resistance. The magnetic field is 0.35 T and the resistance of the U-shaped conductor is 26.0 Ω at a given instant. Calculate: (a) the emf induced; (b) the current flowing in the U-shaped conductor, and (c) the external force needed to keep the rod's velocity constant at that instant.

24. (II) If the U-shaped conductor in Fig. 29–9 has resistivity ρ, whereas that of the moving rod is negligible, derive a formula for the current I as a function of time. Assume the rod has length l, starts at the bottom of the U at $t = 0$, and moves with uniform speed v in the magnetic field B. The cross-sectional area of the rod and all parts of the U is A.

25. (II) A conducting rod rests on two long frictionless parallel rails in a magnetic field **B** ($\perp$ to the rails and rod) as in Fig. 29–35. (a) If the rails are horizontal and the rod is given an initial push, will the rod travel at constant speed even though a magnetic field is present? (b) Suppose at $t = 0$, when the rod has speed $v = v_0$, the two rails are connected electrically by a wire from point a to point b. Assuming the rod has resistance R and the rails have negligible resistance, determine the speed of the rod as a function of time. Discuss your answer.

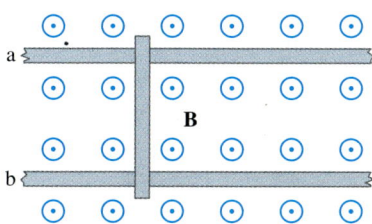

FIGURE 29–35 Problems 25 and 26.

26. (III) Suppose a conducting rod (mass m, resistance R) rests on two frictionless and resistanceless parallel rails a distance l apart in a uniform magnetic field **B** ($\perp$ to the rails and the rod) as in Fig. 29–35. At $t = 0$, the rod is at rest and a source of emf is connected to the points a and b. Determine the speed of the rod as a function of time (a) if the source puts out a constant current I, (b) the source puts out a constant emf $\mathscr{E}_0$. (c) Does the rod reach a terminal speed in either case? If so, what is it?

27. (III) A short section of wire, of length a, is moving with velocity **v**, parallel to a very long wire carrying a current I as shown in Fig. 29–36. The near end of the wire section is a distance b from the long wire. Assuming the vertical wire is very long compared to $a + b$, determine the emf between the ends of the short section. Assume **v** is (a) in the same direction as I, (b) in the opposite direction to I.

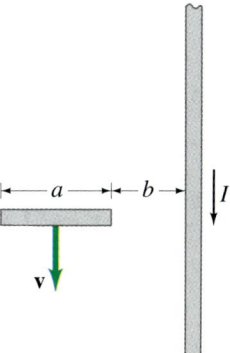

FIGURE 29–36 Problem 27.

Section 29–4

28. (I) The generator of a car idling at 950-rpm produces 12.4 V. What will the output be at a rotation speed of 2500 rpm assuming nothing else changes?

29. (I) Show that the rms output of an ac generator is $V_{\text{rms}} = NAB\omega/\sqrt{2}$.

30. (II) A simple generator has a 420-loop square coil 21.0 cm on a side. How fast must it turn in a 0.350-T field to produce a 120-V peak output?

31. (II) A 350-loop circular armature coil with a diameter of 10.0 cm rotates at 60 rev/s in a uniform magnetic field of strength 0.45 T. What is the rms voltage output of the generator? What would you do to the rotation frequency in order to double the rms voltage output?

* Section 29–5

* **32.** (I) A motor has an armature resistance of 3.75 Ω. If it draws 9.20 A when running at full speed and connected to a 120-V line, how large is the counter emf?

* **33.** (I) The counter emf in a motor is 72 V when operating at 1800 rpm. What would be the counter emf at 2500 rpm if the magnetic field is unchanged?

* **34.** (II) The counter emf in a motor is 100 V when the motor is operating at 1000 rpm. How would you change the motor's magnetic field if you wanted to reduce the counter emf to 75 V when the motor was running at 2500 rpm?

* **35.** (II) What will be the current in the motor of Example 29–6 if the load causes it to run at half speed?

* **36.** (II) A dc generator is rated at 10 kW, 200 V, and 50 A when it rotates at 1000 rpm. The resistance of the armature windings is 0.40 Ω. (a) Calculate the "no-load" voltage at 1000 rpm (when there is no circuit hooked up to the generator). (b) Calculate the full-load voltage (i.e. at 50 A) when the generator is run at 800 rpm. Assume that the magnitude of the magnetic field remains constant.

Section 29-6

37. (I) A transformer is designed to change 120 V into 8500 V, and there are 500 turns in the primary. How many turns are in the secondary, assuming 100 percent efficiency?

38. (I) A transformer has 720 turns in the primary and 120 in the secondary. What kind of transformer is this and, assuming 100 percent efficiency, by what factor does it change the voltage? By what factor does it change the current?

39. (I) A step-up transformer increases 22 V to 120 V. What is the current in the secondary as compared to the primary? Assume 100 percent efficiency.

40. (I) Neon signs require 12 kV for their operation. To operate from a 220-V line, what must be the ratio of secondary to primary turns of the transformer? What would the voltage output be if the transformer were connected backward?

41. (II) A model-train transformer plugs into 120 V ac, and draws 0.65 A while supplying 15 A to the train. (a) What voltage is present across the tracks? (b) Is the transformer step-up or step-down?

42. (II) The output voltage of a 100-W transformer is 12 V and the input current is 26 A. (a) Is this a step-up or a step-down transformer? (b) By what factor is the voltage multiplied?

43. (II) A transformer has 330 primary turns and 1510 secondary turns. The input voltage is 120 V and the output current is 15.0 A. What is the output voltage and input current assuming 100 percent efficiency?

44. (II) If 30 MW of power at 45 kV (rms) arrives at a town from a generator via 3.0-Ω transmission lines, calculate (a) the emf at the generator end of the lines, and (b) the fraction of the power generated that is lost in the lines.

45. (III) If 65 kW is to be transmitted over two 0.100-Ω lines, estimate how much power is saved if the voltage is stepped up from 120 V to 1200 V and then down again, rather than simply transmitting at 120 V. Assume the transformers are each 99 percent efficient.

46. (III) Design a dc transmission line that can transmit 300 MW of electricity 200 km with only a 2 percent loss. The wires are to be made of aluminum and the voltage is 600 kV.

Section 29-7

47. (I) Determine the electric field in the moving rod of Problem 22.

48. (II) In Fig. 29–22c, draw two identical small circles, one placed near the center of the magnetic field region and the other near its edge (but not beyond the edge). Show that the emf around each circle is the same, even though E is greater in the region of the outer circle.

49. (II) The *betatron*, a device used to accelerate electrons to high energy, consists of a circular vacuum tube placed in a magnetic field (Fig. 29–37), into which electrons are injected. The electromagnet produces a field that (1) keeps the electrons in their circular orbit inside the tube, and (2) increases the speed of the electrons when B changes. (a) Explain how the electrons are accelerated. (See Fig. 29–37.) (b) In what direction are the electrons moving in Fig. 29–37 (give directions as if looking down from above)? (c) Should B increase or decrease to accelerate the electrons? (d) The magnetic field is actually 60 Hz ac; show that the electrons can be accelerated only during $\frac{1}{4}$ of a cycle ($\frac{1}{240}$ s). (During this time they make hundreds of thousands of revolutions and acquire very high energy.)

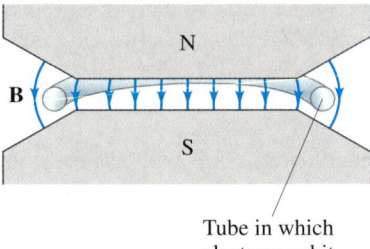

FIGURE 29–37 Problems 49 and 50.

50. (III) Show that the electrons in a betatron, Problem 49 and Fig. 29–37, are accelerated at constant radius if the magnetic field B_0 at the position of the electron orbit in the tube is equal to half the average value of the magnetic field (B_{av}) over the area of the circular orbit at each moment: $B_0 = \frac{1}{2} B_{av}$. (This is the reason the pole faces have a rather odd shape, as indicated in Fig. 29–37.)

51. (III) Find a formula for the net electric field in the moving rod of Problem 26 as a function of time for each case, (a) and (b).

General Problems

52. Suppose you are looking at two current loops in the plane of the page as shown in Fig. 29–38. When the switch is thrown in the left-hand coil, (a) what is the direction of the induced current in the other loop? (b) What is the situation after a "long" time? (c) What is the direction of the induced current in the second loop if the second loop is quickly pulled horizontally to the right?

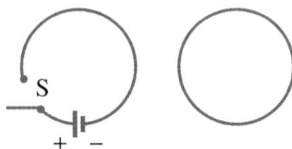

FIGURE 29–38 Problem 52.

53. A simple generator is used to generate a peak output voltage of 24.0 V. The square armature consists of windings that are 7.0 cm on a side and rotates in a field of 0.420 T at a rate of 60 rev/s. How many loops of wire should be wound on the square armature?

54. A square loop 24.0 cm on a side has a resistance of 6.50 Ω. It is initially in a 0.755-T magnetic field, with its plane perpendicular to **B**, but is removed from the field in 40.0 ms. Calculate the electric energy dissipated in this process.

55. Two conducting rails of negligible resistance 30 cm apart rest on a 5.0° ramp. They are joined at the bottom by a 0.60 Ω resistor, and at the top a copper bar of mass 0.040 kg is laid across the rails. The whole apparatus is immersed in a vertical 0.55 T field. What is the terminal (steady) velocity of the bar as it slides frictionlessly down the rails?

56. A pair of power transmission lines each have a 0.80-Ω resistance and carry 700 A over 9.0 km. If the rms input voltage is 42 kV, calculate (a) the voltage at the other end, (b) the power input, (c) power loss in the lines, and (d) the power output.

57. Power is generated at 24 kV at a generating plant which is located 100 km from a town that requires 50 MW of power at 12 kV. The transmission lines from the plant to the town have a total resistance of 0.10 Ω/km. What should the output voltage of the transformer at the generating plant be for an overall transmission efficiency of 98.5 percent, assuming a perfect transformer?

58. A coil with 150 turns, a radius of 5.2 cm and a resistance of 11.0 Ω surrounds a solenoid with 200 turns/cm and a radius of 4.5 cm. The current in the solenoid changes at a constant rate from 0 A to 2.0 A in 0.10 s. Calculate the magnitude and direction of the induced current in the coil.

59. A ring with a radius of 3.0 cm and a resistance of 0.025 Ω is rotated about an axis through its diameter by 90° in a magnetic field of 0.15 T perpendicular to that axis. What is the largest number of electrons that would flow past a fixed point in the ring as this process is accomplished?

60. Calculate the peak output voltage of a simple generator whose square armature windings are 6.60 cm on a side if the armature contains 125 loops and rotates in a field of 0.200 T at a rate of 120 rev/s.

61. A small electric car overcomes a 250-N friction force when traveling 30 km/h. The electric motor is powered by ten 12-V batteries connected in series and is coupled directly to the wheels whose diameters are 50 cm. The 300 armature coils are rectangular, 10 cm by 15 cm, and rotate in a 0.60-T magnetic field. (a) How much current does the motor draw to produce the required torque? (b) What is the back emf? (c) How much power is dissipated in the coils? (d) What percent of the input power is used to drive the car?

62. A **search coil** for measuring B (also called a *flip coil*) is a small coil with N turns, each of cross-sectional area A. It is connected to a so-called **ballistic galvanometer**, which is a device to measure the total charge Q that passes through it in a short time. The flip coil is placed in the magnetic field to be measured with its face perpendicular to the field. It is then quickly rotated 180°. Show that the total charge Q that flows in the induced current during this short "flip" time is proportional to the magnetic field B. In particular, show that B is given by

$$B = \frac{QR}{2NA}$$

where R is the total resistance of the circuit, including that of the coil and that of the ballistic galvanometer which measures the charge Q.

63. The primary windings of a transformer, which has an 80 percent efficiency, are connected to 110 V ac. The secondary windings are connected across a 2.4 Ω, 75 W lightbulb. (a) Calculate the current through the primary windings of the transformer. (b) Calculate the ratio of the number of primary windings of the transformer to the number of secondary windings of the transformer.

64. What is the energy dissipated as a function of time in a circular loop of ten turns of wire having a radius of 10.0 cm and a resistance of 2.0 Ω if the plane of the loop is perpendicular to a magnetic field given by

$$B(t) = B_0 e^{-t/\tau}$$

with $B_0 = 0.50$ T and $\tau = 0.10$ s?

65. A thin metal rod of length L rotates with angular velocity ω about an axis through one end. The rotation axis is perpendicular to the rod and is parallel to a uniform magnetic field **B**. Determine the emf developed between the ends of the rod.

66. High-intensity desk lamps are rated at 40 W but require only 12 V. They contain a transformer that converts 120 V household voltage. (a) Is the transformer step-up or step-down? (b) What is the current in the secondary when the lamp is on? (c) What is the current in the primary? (d) What is the resistance of the bulb when on?

67. Show that the power loss in transmission lines, P_L, is given by $P_L = (P_T)^2 R_L / V^2$, where P_T is the power transmitted to the user, V is the delivered voltage, and R_L is the resistance of the power lines.

* 68. The magnetic field of a "shunt-wound" dc motor is produced by field coils placed in parallel with the armature coils. Suppose that the field coils have a resistance of 36.0 Ω and the armature coils 3.00 Ω. The back emf at full speed is 105 V when the motor is connected to 115 V dc. (a) Draw the equivalent circuit for the situations when the motor is just starting and when it is running full speed. (b) What is the total current drawn by the motor at start up? (c) What is the total current drawn when the motor runs at full speed?

69. Apply Faraday's law, in the form of Eq. 29–8, to show that the static electric field between the plates of a parallel-plate capacitor cannot drop abruptly to zero at the edges, but must, in fact, fringe. Use the path shown dashed in Fig. 29–39.

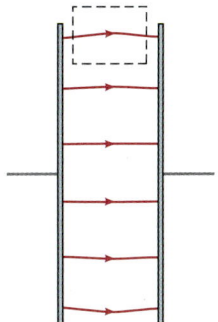

FIGURE 29–39
Problem 69.

70. A circular metal disk of radius R rotates with angular velocity ω about an axis through its center perpendicular to its face. The disk rotates in a uniform magnetic field B whose direction is parallel to the rotation axis. Determine the emf induced between the center and the edges.

71. What is the magnitude and direction of the electric field at each point in the rotating disk of Problem 70?

* 72. Suppose the "magnetic brake" of Fig. 29–17 acts on a circular metal disk of radius R and thickness d, whose electrical resistivity is ρ. The magnetic field **B**, perpendicular to the disk, acts over a small area A whose center is a distance l from the center of the wheel. At a moment when the disk is rotating with angular speed ω about an axis through its center, work out an approximate formula for the torque acting to slow it down.

A spark plug in a car receives a high voltage, which produces a high enough electric field in the air across its gap to pull electrons off the atoms in the air-gasoline mixture and form a spark. The high voltage is produced, from the basic 12 V of the car battery, by an induction coil which is basically a transformer or mutual inductance. Any coil of wire has a self-inductance, and a changing current in it causes an emf to be induced. Such inductors are useful in many circuits.

CHAPTER 30

Inductance; and Electromagnetic Oscillations

We discussed in the last chapter how a changing magnetic flux through a circuit induces an emf in that circuit. Before that we saw that an electric current produces a magnetic field. Combining these two ideas, we expect that a changing current in one circuit ought to induce an emf and a current in a second nearby circuit and even induce an emf in itself. We already saw an example in the previous chapter (transformers), but now we will treat this effect in a more general way in terms of what we will call mutual inductance and self-inductance. The concept of inductance also gives us a springboard to treat energy storage in a magnetic field. This chapter concludes with an analysis of circuits that contain inductance as well as resistance and/or capacitance.

30–1 Mutual Inductance

If two coils of wire are placed near each other, as in Fig. 30–1, a changing current in one will induce an emf in the other. According to Faraday's law, the emf $\mathcal{E}_2$ induced in coil 2 is proportional to the rate of change of flux passing through it. This flux is due to the current I_1 in coil 1, and it is often convenient to express the emf in coil 2 in terms of the current in coil 1.

We let Φ_{21} be the magnetic flux in each loop of coil 2 created by the current in coil 1. If coil 2 contains N_2 closely packed loops, then $N_2 \Phi_{21}$ is the total flux passing through coil 2. If the two coils are fixed in space, $N_2 \Phi_{21}$ is proportional to the current I_1 in coil 1; the proportionality constant is called the **mutual inductance**, M_{21}, defined by

$$M_{21} = \frac{N_2 \Phi_{21}}{I_1}. \qquad (30\text{–}1)$$

The emf $\mathcal{E}_2$ induced in coil 2 due to a changing current in coil 1 is, by Faraday's law,

$$\mathcal{E}_2 = -N_2 \frac{d\Phi_{21}}{dt}.$$

We combine this with Eq. 30–1 rewritten as $\Phi_{21} = M_{21} I_1 / N_2$ (and take its

FIGURE 30–1 A changing current in one coil will induce a current in the second coil.

Mutual inductance

derivative) and obtain

$$\mathcal{E}_2 = -M_{21}\frac{dI_1}{dt}. \quad (30\text{-}2)$$

emf due to mutual inductance

This relates the change in current in coil 1 to the emf it induces in coil 2. The mutual inductance of coil 2 with respect to coil 1, M_{21}, is a "constant" in that it does not depend on I_1; M_{21} depends on "geometric" factors such as the size, shape, number of turns, and relative positions of the two coils, and also on whether iron (or other ferromagnetic material) is present. For example, the farther apart the two coils are in Fig. 30–1, the fewer lines of flux can pass through coil 2, so M_{21} will be less. For some arrangements, the mutual inductance can be calculated (see Example 30–1). More often it is determined experimentally.

Suppose, now, we consider the reverse situation: when a changing current in coil 2 induces an emf in coil 1. In this case,

$$\mathcal{E}_1 = -M_{12}\frac{dI_2}{dt}$$

where M_{12} is the mutual inductance of coil 1 with respect to coil 2. It is possible to show, although we will not prove it here, that $M_{12} = M_{21}$. Hence, for a given arrangement we do not need the subscripts and we can let

$$M = M_{12} = M_{21},$$

so that

$$\mathcal{E}_1 = -M\frac{dI_2}{dt} \quad (30\text{-}3a)$$

and

$$\mathcal{E}_2 = -M\frac{dI_1}{dt}. \quad (30\text{-}3b)$$

The SI unit for mutual inductance is the henry (H), where $1\text{ H} = 1\text{ V}\cdot\text{s}/\text{A} = 1\,\Omega\cdot\text{s}$.

EXAMPLE 30–1 Solenoid and coil. A long thin solenoid of length l and cross-sectional area A contains N_1 closely packed turns of wire. Wrapped around it is an insulated coil of N_2 turns, Fig. 30–2. Assume all the flux from coil 1 (the solenoid) passes through coil 2, and calculate the mutual inductance.

SOLUTION We first determine the flux produced by the solenoid, all of which passes uniformly through coil N_2: The magnetic field inside the solenoid was derived in Chapter 28, Eq. 28–4:

$$B = \mu_0 \frac{N_1}{l} I_1,$$

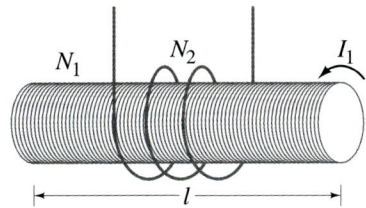

FIGURE 30–2 Example 30–1.

where I_1 is the current in the solenoid. The solenoid is close packed, so we assume all the flux in the solenoid links the coil. Then the flux Φ_{21} through coil 2 is

$$\Phi_{21} = BA = \mu_0 \frac{N_1}{l} I_1 A.$$

Hence the mutual inductance is

$$M = \frac{N_2 \Phi_{21}}{I_1} = \frac{\mu_0 N_1 N_2 A}{l}.$$

Note that we calculated M_{21}; if we had tried to calculate M_{12}, it would have been difficult. Given $M_{12} = M_{21} = M$, we can do the simpler calculation to obtain M. Note also that M depends only on geometric factors, and not on the currents.

> **PHYSICS APPLIED**
> *Transformer is a mutual inductance*

A transformer is an example of mutual inductance in which the coupling is maximized so nearly all flux lines pass through both coils. However, mutual inductance has other uses as well. For example some pacemakers, which are used to regulate the heartbeat so that blood flow is maintained in heart patients, are powered externally. Power in an external coil is transmitted via mutual inductance to a second coil in the pacemaker at the heart. This has the advantage over battery-powered pacemakers in that surgery is not needed to replace a battery when it wears out.

> *Any circuit has some mutual inductance*

Mutual inductance can sometimes be a problem, however. Any changing current in a circuit can induce an emf in another part of the same circuit or in a different circuit even though the conductors are not in the shape of a coil. The mutual inductance M is usually small unless coils with many turns and/or iron cores are involved. However, in situations where small signals are present, problems due to mutual inductance often arise. Shielded cable, in which an inner conductor is surrounded by a cylindrical grounded conductor, is often used to reduce the problem.

30–2 Self-Inductance

The concept of inductance applies also to a single isolated coil of N turns. When a changing current passes through a coil (or solenoid), a changing magnetic flux is produced inside the coil, and this in turn induces an emf in that same coil. This induced emf opposes the change in flux (Lenz's law). For example, if the current through the coil is increasing, the increasing magnetic flux induces an emf that opposes the original current and tends to retard its increase. If the current is decreasing in the coil, the decreasing flux induces an emf in the same direction as the current, thus tending to maintain the original current.

The magnetic flux Φ_B passing through the N turns of a coil is proportional to the current I in the coil, so we define the **self-inductance** L (in analogy to mutual inductance, Eq. 30–1) as

> *Self-inductance*

$$L = \frac{N\Phi_B}{I}. \qquad (30\text{–}4)$$

Then the emf $\mathcal{E}$ induced in a coil of self-inductance L is, from Faraday's law,

> *emf due to self-inductance*

$$\mathcal{E} = -N\frac{d\Phi_B}{dt} = -L\frac{dI}{dt}. \qquad (30\text{–}5)$$

Like mutual inductance, self-inductance is measured in henrys. The magnitude of L depends on the geometry and on the presence of a ferromagnetic material. Self-inductance can be defined, as above, for any circuit or part of a circuit.

Circuits always contain some inductance, but often it is quite small unless the circuit contains a coil of many turns. A coil that has significant self-inductance L is called an **inductor**. It is shown on circuit diagrams by the symbol

—⌒⌒⌒⌒—.   [inductor symbol]

It can serve a useful purpose in certain circuits. Often, inductance is to be avoided in a circuit. Precision resistors are normally wire wound and thus would have inductance as well as resistance. The inductance can be minimized by winding the insulated wire back on itself in the opposite sense so that the current going in opposite directions produces little net magnetic flux; this is called a "noninductive winding."

If an inductor has negligible resistance, it is the inductance (or induced emf) that controls a changing current. If a source of changing or alternating voltage is applied to the coil, this applied voltage will just be balanced by the induced emf of the coil given by Eq. 30–5. Thus we can see from Eq. 30–5 that, for a given $\mathscr{E}$, if the inductance L is large, the change in the current—and therefore the current itself if it is ac—will be small. The greater the inductance, the less the ac current. An inductance thus acts something like a resistance to impede the flow of alternating current. We use the term *reactance* or *impedance* for this quality of an inductor. We will discuss reactance and impedance more fully in Chapter 31, and we shall see that it depends not only on L, but also on the frequency. Here we mention one example of its importance. The resistance of the primary in a transformer is usually quite small, perhaps less than $1\,\Omega$. If resistance alone limited the current in a transformer, tremendous currents would flow when a high voltage was applied. Indeed, a dc voltage applied to a transformer can burn it out. It is the induced emf (or reactance) of the coil that limits the current to a reasonable value.

Common inductors have inductances in the range from about $1\,\mu\text{H}$ to about $1\,\text{H}$ (where $1\,\text{H} = 1\,\text{henry} = 1\,\Omega\cdot\text{s}$).

CONCEPTUAL EXAMPLE 30–2 **Direction of emf in inductor.** (a) Suppose current passes through the coil in Fig. 30–3 from left to right as shown, and is increasing with time. In which direction is the induced emf? (b) If the current passes through the coil in the same direction, but is decreasing in time, what then is the direction of the induced emf?

RESPONSE (a) From Lenz's Law we know that the induced emf must oppose the change in magnetic flux. If the current is increasing, so is the magnetic flux. The induced emf acts to oppose the increasing flux, which means it acts like a source of emf that opposes the outside source of emf driving the current. So the induced emf in the coil acts to oppose I in Fig. 30–3a. In other words the inductor might be thought of as a battery with a positive terminal at point A on the left and negative at point B on the right.
(b) If the current is decreasing, then by Lenz's Law the induced emf acts to bolster the flux—like a source of emf reinforcing the external emf. The induced emf acts to increase I in Fig. 30–3b, so in this situation you can think of the induced emf as a battery with its positive terminal on the right, at point B.

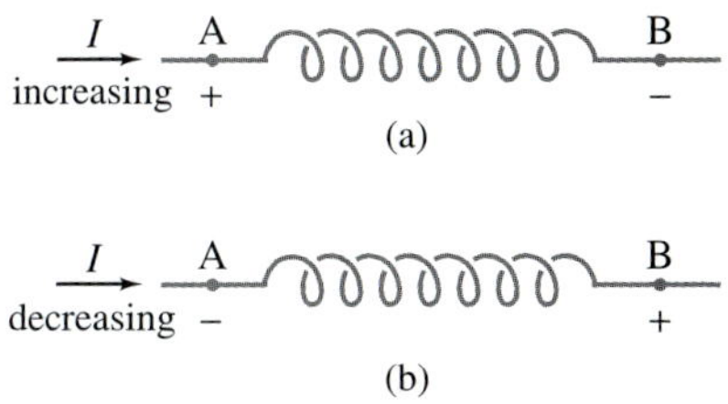

FIGURE 30–3 Example 30–2. The $+$ and $-$ signs refer to the induced emf, due to the changing current, as if points A and B were the terminals of a battery.

EXAMPLE 30–3 **Solenoid inductance.** (a) Determine a formula for the self-inductance L of a tightly wrapped solenoid (a long coil) containing N turns of wire in its length l and whose cross-sectional area is A. (b) Calculate the value of L if $N = 100$, $l = 5.0\,\text{cm}$, $A = 0.30\,\text{cm}^2$ and the solenoid is air filled. (c) Calculate L if the solenoid has an iron core with $\mu = 4000\,\mu_0$.

SOLUTION (a) To determine the inductance L, it is usually simplest to start with Eq. 30–4, so we need to first determine the flux. According to Eq. 28–4, the magnetic field inside a solenoid (ignoring end effects) is constant: $B = \mu_0 n I$ where $n = N/l$. The flux is $\Phi_B = BA = \mu_0 NIA/l$, so

$$L = \frac{N\Phi_B}{I} = \frac{\mu_0 N^2 A}{l}.$$

(b) Since $\mu_0 = 4\pi \times 10^{-7}\,\text{T}\cdot\text{m/A}$

$$L = \frac{(4\pi \times 10^{-7}\,\text{T}\cdot\text{m/A})(100)^2(3.0 \times 10^{-5}\,\text{m}^2)}{(5.0 \times 10^{-2}\,\text{m})} = 7.5\,\mu\text{H}.$$

(c) Here we replace μ_0 by $\mu = 4000\,\mu_0$ so L will be 4000 times larger: $L = 0.030\,\text{H} = 30\,\text{mH}$.

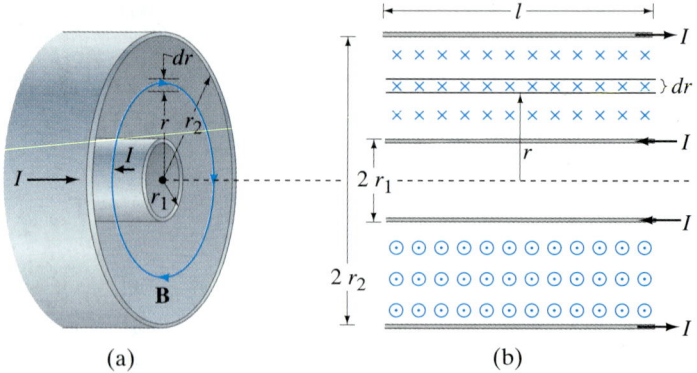

FIGURE 30–4 Example 30–4. Coaxial cable: (a) end view, (b) side view (cross-section).

EXAMPLE 30–4 Coaxial cable inductance. Determine the inductance per unit length of a coaxial cable whose inner conductor has a radius r_1 and the outer conductor has a radius r_2, Fig. 30–4. Assume the conductors are thin, so that the magnetic field within them can be ignored. The conductors carry equal currents I in opposite directions.

SOLUTION Again we need to find the magnetic flux, $\Phi_B = \int \mathbf{B} \cdot d\mathbf{A}$, this time between the conductors. The lines of $\mathbf{B}$ are circles surrounding the inner conductor (only one is shown in Fig. 30–4a). From Ampère's law, $\oint \mathbf{B} \cdot d\mathbf{l} = \mu_0 I$, the magnitude of the field at a distance r from the center, when the inner conductor carries a current I, is (see also Example 28–4):

$$B = \frac{\mu_0 I}{2\pi r}.$$

The magnetic flux through a rectangle of width dr and length l (along the cable, Fig. 30–4b), a distance r from the center, is

$$d\Phi_B = B(l\,dr) = \frac{\mu_0 I}{2\pi r} l\,dr.$$

The total flux in a length l of cable is

$$\Phi_B = \int d\Phi_B = \frac{\mu_0 I l}{2\pi} \int_{r_1}^{r_2} \frac{dr}{r} = \frac{\mu_0 I l}{2\pi} \ln \frac{r_2}{r_1}.$$

Since the current I all flows in one direction in the inner conductor, and the same current I all flows in the opposite direction in the outer conductor, we have only one turn, so $N = 1$. Hence the self-inductance for a length l is

$$L = \frac{\Phi_B}{I} = \frac{\mu_0 l}{2\pi} \ln \frac{r_2}{r_1}.$$

The inductance per unit length is

$$\frac{L}{l} = \frac{\mu_0}{2\pi} \ln \frac{r_2}{r_1}.$$

Note that L depends only on geometric factors and not on the current I.

30–3 Energy Stored in a Magnetic Field

When an inductor of inductance L is carrying a current I which is changing at a rate dI/dt, energy is being supplied to the inductor at a rate

$$P = I\mathscr{E} = LI\frac{dI}{dt}$$

where P stands for power and we used[†] Eq. 30–5. Let us calculate the work needed to increase the current in an inductor from zero to some value I. Using this

[†] No minus sign here because we are supplying power to oppose the emf of the inductor.

last equation the work dW done in a time dt is

$$dW = P\,dt = LI\,dI.$$

Then the total work done to increase the current from zero to I is

$$W = \int dW = \int_0^I LI\,dI = \tfrac{1}{2}LI^2.$$

This work done is equal to the energy U stored in the inductor when it is carrying a current I (and we take $U = 0$ when $I = 0$):

$$U = \tfrac{1}{2}LI^2. \qquad (30\text{–}6)$$

Energy stored in inductor

This can be compared to the energy stored in a capacitor, C, when the potential difference across it is V (see Section 24–4):

$$U = \tfrac{1}{2}CV^2.$$

Just as the energy stored in a capacitor can be considered to reside in the electric field between its plates, so the energy in an inductor can be considered to be stored in its magnetic field. To write the energy in terms of the magnetic field, let us use the result of Example 30–3, that the inductance of an ideal solenoid (end effects ignored) is $L = \mu_0 N^2 A / l$. Now the magnetic field B in a solenoid is related to the current I by $B = \mu_0 NI/l$. Thus

$$U = \tfrac{1}{2}LI^2 = \frac{1}{2}\left(\frac{\mu_0 N^2 A}{l}\right)\left(\frac{Bl}{\mu_0 N}\right)^2$$

$$= \frac{1}{2}\frac{B^2}{\mu_0}Al.$$

We can think of this energy as residing in the volume enclosed by the windings, which is Al. Then the energy per unit volume or *energy density* is

$$u = \text{energy density} = \frac{1}{2}\frac{B^2}{\mu_0}. \qquad (30\text{–}7)$$

Energy density in magnetic field

This formula, which was derived for the special case of a solenoid, can be shown to be valid for any region of space where a magnetic field exists. If a ferromagnetic material is present, μ_0 is replaced by μ. This equation is analogous to that for an electric field, $\tfrac{1}{2}\epsilon_0 E^2$, Eq. 24–6.

EXAMPLE 30–5 Energy stored in a coaxial cable. (*a*) How much energy is being stored per unit length in a coaxial cable whose conductors have radii r_1 and r_2 (Example 30–4, Fig. 30–4) and which carry a current I? (*b*) Where is the energy density highest?

SOLUTION (*a*) The inductance per unit length, as we saw in Example 30–4, is

$$\frac{L}{l} = \frac{\mu_0}{2\pi}\ln\frac{r_2}{r_1}.$$

We want to find the energy stored per unit length, which is

$$\frac{U}{l} = \frac{\tfrac{1}{2}LI^2}{l} = \frac{\mu_0 I^2}{4\pi}\ln\frac{r_2}{r_1}.$$

(*b*) Since $B = \mu_0 I / 2\pi r$, the field is greatest near the surface of the inner conductor, so the energy density, $u = B^2 / 2\mu_0$, will be greatest there.

30–4 LR Circuits

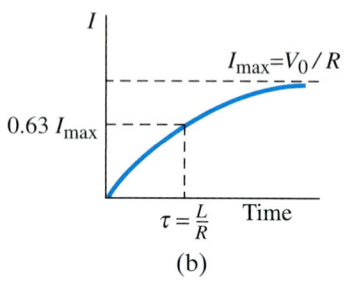

FIGURE 30–5 (a) LR circuit; (b) growth of current when connected to battery.

Any inductor will have some resistance. We represent an inductor by drawing its inductance L and its resistance R separately, as in Fig. 30–5a. The resistance R could also include a separate resistor connected in series. Now we ask, what happens when a battery or other source of dc voltage V_0 is connected in series to such an LR circuit? At the instant the switch connecting the battery is closed, the current starts to flow. The current increases starting from zero. It is opposed by the induced emf in the inductor which means point B in Fig. 30–5 is positive relative to point C. However, as soon as current starts to flow, there is also a voltage $(= IR)$ across the resistance. Hence the voltage applied across the inductance is reduced and the current increases less rapidly. The current thus rises gradually as shown in Fig. 30–5b, and approaches the steady value $I_{max} = V_0/R$ when all the voltage drop is across the resistance.

We can show this analytically by applying Kirchhoff's loop rule to the circuit of Fig. 30–5a. The emf's in the circuit are the battery voltage V_0, and the emf $\mathscr{E} = -L(dI/dt)$ in the inductor opposing the increasing current. Hence the sum of the potential changes around the loop is

$$V_0 - L\frac{dI}{dt} - IR = 0,$$

where I is the current in the circuit at any instant. We rearrange this to obtain

$$L\frac{dI}{dt} + RI = V_0. \quad (30\text{–}8)$$

This is a linear differential equation and can be integrated in the same way we did in Section 26–4 for an RC circuit. We rewrite Eq. 30–8 and then integrate:

$$\int_{I=0}^{I} \frac{dI}{V_0 - IR} = \int_0^t \frac{dt}{L}.$$

Then

$$-\frac{1}{R}\ln\left(\frac{V_0 - IR}{V_0}\right) = \frac{t}{L}$$

or,

Current in LR circuit when battery connected at $t = 0$

$$I = \frac{V_0}{R}\left(1 - e^{-t/\tau}\right) \quad (30\text{–}9)$$

where

Inductive time constant, $\tau = L/R$

$$\tau = \frac{L}{R} \quad (30\text{–}10)$$

is the **time constant** of the LR circuit. The symbol τ represents the time required for the current I to reach $(1 - 1/e) = 0.63$ or 63 percent of its maximum value (V_0/R). Equation 30–9 is plotted in Fig. 30–5b. (Compare to the RC circuit, Section 26–4.)

Now let us flip the switch in Fig. 30–5a so that the battery is taken out of the circuit, and points A and C are connected together as shown in Fig. 30–6 at the moment when the switching occurs (call it $t = 0$) and the current is I_0. Then the differential equation (Eq. 30–8) becomes (since $V_0 = 0$):

$$L\frac{dI}{dt} + RI = 0.$$

We rearrange this equation and integrate:

$$\int_{I_0}^{I} \frac{dI}{I} = -\int_0^t \frac{R}{L}dt$$

where $I = I_0$ at $t = 0$, and $I = I$ at time t.

FIGURE 30–6 The switch is flipped quickly so the battery is removed but we still have a circuit. The current at this moment (call it $t = 0$) is I_0.

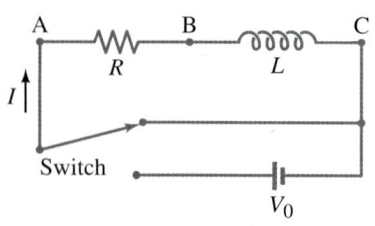

We integrate this last equation to obtain

$$\ln \frac{I}{I_0} = -\frac{R}{L} t$$

or

$$I = I_0 e^{-t/\tau} \quad (30\text{-}11)$$

where again the time constant is $\tau = L/R$. The current thus decays exponentially to zero as shown in Fig. 30–7.

This analysis shows that there is always some "reaction time" when an electromagnet, for example, is turned on or off. We also see that an LR circuit has properties similar to an RC circuit (Section 26–4). Unlike the capacitor case, however, the time constant here is *inversely* proportional to R.

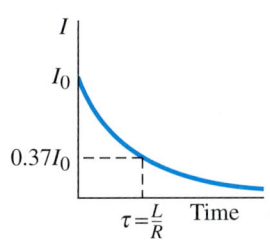

FIGURE 30–7 Decay of the current in Fig. 30–6 in time after the battery is removed from the circuit.

EXAMPLE 30–6 An *LR* circuit. At $t = 0$, a 12.0-V battery is connected in series with a 30-Ω resistor and a 220-mH inductor, as shown in Fig. 30–8. (*a*) What is the current at $t = 0$? (*b*) What is the time constant? (*c*) What is the maximum current? (*d*) How long will it take the current to reach half its maximum possible value? (*e*) At this instant, at what rate is energy being delivered by the battery, and (*f*) at what rate is energy being stored in the inductor's magnetic field?

FIGURE 30–8 Example 30–7.

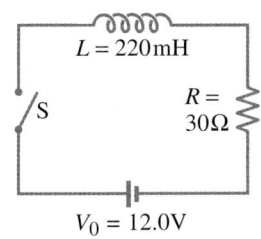

SOLUTION (*a*) The current cannot instantaneously jump from zero to some other value when the switch is closed because the inductor opposes the change ($\mathscr{E}_L = -L(dI/dt)$). Hence just after the switch is closed I is still zero at $t = 0$ and then begins to increase.
(*b*) The time constant is, from Eq. 30–10, $\tau = L/R = (0.22\,\text{H})/(30\,\Omega) = 7.3\,\text{ms}$.
(*c*) The current reaches its maximum steady value after a long time, when $dI/dt = 0$ so $I_{\max} = V_0/R = 12.0\,\text{V}/30\,\Omega = 0.40\,\text{A}$.
(*d*) We set $I = \tfrac{1}{2} I_{\max} = V_0/2R$ in Eq. 30–9, which gives us

$$1 - e^{-t/\tau} = \tfrac{1}{2}$$

or

$$e^{-t/\tau} = 1 - \tfrac{1}{2} = \tfrac{1}{2}.$$

We solve for t:

$$t = \tau \ln 2 = (7.3 \times 10^{-3}\,\text{s})(0.69) = 5.0\,\text{ms}.$$

(*e*) At this instant, $I = I_{\max}/2 = 200\,\text{mA}$, so the power being delivered by the battery is

$$P = IV = (0.20\,\text{A})(12\,\text{V}) = 2.4\,\text{W}.$$

(*f*) From Eq. 30–6, the energy stored in an inductor L at any instant is

$$U = \tfrac{1}{2} L I^2$$

where I is the current in the inductor at that instant. The *rate* at which the energy changes is

$$\frac{dU}{dt} = LI \frac{dI}{dt}.$$

We can differentiate Eq. 30–9 to obtain dI/dt, or use the differential equation, Eq. 30–8, directly:

$$\frac{dU}{dt} = I\left(L \frac{dI}{dt}\right) = I(V_0 - RI)$$

$$= (0.20\,\text{A})[12\,\text{V} - (30\,\Omega)(0.20\,\text{A})] = 1.2\,\text{W}.$$

Since only part of the battery's power is feeding the inductor at this instant, where is the rest going?

30–5 LC Circuits and Electromagnetic Oscillations

In any electric circuit, there can be three basic components: resistance, capacitance, and inductance, in addition to a source of emf. (There can also be more complex components, such as diodes or transistors.) We have previously discussed both RC and LR circuits. Now we look at an LC circuit, one that contains only a capacitance C and an inductance, L, Fig. 30–9. This is an idealized circuit in which we assume there is no resistance in the inductor; in the next Section we include resistance. Let us suppose the capacitor in Fig. 30–9 is initially charged so that one plate has charge Q_0 and the other plate has charge $-Q_0$. Suppose that at $t = 0$, the switch is closed. The capacitor immediately begins to discharge. As it does so, the current I through the inductor increases. We now apply Kirchhoff's loop rule (sum of potential changes around a loop is zero):

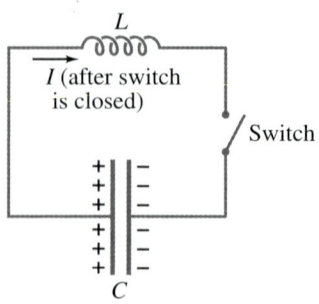

FIGURE 30–9 An LC circuit.

$$-L\frac{dI}{dt} + \frac{Q}{C} = 0.$$

Because charge leaves the positive plate on the capacitor to produce the current I as shown in Fig. 30–9, the charge Q on the (positive) plate of the capacitor is decreasing, so $I = -dQ/dt$. We can then rewrite the above equation as

$$\frac{d^2Q}{dt^2} + \frac{Q}{LC} = 0. \quad (30\text{–}12)$$

This is a familiar differential equation. It has the same form as the equation for simple harmonic motion (Chapter 14, Eq. 14–3). The solution of Eq. 30–12 is

$$Q = Q_0 \cos(\omega t + \phi) \quad (30\text{–}13)$$

where Q_0 and ϕ are constants that depend on the initial conditions. We insert Eq. 30–13 into Eq. 30–12, noting that $d^2Q/dt^2 = -\omega^2 Q_0 \cos(\omega t + \phi)$; thus

$$-\omega^2 Q_0 \cos(\omega t + \phi) + \frac{1}{LC} Q_0 \cos(\omega t + \phi) = 0$$

or

$$\left(-\omega^2 + \frac{1}{LC}\right) \cos(\omega t + \phi) = 0.$$

This relation can be true for all times t only if $(-\omega^2 + 1/LC) = 0$, which tells us that

$$\omega = 2\pi f = \sqrt{\frac{1}{LC}}. \quad (30\text{–}14)$$

LC angular frequency

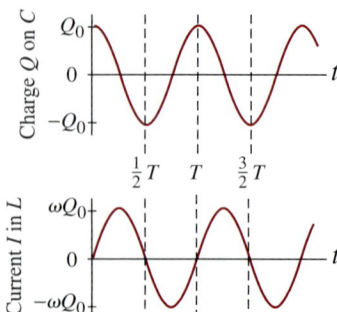

FIGURE 30–10 Charge Q and current I in an LC circuit. The period $T = \frac{1}{f} = \frac{2\pi}{\omega} = 2\pi\sqrt{LC}$.

Equation 30–13 shows that the charge on the capacitor in an LC circuit oscillates sinusoidally. The current in the inductor is

$$I = -\frac{dQ}{dt} = \omega Q_0 \sin(\omega t + \phi)$$
$$= I_0 \sin(\omega t + \phi); \quad (30\text{–}15)$$

so the current too is sinusoidal. The maximum value of I is $I_0 = \omega Q_0 = Q_0/\sqrt{LC}$. Equations 30–14 and 30–15 for Q and I when $\phi = 0$, are plotted in Fig. 30–10.

Now let us look at LC oscillations from the point of view of energy. The energy stored in the electric field of the capacitor at any time t is (see Eq. 24–5):

$$U_E = \frac{1}{2}\frac{Q^2}{C} = \frac{Q_0^2}{2C}\cos^2(\omega t + \phi).$$

The energy stored in the magnetic field of the inductor at the same instant is (Eq. 30–6)

$$U_B = \frac{1}{2}LI^2 = \frac{L\omega^2 Q_0^2}{2}\sin^2(\omega t + \phi) = \frac{Q_0^2}{2C}\sin^2(\omega t + \phi)$$

where we used Eq. 30–14. If we let $\phi = 0$, then at times $t = 0$, $t = \frac{1}{2}T$, $t = T$, and so on (where T is the period $= 1/f = 2\pi/\omega$), we have $U_E = Q_0^2/2C$ and $U_B = 0$. That is, all the energy is stored in the electric field of the capacitor. But at $t = \frac{1}{4}T, \frac{3}{4}T$, and so on, $U_E = 0$ and $U_B = Q_0^2/2C$, and so all the energy is stored in the magnetic field of the inductor. At any time t, the total energy is

$$U = U_E + U_B = \frac{1}{2}\frac{Q^2}{C} + \frac{1}{2}LI^2$$

$$= \frac{Q_0^2}{2C}\left[\cos^2(\omega t + \phi) + \sin^2(\omega t + \phi)\right] = \frac{Q_0^2}{2C}. \quad (30\text{–}16) \quad \textit{Energy}$$

Hence the total energy is constant, and energy is conserved.

What we have in this LC circuit is an **LC oscillator** or **electromagnetic oscillation**. The charge Q oscillates back and forth, from one plate of the capacitor to the other, and repeats this continuously. Likewise, the current oscillates back and forth as well. They are also energy oscillations: when Q is a maximum, the energy is all stored in the electric field of the capacitor; but when Q reaches zero, the current I is a maximum and all the energy is stored in the magnetic field of the inductor. Thus the energy oscillates between being stored in the electric field of the capacitor and in the magnetic field of the inductor. See Fig. 30–11.

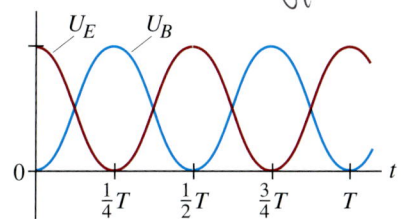

FIGURE 30–11 Energy U_E (red line) and U_B (blue line) stored in the capacitor and the inductor as a function of time. Note how the energy oscillates between electric and magnetic.

EXAMPLE 30–7 LC circuit. A 1200-pF capacitor is fully charged by a 500-V dc power supply. It is disconnected from the power supply and is connected, at $t = 0$, to a 75-mH inductor. Determine: (a) the initial charge on the capacitor; (b) the maximum current; (c) the frequency f and period T of oscillation; and (d) the total energy oscillating in the system.

SOLUTION (a) The 500-V power supply, before being disconnected, charged the capacitor to a charge of

$$Q_0 = CV = (1.2 \times 10^{-9}\,\text{F})(500\,\text{V}) = 6.0 \times 10^{-7}\,\text{C}.$$

(b) The maximum current, $I_{\max}$, is (see Eq. 30–15)

$$I_{\max} = \omega Q_0 = \frac{Q_0}{\sqrt{LC}} = \frac{(6.0 \times 10^{-7}\,\text{C})}{\sqrt{(0.075\,\text{H})(1.2 \times 10^{-9}\,\text{F})}} = 63\,\text{mA}.$$

(c) Equation 30–14 gives us the frequency:

$$f = \frac{\omega}{2\pi} = \frac{1}{(2\pi\sqrt{LC})} = 17\,\text{kHz},$$

and the period T is

$$T = \frac{1}{f} = 6.0 \times 10^{-5}\,\text{s}.$$

(d) Finally the total energy (Eq. 30–16) is

$$U = \frac{Q_0^2}{2C} = \frac{(6.0 \times 10^{-7}\,\text{C})^2}{2(1.2 \times 10^{-9}\,\text{F})} = 1.5 \times 10^{-4}\,\text{W}.$$

30–6 LC Oscillations with Resistance (LRC Circuit)

The LC circuit discussed in the previous Section is an idealization. There is always some resistance R in any circuit, and so we now discuss such a simple LRC circuit, Fig. 30–12.

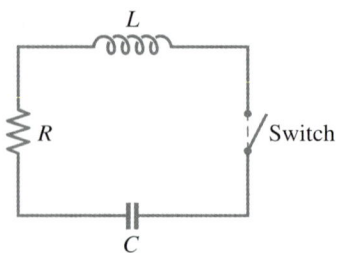

FIGURE 30–12 An LRC circuit.

Suppose again that the capacitor is initially given a charge Q_0 and the battery or other source is then removed from the circuit. The switch is closed at $t = 0$. Since there is now a resistance in the circuit, we expect some of the energy to be converted to thermal energy, and so we don't expect undamped oscillations as in a pure LC circuit. Indeed, if we use Kirchhoff's loop rule around this circuit, we obtain

$$-L\frac{dI}{dt} - IR + \frac{Q}{C} = 0,$$

which is the same equation we had in Section 30–5 with the addition of the voltage drop IR across the resistor. Since $I = -dQ/dt$, as we saw in Section 30–5, this equation becomes

$$L\frac{d^2Q}{dt^2} + R\frac{dQ}{dt} + \frac{1}{C}Q = 0. \tag{30–17}$$

This second-order differential equation in the variable Q has precisely the same form as that for the damped harmonic oscillator, Eq. 14–15:

$$m\frac{d^2x}{dt^2} + b\frac{dx}{dt} + kx = 0.$$

Hence we can analyze our LRC circuit in the same way as for damped harmonic motion, Section 14–7. Our system may undergo damped oscillations, curve A in Fig. 30–13 (underdamped system), or it may be critically damped (curve B), or overdamped (curve C), depending on the relative values of R, L, and C. Using the results of Section 14–7, with m replaced by L, b by R, and k by C^{-1}, we find that the system will be underdamped when

$$R^2 < \frac{4L}{C},$$

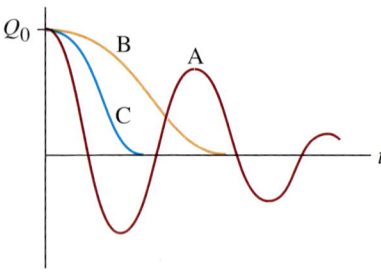

FIGURE 30–13 Charge Q on the capacitor in an LRC circuit as a function of time: curve A is for underdamped oscillation ($R^2 < 4L/C$), curve B is for critically damped, and curve C is for overdamped.

and overdamped for $R^2 > 4L/C$. Critical damping (curve B in Fig. 30–13) occurs when $R^2 = 4L/C$. If R is smaller than $\sqrt{4L/C}$, the angular frequency, ω', will be

Angular frequency (damped LC)

$$\omega' = \sqrt{\frac{1}{LC} - \frac{R^2}{4L^2}} \tag{30–18}$$

(compare to Eq. 14–20). And the charge Q as a function of time will be

$$Q = Q_0 e^{-\frac{R}{2L}t}\cos(\omega' t + \phi) \tag{30–19}$$

where ϕ is a phase constant (compare to Eq. 14–19).

Oscillators are an important element in many electronic devices: radios and television sets use them for tuning, tape recorders use them (the "bias frequency") when recording, and so on. Because some resistance is always present, electrical oscillators generally need a periodic input of power to compensate for the energy converted to thermal energy in the resistance.

EXAMPLE 30–8 Damped oscillations. At $t = 0$, a 40-mH inductor is placed in series with a resistance $R = 3.0\,\Omega$ and a charged capacitor $C = 4.8\,\mu\text{F}$. (a) Show that this circuit will oscillate. (b) Determine the frequency. (c) What is the time required for the charge amplitude to drop to half its starting value? (d) What is the current amplitude? (e) What value of R will make the circuit nonoscillating?

SOLUTION (a) To be underdamped, we must have $R^2 < 4L/C$. Since $R^2 = 9.0\,\Omega^2$ and $4L/C = 4(0.040\,\text{H})/(4.8 \times 10^{-6}\,\text{F}) = 3.3 \times 10^4\,\Omega^2$, this relation is satisfied, so the circuit will oscillate.

(b) We use Eq. 30–18:
$$f' = \frac{\omega'}{2\pi} = \frac{1}{2\pi}\sqrt{\frac{1}{LC} - \frac{R^2}{4L^2}} = 2.3 \times 10^3\,\text{Hz}.$$

(c) From Eq. 30–19, the amplitude will be half when
$$e^{-\frac{R}{2L}t} = \tfrac{1}{2}$$
or
$$t = \frac{2L}{R}\ln 2 = 18\,\text{ms}.$$

(d) We differentiate Eq. 30–19 to obtain $I = -dQ/dt$ (see comment just before Eq. 30–12), noting that $\phi = 0$ since $Q = Q_0$ at $t = 0$:
$$I = -\frac{dQ}{dt} = Q_0 e^{-\frac{R}{2L}t}\left(\frac{R}{2L}\cos\omega' t + \frac{1}{\sqrt{LC}}\sin\omega' t\right)$$
where we approximated $\omega' \approx \sqrt{1/LC}$. Because R is much less than $\sqrt{4L/C}$ (see part a above), we can ignore the $\cos\omega' t$ term, so
$$I \approx \frac{Q_0}{\sqrt{LC}} e^{-\frac{R}{2L}t} \sin\omega' t.$$
Hence the initial current amplitude is $Q_0/\sqrt{LC}$. We are not given Q_0, but if it were, say, $1.0\,\mu\text{C}$, then $I_0 = (1.0 \times 10^{-6}\,\text{C})/\sqrt{(40 \times 10^{-3}\,\text{H})(4.8 \times 10^{-6}\,\text{F})} = 2.3\,\text{mA}$.

(e) To make the circuit critically damped or overdamped, we must use the criterion $R^2 \geq 4L/C = 3.3 \times 10^4\,\Omega^2$. Hence we must have $R \geq 180\,\Omega$.

Summary

A changing current in a coil of wire will induce an emf in a second coil placed nearby. The **mutual inductance**, M, is defined as the proportionality constant between the induced emf $\mathscr{E}_2$ in the second coil and the time rate of change of current in the first:
$$\mathscr{E}_2 = -M\, dI_1/dt.$$
We can also write M as
$$M = \frac{N\Phi_B}{I}$$
where Φ_B is the magnetic flux through one coil (or circuit) with N loops produced by the current I in a second coil or circuit.

Within a single coil, a changing current induces an opposing emf, $\mathscr{E}$, so a coil has a **self-inductance** L defined by
$$\mathscr{E} = -L\, dI/dt.$$
This induced emf acts as an *impedance* to the flow of an alternating current. We can also write L as
$$L = N\frac{\Phi_B}{I}$$
where Φ_B is the flux through the inductance when a current I flows in its N loops.

When the current in an inductance L is I, the energy stored in the inductance is given by
$$U = \tfrac{1}{2}LI^2.$$
This energy can be thought of as being stored in the magnetic field of the inductor. The energy density u in any magnetic field B is given by
$$u = \tfrac{1}{2}\frac{B^2}{\mu},$$
where μ is the magnetic permeability in that region.

When an inductance L and resistor R are connected in series to a source of emf, V_0, the current rises according to an exponential of the form
$$I = \frac{V_0}{R}\left(1 - e^{-t/\tau}\right),$$
where
$$\tau = L/R$$
is the time constant. The current eventually levels out at $I = V_0/R$. If the battery is suddenly switched out of the **LR circuit**, and the circuit remains complete, the current drops exponentially, $I = I_0 e^{-t/\tau}$, with the same time constant τ.

The current in a pure **LC circuit** (or charge on the capacitor) would oscillate sinusoidally. The energy too would oscillate back and forth between electric and magnetic, from the capacitor to the inductor, and back again. The current in a series LRC circuit (or charge on the capacitor), in which the capacitor at some instant is charged, can undergo damped oscillations or undergo critically damped or overdamped oscillations.

Questions

1. In situations where a small signal must travel over a distance, a "shielded cable" is used in which the signal wire is surrounded by an insulator and then enclosed by a cylindrical conductor. Why is a "shield" necessary?
2. What is the advantage of placing the two electric wires carrying ac close together?
3. The primary of a transformer on a telephone pole has a resistance of $0.10\ \Omega$ and the input voltage is 2400 V ac. Can you estimate the current that will flow? Will it be 24,000 A? Explain.
4. A transformer designed for a 120-V ac input will often "burn out" if connected to a 120-V dc source. Explain. (*Hint:* The resistance of the primary coil is usually very low.)
5. How would you arrange two flat circular coils so that their mutual inductance was (*a*) greatest, (*b*) least (without separating them by a great distance)?
6. If the two coils in Fig. 30–1 were connected electrically, would there still be mutual inductance?
7. Suppose the second coil of N_2 turns in Fig. 30–2 were moved so it was near the end of the solenoid. How would this affect the mutual inductance?
8. Would two circuits with mutual inductance also have self-inductance? Explain.
9. Is the self-inductance per unit length of a solenoid greater near its center or near its ends?
10. Is the energy density greatest near the ends of a solenoid or near its center?
11. If you are given a fixed length of wire, how would you shape it to obtain the greatest self-inductance? The least?
12. In a battery, when the current is in the same direction as the emf, the energy of the battery decreases, whereas if the current is in the opposite direction, the energy of the battery increases (as in charging a battery). Is this also true for an inductor?
13. Two solenoids have the same length and circular cross-sectional area. Both consist of tightly packed turns of wire, but one uses thicker wire than the other. Which has the greater inductance? Which has the greater inductive time constant?
14. Does the emf of the battery in Fig. 30–5a affect the time needed for the *LR* circuit to reach (*a*) a given fraction of its maximum possible current, (*b*) a given value of current?
15. A circuit with large inductive time constant carries a steady current. If a switch is opened, there can be a very large (and sometimes dangerous) spark or "arcing over." Explain.
16. At the instant the battery is connected into the *LR* circuit of Fig. 30–5a, the emf in the inductor has its maximum value even though the current is zero. Explain.
17. Explain physically why we might expect the time constant of an *LR* circuit to be proportional to *L* and inversely proportional to *R*, Eq. 30–10.
18. Explain how a solenoid whose ends are connected together can by itself oscillate like an *LC* circuit. Where does the capacitance come from?
19. What keeps an *LC* circuit oscillating even after the capacitor has discharged completely?
20. Is the current in the inductor always the same as that in the resistor of the *LRC* circuit of Fig. 30–12?

Problems

Section 30–1

1. (I) Suppose the second coil of Example 30–1 (Fig. 30–2) has twice the diameter of the solenoid, but is still concentric with it. What then will be the mutual inductance? Assume the solenoid is very long.
2. (II) A 2.44-m-long coil containing 300 loops is wound on an iron core (average $\mu = 2000\mu_0$) along with a second coil of 100 loops. The loops of each coil have a radius of 2.00-cm. If the current in the first coil drops uniformly from 12.0 A to zero in 98.0 ms, determine: (*a*) the mutual inductance *M*; (*b*) the emf induced in the second coil.
3. (II) Determine the mutual inductance per unit length between two long solenoids, one inside the other, whose radii are r_1 and $r_2\,(r_2 < r_1)$ and whose turns per unit length are n_1 and n_2.
4. (III) A small thin coil with N_2 loops, each of area A_2, is placed inside a long solenoid, near its center. The solenoid has N_1 loops in its length l and has area A_1. Determine the mutual inductance as a function of θ, the angle between the plane of the small coil and the axis of the solenoid.
5. (III) A long straight wire and a small rectangular wire loop lie in the same plane, Fig. 30–14. Determine the mutual inductance in terms of l_1, l_2, and *w*. Assume the wire is very long compared to l_1, l_2 and *w*, and that the rest of its circuit is very far away compared to l_1, l_2 and *w*.

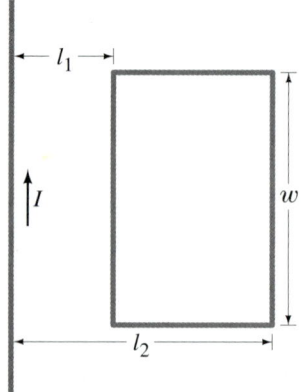

FIGURE 30–14 Problem 5.

Section 30–2

6. (I) If the current in a 180-mH coil changes steadily from 20.0 mA to 38.0 mA in 340 ms, what is the induced emf?

7. (I) Estimate the inductance L of a 0.45-m-long air-filled coil 3.7 cm in diameter containing 20,000 loops.

8. (I) What is the inductance of a coil if it produces an emf of 8.50 V when the current in it changes uniformly from -22.0 mA to $+23.0$ mA in 21.0 ms?

9. (I) What is the inductance of a coaxial cable 22.0 m long if its inner and outer conductors have diameters of 2.0 mm and 3.5 mm?

10. (I) A 35-V emf is induced in a 150-mH coil by a current that rises uniformly from zero to I_0 in 3.0 ms. What is the value of I_0?

11. (II) If the outer conductor of a coaxial cable has radius 3.0 mm, what should be the radius of the inner conductor so that the inductance per unit length does not exceed 40 nH per meter?

12. (II) (a) Show that if two circuits, such as the coils in Fig. 30–1, carry currents I_1 and I_2, the magnetic flux through each is $\Phi_1 = L_1 I_1 + MI_2$ and $\Phi_2 = L_2 I_2 + MI_1$. (b) Determine a formula for the emf induced in each coil in terms of the rate of change of current in the two coils.

13. (II) The wire of a tightly wound solenoid is unwound and used to make another tightly wound solenoid of 3.0 times the diameter. By what factor does the inductance change?

14. (II) A coil has 2.70-Ω resistance and 0.418-H inductance. If the current is 5.00 A and is increasing at a rate of 4.50 A/s, what is the potential difference across the coil at this moment?

15. (II) Ignoring any mutual inductance, what is the equivalent inductance of two inductors connected (a) in series, (b) in parallel?

16. (II) A toroid has a rectangular cross-section as shown in Fig. 30–15. Show that the self-inductance is

$$L = \frac{\mu_0 N^2 h}{2\pi} \ln \frac{r_2}{r_1}$$

where N is the total number of turns and r_1, r_2, and h are the dimensions shown in Fig. 30–15. [*Hint*: Use Ampère's law to get B as a function of r inside the toroid, and integrate.]

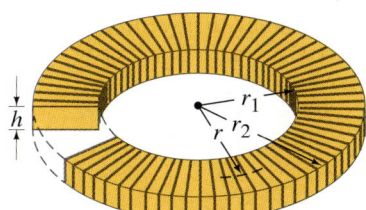

FIGURE 30–15 Problems 16 and 22. A toroid of rectangular cross-section, with N turns carrying a current I.

Section 30–3

17. (I) The magnetic field inside an air-filled solenoid 32.0 cm long and 2.10 cm in diameter is 0.600 T. Approximately how much energy is stored in this field?

18. (I) How much energy is stored in a 400-mH inductor at an instant when the current is 9.0 A?

19. (I) Typical large values for electric and magnetic fields attained in laboratories are about 1.0×10^4 V/m and 2.0 T. (a) Determine the energy density for each field and compare. (b) What magnitude electric field would be needed to produce the same energy density as the 2.0-T magnetic field?

20. (II) What is the energy density at the center of a circular loop of wire carrying a 30-A current if the radius of the loop is 28.0 cm?

21. (II) Calculate the magnetic and electric energy densities at the surface of a 3.0-mm-diameter copper wire carrying a 25-A current.

22. (II) For the toroid of Fig. 30–15, determine the energy density in the magnetic field as a function of $r(r_1 < r < r_2)$ and integrate this over the volume to obtain the total energy stored in the toroid, which carries a current I in each of its N loops.

23. (II) Determine the total energy stored per unit length in the magnetic field between the coaxial cylinders of a coaxial cable by using Eq. 30–7 for the energy density and integrating over the volume. Compare your answer to that obtained in Example 30–5.

Section 30–4

24. (II) After how many time constants does the current in Fig. 30–5 reach within (a) 10 percent, (b) 1.0 percent, and (c) 0.10 percent of its maximum value?

25. (II) How many time constants does it take for the potential difference across the resistor in an LR circuit like that in Fig. 30–6 to drop to 1.0 percent of its original value?

26. (II) In Example 30–6, where is the power delivered by the battery going besides into the inductor? Do a calculation to show that energy is conserved.

27. (II) Determine dI/dt at $t = 0$ (when the battery is connected) for the circuit of Fig. 30–5a and show that if I continued to increase at this rate, it would reach its maximum value in one time constant.

28. (II) It takes 2.56 ms for the current in an LR circuit to increase from zero to half its maximum value. Determine (a) the time constant of the circuit, (b) the resistance of the circuit if $L = 310$ H.

29. (II) (a) Determine the energy stored in the inductor L as a function of time for the LR circuit of Fig. 30–5a. (b) After how many time constants does the stored energy reach 99 percent of its maximum value?

30. (II) In the circuit of Fig. 30–16, determine the current in each resistor (I_1, I_2, I_3) at the moment (a) the switch is closed, (b) a long time after the switch is closed. After the switch has been closed a long time, and reopened, what is each current (c) just after it is opened, (d) after a long time?

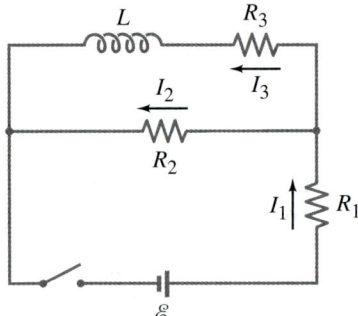

FIGURE 30–16
Problem 30.

Section 30–5

31. (I) The variable capacitor in the tuner of an AM radio has a capacitance of 1800 pF when the radio is tuned to a station at 550 kHz. (a) What must be the capacitance for a station at 1600 kHz? (b) What is the inductance (assumed constant)?

32. (I) (a) If the initial conditions of an LC circuit were $I = I_0$ and $Q = 0$ at $t = 0$, write Q as a function of time. (b) Practically, how could you set up these initial conditions?

33. (I) Use the definitions of the farad and henry to show that $1/\sqrt{LC}$ has units of s^{-1}.

34. (II) A 760-pF capacitor is charged to 135 V and then quickly connected to a 175-mH inductor. Determine (a) the frequency of oscillation, (b) the peak value of the current, and (c) the maximum energy stored in the magnetic field of the inductor.

35. (II) At $t = 0$, $Q = Q_0$ and $I = 0$ in an LC circuit. At the first moment when the energy is shared equally by the inductor and the capacitor, (a) what is the charge on the capacitor? (b) How much time has elapsed (in terms of the period T)?

Section 30–6

36. (II) In an oscillating LRC circuit, how much time does it take for the energy stored in the fields of the capacitor and inductor to fall to half its initial value? (See Fig. 30–12; assume $R \ll \sqrt{4L/C}$.)

37. (II) A damped LC circuit loses 5.5 percent of its electromagnetic energy per cycle to thermal energy. If $L = 65$ mH and $C = 1.00$ μF, what is the value of R?

38. (II) (a) Obtain a differential equation for the current I in the LRC circuit of Fig. 30–12 by differentiating Eq. 30–17. (b) Obtain a general solution for I, assuming the circuit is connnected at $t = 0$ with charge Q_0 on the capacitor. Assume $R \ll \sqrt{4L/C}$. (c) Compare your result to part (d) of Example 30–8, and note any differences. (d) What complications would arise if R were not much less than $\sqrt{4L/C}$, say $R \approx \sqrt{L/C}$?

39. (II) Starting with Eq. 30–19, with $\phi = 0$, show that the current I in a lightly damped LRC circuit is given by

$$I \approx \frac{Q_0}{\sqrt{LC}} e^{-\frac{R}{2L}t} \sin(\omega' t + \delta)$$

where

$$\delta = \tan^{-1} \frac{R}{2L\omega'}.$$

40. (II) Show that the phase constant ϕ in Eq. 30–19 is given by

$$\phi = -\cot^{-1}\left(\frac{4L}{R^2 C}\right)^{\frac{1}{2}}$$

if $I = 0$ at $t = 0$ in the circuit of Fig. 30–12. Assume that $R^2 \ll 4L/C$.

41. (III) How much resistance must be added to a pure LC circuit ($L = 300$ mH, $C = 1800$ pF) to change the oscillator's frequency by 0.10 percent? Will it be increased or decreased?

42. (III) Suppose a battery of fixed voltage V_0 is connected in the LRC circuit of Fig. 30–12. At $t = 0$, the switch is closed. Assuming this is a very overdamped circuit ($R^2 \gg 4L/C$), (a) determine the current as a function of time. (b) Plot I vs. t. (c) Compare your result to a pure RC circuit, for which $L = 0$. Real RC circuits always have some inductance, so the result of this Problem is more realistic than a pure RC circuit.

General Problems

43. An air-filled cylindrical inductor has 3000 turns, and it is 2.5 cm in diameter and 28.2 cm long. Ignoring end effects, what is its inductance? How many turns would you need to generate the same inductance if the core were iron-filled instead? Assume the magnetic permeability of iron is about 1000 times that of free space.

44. At $t = 0$, the current through a 60.0 mH inductor is 50.0 mA and is increasing at the rate of 100 mA/s. What is the initial energy stored in the inductor, and how long does it take for the energy to increase by a factor of 10 from the initial value?

45. A 3000-pF capacitor is charged to 120 V and then quickly connected to an inductor. The frequency of oscillation is observed to be 20 kHz. Determine (a) the inductance, (b) the peak value of the current, and (c) the maximum energy stored in the magnetic field of the inductor.

46. Estimate the mutual inductance M between two small loops, of radius r_1 and r_2, which are separated by a distance l that is large compared to r_1 and r_2, Fig. 30–17. Give M as a function of θ, the angle between the planes of the two coils. Assume the line joining their centers is perpendicular to the plane of coil 1.

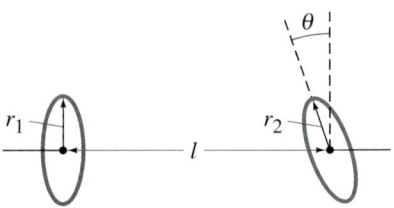

FIGURE 30–17 Problem 46.

47. Approximately how many turns does an air-filled coil have if it is 2.2 cm in diameter and 17.0 cm long and its inductance is 25 mH? How many turns are needed if it has an iron core and $\mu = 10^3 \mu_0$?

48. (a) Show that the self-inductance L of a toroid (Fig. 30–18) of radius r_0 containing N loops each of diameter d is

$$L \approx \frac{\mu_0 N^2 d^2}{8 r_0}$$

if $r_0 \gg d$. Assume the field is uniform inside the toroid; is this actually true? Is this result consistent with L for a solenoid? Should it be? (b) Calculate the inductance L of a large toroid if the diameter of the coils is 2.0 cm and the diameter of the whole ring is 50 cm. Assume the field inside the toroid is uniform. There are a total of 550 loops of wire.

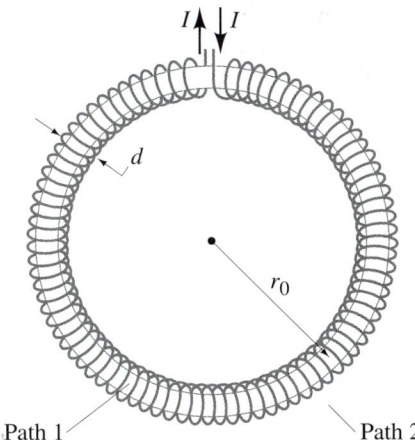

FIGURE 30–18 A toroid, Problem 48.

49. A pair of straight parallel thin wires, such as a lamp cord, each of radius r, are a distance l apart and carry current to a circuit some distance away. Ignoring the field within each wire, show that the inductance per unit length is $(\mu_0/\pi) \ln[(l-r)/r]$.

50. The potential difference across a given coil is 2.55 V at an instant when the current is 360 mA and is changing at a rate of 340 mA/s. At a later instant, the potential difference is 1.82 V while the current is 420 mA and is decreasing at a rate of 180 mA/s. Determine the inductance and resistance of the coil.

51. (a) Show that if two thin coils (inductance L_1 and L_2) are connected in series and placed close to each other, the net self-inductance is

$$L = L_1 + L_2 \pm 2M$$

where M is their mutual inductance. Explain the $\pm$ sign. (b) How can M be made zero (or nearly so)? (c) Determine the net inductance if the two coils are connected in parallel, assuming the mutual inductance is ignorable. If M cannot be ignored, how would it affect your answer?

52. Assuming the Earth's magnetic field averages about 0.50×10^{-4} T near the surface of the Earth, estimate the total energy stored in this field in the first 10 km above the Earth's surface.

53. Show that the fraction of electromagnetic energy lost (to thermal energy) per cycle in a lightly damped $(R^2 \ll 4L/C)$ LRC circuit is approximately

$$\frac{\Delta U}{U} = \frac{2\pi R}{L\omega} = \frac{2\pi}{Q}.$$

The quantity Q, defined as $Q = L\omega/R$, is called the *Q-value*, or *quality factor*, of the circuit and is a measure of the damping present. A high Q-value means smaller damping and less energy input required to maintain oscillations.

54. Two tightly wound solenoids have the same length and circular cross-sectional area. But solenoid 1 uses wire that is half as thick as solenoid 2. (a) What is the ratio of their inductances? (b) What is the ratio of their inductive time constants (assuming no other resistance in the circuits)?

55. (a) For an underdamped LRC circuit, determine a formula for the energy $U = U_E + U_B$ stored in the electric and magnetic fields as a function of time. Give in terms of the initial charge Q_0 on the capacitor. (b) Show how dU/dt is related to the rate energy is transformed in the resistor, I^2R.

56. An electronic device needs to be protected against sudden surges in current. In particular, after the power is turned on the current should rise to no more than 7.5 mA in the first 100 μs. The device has resistance 150 Ω and is designed to operate at 50 mA. How would you protect this device?

57. The circuit shown in Fig. 30–19a can integrate (in the calculus sense) the input voltage V_{in}, if the time constant L/R is large compared with the time during which V_{in} varies. Explain how this integrator works and sketch its output for the square wave signal input shown in Fig. 30–19b. [*Hint*: Write Kirchhoff's loop rule for the circuit. Multiply each term in this differential equation (in I) by a factor $e^{Rt/L}$ to make it easier to integrate.]

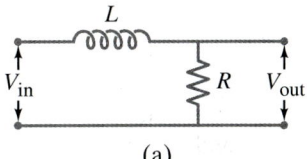

(a)

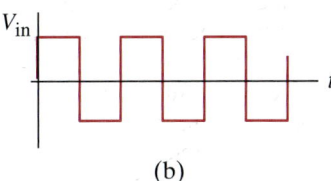

(b)

FIGURE 30–19 Problem 57.

A stereo system is a very complex ac circuit. The power source for home stereo is ac. The audio signal is ac at sound frequencies. The circuits inside a stereo contain resistors, capacitors, and inductors, plus other more complicated elements like diodes and transistors. A loudspeaker also contains an ac circuit, called a crossover (we'll look at this too in this chapter) to divide the signal between woofer and tweeters.

CHAPTER 31

AC Circuits

In earlier chapters we discussed circuits that contain combinations of resistor, capacitor, and inductor (or all three), but only when they are connected to a dc source of emf or to no source as in the discharge of a capacitor in an RC circuit or oscillation of an LC or LRC circuit. See Sections 26–4, 30–4, 30–5, and 30–6. Now we discuss these circuit elements when connected to a source of alternating voltage that produces an alternating current (ac). Such **ac circuits** are important first of all because the output of most generators (Section 29–4) is sinusoidal and most electricity generated and transmitted over wires is ac. Secondly, any voltage that varies in time, no matter how complex it is, can be written as a sum of sine and cosine terms of different frequencies in a Fourier series. Thus, the response of resistors, capacitors, and inductors to a sinusoidal source of emf is of basic importance.

31–1 Introduction: AC Circuits

We briefly discussed alternating currents in Section 25–7, and saw there that for sinusoidally varying currents and voltages, the rms and peak values are related by

$$V_{\rm rms} = \frac{V_0}{\sqrt{2}}, \qquad I_{\rm rms} = \frac{I_0}{\sqrt{2}}.$$

We now examine how a resistor, a capacitor, and an inductor behave when connected to a source of alternating voltage, whose symbol is

[symbol for ac source]

which produces a sinusoidal voltage of frequency f.

We first look at the R, C, and L separately, and then later all of them together. In each case we assume the voltage gives rise to a current

$$I = I_0 \sin 2\pi f t = I_0 \sin \omega t \tag{31-1}$$

where $\omega = 2\pi f$.

31-2 AC Circuit Containing only Resistance R

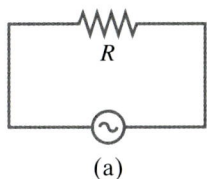

When an ac source is connected to a resistor as in Fig. 31–1a, the current increases and decreases with the alternating voltage, and Kirchhoff's loop rule tells us $V - IR = 0$. Hence

$$V = I_0 R \sin \omega t = V_0 \sin \omega t$$

where $V_0 = I_0 R$ is the peak voltage. Figure 31–1b shows the voltage (red curve) and the current (blue curve). Because the current is zero when the voltage is zero and the current reaches a peak when the voltage does, we say that the current and voltage are **in phase**. Energy is transformed into heat (Section 25–7), at an average rate

$$\overline{P} = \overline{IV} = I_{\text{rms}}^2 R = V_{\text{rms}}^2/R.$$

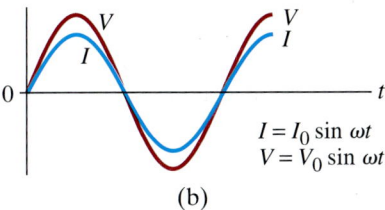

FIGURE 31–1 Resistor connected to ac source. Current is in phase with the voltage across a resistor.

31-3 AC Circuit Containing only Inductance L

In Fig. 31–2a an inductor of inductance L, represented by the symbol —⁓⁓⁓—, is connected to the ac source. We ignore any resistance it might have (it is usually small). The voltage applied to the inductor will be equal to the emf generated in the inductor by the changing current as given by Eq. 30–5. This is because the sum of the electric potential changes around any closed circuit must add up to zero, as Kirchhoff's rule tells us. Thus

$$V - L\frac{dI}{dt} = 0$$

or

$$V = L\frac{dI}{dt} = \omega L I_0 \cos \omega t. \quad (31\text{-}2)$$

Using the identity $\cos \theta = \sin(\theta + 90°)$, we can write

$$V = \omega L I_0 \sin(\omega t + 90°) = V_0 \sin(\omega t + 90°) \quad (31\text{-}3\text{a})$$

where

$$V_0 = I_0 \omega L \quad (31\text{-}3\text{b})$$

is the peak voltage. The current I and voltage V across the inductor as a function of time are graphed in Fig. 31–2b. It is clear from this graph, as well as from Eqs. 31–3, that the current and voltage are out of phase by a quarter cycle, which is equivalent to $\pi/2$ radians or 90°. We see from the graph that

in an inductor, the current lags the voltage by 90°.

That is, the current in an inductor reaches its peaks a quarter cycle after the voltage does. Alternatively, we can say that the voltage leads the current by 90°.

Because the current and voltage are out of phase by 90°, no energy is transformed to other forms of energy in an inductor on the average; in particular, no energy is dissipated as thermal energy. This can be seen as follows from Fig. 31–2b. From point c to d, the voltage is increasing from zero to its maximum. The current, however, is in the opposite direction to the voltage and is approaching zero. The average power over this interval, VI, is negative. From d to e, however, both V and I are positive so VI is positive; this contribution just balances the negative contribution of the previous quarter cycle. Similar considerations apply to the rest of the cycle. Thus, the average power transformed over one or many cycles is zero. We can see that energy from the source passes into the magnetic field of the inductor, where it is stored temporarily. Then the field decreases and the energy is transferred back to the source. None is dissipated in this process. Compare this to a resistor where the current is always in the same direction as the voltage and energy is transferred out of the source and never back into it. (The product VI is never negative.) This energy is not stored in the resistor, but is transformed to thermal energy.

FIGURE 31–2 Inductor connected to an ac source. Current (blue curve) lags voltage (red curve) by a quarter cycle or 90°.

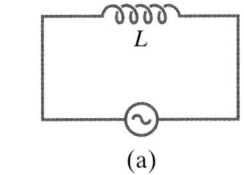

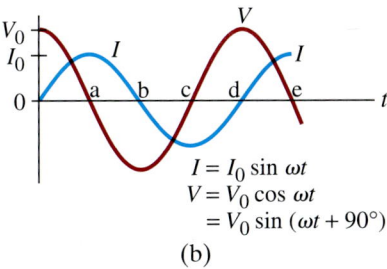

Inductor: current lags voltage

Just as a resistor impedes the flow of charge, so too an inductor impedes the flow of charge in an alternating current due to the back emf produced. For a resistor R, the peak current and peak voltage are related by $V_0 = I_0 R$. We can write a similar relation for an inductor:

$$V_0 = I_0 X_L \quad \begin{bmatrix} \text{maximum values} \\ \text{not an any instant} \end{bmatrix} \quad (31\text{-}4\text{a})$$

where, from Eq. 31–3b

$$X_L = \omega L. \quad (31\text{-}4\text{b})$$

Reactance (impedance) of inductor

The term X_L is called the *inductive reactance,* or *impedance,* of the inductor, and it is easy to show it has units of ohms. Normally, we use the term "reactance" to refer solely to the inductive properties. We then reserve the term "impedance" to include the total "impeding" qualities of the coil—its inductance as well as any resistance it may have (more on this in Section 31–5). In the absence of any resistance (or capacitance), the impedance is the same as the reactance.

The quantities V_0 and I_0 in Eq. 31–4a refer to peak values. (It is also valid for rms values, $V_{\text{rms}} = I_{\text{rms}} X_L$.) Note, however, that although this equation relates the peak values, the peak current and voltage are not reached at the same time; so Eq. 31–4a is *not valid at a particular instant,* as is the case for a resistor ($V = IR$).

Eq. 31–4a NOT valid at any instant

Note from Eq. 31–4b that if $\omega = 2\pi f = 0$ (so the current is dc), there is no back emf and no impedance to the flow of charge.

EXAMPLE 31–1 **Reactance of a coil.** A coil has a resistance $R = 1.00 \, \Omega$ and an inductance of 0.300 H. Determine the current in the coil if (*a*) 120-V dc is applied to it; (*b*) 120-V ac (rms) at 60.0 Hz is applied.

SOLUTION (*a*) There is no inductive reactance ($X_L = 0$ since $f = 0$), so we can write for the resistance:

$$I = \frac{V}{R} = \frac{120 \text{ V}}{1.00 \, \Omega} = 120 \text{ A}.$$

(*b*) The inductive reactance in this case is:

$$X_L = 2\pi f L = (6.28)(60.0 \text{ s}^{-1})(0.300 \text{ H}) = 113 \, \Omega.$$

In comparison to this, the $1.00 \, \Omega$ resistance can be ignored. Thus

$$I_{\text{rms}} \approx \frac{V_{\text{rms}}}{X_L} = \frac{120 \text{ V}}{113 \, \Omega} = 1.06 \text{ A}.$$

[It might be tempting to say that the total impedance is $113 \, \Omega + 1 \, \Omega = 114 \, \Omega$. This might imply that about 1 percent of the voltage drop is across the resistor, or about 1 V; and that across the inductance is 119 V. Although the 1 Vrms across the resistor is accurate, the other statements are *not* true because of the alteration in phase in an inductor. This will be discussed in Section 31–5.]

31–4 AC Circuit Containing only Capacitance C

When a capacitor is connected to a battery, the capacitor plates quickly acquire equal and opposite charges; but no steady current flows in the circuit. A capacitor prevents the flow of a dc current. However, if a capacitor is connected to an alternating source of voltage, as in Fig. 31–3a, an alternating current will flow continuously. This can happen because when the ac voltage is first turned on, charge begins to flow so that one plate acquires a negative charge and the other a positive charge. But when the voltage reverses itself, the charges flow in the opposite direction. Thus, for an alternating applied voltage, an ac current is present in the circuit continuously.

Let us look at this in more detail. By Kirchhoff's loop rule, the applied source voltage must equal the voltage V across the capacitor at any moment:

$$V = \frac{Q}{C}$$

where C is the capacitance and Q is the charge on the capacitor plates. Now the current I at any instant (given as $I = I_0 \sin \omega t$) is

$$I = \frac{dQ}{dt} = I_0 \sin \omega t.$$

Hence the charge Q on the plates at any instant is given by

$$Q = \int_0^t dQ = \int_0^t I_0 \sin \omega t \, dt = -\frac{I_0}{\omega} \cos \omega t.$$

Then the voltage across the capacitor is

$$V = \frac{Q}{C} = -I_0 \left(\frac{1}{\omega C}\right) \cos \omega t.$$

Using the trigonometric identity $\cos \theta = -\sin(\theta - 90°)$ we can rewrite this as

$$V = I_0 \left(\frac{1}{\omega C}\right) \sin(\omega t - 90°) = V_0 \sin(\omega t - 90°) \quad \textbf{(31–5a)}$$

where

$$V_0 = I_0 \left(\frac{1}{\omega C}\right) \quad \textbf{(31–5b)}$$

is the peak voltage. The current $I (= I_0 \sin \omega t)$ and voltage V (Eq. 31–5a) across the capacitor are graphed in Fig. 31–3b. It is clear from this graph, as well as a comparison of Eq. 31–5a with Eq. 31–1, that the current and voltage are out of phase by a quarter cycle or 90° ($\pi/2$ radians):

The current leads the voltage across a capacitor by 90°.

Alternatively we can say that the voltage lags the current by 90°. This is the opposite of what happens for an inductor, where the current lags the voltage by 90°.

Because the current and voltage are out of phase by 90°, the average power dissipated is zero, just as for an inductor. Energy from the source is fed to the capacitor, where it is stored in the electric field between its plates. As the field decreases, the energy returns to the source.

Thus, in an ac circuit, *only a resistance will dissipate energy* to thermal energy.

A relationship between the applied voltage and the current in a capacitor can be written just as for an inductance:

$$V_0 = I_0 X_C \quad \begin{bmatrix}\text{maximum values} \\ \text{not at any instant}\end{bmatrix} \quad \textbf{(31–6a)}$$

where X_C is the **capacitive reactance** (or *impedance*) of the capacitor, and has units of ohms. X_C is given by (see Eq. 31–5b)

$$X_C = \frac{1}{\omega C}. \quad \textbf{(31–6b)}$$

Equation 31–6a relates the peak values of V and I, or the rms values ($V_{\text{rms}} = I_{\text{rms}} X_C$). But it is not valid at a particular instant because I and V are not in phase.

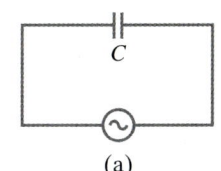

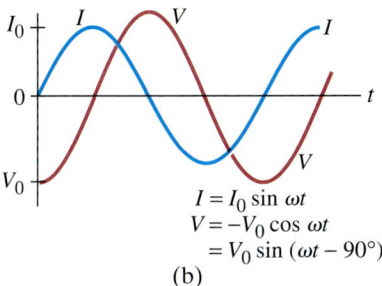

FIGURE 31–3 Capacitor connected to an ac source. Current leads voltage by a quarter cycle, or 90°.

Capacitor: current leads voltage

Only R (not C or L) dissipates electric energy

Reactance (impedance) of capacitor

Note from Eq. 31–6b that for dc conditions, $\omega = 2\pi f = 0$ and X_C becomes infinite. This is as it should be, since a pure capacitor does not pass dc current. Also, note that the reactance of an inductor increases with frequency, but that of a capacitor decreases with frequency.

EXAMPLE 31–2 **Capacitor reactance.** What are the peak and rms currents in the circuit of Fig. 31–3a if $C = 1.0\,\mu\text{F}$ and $V_{\text{rms}} = 120\,\text{V}$? Calculate for (a) $f = 60\,\text{Hz}$, and then for (b) $f = 6.0 \times 10^5\,\text{Hz}$.

SOLUTION (a) $V_0 = \sqrt{2}\,V_{\text{rms}} = 170\,\text{V}$. Then

$$X_C = \frac{1}{2\pi f C} = \frac{1}{(6.28)(60\,\text{s}^{-1})(1.0 \times 10^{-6}\,\text{F})} = 2.7\,\text{k}\Omega.$$

Thus

$$I_0 = \frac{V_0}{X_C} = \frac{170\,\text{V}}{2.7 \times 10^3\,\Omega} = 63\,\text{mA},$$

$$I_{\text{rms}} = \frac{V_{\text{rms}}}{X_C} = \frac{120\,\text{V}}{2.7 \times 10^3\,\Omega} = 44\,\text{mA}.$$

(b) For $f = 6.0 \times 10^5\,\text{Hz}$, X_C will be $0.27\,\Omega$, $I_0 = 630\,\text{A}$, and $I_{\text{rms}} = 440\,\text{A}$. The dependence on f is dramatic.

> **PHYSICS APPLIED**
> *Capacitors as filters*

Capacitors are used for a variety of purposes, some of which have already been described. Two other applications are illustrated in Fig. 31–4. In Fig. 31–4a, circuit A is said to be capacitively coupled to circuit B. The purpose of the capacitor is to prevent a dc voltage from passing from A to B but allowing an ac signal to pass relatively unimpeded. If C is sufficiently large, the ac signal will not be significantly attenuated, whereas dc is filtered out. The capacitor in Fig. 31–4b also passes ac but not dc. In this case, a dc voltage can be maintained between circuits A and B. If the capacitance C is large enough, the capacitor offers little impedance to an ac signal leaving A. Such a signal then passes to ground instead of into B. Thus the capacitor in Fig. 31–4b acts like a *filter* when a constant dc voltage is required; any sharp variation in voltage will pass to ground instead of into circuit B. Capacitors used in these two ways are very common in circuits.

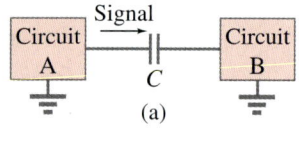
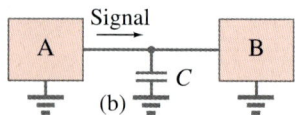

FIGURE 31–4 Two common uses for a capacitor.

FIGURE 31–5 Loudspeaker cross-over.

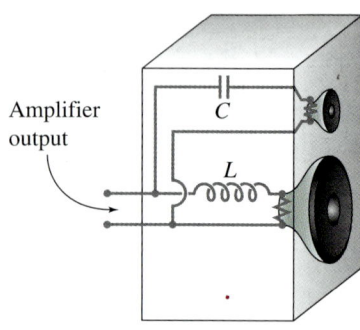

Loudspeakers that have separate speakers for low- and high-frequency sounds can use a simple "cross-over" that uses inductors and capacitors to filter frequencies reaching each speaker. Low-frequency sounds go to a large-diameter speaker, the *woofer*, and high-frequency sounds go to a small-diameter *tweeter*. (Recall standing waves from Chapter 16—small vibrating objects have high resonant frequencies whereas large vibrating objects have lower frequencies.) The simple diagram of Fig. 31–5 shows how the output signal from the amplifier is divided and goes to both tweeter and woofer, in parallel. A capacitor in the tweeter circuit impedes low-frequency signals (much like the capacitor in Fig. 31–4a). An inductor in the woofer circuit impedes high-frequency signals $(X_L = 2\pi f L)$ so mainly low-frequency sounds are emitted by the woofer.

31–5 LRC Series AC Circuit

FIGURE 31–6 An *LRC* circuit.

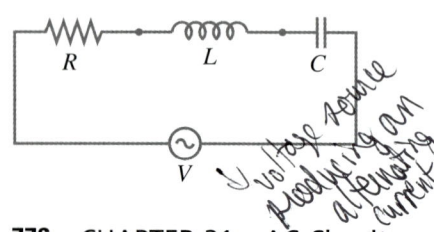

We now examine a circuit containing all three elements in series, a resistor R, an inductor L, and a capacitor C, Fig 31–6. If a given circuit contains only two of these elements, we can still use the results of this Section by setting $R = 0$, $L = 0$, or $C = \infty$ (infinity) as needed. We let V_R, V_L, and V_C represent the voltage across each element at a *given instant* in time; and V_{R0}, V_{L0}, and V_{C0} represent the *maximum* (peak) values of these voltages. The voltage across each of the elements will follow the phase relations we discussed in the previous Sections. That is, V_R will be in phase with the current; V_L will lead the current by 90°; and V_C will lag behind the current by 90°. And at any instant the voltage V supplied by the source will be, by Kirchhoff's loop rule,

$$V = V_R + V_L + V_C. \tag{31-7}$$

However, because the various voltages are not in phase, they do not reach their peak values at the same time, and so the peak voltage of the source, V_0, will *not* equal $V_{R0} + V_{L0} + V_{C0}$. Likewise the rms voltages (which are what ac voltmeters usually read) will not simply add up to give the rms voltage of the source.

Let us now examine the circuit in detail. What we would like to find in particular is the impedance of the circuit as a whole, the peak current I_0 that flows, and the phase difference between the source voltage and the current.

First we note that the current at any instant must be the same at all points in the circuit. Thus, *the currents in each element are in phase with each other, although the voltages are not.* We choose our origin in time ($t = 0$) so that the current I at any time t is

$$I = I_0 \sin \omega t$$

just as we did in the previous Sections of this chapter.

We will analyze the *LRC* circuit using[†] a so-called **phasor diagram**: Arrows (acting somewhat like vectors) are drawn in an xy coordinate system to represent the amplitude of each voltage. The *length of each arrow represents the magnitude of the peak voltage across each element*:

$$V_{R0} = I_0 R, \qquad V_{L0} = I_0 X_L, \qquad V_{C0} = I_0 X_C.$$

The angle of each arrow represents the phase of each voltage, relative to the current, and the arrows rotate at angular frequency ω to take into account the time dependence of the voltages and current. In particular, *the projection of each arrow on the y axis will represent the voltage across each element at a given time.* Let us see how this works.

First we draw the phasor diagram for time $t = 0$. The current at $t = 0$ is $I = I_0 \sin \omega t = 0$, and we draw the arrow representing I_0 along the positive x axis in our phasor diagram, Fig. 31–7a. (Note that the projection of I_0 on the y axis is zero, which corresponds to $I = 0$ at $t = 0$.) The voltage across a resistor is always in phase with the current, so the arrow representing V_{R0} is drawn parallel to I_0 along the x axis as shown. Since the voltage across the inductor, V_L, leads the current by 90°, V_{L0} leads V_{R0} by 90°, and is drawn perpendicular to it as shown. V_C lags the current by 90°, so V_{C0} lags V_{R0} by 90°, and hence V_{C0} is drawn perpendicular to V_{R0}, but downward (Fig. 31–7a). Now if we let this diagram rotate as a whole at angular frequency ω, then we obtain the diagram shown in Fig. 31–7b: after a time t, each arrow has rotated through an angle ωt. Then, as mentioned above, the projections of each arrow on the y axis represent the voltages across each element at the instant t. See Fig. 31–7c. For example, the projection of V_{R0} on the y axis is $V_{R0} \sin \omega t$ ($= I_0 R \sin \omega t = IR$ since $I = I_0 \sin \omega t$). The projections of V_{L0} and V_{C0} on the y axis are $V_L = V_{L0} \cos \omega t = V_L \sin(\omega t + 90°)$, and $V_C = -V_{C0} \cos \omega t = V_{C0} \sin(\omega t - 90°)$. These results are consistent with our earlier results as shown in Figs. 31–1, 31–2, and 31–3. See also Eqs. 31–3a and 31–5a. Maintaining the 90° angle between each vector ensures the correct phase relations. Although these facts show the validity of a phasor diagram, what we are really interested in is how to add the voltages.

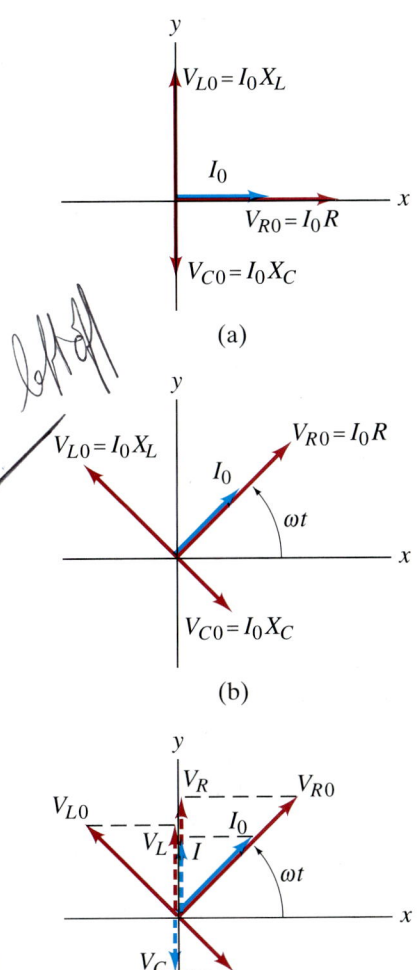

FIGURE 31–7 Phasor diagram for a series *LRC* circuit.

[†]We could instead do our analysis by rewriting Eq. 31–7 as a differential equation (setting $V_C = Q/C$, $V_R = IR = (dQ/dt)R$, and $V_L = L\, dI/dt$) and trying to solve the differential equation. The differential equation we would get would look like Eq. 14–21 in Section 14–8 (on forced vibrations), and would be solved in the same way. Phasor diagrams are easier, and at the same time give us some physical insight into the situation.

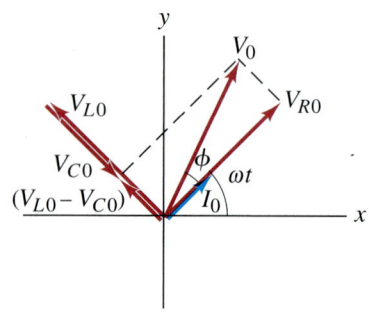

FIGURE 31–8 Phasor diagram for a series LRC circuit showing the sum vector, V_0.

The sum of the projections of the three vectors on the y axis is equal to the projection of their sum. But the sum of the projections represents the instantaneous voltage across the whole circuit, which is the source voltage V. We can then let the vector sum of these vectors be the vector that represents the peak source voltage V_0 on a phasor diagram. This is shown in Fig. 31–8, where it is seen that V_0 makes an angle ϕ with V_{R0} and I_0. As time passes, V_0 rotates with the other vectors, so the instantaneous voltage V (projection of V_0 on y axis) is (see Fig. 31–8):

$$V = V_0 \sin(\omega t + \phi).$$

Thus we see that the voltage from the source is out of phase[†] with the current by an angle ϕ.

From this analysis we can now draw some useful conclusions. First we determine the total **impedance** Z of the circuit, which is defined by the relation

$$V_{\text{rms}} = I_{\text{rms}} Z, \quad \text{or} \quad V_0 = I_0 Z. \quad (31\text{–}8)$$

From Fig. 31–8, we see, using the Pythagorean theorem (V_0 is the hypoteneuse), that

$$\begin{aligned} V_0 &= \sqrt{V_{R0}^2 + (V_{L0} - V_{C0})^2} \\ &= \sqrt{I_0^2 R^2 + (I_0 X_L - I_0 X_C)^2} \\ &= I_0 \sqrt{R^2 + (X_L - X_C)^2}. \end{aligned}$$

Thus, from Eq. 31–8, and then Eqs. 31–4b and 31–6b,

Total impedance of LRC ac circuit

$$Z = \sqrt{R^2 + (X_L - X_C)^2} \quad (31\text{–}9a)$$

$$= \sqrt{R^2 + \left(\omega L - \frac{1}{\omega C}\right)^2}. \quad (31\text{–}9b)$$

This gives the total impedance Z of the circuit. Also from Fig. 31–8 we can find the phase angle ϕ:

Phase angle of LRC ac circuit

$$\tan \phi = \frac{V_{L0} - V_{C0}}{V_{R0}} = \frac{I_0(X_L - X_C)}{I_0 R} = \frac{X_L - X_C}{R} \quad (31\text{–}10a)$$

or

$$\cos \phi = \frac{V_{R0}}{V_0} = \frac{I_0 R}{I_0 Z} = \frac{R}{Z}. \quad (31\text{–}10b)$$

Note that Fig. 31–8 was drawn for the case $X_L > X_C$, and the current lags the source voltage by ϕ. If the reverse is true, $X_L < X_C$, then ϕ in Eq. 31–10 is less than zero, and the current leads the source voltage.

Finally, we can determine the power dissipated in the circuit. We saw earlier that power is dissipated only by a resistance; none is dissipated by inductance or capacitance. Therefore, the average power $\overline{P} = I_{\text{rms}}^2 R$. But from Eq. 31–10b, $R = Z \cos \phi$. Therefore

Power factor

$$\begin{aligned} \overline{P} &= I_{\text{rms}}^2 Z \cos \phi \\ &= I_{\text{rms}} V_{\text{rms}} \cos \phi. \end{aligned} \quad (31\text{–}11)$$

The factor $\cos \phi$ is referred to as the **power factor** of the circuit. For a pure resistor, $\cos \phi = 1$ and $\overline{P} = I_{\text{rms}} V_{\text{rms}}$. For a capacitor or inductor alone, $\phi = -90°$ or $+90°$, respectively, so $\cos \phi = 0$ and no power is dissipated.

[†] As a check, note that if $R = X_C = 0$, then $\phi = 90°$, and V_0 would lead the current by 90°, as it must for an inductor alone. Similarly, if $R = L = 0$, $\phi = -90°$ and V_0 would lag the current by 90°, as it must for a capacitor alone.

The test of this analysis is, of course, in experiment; and experiment is in full agreement with these results.[†]

EXAMPLE 31-3 LRC Circuit. Suppose that $R = 25.0\,\Omega$, $L = 30.0\,\text{mH}$, and $C = 12.0\,\mu\text{F}$ in Fig. 31-6, and that they are connected to a 90.0-V ac (rms) 500-Hz source. Calculate (a) the current in the circuit, (b) the voltmeter readings (rms) across each element, (c) the phase angle ϕ, and (d) the power dissipated in the circuit.

I_{RMS}

SOLUTION (a) First, we find the individual impedances at $f = 500\,\text{Hz} = 500\,\text{s}^{-1}$:

$$X_L = 2\pi f L = 94.2\,\Omega,$$

$$X_C = \frac{1}{2\pi f C} = 26.5\,\Omega.$$

Then

$$Z = \sqrt{R^2 + (X_L - X_C)^2}$$
$$= \sqrt{(25.0\,\Omega)^2 + (94.2\,\Omega - 26.5\,\Omega)^2} = 72.2\,\Omega.$$

From Eq. 31-8,

$$I_{\text{rms}} = \frac{V_{\text{rms}}}{Z} = \frac{90.0\,\text{V}}{72.2\,\Omega} = 1.25\,\text{A}.$$

(b) The rms voltage across each element is

$$(V_R)_{\text{rms}} = I_{\text{rms}} R = (1.25\,\text{A})(25.0\,\Omega) = 31.2\,\text{V}$$

$$(V_L)_{\text{rms}} = I_{\text{rms}} X_L = (1.25\,\text{A})(94.2\,\Omega) = 118\,\text{V}$$

$$(V_C)_{\text{rms}} = I_{\text{rms}} X_C = (1.25\,\text{A})(26.5\,\Omega) = 33.1\,\text{V}.$$

Notice that these do *not* add up to give the source voltage, 90.0 V (rms). Indeed, the rms voltage across the inductance *exceeds* the source voltage. This can happen because the different voltages are out of phase with each other, and at any instant one voltage can be negative, to compensate for a large positive voltage of another. The rms voltages, however, are always positive by definition. Although the rms voltages do not have to add up to the source voltage, the instantaneous voltages at any time do add up, of course, to the source voltage at that instant.

(c) The phase angle ϕ is given by Eq. 31-10b,

$$\cos\phi = \frac{R}{Z} = \frac{25.0\,\Omega}{72.2\,\Omega} = 0.346,$$

so $\phi = 69.7°$. Note that ϕ is positive because $X_L > X_C$ in this case, so $V_{L0} > V_{C0}$ in Fig. 31-8.

(d) $\overline{P} = I_{\text{rms}} V_{\text{rms}} \cos\phi = (1.25\,\text{A})(90.0\,\text{V})(25.0\,\Omega/72.2\,\Omega) = 39.0\,\text{W}.$

[†]At the beginning of this Section, we chose the phase of the current so that $I = I_0 \sin\omega t$. This choice is, of course, arbitrary. (What is physically important is the phase difference, ϕ, between current and voltage.) If we had chosen, instead,

$$V = V_0 \sin\omega t,$$

then the current I would be

$$I = I_0 \sin(\omega t - \phi)$$

where ϕ and I_0 have the same values as given by Eqs. 31-8, 31-9, and 31-10.

31–6 Resonance in AC Circuits

The rms current in an LRC series circuit is given by (see Eqs. 31–8 and 31–9b):

$$I_{\text{rms}} = \frac{V_{\text{rms}}}{Z} = \frac{V_{\text{rms}}}{\sqrt{R^2 + \left(\omega L - \dfrac{1}{\omega C}\right)^2}}. \qquad (31\text{–}12)$$

Because the impedance of inductors and capacitors depends on the frequency f ($= \omega/2\pi$) of the source, the current in an LRC circuit will depend on frequency. From Eq. 31–12 we can see that the current will be maximum at a frequency such that

$$\left(\omega L - \frac{1}{\omega C}\right) = 0.$$

We solve this for ω and call the solution ω_0:

Resonance

$$\omega_0 = \sqrt{\frac{1}{LC}}. \qquad (31\text{–}13)$$

When $\omega = \omega_0$, the circuit is in **resonance**, and $f_0 = \omega_0/2\pi$ is the **resonant frequency** of the circuit. At this frequency, $X_C = X_L$, so the impedance is purely resistive and $\cos\phi = 1$. A graph of I_{rms} versus ω is shown in Fig. 31–9 for particular values of R, L, and C. For smaller R compared to X_L and X_C, the resonance peak will be higher and sharper. When R is very small, the circuit approaches the pure LC circuit we discussed in Section 30–5.

This electrical resonance is analogous to mechanical resonance, which we discussed in Chapter 14. The energy transferred to the system by the source is a maximum at resonance whether it is electrical resonance, the oscillation of a spring, or pushing a child on a swing (Section 14–8). That this is true in the electrical case can be seen from Eq. 31–11. At resonance, $\cos\phi = 1$, and I_{rms} is a maximum. For constant voltage V_{rms}, the power then is a maximum at resonance. A graph of power versus frequency peaks like that for the current, Fig. 31–9.

Electric resonance is used in many circuits. Radio and TV sets, for example, use resonant circuits for tuning in a station. Many frequencies reach the circuit, but a significant current flows only for those at or near the resonant frequency. Either L or C is variable so that different stations can be tuned in.

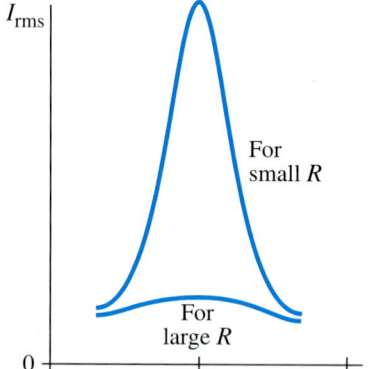

FIGURE 31–9 Current in LRC circuit as a function of frequency, showing resonance peak at $\omega = \omega_0 = \sqrt{1/LC}$.

EXAMPLE 31–4 Radio station oscillator. A radio station is authorized to broadcast on a frequency of 1040 kHz. If you are designing a receiver circuit to pick up this station, and already have a coil with an inductance of 4.0 mH, what capacitance will you need?

SOLUTION The tuned circuit must have a resonant frequency of 1040 kHz. Using an inductor in series with a capacitor will accomplish the job. Since the resonance will be at

$$f_0 = \frac{1}{2\pi}\sqrt{\frac{1}{LC}}$$

you will need to find a capacitor with a value of

$$C = \frac{1}{L(2\pi f_0)^2} = \frac{1}{(4.0 \times 10^{-3}\,\text{H})(2\pi \times 1.04 \times 10^6\,\text{s}^{-1})^2} = 5.85 \times 10^{-12}\,\text{F} = 5.85\,\text{pF}$$

*31-7 Impedance Matching

It is common to connect one electric circuit to a second circuit. For example, a TV antenna is connected to a TV set; an FM tuner is connected to an amplifier; the output of an amplifier is connected to a speaker; electrodes for an ECG or EEG (electrocardiogram and electroencephalogram—electrical traces of heart and brain signals) are connected to an amplifier or a recorder. In many cases it is important that the maximum power be transferred from one to the other, with a minimum of loss. This can be achieved when the output impedance of the one device matches the input impedance of the second.

To show why this is true, we consider simple circuits that contain only resistance. In Fig. 31–10 the source in circuit 1 could represent a power supply, the output of an amplifier, or the signal from an antenna, a laboratory probe, or a set of electrodes. R_1 represents the resistance of this device and includes the internal resistance of the source. R_1 is called the output impedance (or resistance) of circuit 1. The output of circuit 1 is across the terminals a and b which are connected to the input of circuit 2 which may be very complicated. We let R_2 be the equivalent "input resistance" of circuit 2.

The power delivered to circuit 2 is $P = I^2 R_2$ where $I = V/(R_1 + R_2)$. Thus

$$P = I^2 R_2 = \frac{V^2 R_2}{(R_1 + R_2)^2}.$$

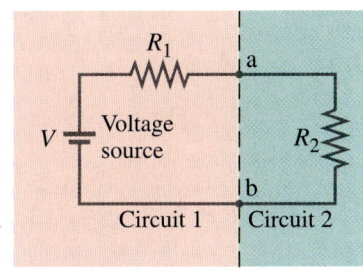

FIGURE 31–10 Output of the circuit on the left is input to the circuit on the right.

We divide the top and bottom of the right side by R_1^2 and find

$$P = \frac{V^2}{R_1} \frac{\left(\dfrac{R_2}{R_1}\right)}{\left(1 + \dfrac{R_2}{R_1}\right)^2}.$$

The question is, if the resistance of the source is R_1, what value should R_2 have so that the maximum power is transferred to circuit 2? To determine this, we take the derivative of P with respect to R_2 and set it equal to zero:

$$0 = \frac{dP}{dR_2} = \frac{V^2}{R_1^2} \frac{(1 - R_2/R_1)}{(1 + R_2/R_1)^3}.$$

This expression can be zero only if $(1 - R_2/R_1) = 0$, or

$$R_2 = R_1.$$

Impedance matching

Thus, the maximum power is transmitted when the *output impedance* of one device *equals the input impedance* of the second. This is called **impedance matching**.

In an ac circuit that contains capacitors and inductors, the different phases are important and the analysis is more complicated. However, the same result holds: to maximize power transfer it is important to match impedances $(Z_2 = Z_1)$.

In addition, one must be aware that it is possible to seriously distort a signal. For example, when a second circuit is connected, it may put the first circuit into resonance, or take it out of resonance for a certain frequency.

Without proper consideration of the impedances involved, one can make measurements that are completely meaningless. These considerations are normally examined by engineers when designing an integrated set of apparatus. It has happened that researchers have connected several components to one another without regard for impedance matching, and made a "new discovery" that later was found (embarrassingly) to be due to impedance mismatch rather than the natural phenomenon they had thought.

In some cases, a transformer is used to alter an impedance, so it can be matched to that of a second circuit. If Z_s is the secondary impedance and Z_p the primary impedance, then $V_s = I_s Z_s$ and $V_p = I_p Z_p$ (I and V are either peak or rms values of current and voltage). Hence

$$\frac{Z_p}{Z_s} = \frac{V_p I_s}{V_s I_p} = \left(\frac{N_p}{N_s}\right)^2$$

where we have used Eqs. 29–5 and 29–6 for a transformer. Thus the impedance can be changed with a transformer.

Some instruments, such as oscilloscopes, require only a signal voltage but very little power. Maximum power transfer is then not important and such instruments can have a high input impedance, which has the advantage that the instrument draws very little current and disturbs the original circuit as little as possible.

*31–8 Three-Phase AC

Transmission lines typically consist of four wires, rather than two as you might have guessed. One of these wires is the ground; the remaining three are used to transmit three-phase ac power which is a superposition of three ac voltages 120° out of phase with each other:

$$V_1 = V_0 \sin \omega t$$
$$V_2 = V_0 \sin(\omega t + 2\pi/3)$$
$$V_3 = V_0 \sin(\omega t + 4\pi/3).$$

(See Fig. 31–11.) Why is three-phase power used? We saw in Fig. 25–20 that single-phase ac (i.e., the voltage V_1 by itself) delivers power to the load in pulses. A much smoother flow of power can be delivered if we use three-phase power. Suppose that each of the three voltages making up the three-phase source is hooked up to a resistor R. Then the power delivered is:

$$P = \frac{1}{R}(V_1^2 + V_2^2 + V_3^2).$$

You can show that this power is a constant equal to $3V_0^2/2R$, which is three times the rms power delivered by a single-phase source. This smooth flow of power makes electrical equipment run smoothly. Although houses use single-phase ac power, most industrial-grade machinery is wired for three-phase power.

FIGURE 31–11 The three voltages, out of phase by 120° ($=\frac{2}{3}\pi$ radians), in a three-phase power line.

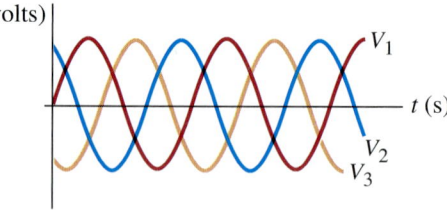

EXAMPLE 31–5 Three-phase circuit. In a three-phase circuit there is 266 V rms between line 1 and ground. What is the rms voltage between lines 2 and 3?

SOLUTION We are given $V_{rms} = V_0/\sqrt{2} = 266$ V. Hence $V_0 = 376$ V. Now $V_3 - V_2 = V_0[\sin(\omega t + 4\pi/3) - \sin(\omega t + 2\pi/3)] = 2V_0 \sin\frac{1}{2}\left(\frac{2\pi}{3}\right)\cos\frac{1}{2}(2\omega t)$ where we used the identity: $\sin A - \sin B = 2\sin\frac{1}{2}(A-B)\cos\frac{1}{2}(A+B)$. The rms voltage is

$$(V_3 - V_2)_{rms} = \frac{1}{\sqrt{2}} 2V_0 \sin\frac{\pi}{3} = \sqrt{2}(376 \text{ V})(0.866) = 460 \text{ Vrms}.$$

Summary

Capacitance and inductance offer *impedance* to the flow of alternating current just as resistance does. This impedance is referred to as **reactance**, X. For capacitance and inductance the reactance is defined, as for resistors, as the proportionality constant between voltage and current (either the rms or peak values). Across a capacitor,

$$V_0 = I_0 X_C,$$

and across an inductor,

$$V_0 = I_0 X_L.$$

The reactance of a capacitor decreases with frequency:

$$X_C = 1/\omega C,$$

where $\omega = 2\pi f$ and f is the frequency.
The reactance of an inductor increases with frequency:

$$X_L = \omega L.$$

Whereas the current through a resistor is always in phase with the voltage across it, this is not true for inductors and capacitors: in an inductor, the current lags the voltage by 90°, and in a capacitor the current leads the voltage by 90°.

In an **LRC series circuit**, the total impedance Z is defined by the equivalent of $V = IR$ for resistance, namely $V_0 = I_0 Z$ or $V_{rms} = I_{rms} Z$; Z is related to $R, C,$ and L by

$$Z = \sqrt{R^2 + (X_L - X_C)^2}.$$

The current in the circuit lags (or leads) the source voltage by an angle ϕ given by $\cos\phi = R/Z$. Only the resistor in an LRC circuit dissipates energy, and at a rate

$$\overline{P} = I_{rms}^2 Z \cos\phi$$

where the factor $\cos\phi$ is referred to as the **power factor**.
An LRC series circuit **resonates** at a frequency given by

$$\omega_0 = \frac{1}{\sqrt{LC}} \quad \text{or} \quad f_0 = \frac{\omega_0}{2\pi} = \frac{1}{2\pi\sqrt{LC}}.$$

The rms current in the circuit is largest when the applied voltage has a frequency equal to f_0. The lower the resistance R, the higher and sharper the resonance peak.

Questions

1. Under what conditions is the impedance in an LRC circuit a minimum?
2. Why can we assume that the current in an LRC circuit will have the same frequency as the applied emf?
3. In an LRC circuit, if $X_L > X_C$, the circuit is said to be predominantly "inductive." And if $X_C > X_L$, the circuit is said to be predominantly "capacitive." Discuss the reasons for these terms. In particular, do they say anything about the relative values of L and C at a given frequency?
4. Do the results of Section 31–5 approach the proper expected results when ω approaches zero? What are the expected results?
5. When an ac generator is connected to an LRC circuit, where does the energy come from ultimately? Where does it go? How do the values of $L, C,$ and R affect the energy supplied by the generator?
6. Discuss the validity of both of Kirchhoff's rules (Section 26–3) when applied to ac circuits that contain several loops.
7. Is it possible for the instantaneous power output of an ac generator connected to an LRC circuit ever to be negative? Explain.
8. Can you tell whether the current in an LRC circuit leads or lags the applied voltage from a knowledge of the power factor, $\cos\phi$?
9. If $\cos\phi$ were less than zero, Eq. 31–11 tells us that $\overline{P} < 0$. Can this happen? Can $\cos\phi$ be made negative? Explain.
10. Does the power factor, $\cos\phi$, depend on frequency? Does the power dissipated in an LRC circuit depend on frequency?
11. What is the significance of the sign of ϕ (+ or −)? Is this a convention, or is it a fixed rule?
12. Describe briefly how the frequency of the source emf affects the impedance of (a) a pure resistance, (b) a pure capacitance, (c) a pure inductance, (d) an LRC circuit near resonance (R small), (e) an LRC circuit far from resonance (R small).
13. Discuss the response of an LRC circuit as $R \to 0$ when the frequency is (a) at resonance, (b) near resonance, (c) far from resonance. Is there energy dissipation in each case? Discuss the transformations of energy that occur in each case.
14. Can you tell whether or not a circuit is in resonance if you are given the value of the power factor, $\cos\phi$?
15. An LRC resonance circuit is often called an *oscillator* circuit. What is it that oscillates?
16. Compare the oscillations of an LRC circuit to the vibration of a mass m on a spring. What do L and C correspond to in the mechanical system?

Problems

Sections 31-1 to 31-4

1. (I) What is the reactance of a $7.2\,\mu\text{F}$ capacitor at a frequency of (a) 60 Hz, (b) 1.0 MHz?

2. (I) At what frequency will a 22.0-mH inductor have a reactance of $660\,\Omega$?

3. (I) At what frequency will a $2.40\text{-}\mu\text{F}$ capacitor have a reactance of $6.70\,\text{k}\Omega$?

4. (I) Plot a graph of the impedance of a $5.8\text{-}\mu\text{F}$ capacitor as a function of frequency from 10 Hz to 1000 Hz.

5. (I) Plot a graph of the impedance of a 5.0-mH inductor as a function of frequency from 100 Hz to 10,000 Hz.

6. (I) Calculate the impedance of, and rms current in, a 36.0-mH radio coil connected to 750-V (rms) 33.3-kHz ac line. Ignore resistance.

7. (II) What is the inductance L of the primary of a transformer whose input is 110 V at 60 Hz and the current drawn is 2.2 A? Assume no current in the secondary.

8. (II) (a) What is the impedance of a well-insulated $0.036\text{-}\mu\text{F}$ capacitor connected to a 22-kV (rms), 600-Hz line? (b) What will be the peak value of the current and its frequency?

9. (II) Instead of starting our analysis of ac circuits with Eq. 31-1, suppose we assumed that the external voltage was given as $V = V_0 \sin \omega t$. Show that when this voltage is (a) connected only to a capacitor C, the current would be $I = \omega C V_0 \cos \omega t = \omega C V_0 \sin(\omega t + 90°)$, and (b) connected only to an inductor L, the current would be $I = -(V_0/\omega L)\cos \omega t = (V_0/\omega L)\sin(\omega t - 90°)$.

10. (II) A current $I = 1.80 \cos 377t$ (I in amps, t in seconds and the "angle" is in radians) flows in a series LR circuit in which $L = 3.85$ mH and $R = 260\,\Omega$. What is the average power dissipation?

11. (II) A capacitor is placed in parallel with some device, B, as in Fig. 31-4b, to filter out stray high-frequency signals, but to allow ordinary 60-Hz ac to pass through with little loss. Suppose that circuit B in Fig. 31-4b is a resistance $R = 400\,\Omega$ connected to ground, and that $C = 0.35\,\mu\text{F}$. What percent of the incoming current will pass through C rather than R if (a) it is 60 Hz; (b) 60,000 Hz?

Section 31-5

12. (I) A $1.20\text{-}\text{k}\Omega$ resistor and a $6.8\text{-}\mu\text{F}$ capacitor are connected in series to an ac source. Calculate the impedance of the circuit if the source frequency is (a) 60 Hz; (b) 60,000 Hz.

13. (I) A $9.0\text{-}\text{k}\Omega$ resistor is in series with a 26.0-mH inductor and an ac source. Calculate the impedance of the circuit if the source frequency is (a) 50 Hz; (b) 30,000 Hz.

14. (I) For a 120-V, 60-Hz voltage, a current of 70 mA passing through the body for 1.0 s could be lethal. What must be the impedance of the body for this to occur?

15. (II) (a) What is the rms current in a series RC circuit if $R = 6.0\,\text{k}\Omega$, $C = 0.80\,\mu\text{F}$, and the rms applied voltage is 120 V at 60 Hz? (b) What is the phase angle between voltage and current? (c) What is the power dissipated by the circuit? (d) What are the voltmeter readings across R and C?

16. (II) (a) What is the rms current in a series LR circuit when a 60.0-Hz, 120-V rms ac voltage is applied, where $R = 765\,\Omega$ and $L = 250$ mH? (b) What is the phase angle between voltage and current? (c) How much power is dissipated? (d) What are the rms voltage readings across R and L?

17. (II) A 35-mH inductor with $2.0\text{-}\Omega$ resistance is connected in series to a $20\text{-}\mu\text{F}$ capacitor and a 60-Hz, 45-V source. Calculate (a) the rms current, (b) the phase angle, and (c) the power dissipated in this circuit.

18. (II) A 40-mH coil whose resistance is $0.80\,\Omega$ is connected to a capacitor C and a 360-Hz source voltage. If the current and voltage are to be in phase, what value must C have?

19. (II) What is the resistance of a coil if its impedance is $335\,\Omega$ and its reactance is $45.5\,\Omega$?

20. (II) In the LRC circuit of Fig. 31-6, suppose $I = I_0 \sin \omega t$ and $V = V_0 \sin(\omega t + \phi)$. Determine the instantaneous power dissipated in the circuit from $P = IV$ using these equations and show that on the average, $\overline{P} = \tfrac{1}{2} V_0 I_0 \cos \phi$, which confirms Eq. 31-11.

21. (II) If $V = V_0 \sin \omega t$, what is the average value of V over (a) a whole cycle, (b) each half cycle? How do these compare to V_{rms}?

22. (II) A circuit consists of a $150\text{-}\Omega$ resistor in series with a 40.0-mH inductor and a 60.0-V ac generator. The power dissipated by the circuit is 15.5 W. What is the frequency of the generator?

23. (II) What is the total impedance, phase angle, and rms current in an LRC circuit connected to a 10.0-kHz, 800-V (rms) source if $L = 32.0$ mH, $R = 8.70\,\text{k}\Omega$, and $C = 5000$ pF?

Section 31-6

24. (I) A 3200-pF capacitor is connected in series to a $26.0\text{-}\mu\text{H}$ coil of resistance $2.00\,\Omega$. What is the resonant frequency of this circuit?

25. (I) What is the resonant frequency of the LRC circuit of Example 31-3? At what rate is energy taken from the generator, on the average, at this frequency?

26. (II) An LRC circuit has $L = 4.15$ mH and $R = 220\,\Omega$. (a) What value must C have to produce resonance at 33.0 kHz? (b) What will be the maximum current at resonance if the peak external voltage is 136 V?

27. (II) What will be the peak current in Problem 26 if the capacitor is chosen so that the resonant frequency is twice the applied frequency of 33.0 kHz?

28. (II) A resonant circuit using a $220\text{-}\mu\text{F}$ capacitor is to resonate at 18.0 kHz. The air-core inductor is to be a solenoid with closely packed coils made from 12.0 m of insulated wire 1.1 mm in diameter. How many loops will the inductor contain?

29. (II) (a) Show that oscillation of charge Q on the capacitor of an LRC circuit has amplitude

$$Q_0 = \frac{V_0}{\sqrt{(\omega R)^2 + \left(\omega^2 L - \dfrac{1}{C}\right)^2}}.$$

(b) At what angular frequency, ω', will Q_0 be a maximum? (c) Compare to a damped harmonic oscillator, and discuss. (See also Question 16 in this chapter.)

30. (II) Show that the width of a sharp resonance peak, defined as the difference in (angular) frequency between the two frequencies where $I = \frac{1}{2}I_0$, is given by $\Delta\omega \approx \sqrt{3}\,R/L$.

31. (II) (a) Determine a formula for the average power $\overline{P}$ dissipated in an LRC circuit in terms of L, R, C, ω, and V_0. (b) At what frequency is the power a maximum? (c) Find an approximate formula for the width of the resonance peak in average power, $\Delta\omega$, which is the difference in the two (angular) frequencies where $\overline{P}$ has half its maximum value. Assume a sharp peak.

*Section 31–7

*32. (I) The output of an ECG amplifier has an impedance of 35 kΩ. It is to be connected to an 8.0-Ω loudspeaker through a transformer. What should be the turns ratio of the transformer?

*33. (I) An audio amplifier has output connections for 4 Ω, 8 Ω, and 16 Ω. If two 8-Ω speakers are to be connected in parallel, to which output terminals should they be connected?

General Problems

34. Suppose circuit B in Fig. 31–4a consists of a resistance $R = 800\,\Omega$, and a capacitance $C = 1.2\,\mu\text{F}$. Will this capacitor act to eliminate 60-Hz ac but pass a high-frequency signal of frequency 60,000 Hz? To check this, determine the voltage drop across R for a 130-mV signal of frequency (a) 60 Hz; (b) 60,000 Hz.

35. A 230-mH coil, whose resistance is 18.5 Ω, is connected to a capacitor C and a 3360-Hz source voltage. If the current and voltage are to be in phase, what value must C have?

36. A circuit contains two elements, but it is not known if they are L, R, or C. The current in this circuit when connected to a 120-V 60-Hz source is 5.6 A and lags the voltage by 50°. What are the two elements and what are their values?

37. An inductance coil operates at 240 V and 60 Hz. It draws 22.8 A. What is the coil's inductance?

38. (a) What is the impedance of a well-insulated 0.038-μF capacitor connected to a 4.0-kV (rms) 700-Hz line? (b) What will be the peak value of the current?

39. A 3.5-kΩ resistor in series with a 620-mH inductor is driven by an ac power supply. At what frequency is the impedance double that of the impedance at 60 Hz?

40. (a) What is the rms current in an RC circuit if $R = 8.80\,\text{k}\Omega$, $C = 1.80\,\mu\text{F}$, and the rms applied voltage is 120 V at 60.0 Hz? (b) What is the phase angle between voltage and current? (c) What is the power dissipated by the circuit? (d) What are the voltmeter readings across R and C?

41. An inductance coil draws 2.5 A dc when connected to a 36-V battery. When connected to a 60-Hz 120-V (rms) source, the current drawn is 3.8 A (rms). Determine the inductance and resistance of the coil.

42. The **Q factor** of a resonance circuit can be defined as the ratio of the voltage across the capacitor (or inductor) to the voltage across the resistor, at resonance. The larger the Q factor, the sharper the resonance curve will be and the sharper the tuning. (a) Show that the Q factor is given by the equation $Q = (1/R)\sqrt{L/C}$. (b) At a resonant frequency $f_0 = 1.0\,\text{MHz}$, what must be the value of L and R to produce a Q factor of 1000? Assume that $C = 0.010\,\mu\text{F}$. (c) What is the Q factor of the circuit of Example 31–3? [Note: see also Problem 53 in Chapter 30.]

43. In a series LRC circuit, the inductance is 20 mH, the capacitance is 50 nF, and the resistance is 200 Ω. At what frequencies is the power factor equal to 0.17?

44. In our analysis of a series LRC circuit, Fig. 31–6, suppose we chose $V = V_0 \sin\omega t$. (a) Construct a phasor diagram, like that of Fig. 31–8, for this case. (b) Write a formula for the current I, defining all terms.

45. A voltage $V = 0.95\sin 754 t$ is applied to an LRC circuit (I is in amperes, t is in seconds and the "angle" is in radians) which has $L = 22.0\,\text{mH}$, $R = 23.2\,\text{k}\Omega$, and $C = 0.30\,\mu\text{F}$. (a) What is the impedance and phase angle? (b) How much power is dissipated in the circuit? (c) What is the rms current and voltage across each element?

46. *Filter circuit.* Figure 31–12 shows a simple filter circuit designed to pass dc voltages with minimal attenuation and to remove, as much as possible, any ac components (such as 60-Hz line voltage that could cause hum in a stereo receiver, for example). Assume $V_{\text{in}} = V_1 + V_2$ where V_1 is dc and $V_2 = V_{20}\sin\omega t$, and that any resistance is very small. (a) Determine the current through the capacitor: give amplitude and phase (assume $R = 0$ and $X_L > X_C$). (b) Show that the ac component of the output voltage, $V_{2\,\text{out}}$, equals $(Q/C) - V_1$, where Q is the charge on the capacitor at any instant, and determine the amplitude and phase of $V_{2\,\text{out}}$. (c) Show that the attenuation of the ac voltage is greatest when $X_C \ll X_L$, and calculate the ratio of the output to input ac voltage in this case. (d) Compare the dc output to input voltage.

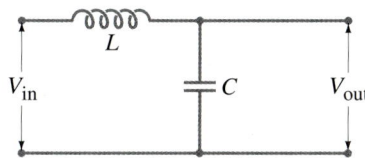

FIGURE 31–12 Problems 46 and 47.

47. Show that if the inductor L in the filter circuit of Fig. 31–12 (Problem 46) is replaced by a large resistor R, there will still be significant attenuation of the ac voltage and little attenuation of the dc voltage if the input dc voltage is high and the current (and power) are low.

48. A resistor R, capacitor C, and inductor L are connected in parallel across an ac generator as shown in Fig. 31–13. The source emf is $V = V_0 \sin \omega t$. Determine the current as a function of time (including amplitude and phase) (a) in the resistor, (b) in the inductor, (c) in the capacitor. (d) What is the total current leaving the source? (Give amplitude I_0 and phase.) (e) Determine the impedance Z defined as $Z = V_0/I_0$. (f) What is the power factor?

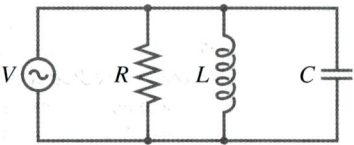

FIGURE 31–13 Problem 48.

49. Suppose a series LRC circuit has two resistors, R_1 and R_2, two capacitors, C_1 and C_2, and two inductors, L_1 and L_2, all in series. Calculate the total impedance of the circuit.

50. Determine the inductance L of the primary of a transformer whose input is 220 V at 60 Hz when the current drawn is 5.8 A. Assume no current in the secondary.

51. You have a small electromagnet that consumes 300 W from a residential circuit operating at 120 V at 60 Hz. Using your ac multimeter, you determine that the unit draws 4.0 A rms. What are the values of the inductance and the internal resistance?

52. An inductor L in series with a resistor R, driven by a sinusoidal voltage source, responds as described by the following differential equation:

$$V_0 \sin \omega t = L \frac{dI}{dt} + RI.$$

Show that a current of the form $I = I_0 \sin(\omega t - \phi)$ flows through the circuit by direct substitution into the differential equation. Determine the amplitude of the current (I_0) and the phase difference ϕ between the current and the voltage source.

53. For the circuit shown in Fig. 31–14, $V = V_0 \sin \omega t$. Calculate the current in each element of the circuit, as well as the total impedance.

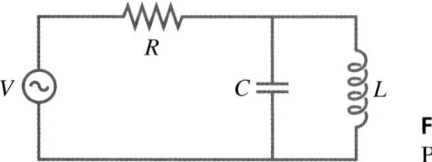

FIGURE 31–14 Problem 53.

54. Show that if the condition $R_1 R_2 = L/C$ is satisfied in the circuit shown in Fig. 31–15, then the potential difference between points a and b is zero for all frequencies.

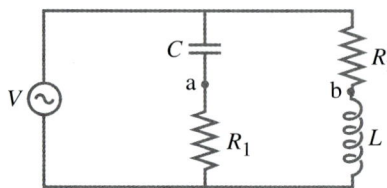

FIGURE 31–15 Problem 54.

These circular disk antennas, each 25 m in diameter, are pointed to receive radio waves from out in space. Radio waves are electromagnetic (EM) waves that have frequencies from a few hundred Hz to about 100 MHz. These antennas are connected together electronically to achieve better detail, and are a part of the Very Large Array in New Mexico searching the heavens for information about the Cosmos.

We will see in this chapter that Maxwell predicted the existence of EM waves from his famous equations. Maxwell's equations themselves are a magnificent summary of electromagnetism. We will also examine how EM waves carry energy and momentum.

CHAPTER 32

Maxwell's Equations and Electromagnetic Waves

The culmination of electromagnetic theory in the nineteenth century was the prediction, and the experimental verification, that waves of electromagnetic fields could travel through space. This achievement opened a whole new world of communication—first the wireless telegraph, then radio and television. And it yielded the spectacular prediction that light is an electromagnetic wave.

The theoretical prediction of electromagnetic waves was the work of the Scottish physicist James Clerk Maxwell (1831–1879; Fig. 32–1), who unified, in one magnificent theory, all the phenomena of electricity and magnetism.

The development of electromagnetic theory in the early part of the nineteenth century by Oersted, Ampère, and others was not actually done in terms of electric and magnetic fields. The idea of the field was introduced somewhat later by Faraday, and was not generally used until Maxwell showed that all electric and magnetic phenomena could be described using only four equations involving electric and magnetic fields. These equations, known as **Maxwell's equations**, are the basic equations for all electromagnetism. They are fundamental in the same sense that Newton's three laws of motion and the law of universal gravitation are for mechanics. In a sense, they are even more fundamental, since they are consistent with the theory of relativity (Chapter 37), whereas Newton's laws are not. Because all of electromagnetism is contained in this set of four equations, Maxwell's equations are considered one of the great triumphs of human intellect.

Before proceeding to a discussion of Maxwell's equations and electromagnetic waves, we first need to discuss a major new prediction of Maxwell's, and, in addition, Gauss's law for magnetism.

FIGURE 32–1 James Clerk Maxwell (1831–1879).

32–1 Changing Electric Fields Produce Magnetic Fields; Ampère's Law and Displacement Current

Ampère's Law

That a magnetic field is produced by an electric current was discovered by Oersted, and the mathematic relation is given by Ampère's law (Eq. 28–3):

$$\oint \mathbf{B} \cdot d\mathbf{l} = \mu_0 I_{\text{encl}}.$$

Is it possible that magnetic fields could be produced in another way as well? For if a changing magnetic field produces an electric field, as discussed in Section 29–7, then perhaps the reverse might be true as well: that *a changing electric field will produce a magnetic field*. If this were true, it would signify a beautiful symmetry in nature.

To back up this idea that a changing electric field might produce a magnetic field, we use an indirect argument that goes something like this. According to Ampère's law, we divide any chosen closed path into short segments $d\mathbf{l}$, take the dot product of each $d\mathbf{l}$ with the magnetic field $\mathbf{B}$ at that segment, and sum (integrate) all these products over the chosen closed path. That sum will equal μ_0 times the total current I that passes through a surface bounded by the path of the line integral. When we applied Ampère's law to the field around a straight wire (Section 28–4), we imagined the current as passing through the circular area enclosed by our circular loop, and that area is the flat surface 1 shown in Fig. 32–2. However, we could just as well use the sackshaped surface 2 in Fig. 32–2 as the surface for Ampère's law, since the same current I passes through it.

Now consider the closed circular path for the situation of Fig. 32–3, where a capacitor is being discharged. Ampère's law works for surface 1 (current I passes through surface 1), but it does not work for surface 2, since no current passes through surface 2. There is a magnetic field around the wire, so the left side of Ampère's law is not zero; yet no current flows through surface 2, so the right side *is* zero. We seem to have a contradiction of Ampère's law.

There is a magnetic field present in Fig. 32–3, however, only if charge is flowing to or away from the capacitor plates. The changing charge on the plates means that the electric field between the plates is changing in time. Maxwell resolved the problem of no current through surface 2 in Fig. 32–3 by proposing that there needs to be an extra term in Ampère's law involving the changing electric field.

Let us see what this term should be by determining it for the changing electric field between the capacitor plates in Fig. 32–3. The charge Q on a capacitor of capacitance C is $Q = CV$ where V is the potential difference between the plates. Also recall that $V = Ed$ where d is the (small) separation of the plates and E is the (uniform) electric field strength between them, and we ignore any fringing of the field. Also, for a parallel-plate capacitor, $C = \epsilon_0 A/d$, where A is the area of each plate (see Chapter 24). We combine these to obtain:

$$Q = CV = \left(\epsilon_0 \frac{A}{d}\right)(Ed) = \epsilon_0 A E.$$

Now if the charge on the plate changes at a rate dQ/dt, the electric field changes at a proportional rate. That is, by differentiating this expression for Q, we have:

$$\frac{dQ}{dt} = \epsilon_0 A \frac{dE}{dt}.$$

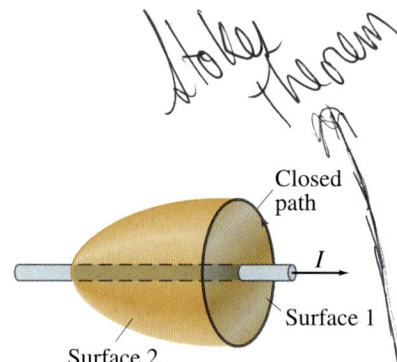

FIGURE 32–2 Ampère's law applied to two different surfaces bounded by the same closed path.

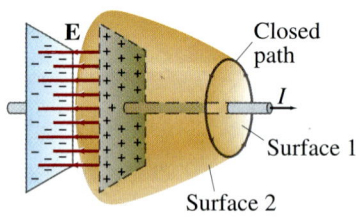

FIGURE 32–3 A capacitor discharging. No conduction current passes through surface 2. An extra term is needed in Ampère's law.

Now dQ/dt is also the current I flowing into or out of the capacitor:

$$I = \frac{dQ}{dt} = \epsilon_0 A \frac{dE}{dt} = \epsilon_0 \frac{d\Phi_E}{dt}$$

where $\Phi_E = EA$ is the **electric flux** through the closed path (surface 2 in Fig. 32–3). In order to make Ampère's law work for surface 2 in Fig. 32–3, as well as for surface 1 (where current I flows), we therefore write:

$$\oint \mathbf{B} \cdot d\mathbf{l} = \mu_0 I_{\text{encl}} + \mu_0 \epsilon_0 \frac{d\Phi_E}{dt}. \tag{32–1}$$

Ampère's law (generalized)

This equation represents the general form of Ampère's law[†], and embodies Maxwell's idea that a magnetic field can be caused not only by an ordinary electric current, but also by a changing electric field or changing electric flux. Although we arrived at it for a special case, Eq. 32–1 has proved valid in general. The last term on the right in Eq. 32–1 is usually very small, and not easy to measure experimentally.

EXAMPLE 32–1 Charging capacitor. A 30-pF air-gap capacitor has circular plates of area $A = 100 \text{ cm}^2$. It is charged by a 70-V battery through a 2.0-Ω resistor. At the instant the battery is connected, the electric field between the plates is changing most rapidly. At this instant, calculate (*a*) the current into the plates, and (*b*) the rate of change of electric field between the plates. (*c*) Determine the magnetic field induced between the plates. Assume **E** is uniform between the plates at any instant and is zero at all points beyond the edges of the plates.

SOLUTION (*a*) In Section 26–4 we discussed RC circuits and saw that the charge on a capacitor being charged, as a function of time, is

$$Q = CV_0(1 - e^{-t/RC}),$$

where V_0 is the voltage of the battery. To find the current at $t = 0$, we differentiate this and substitute the values $V_0 = 70 \text{ V}$, $C = 30 \text{ pF}$, $R = 2.0 \, \Omega$:

$$\left.\frac{dQ}{dt}\right|_{t=0} = \left.\frac{CV_0}{RC} e^{-t/RC}\right|_{t=0} = \frac{V_0}{R} = \frac{70 \text{ V}}{2.0 \, \Omega} = 35 \text{ A}.$$

This is the rate at which charge accumulates on the capacitor and equals the current flowing in the circuit at this instant.
(*b*) The electric field between two closely spaced conductors is given by

$$E = \frac{\sigma}{\epsilon_0} = \frac{Q/A}{\epsilon_0}$$

as we saw in Chapter 21 (see Example 21–12). Hence

$$\frac{dE}{dt} = \frac{dQ/dt}{\epsilon_0 A} = \frac{35 \text{ A}}{(8.85 \times 10^{-12} \text{ C}^2/\text{N}\cdot\text{m}^2)(1.0 \times 10^{-2} \text{ m}^2)} = 4.0 \times 10^{14} \text{ V/m}\cdot\text{s}.$$

[†]Actually, there is a third term on the right for the case when a magnetic field is produced by magnetized materials. This can be accounted for by changing μ_0 to μ, but we will mainly be interested in cases where no magnetic material is present. In the presence of a dielectric, ϵ_0 is replaced by $\epsilon = K\epsilon_0$ (see Section 24–5).

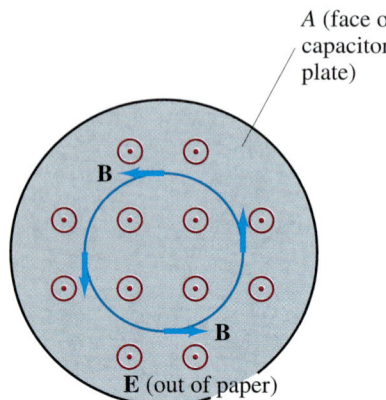

FIGURE 32–4 Frontal view of a circular plate of a parallel-plate capacitor. **E** between plates points out toward viewer; lines of **B** are circles. (Example 32–1.)

(c) Although we will not prove it, we might expect the lines of **B** to be perpendicular to **E** and, because of symmetry, to be circles, as shown in Fig. 32–4; this is the same symmetry we saw for the inverse situation of a changing magnetic field producing an electric field (Section 29–7, see Fig. 29–22). To determine B between the plates, we apply Ampère's law, Eq. 32–1, with the current $I_{encl} = 0$:

$$\oint \mathbf{B} \cdot d\mathbf{l} = \mu_0 \epsilon_0 \frac{d\Phi_E}{dt}.$$

We choose our path to be a circle of radius r, centered at the center of the plate, and thus following a magnetic field line such as the one shown in Fig. 32–4. For $r \leq r_0$ (the radius of plate) the flux through a circle of radius r is $E(\pi r^2)$ since E is assumed uniform within the plates at any moment. So from Ampère's law we have

$$B(2\pi r) = \mu_0 \epsilon_0 \frac{d}{dt}(\pi r^2 E)$$

$$= \mu_0 \epsilon_0 \pi r^2 \frac{dE}{dt}.$$

Hence

$$B = \frac{\mu_0 \epsilon_0}{2} r \frac{dE}{dt}. \qquad [r \leq r_0]$$

We assume $E = 0$ for $r > r_0$, so for points beyond the edge of the plates all the flux is contained within the plates (area $= \pi r_0^2$) and $\Phi_E = E\pi r_0^2$. Thus Ampère's law gives

$$B(2\pi r) = \mu_0 \epsilon_0 \frac{d}{dt}(\pi r_0^2 E)$$

$$= \mu_0 \epsilon_0 \pi r_0^2 \frac{dE}{dt}$$

or

$$B = \frac{\mu_0 \epsilon_0 r_0^2}{2r} \frac{dE}{dt}. \qquad [r \geq r_0]$$

B has its maximum value at $r = r_0$ which, from either relation above (using $r_0 = \sqrt{A/\pi} = 5.6$ cm), is

$$B = \frac{\mu_0 \epsilon_0 r_0}{2} \frac{dE}{dt}$$

$$= \tfrac{1}{2}(4\pi \times 10^{-7}\,\text{T}\cdot\text{m/A})(8.85 \times 10^{-12}\,\text{C}^2/\text{N}\cdot\text{m}^2)(5.6 \times 10^{-2}\,\text{m})(4.0 \times 10^{14}\,\text{V/m}\cdot\text{s})$$

$$= 1.2 \times 10^{-4}\,\text{T}.$$

This is a very small field and lasts only briefly (the time constant $RC = 6.0 \times 10^{-11}$ s) and so would be very difficult to measure.

Let us write the magnetic field B outside the capacitor plates of Example 32–1 in terms of the current I that leaves the plates. The electric field between the plates can be written $E = \sigma/\epsilon_0 = Q/\epsilon_0 A$, as we just saw (part b). Hence B for $r > r_0$ is, with $dQ/dt = I$,

$$B = \frac{\mu_0 \epsilon_0 r_0^2}{2r} \frac{dE}{dt} = \frac{\mu_0 \epsilon_0 r_0^2}{2r} \frac{I}{\epsilon_0 \pi r_0^2} = \frac{\mu_0 I}{2\pi r}.$$

This is the same formula for the field that surrounds a wire (Eq. 28–1). Thus the B field outside the capacitor is the same as that outside the wire. In other words, the magnetic field produced by the changing electric field between the plates is the same as that produced by the current in the wire.

Displacement Current

Maxwell thought it useful to interpret the second term on the right in Eq. 32–1 as being *equivalent* to an electric current. He called it a **displacement current**, I_D. An ordinary current I is then called a **conduction current**. Ampère's law can then be written

$$\oint \mathbf{B} \cdot d\mathbf{l} = \mu_0 (I + I_D)_{\text{encl}} \qquad (32\text{–}2)$$

where

$$I_D = \epsilon_0 \frac{d\Phi_E}{dt}. \qquad (32\text{–}3)$$

Displacement current

The term "displacement current" was based on an old discarded theory; don't let it confuse you: I_D does not represent a flow of electric charge[†] nor is there a displacement.

32–2 Gauss's Law for Magnetism

We are almost in a position to state Maxwell's equations, but first we need to discuss the magnetic equivalent of Gauss's law. As we saw in Chapter 29, for a magnetic field **B**, the *magnetic flux* Φ_B through a surface is defined as

$$\Phi_B = \int \mathbf{B} \cdot d\mathbf{A}$$

where the integral is over the area of either an open or a closed surface. The magnetic flux through a closed surface—that is, a surface which completely encloses a volume—is written

$$\Phi_B = \oint \mathbf{B} \cdot d\mathbf{A}.$$

In the electric case, we saw in Section 22–2 that the electric flux Φ_E through a closed surface is equal to the total net charge Q enclosed by the surface, divided by ϵ_0 (Eq. 22–4):

$$\oint \mathbf{E} \cdot d\mathbf{A} = \frac{Q}{\epsilon_0}.$$

This relation is Gauss's law for electricity.

We can write a similar relation for the magnetic flux. We have seen, however, that in spite of intense searches, no isolated magnetic poles (monopoles)—the magnetic equivalent of single electric charges—have ever been observed. Hence, **Gauss's law for magnetism** is

$$\oint \mathbf{B} \cdot d\mathbf{A} = 0. \qquad (32\text{–}4)$$

In terms of magnetic field lines, this relation tells us that as many lines enter the enclosed volume as leave it. If, indeed, magnetic monopoles do not exist, then there are no "sources" or "sinks" for magnetic field lines to start or stop on, as electric field lines start on positive charges and end on negative charges. Magnetic field lines must then be continuous. Even for a bar magnet, a magnetic field **B** exists inside as well as outside the magnetic material, and the lines of **B** are closed loops as shown in Fig. 32–5.

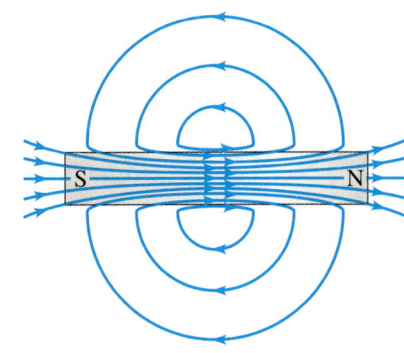

FIGURE 32–5 Magnetic field lines for a bar magnet.

[†]The interpretation of the changing electric field as a displacement current does fit in well with our discussion in Chapter 31 where we saw that an alternating current seems to pass through a capacitor (although charge doesn't). It also means that Kirchhoff's point rule will be valid even at a capacitor plate: for conduction current flows into the plate, but no conduction current flows out of the plate. Instead a "displacement current" flows out of one plate (toward the other).

32–3 Maxwell's Equations

With the extension of Ampère's law given by Eq. 32–1, plus Gauss's law for magnetism (Eq. 32–4), we are now ready to state all four of Maxwell's equations. We have seen them all before in the past dozen chapters. In the absence of dielectric or magnetic materials, **Maxwell's equations** are:

MAXWELL'S

EQUATIONS

$$\oint \mathbf{E} \cdot d\mathbf{A} = \frac{Q}{\epsilon_0} \qquad (32\text{–}5a)$$

$$\oint \mathbf{B} \cdot d\mathbf{A} = 0 \qquad (32\text{–}5b)$$

$$\oint \mathbf{E} \cdot d\mathbf{l} = -\frac{d\Phi_B}{dt} \qquad (32\text{–}5c)$$

$$\oint \mathbf{B} \cdot d\mathbf{l} = \mu_0 I + \mu_0 \epsilon_0 \frac{d\Phi_E}{dt}. \qquad (32\text{–}5d)$$

The first two of Maxwell's equations are simply Gauss's law for electricity (Chapter 22, Eq. 22–4) and Gauss's law for magnetism (Section 32–2, Eq. 32–4). The third is Faraday's law (Chapter 29, Eq. 29–2) and the fourth is Ampère's law as modified by Maxwell (Eq. 32–1). (We dropped the subscripts for simplicity.)

They can be summarized in words: (1) a generalized form of Coulomb's law relating electric field to its sources, electric charges; (2) the same for the magnetic field, except that if there are no magnetic monopoles, magnetic field lines are continuous—they do not begin or end (as electric field lines do on charges); (3) an electric field is produced by a changing magnetic field; (4) a magnetic field is produced by an electric current or by a changing electric field.

Maxwell's equations are the basic equations for all electromagnetism. All of electromagnetism is contained in this set of four equations. They are as fundamental as Newton's three laws of motion and the law of universal gravitation.

In earlier chapters, we have seen that we can treat electric and magnetic fields separately if they do not vary in time. But we cannot treat them independently if they do change in time. For a changing magnetic field produces an electric field; and a changing electric field produces a magnetic field. An important outcome of these relations is the production of electromagnetic waves.

32–4 Production of Electromagnetic Waves

According to Maxwell, a magnetic field will be produced in empty space if there is a changing electric field. From this, Maxwell derived another startling conclusion. If a changing magnetic field produces an electric field, that electric field will itself be changing. This changing electric field will in turn produce a magnetic field, which itself will be changing and so will produce a changing electric field; and so on. When Maxwell manipulated his equations, he found that the net result of these interacting changing fields was a *wave* of electric and magnetic fields that can actually propagate (travel) through space! We now examine, in a simplified way, how such **electromagnetic waves** can be produced. (In Section 32–5 we will examine them more quantitatively.)

Consider two conducting rods that will serve as an "antenna," Fig. 32–6a. Suppose that these two rods are connected by a switch to the opposite terminals of a battery. As soon as the switch is closed, the upper rod quickly becomes positively charged and the lower one negatively charged. Electric field lines are formed as indicated by the red lines in Fig. 32–6b. While the charges are flowing, a current exists whose direction is indicated by the arrows. A magnetic field is therefore produced near the antenna. The magnetic field lines encircle the wires and therefore, in the plane of the page, **B** points into the paper (⊗) on the right and out of the paper (⊙)

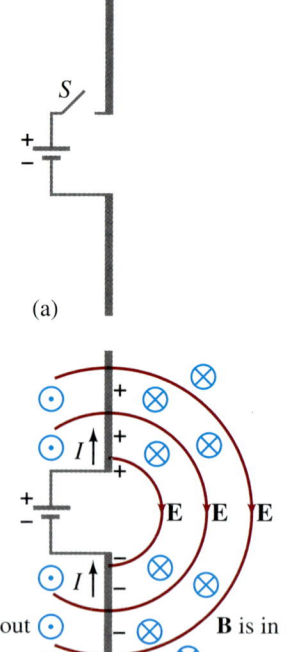

FIGURE 32–6 Fields produced by charge flowing in conductors. It takes time for the **E** and **B** fields to travel outward to distant points.

on the left. Now we ask, how far out do these electric and magnetic fields extend? In the static case, the fields extend outward indefinitely far. However, when the switch in Fig. 32–6 is closed, the fields quickly appear nearby, but it takes time for them to reach distant points. Both electric and magnetic fields store energy, and this energy cannot be transferred to distant points with infinite speed.

Now we look at the situation of Fig. 32–7 where our antenna is connected to an ac generator. In Fig. 32–7a, the connection has just been completed. Charge starts building up and fields form just as in Fig. 32–6. The + and − signs in Fig. 32–7a indicate the net charge on each rod. The black arrows indicate the direction of the current. The electric field is represented by the red lines in the plane of the page; and the magnetic field, according to the right-hand rule, is into (⊗) or out of (⊙) the page. In Fig. 32–7b, the voltage of the ac generator has reversed in direction; the current is reversed and the new magnetic field is in the opposite direction. Because the new fields have changed direction, the old lines fold back to connect up to some of the new lines and form closed loops as shown.† The old fields, however, don't suddenly disappear; they are on their way to distant points. Indeed, because a changing magnetic field produces an electric field, and a changing electric field produces a magnetic field, this combination of changing electric and magnetic fields moving outward is self-supporting, no longer depending on the antenna charges.

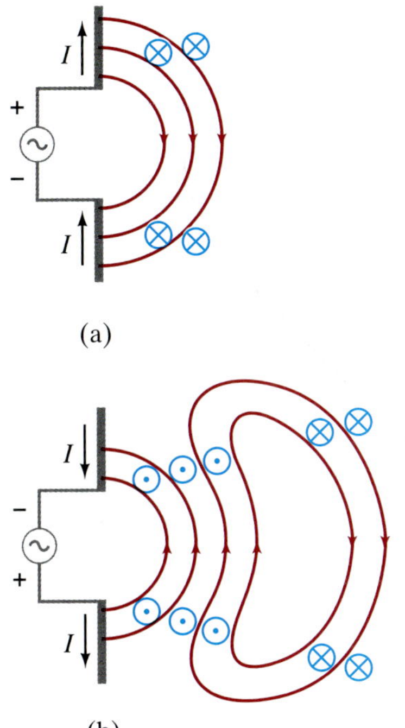

The fields not far from the antenna, referred to as the *near field*, become quite complicated, but we are not so interested in them. We are instead mainly interested in the fields far from the antenna (they are generally what we detect), which we refer to as the **radiation field**. The electric field lines form loops, as shown in Fig. 32–8, and continue moving outward. The magnetic field lines also form closed loops, but are not shown since they are perpendicular to the page. Although the lines are shown only on the right of the source, fields also travel in other directions. The field strengths are greatest in directions perpendicular to the oscillating charges; and they drop to zero along the direction of oscillation—above and below the antenna in Fig. 32–8.

The magnitudes of both **E** and **B** in the radiation field are found to decrease with distance as $1/r$. (Compare this to the static electric field given by Coulomb's law where **E** decreases as $1/r^2$.) The energy carried by the electromagnetic wave is proportional (as for any wave, Chapter 15) to the square of the amplitude, E^2 or B^2, as will be discussed further in Section 32–7, so the intensity of the wave decreases as $1/r^2$.

Several things about the radiation field can be noted from Fig. 32–8. First, *the electric and magnetic fields at any point are perpendicular to each other, and to the direction of motion*. Second, we can see that the fields alternate in direction (**B** is into the page at some points and out of the page at others, similarly for **E**). Thus, the field strengths vary from a maximum in one direction, to zero, to a maximum in the other direction. The electric and magnetic fields are "in phase": that is, they each are zero at the same points and reach their maxima at the same points in space. Finally, very far from the antenna (Fig. 32–8b) the field lines are quite flat over a reasonably large area, and the waves are referred to as **plane waves**.

†We are considering waves traveling through empty space, so there are no charges for lines of **E** to start or stop on, so they form closed loops. Magnetic field lines always form closed loops since there seem to be no single (separate) magnetic poles.

FIGURE 32–7 Sequence showing electric and magnetic fields that spread outward from oscillating charges on two conductors connected to an ac source (see the text).

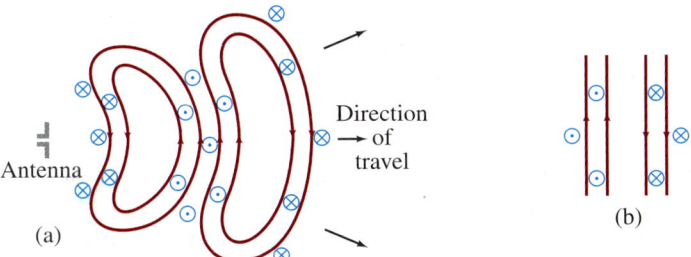

FIGURE 32–8 (a) The radiation fields (far from the antenna) produced by a sinusoidal signal on the antenna. The closed loops represent electric field lines. The magnetic field lines, perpendicular to the page and represented by ⊗ and ⊙, also form closed loops. (b) Very far from the antenna the wave fronts (field lines) are essentially flat over a fairly large area, and are referred to as *plane waves*.

If the source voltage varies sinusoidally, then the electric and magnetic field strengths in the radiation field will also vary sinusoidally. The sinusoidal character of the waves is diagrammed in Fig. 32–9, which shows the field *strengths* plotted as a function of position. Notice that **B** and **E** are perpendicular to each other and to the direction of travel, as pointed out above.

We call these waves electromagnetic (EM) waves. They are *transverse* waves and resemble other types of waves (Chapter 15). However, EM waves are always waves of *fields*, not of matter like waves on water or a rope. Because they are fields, EM waves can propagate in empty space.

We have seen in the above analysis that EM waves are produced by electric charges that are oscillating in an antenna, and hence are undergoing acceleration. In fact, we can say in general that

accelerating electric charges give rise to electromagnetic waves.

Electromagnetic waves can be produced in other ways as well, requiring description at the atomic and nuclear levels, as we will discuss later.

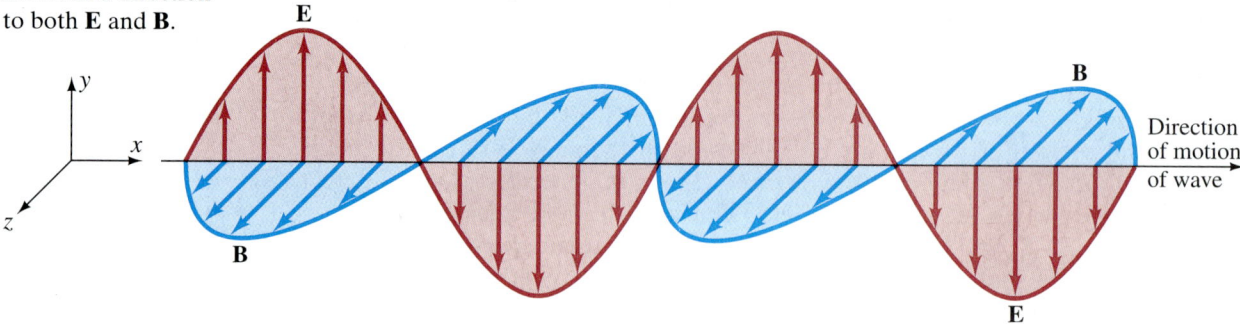

FIGURE 32–9 Electric and magnetic field strengths in an electromagnetic wave. **E** and **B** are at right angles to each other. The entire pattern moves in a direction perpendicular to both **E** and **B**.

32–5 Electromagnetic Waves, and Their Speed, from Maxwell's Equations

Let us now examine how the existence of EM waves follows from Maxwell's equations. We will see that Maxwell's prediction of the existence of EM waves was startling. Equally startling was the speed at which they were predicted to travel.

We begin by considering a region of free space, where there are *no charges or conduction currents*—that is, far from the source so that the wave fronts (the field lines in Fig. 32–8) are essentially flat over a reasonable area. They are then called **plane waves**, as we saw, meaning that at any instant, **E** and **B** are uniform over a reasonably large plane perpendicular to the direction of propagation. We also assume, in a particular coordinate system, that the wave is traveling in the x direction with velocity $\mathbf{v} = v\mathbf{i}$, that **E** is parallel to the y axis, and that **B** is parallel to the z axis, as in Fig. 32–9.

Maxwell's equations, with $Q = I = 0$, become

Maxwell's equations in vacuum

$$\oint \mathbf{E} \cdot d\mathbf{A} = 0 \tag{32-6a}$$

$$\oint \mathbf{B} \cdot d\mathbf{A} = 0 \tag{32-6b}$$

$$\oint \mathbf{E} \cdot d\mathbf{l} = -\frac{d\Phi_B}{dt} \tag{32-6c}$$

$$\oint \mathbf{B} \cdot d\mathbf{l} = \mu_0 \epsilon_0 \frac{d\Phi_E}{dt}. \tag{32-6d}$$

Notice the beautiful symmetry of these equations. The term on the right in the last equation, conceived by Maxwell, is essential for this symmetry. It is also essential if electromagnetic waves are to be produced, as we will now see.

If the wave is sinusoidal with wavelength λ and frequency f, then, as we saw in Chapter 15, Section 15–4, such a traveling wave can be written as

$$E = E_y = E_0 \sin(kx - \omega t)$$
$$B = B_z = B_0 \sin(kx - \omega t) \tag{32-7}$$

where

$$k = \frac{2\pi}{\lambda}, \quad \omega = 2\pi f, \quad \text{and} \quad f\lambda = \frac{\omega}{k} = v, \tag{32-8}$$

with v being the speed of the wave. Although visualizing the wave as sinusoidal is helpful, we will not have to assume this in most of what follows.

Consider now a small rectangle in the plane of the electric field as shown in Fig. 32–10. This rectangle has a finite height Δy, and a very thin width which we take to be the infinitesimal distance dx. In order to show that $\mathbf{E}$, $\mathbf{B}$, and $\mathbf{v}$ are in the orientation shown, we apply Lenz's law to this rectangular loop. The changing magnetic flux through this loop is related to the electric field around the loop by Faraday's law (Maxwell's third equation, Eq. 32–6c). For the case shown, B through the loop is decreasing in time (the wave is moving to the right). So the electric field must be in a direction to oppose this change, meaning E must be greater on the right side of the loop than on the left, as shown (so it could produce a counterclockwise current whose magnetic field would act to oppose the change in Φ_B—but of course there is no current). This brief argument shows that the orientation of $\mathbf{E}$, $\mathbf{B}$, and $\mathbf{v}$ are in the correct relation as shown. That is, $\mathbf{v}$ is in the direction of $\mathbf{E} \times \mathbf{B}$. Now let us apply Faraday's law, which is Maxwell's third equation (Eq. 32–6c),

$$\oint \mathbf{E} \cdot d\mathbf{l} = -\frac{d\Phi_B}{dt}$$

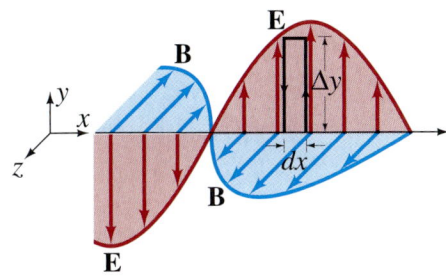

FIGURE 32–10 Applying Faraday's law to the rectangle $(\Delta y)(dx)$.

to the rectangle of height Δy and width dx shown in Fig. 32–10. First we consider $\oint \mathbf{E} \cdot d\mathbf{l}$. Along the short top and bottom sections of length dx, $\mathbf{E}$ is perpendicular to $d\mathbf{l}$, so $\mathbf{E} \cdot d\mathbf{l} = 0$. Along the vertical sides, we let E be the electric field along the left side, and on the right side where it will be slightly larger, it is $E + dE$. Thus, if we take our loop counterclockwise,

$$\oint \mathbf{E} \cdot d\mathbf{l} = (E + dE)\Delta y - E\Delta y = dE\,\Delta y.$$

For the right side of Faraday's law, the magnetic flux through the loop changes as

$$\frac{d\Phi_B}{dt} = \frac{dB}{dt} dx\,\Delta y,$$

since the area of the loop, $(dx)(\Delta y)$, is not changing. Thus, Faraday's law gives us

$$dE\,\Delta y = -\frac{dB}{dt} dx\,\Delta y$$

or

$$\frac{dE}{dx} = -\frac{dB}{dt}.$$

Actually, both E and B are functions of position x and time t. We should therefore use partial derivatives:

$$\frac{\partial E}{\partial x} = -\frac{\partial B}{\partial t} \tag{32-9}$$

where $\partial E/\partial x$ means the derivative of E with respect to x while t is held fixed, and $\partial B/\partial t$ is the derivative of B with respect to t while x is kept fixed.

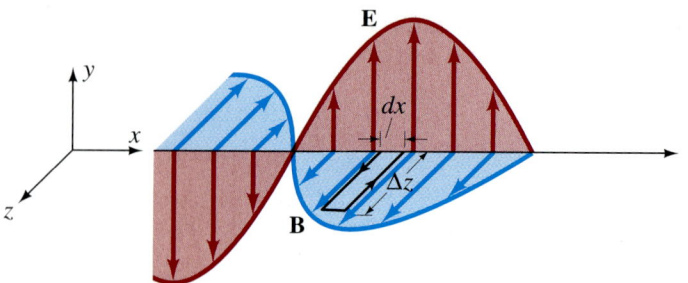

FIGURE 32–11 Applying Maxwell's fourth equation to the rectangle $(\Delta z)(dx)$.

We can obtain another important relation between E and B in addition to Eq. 32–9. To do so, we consider now a small rectangle in the plane of **B**, whose length and width are Δz and dx as shown in Fig. 32–11. To this rectangular loop we apply Maxwell's fourth equation (the extension of Ampère's law):

$$\oint \mathbf{B} \cdot d\mathbf{l} = \mu_0 \epsilon_0 \frac{d\Phi_E}{dt}$$

where we have taken $I = 0$ since we assume the absence of conduction currents. Along the short sides (dx), $\mathbf{B} \cdot d\mathbf{l}$ is zero since **B** is perpendicular to $d\mathbf{l}$. Along the longer sides (Δz), we let B be the magnetic field along the left side of length Δz, and $B + dB$ be the field along the right side. We again integrate counterclockwise, so

$$\oint \mathbf{B} \cdot d\mathbf{l} = B\,\Delta z - (B + dB)\,\Delta z = -dB\,\Delta z.$$

The right side of Maxwell's fourth equation is

$$\mu_0 \epsilon_0 \frac{d\Phi_E}{dt} = \mu_0 \epsilon_0 \frac{dE}{dt} dx\, \Delta z.$$

Equating the two expressions, we obtain

$$-dB\,\Delta z = \mu_0 \epsilon_0 \frac{dE}{dt} dx\, \Delta z$$

or

$$\frac{\partial B}{\partial x} = -\mu_0 \epsilon_0 \frac{\partial E}{\partial t} \quad (32\text{–}10)$$

where we have replaced dB/dx and dE/dt by the proper partial derivatives as before.

We can use Eqs. 32–9 and 32–10 to obtain a relation between the magnitudes of **E** and **B**, and the speed v. Let E and B be given by Eqs. 32–7 as a function of x and t. When we apply Eq. 32–9, taking the derivatives of E and B as given by Eqs. 32–7, we obtain

$$kE_0 \cos(kx - \omega t) = \omega B_0 \cos(kx - \omega t)$$

or

$$\frac{E_0}{B_0} = \frac{\omega}{k} = v,$$

since $v = \omega/k$ (see Eq. 32–8 or 15–12). Since E and B are in phase, we see that E and B are related by

$$\frac{E}{B} = v \quad (32\text{–}11)$$

at any point in space, where v is the velocity of the wave.

Now we apply Eq. 32–10 to the sinusoidal fields (Eqs. 32–7) and we obtain

$$kB_0 \cos(kx - \omega t) = \mu_0 \epsilon_0 \omega E_0 \cos(kx - \omega t)$$

or

$$\frac{B_0}{E_0} = \frac{\mu_0 \epsilon_0 \omega}{k} = \mu_0 \epsilon_0 v.$$

But $B_0/E_0 = 1/v$ from Eq. 32–11, so

$$\mu_0 \epsilon_0 v = \frac{1}{v}$$

or

$$v = \frac{1}{\sqrt{\epsilon_0 \mu_0}}. \qquad (32\text{–}12)$$

Thus the speed of electromagnetic waves in free space is a constant, independent of the wavelength or frequency. If we put in values for ϵ_0 and μ_0 we find

$$v = \frac{1}{\sqrt{\epsilon_0 \mu_0}} = \frac{1}{\sqrt{(8.85 \times 10^{-12}\,\text{C}^2/\text{N}\cdot\text{m}^2)(4\pi \times 10^{-7}\,\text{T}\cdot\text{m/A})}}$$

$$= 3.00 \times 10^8\,\text{m/s}.$$

Speed of EM waves is speed of light

This is a remarkable result. For this is precisely equal to the measured speed of light!

* Derivation of Speed of Light (General)

We can derive the speed of EM waves without having to assume sinusoidal waves by combining Eqs. 32–9 and 32–10 as follows. We take the derivative with respect to t of Eq. 32–10

$$\frac{\partial^2 B}{\partial t\, \partial x} = -\mu_0 \epsilon_0 \frac{\partial^2 E}{\partial t^2}.$$

We next take the derivative of Eq. 32–9 with respect to x:

$$\frac{\partial^2 E}{\partial x^2} = -\frac{\partial^2 B}{\partial t\, \partial x}.$$

Since $\partial^2 B/\partial t\, \partial x$ appears in both relations, we obtain

$$\frac{\partial^2 E}{\partial t^2} = \frac{1}{\mu_0 \epsilon_0} \frac{\partial^2 E}{\partial x^2}. \qquad (32\text{–}13\text{a})$$

By taking other derivatives of Eqs. 32–9 and 32–10 we obtain the same relation for B:

$$\frac{\partial^2 B}{\partial t^2} = \frac{1}{\mu_0 \epsilon_0} \frac{\partial^2 B}{\partial x^2}. \qquad (32\text{–}13\text{b})$$

Both of Eqs. 32–13 have the form of the *wave equation* for a plane wave traveling in the x direction,

$$\frac{\partial^2 y}{\partial t^2} = v^2 \frac{\partial^2 y}{\partial x^2},$$

as discussed in Section 15–5 (Eq. 15–16). We see that the velocity v is given by

$$v^2 = \frac{1}{\mu_0 \epsilon_0}$$

in agreement with Eq. 32–12. Thus we see that a natural outcome of Maxwell's equations is that E and B obey the wave equation for waves traveling with speed $v = 1/\sqrt{\mu_0 \epsilon_0}$. It was on this basis that Maxwell predicted the existence of electromagnetic waves.

32–6 Light as an Electromagnetic Wave and the Electromagnetic Spectrum

The calculations in Section 32–5 gave the result that Maxwell himself determined: that the speed of EM waves is 3.00×10^8 m/s, the same as the measured speed of light.

Light had been shown some 60 years previously to behave like a wave (we'll discuss this in Chapter 35). But nobody knew what kind of wave it was—that is, what is it that is oscillating in a light wave? Maxwell, on the basis of the calculated speed of EM waves, argued that light must be an electromagnetic wave. This idea soon came to be generally accepted by scientists, but not fully until after EM waves were experimentally detected. EM waves were first generated and detected experimentally by Heinrich Hertz (1857–1894) in 1887, eight years after Maxwell's death. Hertz used a spark-gap apparatus in which charge was made to rush back and forth for a short time, generating waves whose frequency was about 10^9 Hz. He detected them some distance away using a loop of wire in which an emf was produced when a changing magnetic field passed through. These waves were later shown to travel at the speed of light, 3.00×10^8 m/s, and to exhibit all the characteristics of light such as reflection, refraction, and interference. The only difference was that they were not visible. Hertz's experiment was a strong confirmation of Maxwell's theory.

The wavelengths of visible light were measured in the first decade of the nineteenth century, long before anyone imagined that light was an electromagnetic wave. The wavelengths were found to lie between 4.0×10^{-7} m and 7.5×10^{-7} m; or 400 nm to 750 nm ($1 \text{ nm} = 10^9$ m). The frequencies of visible light can be found using Eq. 15–1, which we rewrite here:

$$f\lambda = c, \qquad (32\text{--}14)$$

where f and λ are the frequency and wavelength, respectively, of the wave. Here, *c is symbol for speed of light*
c is the speed of light, 3.00×10^8 m/s; it gets the special symbol c because of its universality for all EM waves in free space. Equation 32–14 tells us that the frequencies of visible light are between 4.0×10^{14} Hz and 7.5×10^{14} Hz. (Recall that $1 \text{ Hz} = 1$ cycle per second $= 1 \text{ s}^{-1}$.)

But visible light is only one kind of EM wave. As we have seen, Hertz produced EM waves of much lower frequency, about 10^9 Hz. These are called **radio waves**, since frequencies in this range are used today to transmit radio and TV signals. Electromagnetic waves, or EM radiation as we sometimes call it, have been produced or detected over a wide range of frequencies. They are usually categorized as shown in Fig. 32–12, which is known as the **electromagnetic spectrum**.

EM spectrum

Radio waves and microwaves can be produced in the laboratory using electronic equipment, as we saw in Fig. 32–7. Higher-frequency waves are very difficult to produce electronically. These and other types of EM waves are produced in natural processes, as emission from atoms, molecules, and nuclei (more on this later).

FIGURE 32–12 Electromagnetic spectrum.

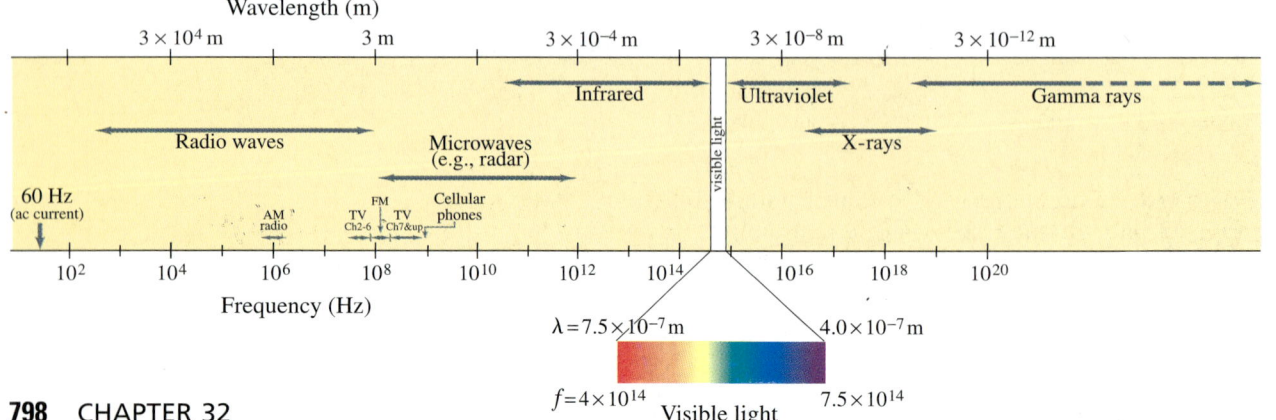

EM waves can be produced by the acceleration of electrons or other charged particles, such as electrons accelerating in the antenna of Fig. 32–7. Another example is X-rays, which are produced (Chapter 36) when fast-moving electrons are rapidly decelerated upon striking a metal target. Even the visible light emitted by an ordinary incandescent light is due to electrons undergoing acceleration within the hot filament. We will meet various types of EM waves later. However, it is worth mentioning here that infrared (IR) radiation (EM waves whose frequency is just less than that of visible light) is mainly responsible for the heating effect of the Sun. The Sun emits not only visible light but substantial amounts of IR and UV (ultraviolet) as well. The molecules of our skin tend to "resonate" at infrared frequencies, so it is these that are preferentially absorbed and thus warm us up. We humans experience EM waves differently depending on their wavelengths: Our eyes detect wavelengths between 4 and 7×10^{-7} m (visible light), whereas our skin detects longer wavelengths (IR). Many EM wavelengths we don't detect directly at all.

EXAMPLE 32–2 Wavelengths of EM waves. Calculate the wavelength: (*a*) of a 60-Hz EM wave, (*b*) of a 93.3-MHz FM radio wave, and (*c*) of a beam of visible red light from a laser at frequency 4.74×10^{14} Hz.

SOLUTION (*a*) Since $c = \lambda f$,

$$\lambda = \frac{c}{f} = \frac{3.0 \times 10^8 \text{ m/s}}{60 \text{ s}^{-1}} = 5.0 \times 10^6 \text{ m},$$

or 5000 km. 60 Hz is the frequency of ac current in the United States, and, as we see here, one wavelength stretches all the way across the country.

(*b*) $$\lambda = \frac{3.00 \times 10^8 \text{ m/s}}{93.3 \times 10^6 \text{ s}^{-1}} = 3.22 \text{ m}.$$

The length of an FM antenna is about half this $(\frac{1}{2}\lambda)$.

(*c*) $$\lambda = \frac{3.00 \times 10^8 \text{ m/s}}{4.74 \times 10^{14} \text{ s}^{-1}} = 6.33 \times 10^{-7} \text{ m} (= 633 \text{ nm}).$$

EXAMPLE 32–3 Determining E and B in EM waves. Assume the 60 Hz EM wave in Example 32–2 is a sinusoidal wave propagating in the z direction with **E** pointing in the x direction, and $E_0 = 2.0$ V/m. Write vector expressions for **E** and **B** as functions of position and time.

SOLUTION From Eq. 32–8 we have

$$k = 2\pi/\lambda = \frac{2\pi}{5.0 \times 10^6 \text{ m}} = 1.26 \times 10^{-6} \text{ m}^{-1}$$

$$\omega = 2\pi f = 2\pi(60 \text{ Hz}) = 3.77 \times 10^2 \text{ rad/s}.$$

From Eq. 32–11 with $v = c$, we find that

$$B_0 = \frac{E_0}{c} = \frac{2.0 \text{ V/m}}{3.0 \times 10^8 \text{ m/s}} = 6.7 \times 10^{-9} \text{ T}.$$

The direction of propagation is that of $\mathbf{E} \times \mathbf{B}$, as in Fig. 32–9. With **E** pointing in the x direction, and the wave propagating in the z direction, **B** must point in the y direction. Using Eqs. 32–7 we find:

$$\mathbf{E} = \mathbf{i}(2.0 \text{ V/m}) \sin[(1.26 \times 10^{-6} \text{ m}^{-1})z - (3.77 \times 10^2 \text{ rad/s})t]$$

$$\mathbf{B} = \mathbf{j}(6.67 \times 10^{-9} \text{ T}) \sin[(1.26 \times 10^{-6} \text{ m}^{-1})z - (3.77 \times 10^2 \text{ rad/s})t]$$

Electromagnetic waves can travel along transmission lines as well as in empty space. When a source of emf is connected up to a transmission line—be it two parallel wires or a coaxial cable (Fig. 32–13)—the electric field within the wires is not set up immediately at all points along them, just as we saw in Section 32–4 with reference to Fig. 32–7. Indeed, it can be shown that if the wires are separated by air, the electrical signal travels along the wires at the speed $c = 3.0 \times 10^8 \, \text{m/s}$. For example, when you flip a light switch, the light actually goes on a tiny fraction of a second later. If the wires are in a medium whose electric permittivity is ϵ and magnetic permeability is μ, the speed is not given by Eq. 32–12, but by

$$v = \frac{1}{\sqrt{\epsilon \mu}}.$$

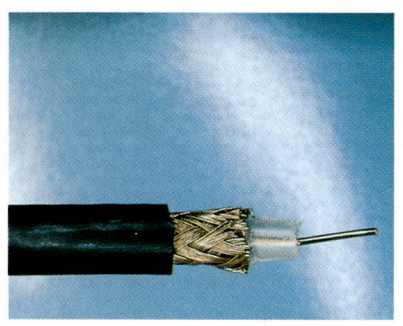

FIGURE 32–13 Coaxial cable.

32–7 Energy in EM Waves; the Poynting Vector

Electromagnetic waves carry energy from one region of space to another. This energy is associated with the moving electric and magnetic fields. In Section 24–4, we saw that the energy density (J/m^3) stored in an electric field E is $u_E = \frac{1}{2}\epsilon_0 E^2$, where u_E is the energy per unit volume. The energy stored in a magnetic field B, as we discussed in Section 30–3, is given by $u_B = \frac{1}{2} B^2/\mu_0$. Thus, the total energy stored per unit volume in a region of space where there is an electromagnetic wave is

$$u = \frac{1}{2}\epsilon_0 E^2 + \frac{1}{2}\frac{B^2}{\mu_0}. \tag{32–15}$$

In this equation, E and B represent the electric and magnetic field strengths of the wave at any instant in a small region of space. We can write Eq. 32–15 in terms of the E field only, since from Eq. 32–12 we have $\sqrt{\epsilon_0 \mu_0} = 1/c$, and from Eq. 32–11, $B = E/c$. We insert these into Eq. 32–15 to obtain

$$u = \frac{1}{2}\epsilon_0 E^2 + \frac{1}{2}\frac{\epsilon_0 \mu_0 E^2}{\mu_0}$$

$$= \epsilon_0 E^2. \tag{32–16a}$$

Notice that the energy density associated with the B field is equal to that associated with the E field, so each contributes half to the total energy. We can also write the energy density in terms of the B field only:

$$u = \epsilon_0 E^2 = \epsilon_0 c^2 B^2 = \frac{\epsilon_0 B^2}{\epsilon_0 \mu_0},$$

so

$$u = \frac{B^2}{\mu_0}. \tag{32–16b}$$

We can also write u in one term containing both E and B:

$$u = \epsilon_0 E^2 = \epsilon_0 E c B = \frac{\epsilon_0 EB}{\sqrt{\epsilon_0 \mu_0}},$$

or

$$u = \sqrt{\frac{\epsilon_0}{\mu_0}} \, EB. \tag{32–16c}$$

Equations 32–16 give the energy density in any region of space at any instant.

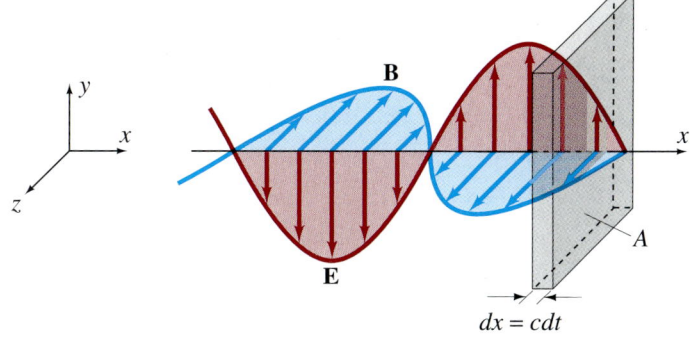

FIGURE 32–14 Electromagnetic wave carrying energy through area A.

Now let us determine the energy the wave transports per unit time per unit area. This is given by a vector **S**, which is called the **Poynting vector**.[†] The units of **S** are W/m². The direction of **S** is the direction in which the energy is transported, which is the direction in which the wave is moving. Let us imagine the wave is passing through an area A perpendicular to the x axis as shown in Fig. 32–14. In a short time dt, the wave moves to the right a distance $dx = c\,dt$ where c is the wave speed. The energy that passes through A in the time dt is the energy that occupies the volume $dV = A\,dx = Ac\,dt$. The energy density u is $u = \epsilon_0 E^2$ where E is the electric field in this volume at the given instant. So the total energy dU contained in this volume dV is the energy density u times the volume: $dU = u\,dV = (\epsilon_0 E^2)(Ac\,dt)$. Therefore the energy crossing the area A per time dt is

$$S = \frac{1}{A}\frac{dU}{dt} = \epsilon_0 c E^2. \tag{32-17}$$

Since $E = cB$ and $c = 1/\sqrt{\epsilon_0 \mu_0}$, this can also be written:

$$S = \epsilon_0 c E^2 = \frac{cB^2}{\mu_0} = \frac{EB}{\mu_0}.$$

The direction of **S** is along **v**, perpendicular to **E** and **B**, so the Poynting vector **S** can be written

$$\mathbf{S} = \frac{1}{\mu_0}(\mathbf{E} \times \mathbf{B}). \tag{32-18}$$

Poynting vector

Equation 32–17 or 32–18 gives the energy transported per unit area per unit time at any *instant*. We often want to know the *average* over an extended period of time since the frequencies are usually so high we don't detect the rapid time variation. If E and B are sinusoidal, then $\overline{E^2} = E_0^2/2$, just as for electric currents and voltages (Section 25–7), where E_0 is the *maximum* value of E. Thus we can write for the magnitude of the Poynting vector, on the average,

$$\overline{S} = \frac{1}{2}\epsilon_0 c E_0^2 = \frac{1}{2}\frac{c}{\mu_0}B_0^2 = \frac{E_0 B_0}{2\mu_0}, \tag{32-19a}$$

where B_0 is the maximum value of B. This time averaged value of **S** is the **intensity**, defined as the average power transferred across unit area (Section 15–3). We can also write

Intensity

$$\overline{S} = \frac{E_{\text{rms}} B_{\text{rms}}}{\mu_0} \tag{32-19b}$$

where E_{rms} and B_{rms} are the rms values ($E_{\text{rms}} = \sqrt{\overline{E^2}}$, $B_{\text{rms}} = \sqrt{\overline{B^2_{\text{rms}}}}$).

[†] After J. H. Poynting (1852–1914).

EXAMPLE 32–4 **E and B from the Sun.** Radiation from the Sun reaches the Earth (above the atmosphere) at a rate of about 1350 W/m². Assume that this is a single EM wave and calculate the maximum values of E and B.

SOLUTION Since $\overline{S} = 1350 \text{ W/m}^2 = \epsilon_0 c E_0^2/2$, then

$$E_0 = \sqrt{\frac{2\overline{S}}{\epsilon_0 c}}$$

$$= \sqrt{\frac{2(1350 \text{ W/m}^2)}{(8.85 \times 10^{-12} \text{ C}^2/\text{N} \cdot \text{m}^2)(3.0 \times 10^8 \text{ m/s})}}$$

$$= 1.01 \times 10^3 \text{ V/m}.$$

From Eq. 32–11, $B = E/c$, so

$$B_0 = \frac{E_0}{c} = \frac{1.01 \times 10^3 \text{ V/m}}{3.00 \times 10^8 \text{ m/s}} = 3.37 \times 10^{-6} \text{ T}.$$

This Example illustrates that B has a small numerical value compared to E. This is because of the different units for E and B and the way these units are defined. But, as we saw earlier, B contributes the same energy to the wave as E does.

*32–8 Radiation Pressure

If electromagnetic waves carry energy, as we have just seen, then we might expect them to also carry linear momentum. When an electromagnetic wave encounters the surface of a material object and is absorbed or reflected, there will be a force exerted on the surface as a result of the momentum transfer $(F = dp/dt)$ just as when a moving object strikes a surface. The force per unit area exerted by the waves is called **radiation pressure**, and its existence was predicted by Maxwell. He showed that if a beam of EM radiation (light, for example) is completely absorbed by an object, then the momentum transferred is,

$$\Delta p = \frac{\Delta U}{c}$$

where ΔU is the energy absorbed by the object in a time Δt and c is the speed of light. If instead, the radiation is fully reflected (suppose the object is a mirror) then the momentum transferred is twice as great, just as when a ball bounces elastically off a surface (Chapter 7):

$$\Delta p = \frac{2\Delta U}{c}.$$

If a surface absorbs some of the energy, and reflects some of it, then $\Delta p = k \Delta U/c$, where k is a constant between 1 and 2.

Using Newton's second law we can calculate the force and the pressure exerted by radiation on the object. The force F is given by

$$F = \frac{dp}{dt}.$$

The average rate that energy is delivered to the object is related to the Poynting vector by

$$\frac{dU}{dt} = \overline{S}A,$$

where A is the cross-sectional area of the object which intercepts the radiation.

The radiation pressure P (assuming full absorption) is given by

$$P = \frac{F}{A} = \frac{1}{A}\frac{dp}{dt} = \frac{1}{Ac}\frac{dU}{dt} = \frac{\overline{S}}{c}.$$

Radiation pressure (absorbed)

If the light is fully reflected, the pressure is twice as great:

$$P = \frac{2\overline{S}}{c}.$$

Radiation pressure (reflected)

EXAMPLE 32–5 ESTIMATE Solar pressure. Radiation from the sun that reaches the Earth's surface transports energy at a rate of about 1000 W/m². Estimate the pressure and force exerted by the Sun on your outstretched hand.

SOLUTION The radiation is partially reflected and partially absorbed, so let us estimate simply $P = \overline{S}/c$, which gives

$$P \approx \frac{\overline{S}}{c} = \frac{1000 \text{ W/m}^2}{3 \times 10^8 \text{ m/s}} \approx 3 \times 10^{-6} \text{ N/m}^2.$$

An estimate of the area of your outstretched hand might be about 10 cm by 20 cm, so $A = 0.02 \text{ m}^2$. Then the force is

$$F = PA \approx (3 \times 10^{-6} \text{ N/m}^2)(0.02 \text{ m}^2) \approx 6 \times 10^{-8} \text{ N}.$$

These numbers are tiny. The force of gravity on your hand, for comparison, is maybe a half pound, or with $m = 0.2 \text{ kg}$, $mg \approx (0.2 \text{ kg})(9.8 \text{ m/s}^2) \approx 2 \text{ N}$. We see that the radiation pressure on your hand is imperceptible.

Although you can not directly feel the effects of radiation pressure, the phenomenon is quite dramatic when applied to atoms irradiated by a laser beam. An atom has a mass on the order of 10^{-27} kg, and a laser can deliver energy at a rate of 1000 W/m². This is the same intensity used in Example 32–5, above, but here a radiation pressure of 10^{-6} N/m² would be very significant on a molecule whose mass might be 10^{-23} to 10^{-26} kg. It is possible, in fact, to move atoms and molecules around by steering them with a laser beam, in a device called "optical tweezers." Optical tweezers have some remarkable applications. They are of great interest to biologists, especially since optical tweezers can manipulate live organisms without damaging them. Optical tweezers have been used to measure the elastic properties of DNA by pulling each end of the molecule with a tweezers.

➡ PHYSICS APPLIED
Optical tweezers

*32–9 Radio and Television

Electromagnetic waves offer the possibility of transmitting information over long distances. Among the first to realize this and put it into practice was Guglielmo Marconi (1874–1937) who, in the 1890s, invented and developed the wireless telegraph. With it, messages could be sent hundreds of kilometers at the speed of light without the use of wires. The first signals were merely long and short pulses that could be translated into words by a code, such as the "dots" and "dashes" of the Morse code. The next decade saw the development of vacuum tubes. Out of this early work radio and television were born. We now discuss briefly (1) how radio and TV signals are transmitted, and (2) how they are received at home.

➡ PHYSICS APPLIED
"Wireless" transmission

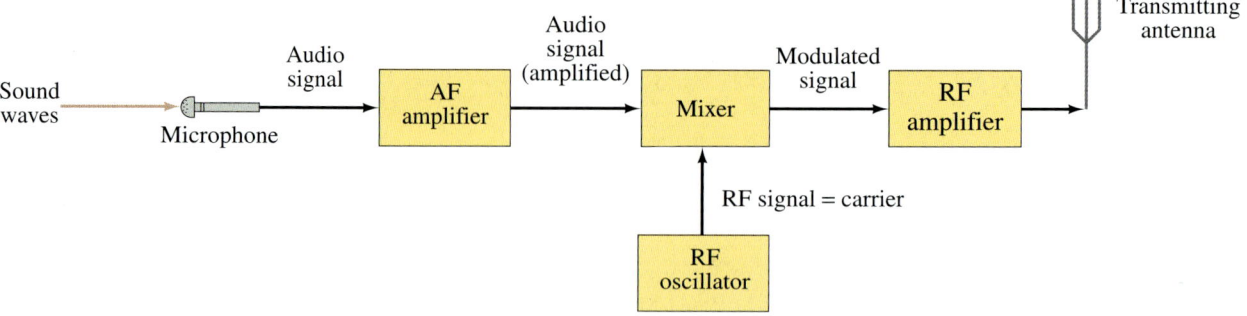

FIGURE 32–15 Block diagram of a radio transmitter.

Transmission of radio waves

The process by which a radio station transmits information (words and music) is outlined in Fig. 32–15. The audio (sound) information is changed into an electrical signal of the same frequencies by, say, a microphone or tape recorder head. This electrical signal is called an audiofrequency (AF) signal, since the frequencies are in the audio range (20 to 20,000 Hz). The signal is amplified electronically and is then mixed with a radio-frequency (RF) signal called its **carrier frequency**. AM radio stations have carrier frequencies from about 530 kHz to 1600 kHz. For example, "710 on your dial" means a station whose carrier frequency is 710 kHz. FM radio stations have much higher carrier frequencies, between 88 MHz and 108 MHz. The carrier frequencies for TV stations in the United States lie between 54 MHz and 88 MHz for channels 2 through 6, and between 174 MHz and 216 MHz for channels 7 through 13; UHF (ultra-high-frequency) stations have even higher carrier frequencies, between 470 MHz and 890 MHz.

Carrier frequency

PHYSICS APPLIED
AM and FM

The mixing of the audio and carrier frequencies is done in two ways. In **amplitude modulation** (AM), the amplitude of the higher frequency carrier wave is made to vary in proportion to the amplitude of the audio signal, as shown in Fig. 32–16. It is called "amplitude modulation" because the amplitude of the carrier is altered ("modulate" means to change or alter). In **frequency modulation** (FM), the *frequency* of the carrier wave is made to change in proportion to the audio signal's amplitude, as shown in Fig. 32–17.

The mixed signal is amplified further and is then sent to the antenna, where the complex mixture of frequencies is sent out in the form of EM waves.

A television transmitter works in a similar way, using frequency modulation, except that both audio and video signals are mixed with carrier frequencies.

FIGURE 32–16 In amplitude modulation (AM), the amplitude of the carrier signal is made to vary in proportion to the audio signal's amplitude.

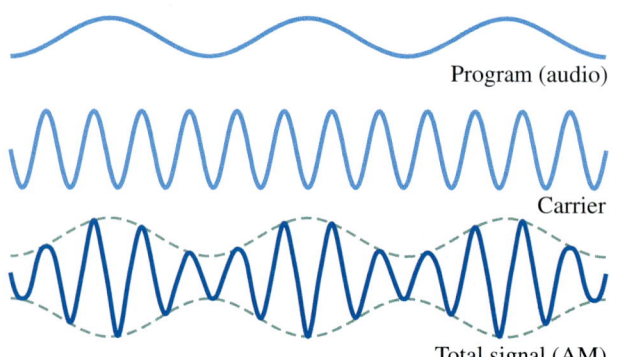

FIGURE 32–17 In frequency modulation (FM), the frequency of the carrier signal is made to change in proportion to the audio signal's amplitude. This method is used by FM radio and television.

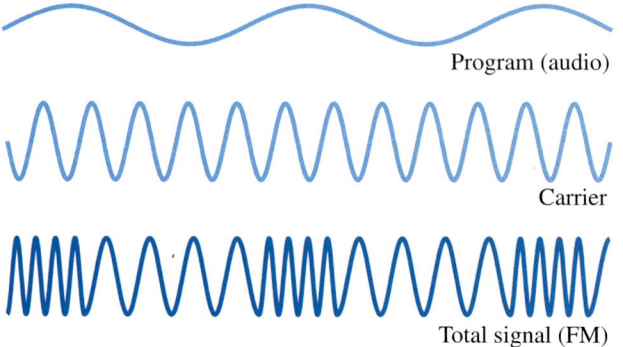

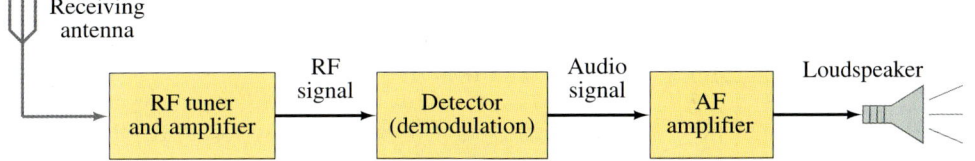

FIGURE 32–18 Block diagram of a simple radio receiver.

Now let us look at the other end of the process, the reception of radio and TV programs at home. A simple radio receiver is diagrammed in Fig. 32–18. The EM waves sent out by all stations are received by the antenna. One kind of antenna consists of one or more conducting rods; the electric field in the EM waves exerts a force on the electrons in the conductor, causing them to move back and forth at the frequencies of the waves (Fig. 32–19a). A second type of antenna consists of a tubular coil of wire, often found in AM radios, or the simple loop of a UHF television antenna. These antennas detect the magnetic field of the wave, for the changing B field induces an emf in the coil (Fig. 32–19b).

→ **PHYSICS APPLIED**
Radio and TV receivers

Antennas

FIGURE 32–19 Antennas. (a) Electric field of EM wave produces a current in an antenna consisting of straight wire or rods. (b) Changing magnetic field induces an emf and current in a loop antenna.

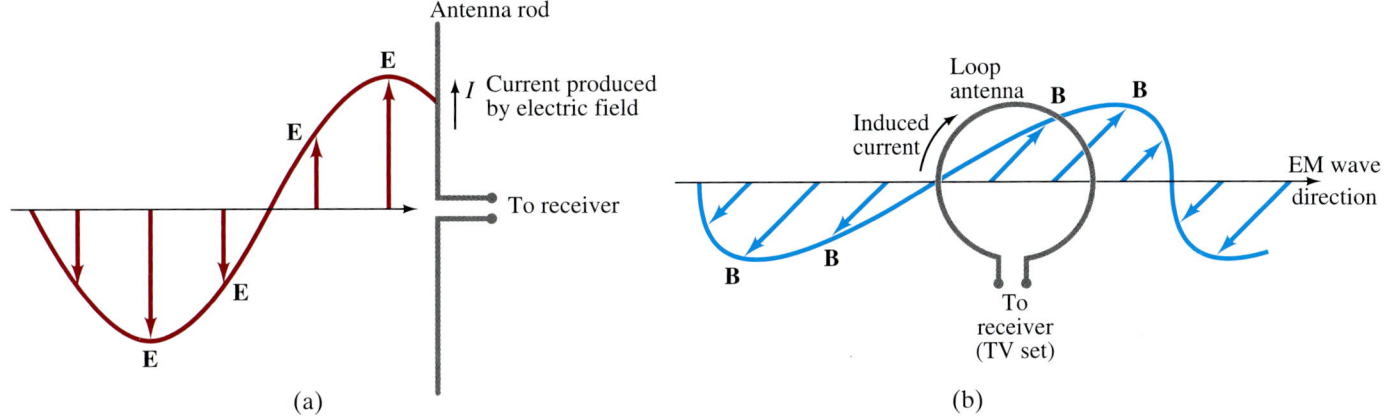

The signal the antenna detects and sends to the receiver is very small and contains frequencies from many different stations. The receiver selects out a particular RF frequency (actually a narrow range of frequencies) corresponding to a particular station using a resonant LC circuit (Sections 30–5 and 31–6) with a variable capacitor or inductor. A simple example is shown in Fig. 32–20. A particular station is "tuned-in" by adjusting L and C so that the resonant frequency of the circuit equals that of the station's carrier frequency. The signal, containing both audio and carrier frequencies, next goes to the *detector* (Fig. 32–18) where "demodulation" takes place—that is, the RF carrier frequency is separated from the audio signal. The audio signal is amplified and sent to a loudspeaker or headphones.

Modern receivers have more stages than those shown. Various means are used to increase the sensitivity and selectivity (ability to detect weak signals and distinguish them from other stations), and to minimize distortion of the original signal.[†]

A television receiver does similar things to both the audio and the video signals. The audio signal goes finally to the loudspeaker, and the video signal to the picture tube, a *cathode ray tube* (CRT) whose operation was discussed in Sections 23–9 and 27–7.

FIGURE 32–20 Simple tuning stage of a radio.

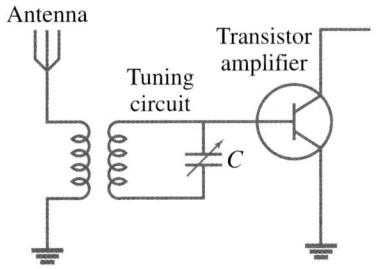

[†]For *FM stereo broadcasting*, two signals are carried by the carrier wave. One of these contains frequencies up to about 15 kHz, which includes most audio frequencies. The other signal includes the same range of frequencies, but 19 kHz is added to it. A stereo receiver subtracts this 19,000-Hz signal and distributes the two signals to the left and right channels. The first signal actually consists of the sum of left and right channels $(L + R)$, so mono radios detect all the sound. The second signal is the difference between left and right $(L - R)$. Hence the receiver must add and subtract the two signals to get pure left and right signal for each channel.

*SECTION 32–9 Radio and Television

EXAMPLE 32-6 Tuning a station. An FM radio station transmits at 100 MHz. Calculate (a) the transmitting wavelength, and (b) the value of the capacitance in the tuning circuit if $L = 0.40\,\mu\text{H}$.

SOLUTION (a) The carrier frequency is $f = 100\,\text{MHz} = 1.0 \times 10^8\,\text{s}^{-1}$, so
$$\lambda = \frac{c}{f} = \frac{(3.0 \times 10^8\,\text{m/s})}{(1.0 \times 10^8\,\text{s}^{-1})} = 3.0\,\text{m}.$$

The wavelengths of other FM signals (88 MHz to 108 MHz) are close to this. FM antennas are typically 1.5 m long, or about a half wavelength. This length is chosen so that the antenna reacts in a resonant fashion and thus is more sensitive to FM frequencies.

(b) According to Eq. 30–14 or 31–13, the resonant frequency is $f_0 = 1/(2\pi\sqrt{LC})$. Therefore,
$$C = \frac{1}{4\pi^2 f_0^2 L} = \frac{1}{4(3.14)^2(1.0 \times 10^8\,\text{s}^{-1})^2(4.0 \times 10^{-7}\,\text{H})}$$
$$= 6.3\,\text{pF}.$$

Of course, the capacitor or inductor is variable, so other stations can be selected too.

The various regions of the radio-wave spectrum are assigned by governmental agencies for various purposes. Besides those mentioned above, there are "bands" assigned for use by ships, airplanes, police, military, amateurs, satellites and space, and radar. Cellular telephones, for example, occupy a band from 824 MHz to 894 MHz.

Summary

James Clerk Maxwell synthesized an elegant theory in which all electric and magnetic phenomena could be described using four equations, now called **Maxwell's equations**. They are based on earlier ideas, but Maxwell added one more—that a changing electric field produces a magnetic field. Maxwell's equations are

$$\oint \mathbf{E} \cdot d\mathbf{A} = \frac{Q}{\epsilon_0} \quad \oint \mathbf{E} \cdot d\mathbf{l} = -\frac{d\Phi_B}{dt}$$

$$\oint \mathbf{B} \cdot d\mathbf{A} = 0 \quad \oint \mathbf{B} \cdot d\mathbf{l} = \mu_0 I + \mu_0 \epsilon_0 \frac{d\Phi_E}{dt}.$$

The two on the left are Gauss's laws for electricity and for magnetism; the two on the right are Faraday's law and Ampère's law (as extended by Maxwell).

Maxwell's theory predicted that transverse **electromagnetic (EM) waves** would be produced by accelerating electric charges, and that these waves would propagate through space at the speed of light c, given by

$$c = \frac{1}{\sqrt{\epsilon_0 \mu_0}}.$$

The oscillating electric and magnetic fields in an EM wave are perpendicular to each other and to the direction of propagation.

After EM waves were experimentally detected in the late 1800s, the idea that light is an EM wave (although of much higher frequency than those detected directly) became generally accepted. The **electromagnetic spectrum** includes EM waves of a wide variety of wavelengths, from microwaves and radio waves to visible light to X-rays and gamma rays, all of which travel through space at a speed $c = 3.00 \times 10^8\,\text{m/s}$.

The energy carried by EM waves can be described by the **Poynting vector**

$$\mathbf{S} = \frac{1}{\mu_0} \mathbf{E} \times \mathbf{B}$$

which gives the rate energy is carried across unit area per unit time when the electric and magnetic fields in an EM wave in free space are $\mathbf{E}$ and $\mathbf{B}$.

Questions

1. Suppose you are looking along the same direction as an electric field **E** that is increasing. Will the induced magnetic field be clockwise or counterclockwise? What if **E** points toward you and is decreasing?
2. What is the direction of the displacement current in Fig. 32–3? (Note: The capacitor is discharging.)
3. Why is it that the magnetic field of a displacement current in a capacitor is so much harder to detect than the magnetic field of a conduction current?
4. Are there any good reasons for calling the term $\mu_0 \epsilon_0 d\Phi_E/dt$ in Eq. 32–1 a "current"? Explain.
5. The electric field in an EM wave traveling north oscillates in an east–west plane. Describe the direction of the magnetic field vector in this wave.
6. Is sound an electromagnetic wave? If not, what kind of wave is it?
7. Can EM waves travel through a perfect vacuum? Can sound waves?
8. When you flip a light switch, does the overhead light go on immediately? Explain.
9. Are the wavelengths of radio and television signals longer or shorter than those detectable by the human eye?
10. What does the result of Example 32–2 tell you about the phase of a 60-Hz ac current that starts at a power plant as compared to its phase at a house 200 km away?
11. When you connect two loudspeakers to the output of a stereo amplifier, should you be sure the lead wires are equal in length so that there will not be a time lag between speakers? Explain.
12. In the electromagnetic spectrum, what type of EM wave would have a wavelength of 10^3 km? 1 km? 1 m? 1 cm? 1 mm? 1 μm?
*13. A lost person may signal by flashing a flashlight on and off using Morse code. This is actually a modulated EM wave. Is it AM or FM? What is the frequency of the carrier, approximately?
*14. Can two radio or TV stations broadcast on the same carrier frequency? Explain.
*15. If a radio transmitter has a vertical antenna, should a receiver's antenna (rod type) be vertical or horizontal to obtain best reception?
*16. The carrier frequencies of FM broadcasts are much higher than for AM broadcasts. On the basis of what you learned about diffraction in Chapter 15, explain why AM signals can be detected more readily than FM signals behind low hills or buildings.
17. Discuss how cordless telephones make use of EM waves. What about cellular telephones?

Problems

Section 32–1

1. (I) Determine the rate at which the electric field changes between the round plates of a capacitor, 6.0 cm in diameter, if the plates are spaced 1.3 mm apart and the voltage across them is changing at a rate of 120 V/s.
2. (I) Calculate the displacement current I_D between the square plates, 3.8 cm on a side, of a capacitor if the electric field is changing at a rate of 2.0×10^6 V/m·s.
3. (II) At a given instant, a 1.8-A current flows in the wires connected to a parallel-plate capacitor. What is the rate at which the electric field is changing between the plates if the square plates are 1.60 cm on a side?
4. (II) A 1200-nF capacitor with circular parallel plates 2.0 cm in diameter is accumulating charge at the rate of 35.0 mC/s at some instant in time. What will be the induced magnetic field strength 10.0 cm radially outward from the center of the plates? What will be the value of the field strength after the capacitor is fully charged?
5. (II) Show that the displacement current through a parallel-plate capacitor can be written $I_D = CdV/dt$, where V is the voltage across the capacitor at any instant.
6. (II) Suppose an air-gap capacitor has circular plates of radius $R = 2.5$ cm and separation $d = 2.0$ mm. A 96.0-Hz emf, $\mathcal{E} = \mathcal{E}_0 \cos \omega t$, is applied to the capacitor. The maximum displacement current is 35 μA. Determine (a) the maximum conduction current I, (b) the value of $\mathcal{E}_0$, (c) the maximum value of $d\Phi_E/dt$ between the plates. Neglect fringing.

Section 32–3

7. (II) If magnetic monopoles existed, which of Maxwell's equations would be altered, and what would be their new form? Let Q_m be the strength of a magnetic monopole, analogous to electric charge Q.

Section 32–5

8. (I) If the magnetic field in a traveling EM wave has a peak value of 17.5 nT, what is the peak value of the electric field strength?
9. (I) If the electric field in an EM wave has a peak value of 0.43×10^{-4} V/m, what is the peak value of the magnetic field strength?
10. (I) In an EM wave traveling west, the B field oscillates vertically and has a frequency of 80.0 kHz and an rms strength of 6.75×10^{-9} T. What is the frequency and rms strength of the electric field and what is its direction?
11. (II) The electric field of a plane EM wave is given by $E_x = E_0 \cos(kz + \omega t)$, $E_y = E_z = 0$. Determine (a) the magnitude and direction of **B**, and (b) the direction of propagation.

Section 32-6

12. (I) What is the frequency of a microwave whose wavelength is 1.80 cm?

13. (I) (a) What is the wavelength of a 27.75×10^9 Hz radar signal? (b) What is the frequency of an X-ray with wavelength 0.10 nm?

14. (I) An EM wave has a wavelength of 850 nm. What is its frequency, and how would we classify it?

15. (I) An EM wave has frequency 9.56×10^{14} Hz. What is its wavelength, and how would we classify it?

16. (I) A light-year is a measure of distance (not time). How many meters does light travel in a year?

17. (II) How long would it take a message sent as radio waves from Earth to reach (a) Mars, (b) a spacecraft near Saturn?

18. (II) Pulsed lasers used for science and medicine produce very short bursts of electromagnetic energy. If the laser light wavelength is 1062 nm (this corresponds to a Neodymium-YAG laser), and the pulse lasts for 30 picoseconds, how many wavelengths are found within the laser pulse? How short would the pulse need to be to fit only one wavelength?

Section 32-7

19. (I) The **E** field in an EM wave has a peak of 36.5 mV/m. What is the average rate at which this wave carries energy across unit area per unit time?

20. (II) The magnetic field in a traveling EM wave has an rms strength of 32.5 nT. How long does it take to deliver 335 J of energy to 1.00 cm² of a wall that it hits perpendicularly?

21. (II) How much energy is transported across a 1.00 cm² area per hour by an EM wave whose E field has an rms strength of 28.6 mV/m?

22. (II) A spherically spreading EM wave comes from a 1000-W source. At a distance of 10.0 m, what is the intensity, and what is the rms value of the electric field?

23. (II) What is the energy contained in a 1.00-m³ volume near the Earth's surface due to radiant energy from the Sun? See Example 32–4.

24. (II) A 12.8-mW laser puts out a narrow beam 2.00 mm in diameter. What are the average (rms) values of E and B in the beam?

25. (II) Estimate the average power output of the Sun, given that about 1350 W/m² reaches the upper atmosphere of the Earth.

26. (II) A high-energy pulsed laser emits a 1.0-ns-long pulse of average power 2.5×10^{11} W. The beam is 2.2×10^{-3} m in radius. Determine (a) the energy delivered in each pulse, and (b) the rms value of the electric field.

27. (III) (a) Show that the Poynting vector **S** points radially inward toward the center of a circular parallel-plate capacitor when it is being charged as in Example 32–1. (b) Integrate **S** over the cylindrical boundary of the capacitor gap to show that the rate at which energy enters the capacitor is equal to the rate at which electrostatic energy is being stored in the electric field of the capacitor (Section 24–4). Ignore fringing of **E**.

*Section 32-8

*28. (II) Estimate the radiation pressure due to a 100 W bulb at a distance of 8.0 cm from the center of the bulb. Estimate the force exerted on your fingertip if you place it at this point.

*29. (III) When the Sun became hot and luminous long ago, it is believed that it ejected dust particles and individual atoms out of the solar system using radiation pressure. Calculate how small the dust particles had to be in order to be ejected by comparing the radiation force with the gravitational force on the particles. Assume the particles are spherical, have a density of 2.0×10^3 kg/m³, and totally absorb the Sun's radiation. The Sun has an average power output of 3.8×10^{26} W.

*30. (III) We discussed in the text two different equations for the momentum transfer due to radiation, one valid for total absorption, the other valid for total reflection. The latter equation is identical to the former with the exception of a factor of two. Give a simple mechanical argument for this factor of two by treating light as massless particles (photons) each with momentum p. In the case of absorption, light particles collide completely inelastically with a massive object, whereas in the case of reflection they collide elastically. Analyze these collisions and explain the factor of two.

*Section 32-9

*31. (I) The variable capacitor in the tuner of an AM radio has a capacitance of 2400 pF when the radio is tuned to a station at 550 kHz. What must the capacitance be for a station at the other end of the dial, 1550 kHz?

*32. (I) The oscillator of a 96.1-MHz FM station has an inductance of 1.8 μH. What value must the capacitance be?

*33. (II) A certain FM radio tuning circuit has a fixed capacitor $C = 840$ pF. Tuning is done by a variable inductance. What range of values must the inductance have to tune stations from 88 MHz to 108 MHz?

*34. (II) An amateur radio operator wishes to build a receiver that can tune a range from 14.0 MHz to 15.0 MHz. A variable capacitor has a minimum capacitance of 92 pF. (a) What is the required value of the inductance? (b) What is the maximum capacitance used on the variable capacitor?

General Problems

35. (a) What is the wavelength of an AM station at 680 kHz? (b) An FM station broadcasts at 100.7 MHz. What is the wavelength of this wave?

36. Compare 940 on the AM dial to 94 on the FM dial. Which has the longer wavelength, and by what factor is it larger?

37. Television broadcast frequencies range between 54.0 MHz for Channel 2 to 806 MHz for Channel 69 (and beyond). What are the wavelengths for these two channels?

38. If the Sun were to disappear or somehow radically change its output, how long would it take for us on Earth to learn about it?

39. (a) How long did it take for a message sent from Earth to reach the first astronauts on the Moon? (b) How long will it take for a message from Earth to reach the first astronauts who arrive on Mars; assume Mars is at its closest approach to Earth $(78 \times 10^6 \, \text{km})$?

40. A radio voice signal from the Apollo crew on the Moon is beamed to a listening crowd from a radio speaker. If you are standing 50 m from the loudspeaker, what is the total time lag between when you hear the sound and when the sound left the Moon?

41. A point source emits light energy uniformly in all directions at an average rate P_0 with a single frequency f. Show that the peak electric field in the wave is given by

$$E_0 = \sqrt{\frac{\mu_0 c P_0}{2\pi r^2}}.$$

42. What are E_0 and B_0 2.00 m from a 100-W light source? Assume the bulb emits radiation of a single frequency uniformly in all directions.

43. Estimate the rms electric field in the sunlight that hits Mars, knowing that the Earth receives about 1350 W/m² and that Mars is 1.52 times farther away from the Sun (on average) than is the Earth.

44. At a given instant in time, a traveling EM wave is noted to have its maximum magnetic field pointing west and its maximum electric field pointing south. In which direction is the wave traveling? If the rate of energy flow is 500 W/m², what are the maximum values for the two fields?

45. Who will hear the voice of a singer first: a person in the balcony 50 m away from the stage (see Fig. 32–21), or a person 3000 km away at home whose ear is next to the radio? Roughly how much sooner? Assume the microphone is a few centimeters from the singer and the temperature is 20°C.

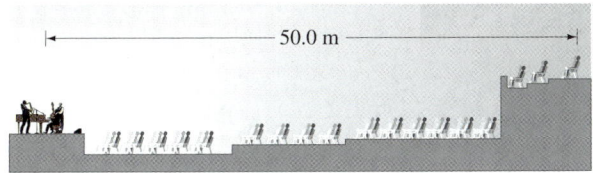

FIGURE 32–21 Problem 45.

46. A 1.80-m-long FM receiving antenna is oriented parallel to the electric field of an EM wave. How large must the **E** field strength be to produce a 1.0-mV rms voltage between the ends of the antenna? What is the rate of energy transport per unit area?

47. Suppose a 50-kW radio station emits EM waves uniformly in all directions. (a) How much energy per second crosses a 1.0-m² area 100 m from the transmitting antenna? (b) What is the rms magnitude of the **E** field at this point, assuming the station is operating at full power? (c) What is the voltage induced in a 1.0-m-long vertical car antenna at this distance?

48. Repeat Problem 47 for a distance of 100 km from the station.

49. Referring to Problem 47, what is the maximum power level of the radio station so as to avoid electrical breakdown of air at a distance of 1.0 m from the antenna? Assume the antenna is a point source. Air breaks down in an electric field of about 3×10^6 V/m. [Hint: See Problem 41.]

50. How large an emf (rms) will be generated in an antenna that consists of a 380-loop circular coil of wire 2.2 cm in diameter if the EM wave has a frequency of 810 kHz and is transporting energy at an average rate of 1.0×10^{-4} W/m² at the antenna? [Hint: You can use Eq. 29–4 for a generator, since it could be applied to an observer moving with the coil so that the magnetic field is oscillating with the frequency $f = \omega/2\pi$.]

*51. The variable capacitance of a radio tuner consists of six plates connected together placed alternately between six other plates, also connected together (Fig. 32–22). Each plate is separated from its neighbor by 1.1 mm of air. One set of plates can move so that the area of overlap varies from 1.0 cm² to 9.0 cm². (a) Are these capacitors connected in series or in parallel? (b) Determine the range of capacitance values. (c) What value of inductor is needed if the radio is to tune AM stations from 550 kHz to 1600 kHz?

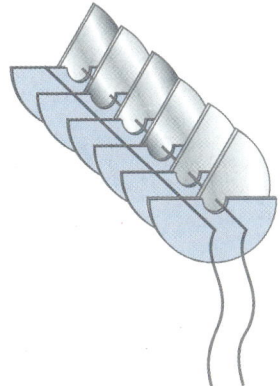

FIGURE 32–22 Problem 51.

52. A cylindrical conductor of radius r and conductivity σ carries a steady current I distributed uniformly across its cross-section. (a) Determine **E** inside the conductor. (b) Determine **B** just outside the conductor. (c) Determine the Poynting vector **S** at the surface of the conductor and show that it is normal to the surface and points inward. (d) Integrate over **S** to show that the rate at which electromagnetic energy enters the sides of the conductor is equal to the rate at which energy is dissipated, $I^2 R$. Thus we can think of the energy as entering the conductor in the form of electromagnetic fields from the sides rather than through the ends.

APPENDIX

Mathematical Formulas

A–1 Quadratic Formula

If
$$ax^2 + bx + c = 0$$
then
$$x = \frac{-b \pm \sqrt{b^2 - 4ac}}{2a}$$

A–2 Binomial Expansion

$$(1 \pm x)^n = 1 \pm nx + \frac{n(n-1)}{2!}x^2 \pm \frac{n(n-1)(n-2)}{3!}x^3 + \cdots$$

$$(x + y)^n = x^n\left(1 + \frac{y}{x}\right)^n = x^n\left(1 + n\frac{y}{x} + \frac{n(n-1)}{2!}\frac{y^2}{x^2} + \cdots\right)$$

A–3 Other Expansions

$$e^x = 1 + x + \frac{x^2}{2!} + \frac{x^3}{3!} + \cdots$$

$$\ln(1 + x) = x - \frac{x^2}{2} + \frac{x^3}{3} - \frac{x^4}{4} + \cdots$$

$$\sin\theta = \theta - \frac{\theta^3}{3!} + \frac{\theta^5}{5!} - \cdots$$

$$\cos\theta = 1 - \frac{\theta^2}{2!} + \frac{\theta^4}{4!} - \cdots$$

$$\tan\theta = \theta + \frac{\theta^3}{3} + \frac{2}{15}\theta^5 + \cdots \qquad |\theta| < \frac{\pi}{2}$$

In general:
$$f(x) = f(0) + \left(\frac{df}{dx}\right)_0 x + \left(\frac{d^2f}{dx^2}\right)_0 \frac{x^2}{2!} + \cdots$$

A–4 Areas and Volumes

Object	Surface area	Volume
Circle, radius r	πr^2	—
Sphere, radius r	$4\pi r^2$	$\frac{4}{3}\pi r^3$
Right circular cylinder, radius r, height h	$2\pi r^2 + 2\pi r h$	$\pi r^2 h$
Right circular cone, radius r, height h	$\pi r^2 + \pi r \sqrt{r^2 + h^2}$	$\frac{1}{3}\pi r^2 h$

A–5 Plane Geometry

1.
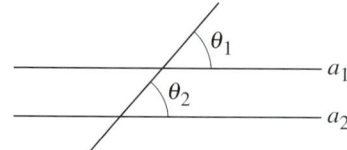

FIGURE A–1

If line a_1 is parallel to line a_2, then $\theta_1 = \theta_2$.

2.
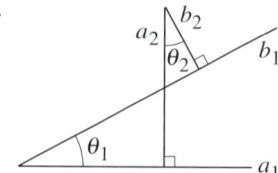

FIGURE A–2

If $a_1 \perp a_2$ and $b_1 \perp b_2$, then $\theta_1 = \theta_2$.

3. The sum of the angles in any plane triangle is 180°.

4. *Pythagorean theorem:*

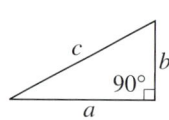

FIGURE A–3

In any right triangle (one angle = 90°) of sides a, b, and c:
$$a^2 + b^2 = c^2$$
where c is the length of the hypotenuse (opposite the 90° angle).

A–6 Trigonometric Functions and Identities

(See Fig. A–4.)

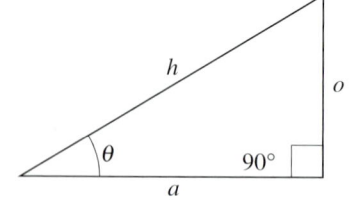

FIGURE A–4

$$\sin\theta = \frac{o}{h} \qquad \csc\theta = \frac{1}{\sin\theta} = \frac{h}{o}$$

$$\cos\theta = \frac{a}{h} \qquad \sec\theta = \frac{1}{\cos\theta} = \frac{h}{a}$$

$$\tan\theta = \frac{o}{a} = \frac{\sin\theta}{\cos\theta} \qquad \cot\theta = \frac{1}{\tan\theta} = \frac{a}{o}$$

$$a^2 + o^2 = h^2 \qquad \text{[Pythagorean theorem]}.$$

Figure A–5 shows the signs (+ or −) that cosine, sine, and tangent take on for angles θ in the four quadrants (0° to 360°). Note that angles are measured counterclockwise from the x axis as shown; negative angles are measured from *below* the x axis, clockwise: for example, $-30° = +330°$, and so on.

FIGURE A–5

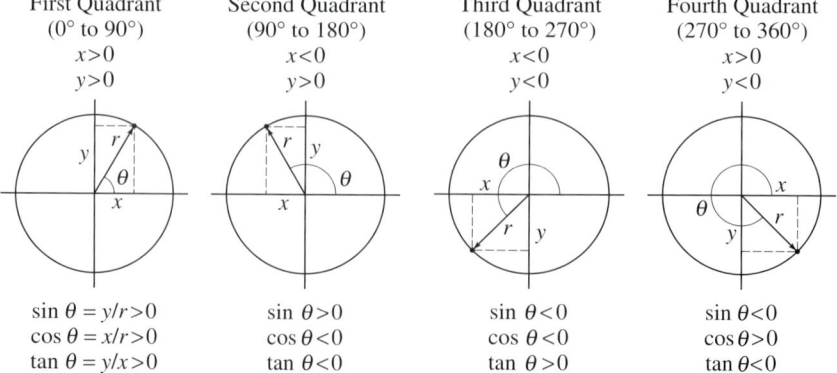

The following are some useful identities among the trigonometric functions:

$$\sin^2\theta + \cos^2\theta = 1, \quad \sec^2\theta - \tan^2\theta = 1, \quad \csc^2\theta - \cot^2\theta = 1$$

$$\sin 2\theta = 2\sin\theta\cos\theta$$

$$\cos 2\theta = \cos^2\theta - \sin^2\theta = 2\cos^2\theta - 1 = 1 - 2\sin^2\theta$$

$$\tan 2\theta = \frac{2\tan\theta}{1 - \tan^2\theta}$$

$$\sin(A \pm B) = \sin A \cos B \pm \cos A \sin B$$

$$\cos(A \pm B) = \cos A \cos B \mp \sin A \sin B$$

$$\tan(A \pm B) = \frac{\tan A \pm \tan B}{1 \mp \tan A \tan B}$$

$$\sin(180° - \theta) = \sin\theta$$
$$\cos(180° - \theta) = -\cos\theta$$
$$\sin(90° - \theta) = \cos\theta$$
$$\cos(90° - \theta) = \sin\theta$$
$$\cos(-\theta) = \cos\theta$$
$$\sin(-\theta) = -\sin\theta$$
$$\tan(-\theta) = -\tan\theta$$

$$\sin\tfrac{1}{2}\theta = \sqrt{\frac{1 - \cos\theta}{2}}, \quad \cos\tfrac{1}{2}\theta = \sqrt{\frac{1 + \cos\theta}{2}}, \quad \tan\tfrac{1}{2}\theta = \sqrt{\frac{1 - \cos\theta}{1 + \cos\theta}}$$

$$\sin A \pm \sin B = 2\sin\left(\frac{A \pm B}{2}\right)\cos\left(\frac{A \mp B}{2}\right).$$

For any triangle (see Fig. A–6):

$$\frac{\sin\alpha}{a} = \frac{\sin\beta}{b} = \frac{\sin\gamma}{c} \qquad \text{[Law of sines]}$$

$$c^2 = a^2 + b^2 - 2ab\cos\gamma. \qquad \text{[Law of cosines]}$$

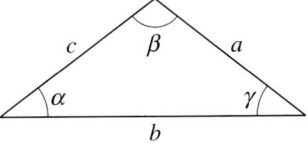

FIGURE A–6

A–7 Logarithms

The following identities apply to common logs (base 10), natural logs (base e) which are often abbreviated ln, or logs to any other base.

$$\log(ab) = \log a + \log b$$

$$\log\left(\frac{a}{b}\right) = \log a - \log b$$

$$\log a^n = n\log a.$$

A–8 Vectors

Vector addition is covered in Sections 3–2 to 3–5.
Vector multiplication is covered in Sections 3–3, 7–2 and 11–1.

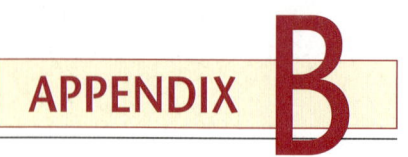

APPENDIX B

Derivatives and Integrals

B–1 Derivatives: General Rules

(See also Section 2–3.)

$$\frac{dx}{dx} = 1$$

$$\frac{d}{dx}[af(x)] = a\frac{df}{dx} \quad (a = \text{constant})$$

$$\frac{d}{dx}[f(x) + g(x)] = \frac{df}{dx} + \frac{dg}{dx}$$

$$\frac{d}{dx}[f(x)g(x)] = \frac{df}{dx}g + f\frac{dg}{dx}$$

$$\frac{d}{dx}[f(y)] = \frac{df}{dy}\frac{dy}{dx} \quad \text{[chain rule]}$$

$$\frac{dx}{dy} = \frac{1}{\left(\dfrac{dy}{dx}\right)} \quad \text{if } \frac{dy}{dx} \neq 0.$$

B–2 Derivatives: Particular Functions

$$\frac{da}{dx} = 0 \quad (a = \text{constant})$$

$$\frac{d}{dx}x^n = nx^{n-1}$$

$$\frac{d}{dx}\sin ax = a\cos ax$$

$$\frac{d}{dx}\cos ax = -a\sin ax$$

$$\frac{d}{dx}\tan ax = a\sec^2 ax$$

$$\frac{d}{dx}\ln ax = \frac{1}{x}$$

$$\frac{d}{dx}e^{ax} = ae^{ax}$$

B–3 Indefinite Integrals: General Rules

(See also Section 7–3.)

$$\int dx = x$$

$$\int af(x)\,dx = a\int f(x)\,dx \quad (a = \text{constant})$$

$$\int [f(x) + g(x)]\,dx = \int f(x)\,dx + \int g(x)\,dx$$

$$\int u\,dv = uv - \int v\,du \quad \text{[integration by parts]}$$

B–4 Indefinite Integrals: Particular Functions

(An arbitrary constant can be added to the right side of each equation.)

$$\int a\,dx = ax \quad (a = \text{constant})$$

$$\int x^m\,dx = \frac{1}{m+1}x^{m+1} \quad (m \ne -1)$$

$$\int \sin ax\,dx = -\frac{1}{a}\cos ax$$

$$\int \cos ax\,dx = \frac{1}{a}\sin ax$$

$$\int \tan ax\,dx = \frac{1}{a}\ln|\sec ax|$$

$$\int \frac{1}{x}\,dx = \ln x$$

$$\int e^{ax}\,dx = \frac{1}{a}e^{ax}$$

$$\int \frac{dx}{x^2 + a^2} = \frac{1}{a}\tan^{-1}\frac{x}{a}$$

$$\int \frac{dx}{x^2 - a^2} = \frac{1}{2a}\ln\left(\frac{x-a}{x+a}\right) \quad (x^2 > a^2)$$

$$= -\frac{1}{2a}\ln\left(\frac{a+x}{a-x}\right) \quad (x^2 < a^2)$$

$$\int \frac{dx}{\sqrt{x^2 \pm a^2}} = \ln(x + \sqrt{x^2 \pm a^2})$$

$$\int \frac{dx}{(x^2 \pm a^2)^{\frac{3}{2}}} = \frac{\pm x}{a^2\sqrt{x^2 \pm a^2}}$$

$$\int \frac{x\,dx}{(x^2 \pm a^2)^{\frac{3}{2}}} = \frac{-1}{\sqrt{x^2 \pm a^2}}$$

$$\int \sin^2 ax\,dx = \frac{x}{2} - \frac{\sin 2ax}{4a}$$

$$\int xe^{-ax}\,dx = -\frac{e^{-ax}}{a^2}(ax + 1)$$

$$\int x^2 e^{-ax}\,dx = -\frac{e^{-ax}}{a^3}(a^2x^2 + 2ax + 2)$$

B–5 A few Definite Integrals

$$\int_0^\infty x^n e^{-ax}\,dx = \frac{n!}{a^{n+1}}$$

$$\int_0^\infty e^{-ax^2}\,dx = \sqrt{\frac{\pi}{4a}}$$

$$\int_0^\infty xe^{-ax^2}\,dx = \frac{1}{2a}$$

$$\int_0^\infty x^2 e^{-ax^2}\,dx = \sqrt{\frac{\pi}{16a^3}}$$

$$\int_0^\infty x^3 e^{-ax^2}\,dx = \frac{1}{2a^2}$$

$$\int_0^\infty x^{2n} e^{-ax^2}\,dx = \frac{1 \cdot 3 \cdot 5 \cdots (2n-1)}{2^{n+1}a^n}\sqrt{\frac{\pi}{a}}$$

APPENDIX C

Gravitational Force due to a Spherical Mass Distribution

In Chapter 6 (Section 6–1), we stated that the gravitational force exerted by or on a uniform sphere acts as if all the mass of the sphere were concentrated at its center, if the other mass is outside the sphere. In other words, the gravitational force that a uniform sphere exerts on a particle outside it is

$$F = G\frac{mM}{r^2},\qquad [m\text{ outside sphere of mass }M]$$

where m is the mass of the particle, M the mass of the sphere, and r the distance of m from the center of the sphere. Now we will derive this result. We will use the concepts of infinitesimally small quantities and integration.

First we consider a very thin, uniform spherical shell (like a thin-walled basketball) of mass M whose thickness t is small compared to its radius R (Fig. C–1). The force on a particle of mass m at a distance r from the center of the shell can be calculated as the vector sum of the forces due to all the particles of the shell. We imagine the shell divided up into thin (infinitesimal) circular strips so that all points on a strip are equidistant from our particle m. One of these circular strips, labeled AB, is shown in Fig. C–1. It is $R\,d\theta$ wide, t thick, and has a radius $R\sin\theta$. The force on our particle m due to a tiny piece of the strip at point A is represented by the vector $\mathbf{F}_A$ shown. The force due to a tiny piece of the strip at point B, which is diametrically opposite A, is the force $\mathbf{F}_B$. We take the two pieces at A and B to be of equal mass, so $F_A = F_B$. The horizontal components of $\mathbf{F}_A$ and $\mathbf{F}_B$ are each equal to

$$F_A \cos\phi$$

and point toward the center of the shell. The vertical components of $\mathbf{F}_A$ and $\mathbf{F}_B$ are of equal magnitude and point in opposite directions, and so cancel. Since for every point on the strip there is a corresponding point diametrically opposite (as with A and B), we see that the net force due to the entire strip points toward the center of the shell. Its magnitude will be

$$dF = G\frac{m\,dM}{l^2}\cos\phi,$$

where dM is the mass of the entire circular strip and l is the distance from all

FIGURE C–1 Calculating the gravitational force on a particle of mass m due to a uniform spherical shell of radius R and mass M.

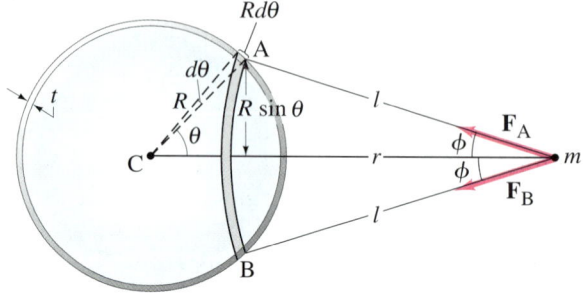

points on the strip to m, as shown. We write dM in terms of the density ρ; by density we mean the mass per unit volume (Section 13–1). Hence, $dM = \rho\, dV$, where dV is the volume of the strip and equals $(2\pi R \sin\theta)(t)(R\, d\theta)$. Then the force dF due to the circular strip shown is

$$dF = G\frac{m\rho 2\pi R^2 t \sin\theta\, d\theta}{l^2}\cos\phi. \tag{C-1}$$

To get the total force F that the entire shell exerts on the particle m, we must integrate over all the circular strips: that is, from $\theta = 0°$ to $\theta = 180°$. But our expression for dF contains l and ϕ, which are functions of θ. From Fig. C–1 we can see that

$$l\cos\phi = r - R\cos\theta.$$

Furthermore, we can write the law of cosines for triangle CmA:

$$\cos\theta = \frac{r^2 + R^2 - l^2}{2rR}. \tag{C-2}$$

With these two expressions we can reduce our three variables (l, θ, ϕ) to only one, which we take to be l. We do two things with Eq. C–2: (1) We put it into the equation for $l\cos\phi$ above:

$$\cos\phi = \frac{1}{l}(r - R\cos\theta) = \frac{r^2 + l^2 - R^2}{2rl};$$

and (2) we take the differential of both sides of Eq. C–2 (because $\sin\theta\, d\theta$ appears in the expression for dF, Eq. C–1):

$$-\sin\theta\, d\theta = -\frac{2l\, dl}{2rR} \quad\text{or}\quad \sin\theta\, d\theta = \frac{l\, dl}{rR},$$

since r and R are considered constants when summing over the strips. Now we insert these into Eq. C–1 for dF and find

$$dF = Gm\rho\pi t\frac{R}{r^2}\left(1 + \frac{r^2 - R^2}{l^2}\right)dl.$$

Now we integrate to get the net force on our thin shell of radius R. To integrate over all the strips ($\theta = 0°$ to $180°$), we must go from $l = r - R$ to $l = r + R$ (see Fig. C–1). Thus,

$$F = Gm\rho\pi t\frac{R}{r^2}\left[l - \frac{r^2 - R^2}{l}\right]_{l=r-R}^{l=r+R}$$

$$= Gm\rho\pi t\frac{R}{r^2}(4R).$$

The volume V of the spherical shell is its area $(4\pi R^2)$ times the thickness t. Hence the mass $M = \rho V = \rho 4\pi R^2 t$, and finally

$$F = G\frac{mM}{r^2}. \quad \begin{bmatrix}\text{particle of mass } m \text{ outside a}\\ \text{thin uniform spherical shell of mass } M\end{bmatrix}$$

This result gives us the force a thin shell exerts on a particle of mass m a distance r from the center of the shell, and *outside* the shell. We see that the force is the same as that between m and a particle of mass M at the center of the shell. In other words, for purposes of calculating the gravitational force exerted on or by a uniform spherical shell, we can consider all its mass concentrated at its center.

What we have derived for a shell holds also for a solid sphere, since a solid sphere can be considered as made up of many concentric shells, from $R = 0$ to $R = R_0$, where R_0 is the radius of the solid sphere. Why? Because if each shell has mass dM, we write for each shell, $dF = Gm\,dM/r^2$, where r is the distance from the center C to mass m and is the same for all shells. Then the total force equals the sum or integral over dM, which gives the total mass M. Thus the result

$$F = G\frac{mM}{r^2} \qquad \begin{bmatrix} \text{particle of mass } m \text{ outside} \\ \text{solid sphere of mass } M \end{bmatrix} \quad \text{(C-3)}$$

is valid for a solid sphere of mass M even if the density varies with distance from the center. (It is not valid if the density varies within each shell—that is, depends not only on R.) Thus the gravitational force exerted on or by spherical objects, including nearly spherical objects like the Earth, Sun, and Moon, can be considered to act as if the objects were point particles.

This result, Eq. C-3, is true only if the mass m is outside the sphere. Let us next consider a point mass m that is located inside the spherical shell of Fig. C-1. Here, r would be less than R, and the integration over l would be from $l = R - r$ to $l = R + r$, so

$$\left[l - \frac{r^2 - R^2}{l} \right]_{R-r}^{R+r} = 0.$$

Thus the force on any mass inside the shell would be zero. This result has particular importance for the electrostatic force, which is also an inverse square law. For the gravitational situation, we see that at points within a solid sphere, say 1000 km below the earth's surface, only the mass up to that radius contributes to the net force. The outer shells beyond the point in question contribute no net gravitational effect.

The results we have obtained here can also be reached using the gravitational analog of Gauss's law for electrostatics (Chapter 22).

Selected Isotopes

(1) Atomic Number Z	(2) Element	(3) Symbol	(4) Mass Number A	(5) Atomic Mass†	(6) % Abundance (or Radioactive Decay Mode)	(7) Half-life (if radioactive)
0	(Neutron)	n	1	1.008665	β^-	10.4 min
1	Hydrogen	H	1	1.007825	99.985%	
	Deuterium	D	2	2.014102	0.015%	
	Tritium	T	3	3.016049	β^-	12.33 yr
2	Helium	He	3	3.016029	0.000137%	
			4	4.002603	99.999863%	
3	Lithium	Li	6	6.015122	7.5%	
			7	7.016004	92.5%	
4	Beryllium	Be	7	7.016929	EC, γ	53.12 days
			9	9.012182	100%	
5	Boron	B	10	10.012937	19.9%	
			11	11.009305	80.1%	
6	Carbon	C	11	11.011434	β^+, EC	20.39 min
			12	12.000000	98.90%	
			13	13.003355	1.10%	
			14	14.003242	β^-	5730 yr
7	Nitrogen	N	13	13.005739	β^+	9.965 min
			14	14.003074	99.63%	
			15	15.000108	0.37%	
8	Oxygen	O	15	15.003065	β^+, EC	122.24 s
			16	15.994915	99.76%	
			18	17.999160	0.20%	
9	Fluorine	F	19	18.998403	100%	
10	Neon	Ne	20	19.992440	90.48%	
			22	21.991386	9.25%	
11	Sodium	Na	22	21.994437	β^+, EC, γ	2.6019 yr
			23	22.989770	100%	
			24	23.990963	β^-, γ	14.9590 h
12	Magnesium	Mg	24	23.985042	78.99%	
13	Aluminum	Al	27	26.981538	100%	

†The masses given in column (5) are those for the neutral atom, including the Z electrons.

(1) Atomic Number Z	(2) Element	(3) Symbol	(4) Mass Number A	(5) Atomic Mass†	(6) % Abundance (or Radioactive Decay Mode)	(7) Half-life (if radioactive)
14	Silicon	Si	28	27.976927	92.23%	
			31	30.975363	β^-, γ	157.3 min
15	Phosphorus	P	31	30.973762	100%	
			32	31.973907	β^-	14.262 days
16	Sulfur	S	32	31.972071	95.02%	
			35	34.969032	β^-	87.32 days
17	Chlorine	Cl	35	34.968853	75.77%	
			37	36.965903	24.23%	
18	Argon	Ar	40	39.962383	99.600%	
19	Potassium	K	39	38.963707	93.2581%	
			40	39.963999	0.0117% β^-, EC, γ, β^+	1.28×10^9 yr
20	Calcium	Ca	40	39.962591	96.941%	
21	Scandium	Sc	45	44.955910	100%	
22	Titanium	Ti	48	47.947947	73.8%	
23	Vanadium	V	51	50.943964	99.750%	
24	Chromium	Cr	52	51.940512	83.79%	
25	Manganese	Mn	55	54.938049	100%	
26	Iron	Fe	56	55.934942	91.72%	
27	Cobalt	Co	59	58.933200	100%	
			60	59.933822	β^-, γ	5.2714 yr
28	Nickel	Ni	58	57.935348	68.077%	
			60	59.930791	26.233%	
29	Copper	Cu	63	62.929601	69.17%	
			65	64.927794	30.83%	
30	Zinc	Zn	64	63.929147	48.6%	
			66	65.926037	27.9%	
31	Gallium	Ga	69	68.925581	60.108%	
32	Germanium	Ge	72	71.922076	27.66%	
			74	73.921178	35.94%	
33	Arsenic	As	75	74.921596	100%	
34	Selenium	Se	80	79.916522	49.61%	
35	Bromine	Br	79	78.918338	50.69%	
36	Krypton	Kr	84	83.911507	57.0%	
37	Rubidium	Rb	85	84.911789	72.17%	
38	Strontium	Sr	86	85.909262	9.86%	
			88	87.905614	82.58%	
			90	89.907737	β^-	28.79 yr
39	Yttrium	Y	89	88.905848	100%	
40	Zirconium	Zr	90	89.904704	51.45%	
41	Niobium	Nb	93	92.906377	100%	
42	Molybdenum	Mo	98	97.905408	24.13%	

†The masses given in column (5) are those for the neutral atom, including the Z electrons.

(1) Atomic Number Z	(2) Element	(3) Symbol	(4) Mass Number A	(5) Atomic Mass†	(6) % Abundance (or Radioactive Decay Mode)	(7) Half-life (if radioactive)
43	Technetium	Tc	98	97.907216	β^-, γ	4.2×10^6 yr
44	Ruthenium	Ru	102	101.904349	31.6%	
45	Rhodium	Rh	103	102.905504	100%	
46	Palladium	Pd	106	105.903483	27.33%	
47	Silver	Ag	107	106.905093	51.839%	
			109	108.904756	48.161%	
48	Cadmium	Cd	114	113.903358	28.73%	
49	Indium	In	115	114.903878	95.7%; β^-, γ	4.41×10^{14} yr
50	Tin	Sn	120	119.902197	32.59%	
51	Antimony	Sb	121	120.903818	57.36%	
52	Tellurium	Te	130	129.906223	33.80%	7.9×10^{20} yr
53	Iodine	I	127	126.904468	100%	
			131	130.906124	β^-, γ	8.0207 days
54	Xenon	Xe	132	131.904154	26.9%	
			136	135.907220	8.9%	
55	Cesium	Cs	133	132.905446	100%	
56	Barium	Ba	137	136.905821	11.23%	
			138	137.905241	71.70%	
57	Lanthanum	La	139	138.906348	99.9098%	
58	Cerium	Ce	140	139.905434	88.48%	
59	Praseodymium	Pr	141	140.907647	100%	
60	Neodymium	Nd	142	141.907718	27.13%	
61	Promethium	Pm	145	144.912744	EC, γ, α	17.7 yr
62	Samarium	Sm	152	151.919728	26.7%	
63	Europium	Eu	153	152.921226	52.2%	
64	Gadolinium	Gd	158	157.924101	24.84%	
65	Terbium	Tb	159	158.925343	100%	
66	Dysprosium	Dy	164	163.929171	28.2%	
67	Holmium	Ho	165	164.930319	100%	
68	Erbium	Er	166	165.930290	33.6%	
69	Thulium	Tm	169	168.934211	100%	
70	Ytterbium	Yb	174	173.938858	31.8%	
71	Lutecium	Lu	175	174.940767	97.4%	
72	Hafnium	Hf	180	179.946549	35.100%	
73	Tantalum	Ta	181	180.947996	99.988%	
74	Tungsten (wolfram)	W	184	183.950933	30.67%	
75	Rhenium	Re	187	186.955751	62.60%; β^-	4.35×10^{10} yr
76	Osmium	Os	191	190.960927	β^-, γ	15.4 days
			192	191.961479	41.0%	
77	Iridium	Ir	191	190.960591	37.3%	
			193	192.962923	62.7%	
78	Platinum	Pt	195	194.964774	33.8%	

†The masses given in column (5) are those for the neutral atom, including the Z electrons.

(1) Atomic Number Z	(2) Element	(3) Symbol	(4) Mass Number A	(5) Atomic Mass†	(6) % Abundance (or Radioactive Decay Mode)	(7) Half-life (if radioactive)
79	Gold	Au	197	196.966551	100%	
80	Mercury	Hg	199	198.968262	16.87%	
			202	201.970625	29.86%	
81	Thallium	Tl	205	204.974412	70.476%	
82	Lead	Pb	206	205.974449	24.1%	
			207	206.975880	22.1%	
			208	207.976635	52.4%	
			210	209.984173	β^-, γ, α	22.3 yr
			211	210.988731	β^-, γ	36.1 min
			212	211.991887	β^-, γ	10.64 h
			214	213.999798	β^-, γ	26.8 min
83	Bismuth	Bi	209	208.980383	100%	
			211	210.987258	α, γ, β^-	2.14 min
84	Polonium	Po	210	209.982857	α, γ	138.376 days
			214	213.995185	α, γ	164.3 μs
85	Astatine	At	218	218.008681	α, β^-	1.5 s
86	Radon	Rn	222	222.017570	α, γ	3.8235 days
87	Francium	Fr	223	223.019731	β^-, γ, α	21.8 min
88	Radium	Ra	226	226.025402	α, γ	1600 yr
89	Actinium	Ac	227	227.027746	β^-, γ, α	21.773 yr
90	Thorium	Th	228	228.028731	α, γ	1.9116 yr
			232	232.038050	100%; α, γ	1.405×10^{10} yr
91	Protactinium	Pa	231	231.035878	α, γ	3.276×10^4 yr
92	Uranium	U	232	232.037146	α, γ	68.9 yr
			233	233.039628	α, γ	1.592×10^5 yr
			235	235.043923	0.720%, α, γ	7.038×10^8 yr
			236	236.045561	α, γ	2.342×10^7 yr
			238	238.050782	99.2745%; α, γ	4.468×10^9 yr
			239	239.054287	β^-, γ	23.45 min
93	Neptunium	Np	237	237.048166	α, γ	2.144×10^6 yr
			239	239.052931	β^-, γ	2.3565 d
94	Plutonium	Pu	239	239.052157	α, γ	24,110 yr
			244	244.064197	α	8.08×10^7 yr
95	Americium	Am	243	243.061373	α, γ	7370 yr
96	Curium	Cm	247	247.070346	α, γ	1.56×10^7 yr
97	Berkelium	Bk	247	247.070298	α, γ	1380 yr
98	Californium	Cf	251	251.079580	α, γ	898 yr
99	Einsteinium	Es	252	252.082972	α, EC, γ	472 d
100	Fermium	Fm	257	257.095099	α, γ	101 d
101	Mendelevium	Md	258	258.098425	α, γ	51.5 d
102	Nobelium	No	259	259.10102	α, EC	58 min
103	Lawrencium	Lr	262	262.10969	α, EC, fission	216 min

†The masses given in column (5) are those for the neutral atom, including the Z electrons.

(1) Atomic Number Z	(2) Element	(3) Symbol	(4) Mass Number A	(5) Atomic Mass†	(6) % Abundance (or Radioactive Decay Mode)	(7) Half-life (if radioactive)
104	Rutherfordium	Rf	261	261.10875	α	65 s
105	Dubnium	Db	262	262.11415	α, fission, EC	34 s
106	Seaborgium	Sg	266	266.12193	α, fission	21 s
107	Bohrium	Bh	264	264.12473	α	0.44 s
108	Hassium	Hs	269	269.13411	α	9 s
109	Meitnerium	Mt	268	268.13882	α	0.07 s
110			271	271.14608	α	0.06 s
111			272	272.15348	α	1.5 ms
112			277	277	α	0.24 ms
114			289	289	α	20 s
116			289	289	α	0.6 ms
118			293	293	α	0.1 ms

†The masses given in column (5) are those for the neutral atom, including the Z electrons.

Answers to Odd-Numbered Problems

CHAPTER 1

1. (a) 1×10^{10} yr; (b) 3×10^{17} s.
3. (a) 1.156×10^3; (b) 2.18×10^1;
   (c) 6.8×10^{-3}; (d) 2.7635×10^1;
   (e) 2.19×10^{-1}; (f) 2.2×10^1.
5. 7.7%.
7. (a) 4%; (b) 0.4%; (c) 0.07%.
9. 1.0×10^5 s.
11. 9%.
13. (a) 0.286 6 m; (b) 0.000 085 V;
    (c) 0.000 760 kg;
    (d) 0.000 000 000 060 0 s;
    (e) 0.000 000 000 000 022 5 m;
    (f) 2,500,000,000 volts.
15. 1.8 m.
17. (a) 0.111 yd^2; (b) 10.76 ft^2.
19. (a) 3.9×10^{-9} in;
    (b) 1.0×10^8 atoms.
21. (a) 0.621 mi/h;
    (b) 1 m/s = 3.28 ft/s; (c) 0.278 m/s.
23. (a) 9.46×10^{15} m;
    (b) 6.31×10^4 AU; (c) 7.20 AU/h.
25. (a) 10^3; (b) 10^4; (c) 10^{-2}; (d) 10^9.
27. $\approx 20\%$.
29. 1×10^5 cm^3.
31. (a) $\approx$600 dentists.
33. $\approx 3 \times 10^8$ kg/yr.
35. 51 km.
37. $A = [L/T^4] = $ m/s^4,
    $B = [L/T^2] = $ m/s^2.
39. (a) 0.10 nm; (b) 1.0×10^5 fm;
    (c) 1.0×10^{10} Å; (d) 9.5×10^{25} Å.
41. (a) 3.16×10^7 s; (b) 3.16×10^{16} ns;
    (c) 3.17×10^{-8} yr.
43. (a) 1,000 drivers.
45. 1×10^{11} gal/yr.
47. 9 cm.
49. 4×10^5 t.
51. $\approx$4 yr.
53. 1.9×10^2 m.
55. (a) 3%, 3%; (b) 0.7%, 0.2%.

CHAPTER 2

1. 5.0 h.
3. 61 m.
5. 0.78 cm/s (toward +x).
7. $\approx$300 m/s.
9. (a) 10.1 m/s;
   (b) +3.4 m/s, away from trainer.
11. (a) 0.28 m/s; (b) 1.2 m/s;
    (c) 0.28 m/s; (d) 1.6 m/s;
    (e) −1.0 m/s.
13. (a) 13.4 m/s;
    (b) +4.5 m/s, away from master.
15. 24 s.
17. 55 km/h, 0.
19. 6.73 m/s.
21. 5.2 s
23. −7.0 m/s^2, 0.72.
25. (a) 4.7 m/s^2; (b) 2.2 m/s^2;
    (c) 0.3 m/s^2; (d) 1.6 m/s^2.
27. $v = (6.0 \text{ m/s}) + (17 \text{ m/s}^2)t$,
    $a = 17$ m/s^2.
29. 1.5 m/s^2, 99 m.
31. 1.7×10^2 m.
33. 4.41 m/s^2, $t = 2.61$ s.
35. 55.0 m.
37. (a) 2.3×10^2 m; (b) 31 s;
    (c) 15 m, 13 m.
39. (a) 103 m; (b) 64 m.
41. 31 m/s.
43. (b) 3.45 s.
45. 32 m/s (110 km/h).
47. 2.83 s.
49. (a) 8.81 s; (b) 86.3 m/s.
51. 1.44 s.
53. 15 m/s.
55. 5.44 s.
59. 0.035 s.
61. 1.8 m above the top of the window.
63. 52 m.
65. 19.8 m/s, 20.0 m.
67. (a) $v = (g/k)(1 - e^{-kt})$;
    (b) $v_{\text{term}} = g/k$.
69. $6h_{\text{Earth}}$.
71. 1.3 m.
73. (b) $H_{50} = 9.8$ m; (c) $H_{100} = 39$ m.
75. (a) 1.3 m; (b) 6.1 m/s; (c) 1.2 s.
77. (a) 3.88 s; (b) 73.9 m; (c) 38.0 m/s, 48.4 m/s.
79. (a) 52 min; (b) 31 min.
81. (a) $v_0 = 26$ m/s; (b) 35 m; (c) 1.2 s; (d) 4.1 s.
83. (a) 4.80 s; (b) 37.0 m/s; (c) 75.2 m.
85. She should decide to stop!
87. $\Delta v_{0\text{down}} = 0.8$ m/s, $\Delta v_{0\text{up}} = 0.9$ m/s.
89. 29.0 m.

CHAPTER 3

1. 263 km, 13° S of W.
3. $\mathbf{V}_{\text{wrong}} = \mathbf{V}_2 - \mathbf{V}_1$.
5. 13.6 m, 18° N of E,

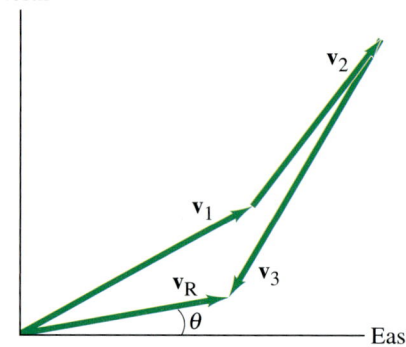

7. (a)

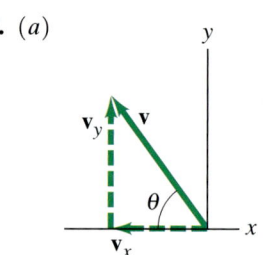

(b) $V_x = -11.7$, $V_y = 8.16$;
(c) 14.3, 34.9° above $-x$-axis.

9. (a) $V_N = 476$ km/h,
   $V_W = 421$ km/h;
   (b) $d_N = 1.43 \times 10^3$ km,
   $d_W = 1.26 \times 10^3$ km.
11. (a) 4.2, 45° below +x-axis;
    (b) 5.1, 79° below +x-axis.
13. (a) 53.7, 1.40° above −x-axis;
    (b) 53.7, 1.40° below +x-axis.
15. (a) 94.5, 11.8° below −x-axis;
    (b) 150, 35.3° below +x-axis.
17. (a) $A_x = \pm 82.9$; (b) 166.6, 12.1° above −x-axis.
19. $(7.60 \text{ m/s})\mathbf{i} - (4.00 \text{ m/s})\mathbf{k}$; 8.59 m/s.
21. (a) Unknown; (b) 4.11 m/s^2, 33.2° north of east; (c) unknown.

23. (a) $\mathbf{v} = (4.0\,\text{m/s}^2)t\mathbf{i} + (3.0\,\text{m/s}^2)t\mathbf{j}$;
    (b) $(5.0\,\text{m/s}^2)t$;
    (c) $\mathbf{r} = (2.0\,\text{m/s}^2)t^2\mathbf{i} + (1.5\,\text{m/s}^2)t^2\mathbf{j}$;
    (d) $\mathbf{v} = (8.0\,\text{m/s})\mathbf{i} + (6.0\,\text{m/s})\mathbf{j}$,
    $|\mathbf{v}| = 10.0\,\text{m/s}$,
    $\mathbf{r} = (8.0\,\text{m})\mathbf{i} + (6.0\,\text{m})\mathbf{j}$.

25. (a) $-(18.0\,\text{m/s})\sin(3.0\,\text{s}^{-1})t\,\mathbf{i} + (18.0\,\text{m/s})\cos(3.0\,\text{s}^{-1})t\,\mathbf{j}$;
    (b) $-(54.0\,\text{m/s}^2)\cos(3.0\,\text{s}^{-1})t\,\mathbf{i} - (54.0\,\text{m/s}^2)\sin(3.0\,\text{s}^{-1})t\,\mathbf{j}$;
    (c) circle; (d) $a = (9.0\,\text{s}^{-2})r$, $180°$.

27. 44 m, 6.3 m.
29. 38° and 52°.
31. 1.95 s.
33. 22 m.

35.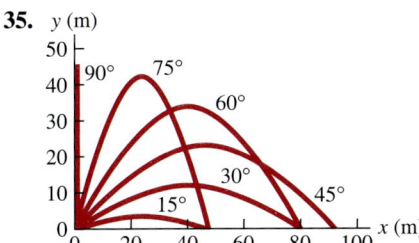

37. 5.71 s.
39. (a) 65.7 m; (b) 7.32 s; (c) 267 m; (d) 42.2 m/s, 30.1° above the horizontal.
43. Unsuccessful, 34.7 m.
45. (a) $\mathbf{v}_0 = 3.42$ m/s, 47.5° above the horizontal; (b) 5.32 m above the water; (c) $\mathbf{v}_f = 10.5$ m/s, 77° below the horizontal.
47. $\theta = \tan^{-1}(gt/v_0)$ below the horizontal.
49. $\theta = \frac{1}{2}\tan^{-1}(-\cot\phi)$.
51. $7.29g$ up.
53. 5.9×10^{-3} m/s² toward the Sun.
55. $0.94g$.
59. 2.7 m/s, 22° from the river bank.
61. 23.1 s.
63. 1.41 m/s.
65. (a) 1.82 m/s; (b) 3.22 m/s.
67. (a) 60 m; (b) 75 s.
69. 58 km/h, 31°, 58 km/h opposite to $\mathbf{v}_{12}$.
71. 0.0889 m/s².
73. $D_x = 60$ m, $D_y = -35$ m, $D_z = -12$ m; 70 m; $\theta_h = 30°$ from the x-axis toward the $-y$-axis, $\theta_v = 9.8°$ below the horizontal.
75. 7.0 m/s.

77. $\pm 28.5°, \pm 25.2$.
79. 170 km/h, 41.5° N of E.
81. 1.6 m/s².
83. 2.7 s, 1.9 m/s.
85. (a) $Dv/(v^2 - u^2)$; (b) $D/(v^2 - u^2)^{1/2}$.
87. 54.6° below the horizontal.
89. (a) 464 m/s; (b) 355 m/s.
91. Row at an angle of 23° upstream and run 243 m in a total time of 20.7 min.
93. 1.8×10^3 rev/day.

CHAPTER 4

3. 6.9×10^2 N.
5. (a) 5.7×10^2 N; (b) 99 N; (c) 2.1×10^2 N; (d) 0.
7. 107 N.
9. -9.3×10^5 N, 25% of the weight of the train.
11. $m > 1.9$ kg.
13. 2.1×10^2 N.
15. -1.40 m/s² (down).
17. a (downward) ≥ 1.2 m/s².
19. $a_{\max} = 0.557$ m/s².
21. (a) 2.2 m/s²; (b) 18 m/s; (c) 93 s.
23. 3.0×10^3 N downward.
25. (a) 1.4×10^2 N; (b) 14.5 m/s.
27. Southwesterly direction.

29.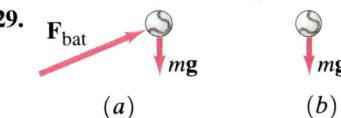

31. (a) 1.13 m/s², 52.2° below $-x$-axis; (b) 0.814 m/s², 42.3° above $+x$-axis.
33. (a) $m_1 g - F_T = m_1 a$, $F_T - m_2 g = m_2 a$.
35. (a) lower bucket = 34 N, upper bucket = 68 N; (b) lower bucket = 40 N, upper bucket = 80 N.
37. 1.4×10^3 N.
39. $F_B = 6890$ N, $F_A + F_B = 8860$ N.

41. (a) 2.2 m up the plane; (b) 2.2 s.
43. $\frac{5}{2}(F_0/m)t_0^2$.
47. (a)

(b) $a = m_2 g/(m_1 + m_2)$,
$F_T = m_1 m_2 g/(m_1 + m_2)$.

49. $a = [m_2 + m_C(\ell_2/\ell)]g / (m_1 + m_2 + m_C)$.
51. 1.74 m/s², $F_{T1} = 22.6$ N, $F_{T2} = 20.9$ N.
53. $(m + M)g\tan\theta$.
55. $F_{T1} = [4m_1 m_2 m_3 / (m_1 m_3 + m_2 m_3 + 4m_1 m_2)]g$,
$F_{T3} = [8m_1 m_2 m_3 / (m_1 m_3 + m_2 m_3 + 4m_1 m_2)]g$,
$a_1 = [(m_1 m_3 - 3m_2 m_3 + 4m_1 m_2) / (m_1 m_3 + m_2 m_3 + 4m_1 m_2)]g$,
$a_2 = [(-3m_1 m_3 + m_2 m_3 + 4m_1 m_2) / (m_1 m_3 + m_2 m_3 + 4m_1 m_2)]g$,
$a_3 = [(m_1 m_3 + m_2 m_3 - 4m_1 m_2) / (m_1 m_3 + m_2 m_3 + 4m_1 m_2)]g$.
57. $v = \{[2m_2 \ell_2 + m_C(\ell_2^2/\ell)]g / (m_1 + m_2 + m_C)\}^{1/2}$.
59. 2.0×10^{-2} N.
61. 4.3 N.
63. 1.5×10^4 N.
65. 1.2 s, no change.
67. (a) 2.45 m/s² (up the incline); (b) 0.50 kg; (c) 7.35 N, 4.9 N.
69. 1.3×10^2 N.
71. 8.8°.
73. 82 m/s (300 km/h).
75. (a) $F = \frac{1}{2}Mg$; (b) $F_{T1} = F_{T2} = \frac{1}{2}Mg$, $F_{T3} = \frac{3}{2}Mg$, $F_{T4} = Mg$.
77. -8.3×10^2 N.
79. (a) 0.606 m/s²; (b) 150 kN.

CHAPTER 5

1. 35 N, no force.
3. (a)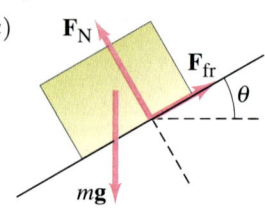

   (b)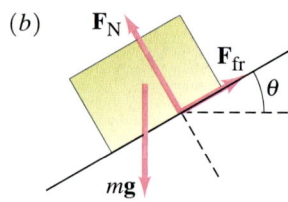

   (c)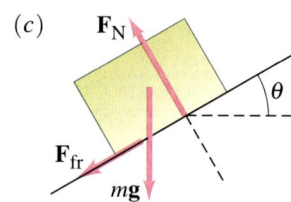

5. 0.20.
7. 69 N, $\mu_k = 0.54$.
9. 8.0 kg.
11. 1.3 m.
13. 1.3×10^3 N.
15. (a) 0.58; (b) 5.7 m/s; (c) 15 m/s.
17. (a) 1.4 m/s^2; (b) 5.4×10^2 N; (c) 1.41 m/s^2, 2.1×10^2 N.
19. (a) 86 cm up the plane; (b) 1.5 s.
21. (a) 2.8 m/s^2; (b) 2.1 N.
23. $a = \{\sin\theta - [(\mu_1 m_1 + \mu_2 m_2)/(m_1 + m_2)]\cos\theta\}g$,
    $F_T = [m_1 m_2(\mu_2 - \mu_1)/(m_1 + m_2)]g\cos\theta$.
25. (a) $\mu_k = (v_0^2/2gd\cos\theta) - \tan\theta$; (b) $\mu_s \geq \tan\theta$.
27. (a) 2.0 m/s^2 up the plane; (b) 5.4 m/s^2 up the plane.
29. $\mu_k = 0.40$.
31. (a) $c = 14$ kg/m; (b) 5.7×10^2 N.
33. $F_{\min} = (m + M)g(\sin\theta + \mu\cos\theta)/(\cos\theta - \mu_s\sin\theta)$.
35. $v_{\max} = 21$ m/s, independent of the mass.
37. (a) 0.25 m/s^2 toward the center; (b) 6.3 N toward the center.
39. Yes, $v_{\text{top, min}} = (gR)^{1/2}$.
41. 0.34.
43. 2.1×10^2 N.
45. 5.91°, 14.3 N.
47. (a) 5.8×10^3 N; (b) 4.1×10^2 N; (c) 31 m/s.
49. $F_T = 2\pi mRf^2$.
51. 66 km/h $< v <$ 123 km/h.
53. (a) $(1.6$ m/s$^2)$**i**;
    (b) $(0.98$ m/s$^2)$**i** $- (1.7$ m/s$^2)$**j**;
    (c) $-(4.9$ m/s$^2)$**i** $- (1.6$ m/s$^2)$**j**.
55. (a) 9.0 m/s^2; (b) 15 m/s^2.
57. $\tau = m/b$.
59. (a) $v = (mg/b) + [v_0 - (mg/b)]e^{-bt/m}$;
    (b) $v = -(mg/b) + [v_0 + (mg/b)]e^{-bt/m}$,
    $v \geq 0$.
61. (b) 1.8°.
63. 10 m.
65. $\mu_s = 0.41$.
67. 2.3.
69. 101 N, $\mu_k = 0.719$.
71. (b) Will slide.
73. Emerges with a speed of 13 m/s.
75. 27.6 m/s, 0.439 rev/s.
77. $\Sigma F_{\tan} = 3.3 \times 10^3$ N, $\Sigma F_R = 2.0 \times 10^3$ N.
79. (a) $F_{NC} > F_{NB} > F_{NA}$; (b) heaviest at C, lightest at A; (c) $v_{A\max} = (gR)^{1/2}$.
81. (a) 1.23 m/s; (b) 3.01 m/s.
83. $\phi = 31°$.
85. (a) $r = v^2/g\cos\theta$; (b) 92 m.
87. (a) 59 s; (b) greater normal force.
89. 29.2 m/s.
91. 302 m, 735 m.
93. $g(1 - \mu_s\tan\phi)/4\pi^2 f^2(\tan\phi + \mu_s) < r < g(1 + \mu_s\tan\phi)/4\pi^2 f^2(\tan\phi - \mu_s)$.

CHAPTER 6

1. 1.52×10^3 N.
3. 1.6 m/s^2.
5. $g_h = 0.91 g_{\text{surface}}$.
7. 1.9×10^{-8} N toward center of square.
9. $Gm^2\{(2/x_0^2) + [3x_0/(x_0^2 + y_0^2)^{3/2}]\}$**i** $+ Gm^2\{[3y_0/(x_0^2 + y_0^2)^{3/2}] + (4/y_0^2)\}$**j**.
11. 1.26.
13. 3.46×10^8 m from Earth's center.
15. (b) g decreases with an increase in height; (c) 9.493 m/s^2.
19. 7.56×10^3 m/s.
21. 2.0 h.
23. (a) 56 kg; (b) 56 kg; (c) 75 kg; (d) 38 kg; (e) 0.
25. (a) 22 N (toward the Moon); (b) -1.7×10^2 N (away from the Moon).
27. (a) Gravitational force provides required centripetal acceleration; (b) 9.6×10^{26} kg.
29. 7.9×10^3 m/s.
31. $v = (Gm/L)^{1/2}$.
33. 0.0587 days (1.41 h).
35. 1.6×10^2 yr.
37. 2×10^8 yr.
39. $r_{\text{Europa}} = 6.71 \times 10^5$ km, $r_{\text{Ganymede}} = 1.07 \times 10^6$ km, $r_{\text{Callisto}} = 1.88 \times 10^6$ km.
41. 9.0 Earth-days.
43. (a) 2.1×10^2 A.U. (3.1×10^{13} m); (b) 4.2×10^2 A.U.; (c) 4.2×10^2.
45. (a) 5.9×10^{-3} N/kg; (b) not significant.
47. 2.7×10^3 km.
49. 6.7×10^{12} m/s^2.
51. 4.4×10^7 m/s^2.
53. $G' = 1 \times 10^{-4}$ N·m^2/kg$^2 \approx 10^6 G$.
55. 5 h 35 min, 19 h 50 min.
57. (a) 10 h; (b) 6.5 km; (c) 4.2×10^{-3} m/s^2.
59. 5.4×10^{12} m, in the Solar System, Pluto.
61. $2.3 g_{\text{Earth}}$.
63. $m_P = g_P r^2/G$.
67. 7.9×10^3 m/s.

CHAPTER 7

1. 6.86×10^3 J.
3. 1.27×10^4 J.
5. 8.1×10^3 J.
7. 1 J = 1×10^7 erg = 0.738 ft·lb.
9. 1.0×10^4 J.
13. (a) 3.6×10^2 N; (b) -1.3×10^3 J; (c) -4.6×10^3 J; (d) 5.9×10^3 J; (e) 0.
15. $W_{FN} = W_{mg} = 0$, $W_{FP} = -W_{fr} = 2.0 \times 10^2$ J.
21. (a) -16.1; (b) -238; (c) -3.9.
23. **C** $= -1.3$**i** $+ 1.8$**j**.
25. $\theta_x = 42.7°, \theta_y = 63.8°, \theta_z = 121°$.
27. 95°, $-35°$ from x-axis.
31. 0.089 J.
33. 2.3×10^3 J.
35. 2.7×10^3 J.
37. $(kX^2/2) + (aX^4/4) + (bX^5/5)$.

39. (a) 5.0×10^{10} J.
41. (a) $\sqrt{3}$; (b) $\frac{1}{4}$.
43. -5.02×10^5 J.
45. 3.0×10^2 N in the direction of the motion of the ball.
47. 24 m/s (87 km/h or 54 mi/h), the mass cancels.
49. (a) 72 J; (b) -35 J; (c) 37 J.
51. 10.2 m/s.
53. $\mu_k = F/2mg$.
55. (a) 6.5×10^2 J; (b) -4.9×10^2 J; (c) 0; (d) 4.0 m/s.
57. (a) 1.66×10^5 J; (b) 21.0 m/s; (c) 2.13 m.
59. $v_p = 2.0 \times 10^7$ m/s, $v_{pc} = 2.0 \times 10^7$ m/s; $v_e = 2.9 \times 10^8$ m/s, $v_{ec} = 8.4 \times 10^8$ m/s.
61. 1.74×10^3 J.
63. (a) 15 J; (b) 4.2×10^2 J; (c) -1.8×10^2 J; (d) -2.5×10^2 J; (e) 0; (f) 10 J.
65. (a) 12 J; (b) 10 J; (c) -2.1 J.
67. 86 kJ, $\theta = 42°$.
69. $(A/k)e^{-(0.10\,\mathrm{m})k}$.
71. 1.5 N.
73. 5.0×10^3 N/m.
75. (a) $6.6°$; (b) $10.3°$.

CHAPTER 8

1. 0.924 m.
3. 2.2×10^3 J.
5. (a) 51.7 J; (b) 15.1 J; (c) 51.7 J.
7. (a) Conservative; (b) $\frac{1}{2}kx^2 - \frac{1}{4}ax^4 - \frac{1}{5}bx^5 +$ constant.
9. (a) $\frac{1}{2}k(x^2 - x_0^2)$; (b) same.
11. 45.4 m/s.
13. 6.5 m/s.
15. (a) 1.0×10^2 N/m; (b) 22 m/s^2.
17. (a) 8.03 m/s; (b) 3.44 m.
19. (a) $v_{\max} = [v_0^2 + (kx_0^2/m)]^{1/2}$; (b) $x_{\max} = [x_0^2 + (mv_0^2/k)]^{1/2}$.
21. (a) 2.29 m/s; (b) 1.98 m/s; (c) 1.98 m/s; (d) $F_{Ta} = 0.87$ N, $F_{Tb} = 0.80$ N, $F_{Tc} = 0.80$ N; (e) $v_a = 2.59$ m/s, $v_b = 2.31$ m/s, $v_c = 2.31$ m/s.
23. $k = 12Mg/h$.
25. 4.5×10^6 J.
27. (a) 22 m/s; (b) 2.9×10^2 m.
29. 13 m/s.
31. 0.23.
33. 0.40.
35. (a) 0.13 m; (b) 0.77; (c) 0.46 m/s.
37. (a) $K = GM_E m_S/2r_S$; (b) $U = -GM_E m_S/r_S$; (c) $-\frac{1}{2}$.
39. (a) 6.2×10^5 m/s; (b) 4.2×10^4 m/s, $v_{\mathrm{esc}}/v_{\mathrm{orbit}} = \sqrt{2}$.
45. (a) 1.07×10^4 m/s; (b) 1.17×10^4 m/s; (c) 1.12×10^4 m/s.
47. (a) $dv_{\mathrm{esc}}/dr = -\frac{1}{2}(2GM_E/r^3)^{1/2}$ $= -v_{\mathrm{esc}}/2r$; (b) 1.09×10^4 m/s.
49. $GmM_E/12r_E$.
51. 1.1×10^4 m/s.
55. 5.4×10^2 N.
57. (a) 1.0×10^3 J; (b) 1.0×10^3 W.
59. 2.1×10^4 W, 28 hp.
61. 4.8×10^2 W.
63. 1.2×10^3 W.
65. 1.8×10^6 W.
67. (a) -25 W; (b) $+4.3 \times 10^3$ W; (c) $+1.5 \times 10^3$ W.
69. (a) 80 J; (b) 60 J; (c) 80 J; (d) 5.7 m/s at $x = 0$; (e) 32 m/s^2 at $x = \pm x_0$.
71. $a^2/4b$.
73. 8.0 m/s.
75. 32.5 hp.
77. (a) 28 m/s; (b) 1.2×10^2 m.
79. (a) $(2gL)^{1/2}$; (b) $(1.2gL)^{1/2}$.
81. (a) 1.1×10^6 J; (b) 60 W (0.081 hp); (c) 4.0×10^2 W (0.54 hp).
83. (a) 40 m/s; (b) 2.6×10^5 W.
87. (a) $29°$; (b) 6.4×10^2 N; (c) 9.2×10^2 N.
89. (a) $-\dfrac{U_0}{r}\left(\dfrac{r_0}{r} + 1\right)e^{-r/r_0}$; (b) 0.030; (c) $F(r) = -C/r^2$, 0.11.
91. 6.7 hp.
93. (a) 2.8 m; (b) 1.5 m; (c) 1.5 m.
95. 76 hp.
97. (a) 5.00×10^3 m/s; (b) 2.89×10^3 m/s.

CHAPTER 9

1. 6.0×10^7 N, up.
3. (a) 0.36 kg·m/s; (b) 0.12 kg·m/s.
5. $(26\,\mathrm{N}\cdot\mathrm{s})\mathbf{i} - (28\,\mathrm{N}\cdot\mathrm{s})\mathbf{j}$.
7. (a) $(8h/g)^{1/2}$; (b) $(2gh)^{1/2}$; (c) $-(8m^2gh)^{1/2}$ (up); (d) mg (down), a surprising result.
9. 3.4×10^4 kg.
11. 4.4×10^3 m/s.
13. -0.667 m/s (opposite to the direction of the package).
15. 2, lesser kinetic energy has greater mass.
17. $\frac{3}{2}v_0\mathbf{i} - v_0\mathbf{j}$.
19. 1.1×10^{-22} kg·m/s, $36°$ from the direction opposite to the electron's.
21. (a) $(100\,\mathrm{m/s})\mathbf{i} + (50\,\mathrm{m/s})\mathbf{j}$; (b) 3.3×10^5 J.
23. 130 N, not large enough.
25. 1.1×10^3 N.
27. (a) $2mv/\Delta t$; (b) $2mv/t$.
29. (a)

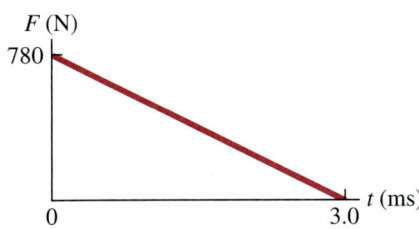

(b) 1.2 N·s; (c) 1.2 N·s; (d) 3.9 g.
31. (a) $(0.84\,\mathrm{N}) + (1.2\,\mathrm{N/s})t$; (b) 18.5 N; (c) $(0.12\,\mathrm{kg/s})\{[(49\,\mathrm{m^2/s^2}) - (1.18\,\mathrm{m^2/s^3})t]^{1/2} + (9.80\,\mathrm{m/s^2})t\}$, 18.3 N.
33. $v_1' = -1.40$ m/s (rebound), $v_2' = 2.80$ m/s.
35. (a) 2.7 m/s; (b) 0.84 kg.
37. 3.2×10^3 m/s.
39. (a) 1.00; (b) 0.89; (c) 0.29; (d) 0.019.
41. (a) 0.32 m; (b) -3.1 m/s (rebound), 4.9 m/s; (c) Yes.
43. (a) $+M/(m + M)$; (b) 0.964.
45. $141°$.
47. (b) $e = (h'/h)^{1/2}$.
49. (a) $v_1' = v_2' = 1.9$ m/s; (b) $v_1' = -1.6$ m/s, $v_2' = 7.9$ m/s; (c) $v_1' = 0$, $v_2' = 5.2$ m/s; (d) $v_1' = 3.1$ m/s, $v_2' = 0$; (e) $v_1' = -4.0$ m/s, $v_2' = 12$ m/s; result for (c) is reasonable, result for (d) is not reasonable, result for (e) is not reasonable.
51. $61°$ from first eagles's direction, 6.8 m/s.
53. (a) $30°$; (b) $v_2' = v/\sqrt{3}$, $v_1' = v/\sqrt{3}$; (c) $\frac{2}{3}$.

55. $\theta_1' = 76°$, $v_n' = 5.1 \times 10^5$ m/s,
    $v_{He}' = 1.8 \times 10^5$ m/s.
59. 6.5×10^{-11} m from the carbon atom.
61. 0.030 nm above center of H triangle.
63. $x_{CM} = 1.10$ m (East),
    $y_{CM} = -1.10$ m (South).
65. $x_{CM} = 0$, $y_{CM} = 2r/\pi$.
67. $x_{CM} = 0$, $y_{CM} = 0$,
    $z_{CM} = 3h/4$ above the point.
69. (a) 4.66×10^6 m.
71. (a) $x_{CM} = 4.6$ m; (b) 4.3 m;
    (c) 4.6 m.
73. $mv/(M + m)$ up, balloon will also stop.
75. 55 m.
77. 0.899 hp.
79. (a) 2.3×10^3 N; (b) 2.8×10^4 N;
    (c) 1.1×10^4 hp.
81. A "scratch shot".
83. 1.4×10^4 N, 43.3°.
85. 5.1×10^2 m/s.
87. $m_2 = 4.00m$.
89. 50%.
91. (a) No; (b) $v_1/v_2 = -m_2/m_1$;
    (c) m_2/m_1; (d) does not move;
    (e) center of mass will move.
93. 8.29 m/s.
95. (a) 2.5×10^{-13} m/s; (b) 1.7×10^{-17};
    (c) 0.19 J.
97. $m \leq M/3$.
99. 29.6 km/s.
101. (a) 2.3 N·s; (b) 4.5×10^2 N.
103. (a) Inelastic collision; (b) 0.10 s;
    (c) -1.4×10^5 N.
105. 0.28 m, 1.1 m.

CHAPTER 10

1. (a) $\pi/6$ rad = 0.524 rad;
    (b) $19\pi/60$ = 0.995 rad;
    (c) $\pi/2$ = 1.571 rad;
    (d) 2π = 6.283 rad;
    (e) $7\pi/3$ = 7.330 rad.
3. 2.3×10^3 m.
5. (a) 0.105 rad/s;
    (b) 1.75×10^{-3} rad/s;
    (c) 1.45×10^{-4} rad/s; (d) zero.
7. (a) 464 m/s; (b) 185 m/s;
    (c) 355 m/s.
9. (a) 262 rad/s;
    (b) 46 m/s, 1.2×10^4 m/s² radial.
11. 7.4 cm.

13. (a) 1.75×10^{-4} rad/s²;
    (b) $a_R = 1.17 \times 10^{-2}$ m/s²,
    $a_{tan} = 7.44 \times 10^{-4}$ m/s².
15. (a) 0.58 rad/s2; (b) 12 s.
17. (a) $(1.67\,\text{rad/s}^4)t^3 - (1.75\,\text{rad/s}^3)t^2$;
    (b) $(0.418\,\text{rad/s}^4)t^4 - (0.583\,\text{rad/s}^3)t^3$;
    (c) 6.4 rad/s, 2.0 rad.
19. (a) ω_1 is in the $-x$-direction,
    ω_2 is in the $+z$-direction;
    (b) $\omega = 61.0$ rad/s, 35.0° above $-x$-axis;
    (c) $-(1.75 \times 10^3\,\text{rad/s}^2)\mathbf{j}$.
21. (a) 35 m·N; (b) 30 m·N.
23. 1.2 m·N (clockwise).
25. 3.5×10^2 N, 2.0×10^3 N.
27. 53 m·N.
29. (a) 3.5 kg·m²; (b) 0.024 m·N.
31. 2.25×10^3 kg·m², 8.8×10^3 m·N.
33. 9.5×10^4 m·N.
35. 10 m/s.
37. (a)

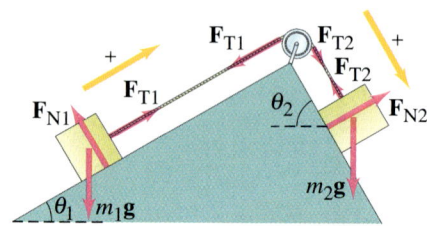

(b) $F_{T1} = 47$ N, $F_{T2} = 75$ N;
(c) 7.0 m·N, 1.7 kg·m².
39. Thin hoop (through center):
    $k = R_0$;
    Thin hoop (through diameter):
    $k = [(R_0^2/2) + (w^2/12)]^{1/2}$;
    Solid cylinder (through center):
    $k = R/\sqrt{2}$;
    Hollow cylinder (through center):
    $k = [(R_1^2 + R_2^2)/2]^{1/2}$;
    Uniform sphere (through center):
    $k = (2r_0^2/5)^{1/2}$;
    Rod (through center): $k = \ell/\sqrt{12}$;
    Rod (through end): $k = \ell/\sqrt{3}$;
    Plate (through center):
    $k = [(\ell^2 + w^2)/12]^{1/2}$.
41. (a) 4.18 rad/s²; (b) 8.37 m/s²;
    (c) 421 m/s²; (d) 3.07×10^3 N;
    (e) 1.14°.
43. (a) $I_a = Ms^2/12$; (b) $I_b = Ms^2/12$.
45. (a) $5.30MR_0^2$; (b) -15%.
47. (a) $9MR_0^2/16$; (b) $MR_0^2/4$;
    (c) $5MR_0^2/4$.
51. (b) $M\ell^2/12$, $Mw^2/12$.

53. 0.38 rev/s.
55. (a) As moment of inertia increases, angular velocity must decrease;
    (b) 1.6.
57. (a) 7.1×10^{33} kg·m²/s;
    (b) 2.7×10^{40} kg·m²/s.
59. 0.45 rad/s, 0.80 rad/s.
61. 2.33×10^4 J.
63. 5×10^9, loss of gravitational potential energy.
65. 1.4 m/s.
67. (a) 2.5 kg·m²; (b) 0.58 kg·m²;
    (c) 0.35 s; (d) -72 J; (e) rotating.
69. 12.4 m/s.
71. 1.4×10^2 J.
73. (a) 4.48 m/s; (b) 1.21 J;
    (c) $\mu_s \geq 0.197$.
75. $v = [10g(R_0 - r_0)/7]^{1/2}$.
77. (a) 4.5×10^5 J; (b) 0.18 (18%);
    (c) 1.71 m/s²; (d) 6.4%.
79. (a) $12v_0^2/49\mu_k g$;
    (b) $v = 5v_0/7$, $\omega = 5v_0/7R$.
81. (a) 4.5 m/s², 19 rad/s²; (b) 5.8 m/s;
    (c) 15.3 J; (d) 1.4 J;
    (e) $K = 16.7$ J, $\Delta E = 0$;
    (f) $a = 4.5$ m/s², $v = 5.8$ m/s, 14.1 J.
83. $\theta_{Sun} = 9.30 \times 10^{-3}$ rad (0.53°),
    $\theta_{Moon} = 9.06 \times 10^{-3}$ rad (0.52°).
85. $\omega_1/\omega_2 = R_2/R_1$.
87. $\ell/2, \ell/2$.
89. (a) $-(I_W/I_P)\omega_W$ (down);
    (b) $-(I_W/2I_P)\omega_W$ (down);
    (c) $(I_W/I_P)\omega_W$ (up); (d) 0.
91. (a) $\omega_R/\omega_F = N_F/N_R$; (b) 4.0;
    (c) 1.5.
93. (a) 1.5×10^2 rad/s²;
    (b) 1.2×10^3 N.
95. (a) 0.070 rad/s²; (b) 40 rpm.
97. 7.9 N.
99. (b) 2.2×10^3 rad/s; (c) 24 min.
101. (a) 2.9 m; (b) 3.6 m.
103. (a) 1.2 rad/s; (b) 2.0×10^3 J,
    1.2×10^3 J, loss of 8.0×10^2 J,
    decrease of 40%.
105. (a) 1.7 m/s; (b) 0.84 m/s.
107. (a) $h_{min} = 2.7R_0$;
    (b) $h_{min} = 2.7R_0 - 1.7r_0$.
109. (a) 0.84 m/s; (b) 0.96.

CHAPTER 11

7. (a) $-7.0\mathbf{i} - 14.0\mathbf{j} + 19.3\mathbf{k}$; (b) 164°.
11. $-(30.3\,\text{m·N})\mathbf{k}$ (in $-z$-direction).
13. $(18\,\text{m·kN})\mathbf{i} \pm (14\,\text{m·kN})\mathbf{j}$
    $\mp (19\,\text{m·kN})\mathbf{k}$.

19. $(55\mathbf{i} - 90\mathbf{j} + 42\mathbf{k})$ kg·m²/s.
21. (a) $[(7m/9) + (M/6)]\ell^2\omega^2$;
    (b) $[(14m/9) + (M/3)]\ell^2\omega$.
23. 2.30 m/s².
25. (a) $L = [R_0 M_1 + R_0 M_2 + (I/R_0)]v$;
    (b) $a = M_2 g / [M_1 + M_2 + (I/R_0^2)]$.
27. Rod rotates at 7.8 rad/s about the center of mass, which moves with constant velocity of 0.21 m/s.
31. $F_1 = [(d + r\cos\phi)/2d]m_1 r\omega^2 \sin\phi$,
    $F_2 = [(d - r\cos\phi)/2d]m_1 r\omega^2 \sin\phi$.
33. 16 N, −7.5 N.
35. $3m^2v^2/g(3m + 4M)(m + M)$.
37. $(1 - 4.7 \times 10^{-13})\omega_E$.
39. (a) 14 m/s; (b) 6.8 rad/s.
41. 1.02×10^{-3} kg·m².
43. 2.2 rad/s (0.35 rev/s).
45. $\tan^{-1}(r\omega^2/g)$.
47. (a) g, along a radial line; (b) 0.998g, 0.0988° south from a radial line;
    (c) 0.997g, along a radial line.
49. North or south direction.
51. (a) South; (b) $\omega D^2 \sin\lambda / v_0$;
    (c) 0.46 m.
53. (a) $(-9.0\mathbf{i} + 12\mathbf{j} - 8.0\mathbf{k})$ kg·m²/s;
    (b) $(9.0\mathbf{j} - 6.0\mathbf{k})$ m·N.
55. (a) Turn in the direction of the lean;
    (b) $\Delta L = 0.98$ kg·m²/s,
    $\Delta L = 0.18 L_0$.
57. (a) 1.8×10^3 kg·m²/s²;
    (b) 1.8×10^3 m·N; (c) 2.1×10^3 W.
59. $v_{CM} = (3g\ell/4)^{1/2}$.
61. $(19 \text{ m/s})(1 - \cos\theta)^{1/2}$.
63. (a) 2.3×10^4 rev/s;
    (b) 5.7×10^3 rev/s.

CHAPTER 12

1. 379 N, 141°.
3. 1.6×10^3 m·N.
5. 6.52 kg.
7. 2.84 m from the adult.
9. 0.32 m.
11. $F_{T1} = 3.4 \times 10^3$ N,
    $F_{T2} = 3.9 \times 10^3$ N.
13. $F_1 = -2.94 \times 10^3$ N (down),
    $F_2 = 1.47 \times 10^4$ N.
15. Top hinge: $F_{Ax} = 55.2$ N,
    $F_{Ay} = 63.7$ N; bottom hinge:
    $F_{Bx} = -55.2$ N, $F_{By} = 63.7$ N.

17. (a)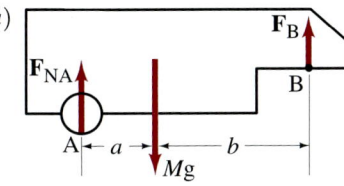
    (b) 1.5×10^4 N; (c) 6.7×10^3 N.
19. $F_T = 1.4 \times 10^3$ N (up),
    $F_{bone} = 2.1 \times 10^3$ N (down).
21. 2.7×10^3 N.
23. 89.5 cm from the feet.
25. $F_1 = 5.8 \times 10^3$ N, $F_2 = 5.6 \times 10^3$ N.
27. (a) 2.1×10^2 N; (b) 2.0×10^3 N.
29. 7.1×10^2 N.
31. $F_T = 2.5 \times 10^2$ N,
    $F_{AH} = 2.5 \times 10^2$ N,
    $F_{AV} = 2.0 \times 10^2$ N.
33. (a) 1.00 N; (b) 1.25 N.
35. $\theta_{max} = 40°$, same.
37. (a) $F_T = 182$ N; (b) $F_{N1} = 352$ N,
    $F_{N2} = 236$ N; (c) $F_B = 298$ N, 52.4°.
39. 1.0×10^2 N.
41. (a) 1.2×10^5 N/m²; (b) 2.4×10^{-6}.
43. (a) 1.3×10^5 N/m²; (b) 6.5×10^{-7};
    (c) 0.0062 mm.
45. 9.6×10^6 N/m².
47. 9.0×10^7 N/m², 9.0×10^2 atm.
49. 2.2×10^7 N.
51. (a) 1.1×10^2 m·N; (b) wall;
    (c) all three.
53. 3.9×10^2 N, thicker strings, maximum strength is exceeded.
55. (a) 4.4×10^{-5} m²; (b) 2.7 mm.
57. 1.2 cm.
61. (a) $F_T = 129$ kN;
    $F_A = 141$ kN, 23.5°;
    (b) $F_{DE} = 64.7$ kN (tension),
    $F_{CE} = 32.3$ kN (compression),
    $F_{CD} = 64.7$ kN (compression),
    $F_{BD} = 64.7$ kN (tension),
    $F_{BC} = 64.7$ kN (tension),
    $F_{AC} = 97.0$ kN (compression),
    $F_{AB} = 64.7$ kN (compression).
63. (a) 4.8×10^{-2} m²; (b) 6.8×10^{-2} m².
65. $F_{AB} = 5.44 \times 10^4$ N (compression),
    $F_{ACx} = 2.72 \times 10^4$ N (tension),
    $F_{BC} = 5.44 \times 10^4$ N (tension),
    $F_{BD} = 5.44 \times 10^4$ N (compression),
    $F_{CD} = 5.44 \times 10^4$ N (tension),
    $F_{CE} = 2.72 \times 10^4$ N (tension),
    $F_{DE} = 5.44 \times 10^4$ N (compression).
67. 12 m.
69. $M_C = 0.191$ kg, $M_D = 0.0544$ kg,
    $M_A = 0.245$ kg.

71. (a) $Mg[h/(2R - h)]^{1/2}$;
    (b) $Mg[h(2R - h)]^{1/2}/(R - h)$.
73. $\theta_{max} = 29°$.
75. 6, 2.0 m apart.
77. 3.8.
79. 5.0×10^5 N, 3.2 m.
81. (a) 600 N; (b) $F_A = 0$, $F_B = 1200$ N;
    (c) $F_A = 150$ N, $F_B = 1050$ N;
    (d) $F_A = 750$ N, $F_B = 450$ N.
83. 6.5×10^2 N.
85. 0.67 m.
87. Right end is safe, left end is not safe, 0.10 m.
89. (a) $F_L = 3.3 \times 10^2$ N up,
    $F_R = 2.3 \times 10^2$ N down;
    (b) 65 cm from right hand;
    (c) 123 cm from right hand.
91. $\theta \geq 40°$.
93. (b) beyond the table;
    (c) $D = L \sum_{i=1}^{n} \frac{1}{2i}$; (d) 32 bricks.
95. $F_{TB} = 134$ N, $F_{TA} = 300$ N.
97. $2.6w$, 31° above horizontal.

CHAPTER 13

1. 3×10^{11} kg.
3. 4.3×10^2 kg.
5. 0.8477.
7. (a) 3×10^7 N/m²;
   (b) 2×10^5 N/m².
9. 1.1 m.
11. 8.28×10^3 kg.
13. 1.2×10^5 N/m²,
    2.3×10^7 N (down),
    1.2×10^5 N/m².
15. 6.54×10^2 kg/m³.
17. 3.36×10^4 N/m² (0.331 atm).
19. (a) 1.41×10^5 Pa; (b) 9.8×10^4 Pa.
21. (a) 0.34 kg; (b) 1.5×10^4 N (up).
23. (c) $\geq 0.38h$, no.
27. 4.70×10^3 kg/m³.
29. 8.5×10^2 kg.
31. Copper.
33. (a) 1.14×10^6 N; (b) 4.0×10^5 N.
35. (b) Above the center of gravity.
37. 0.88.
39. 7.9×10^2 kg.
43. 4.1 m/s.
45. 9.5 m/s.
47. 1.5×10^5 N/m² = 1.5 atm.
49. 4.11×10^{-3} m³/s.
51. 1.7×10^6 N.

59. (a) $2[h_1(h_2 - h_1)]^{1/2}$;
    (b) $h'_1 = h_2 - h_1$.
61. 0.072 Pa·s.
63. 4.0×10^3 Pa.
65. 11 cm.
67. (a) Laminar; (b) 3200, turbulent.
69. 1.9 m.
71. 9.1×10^{-3} N.
73. (a) $\gamma = F/4\pi r$; (b) 0.024 N/m.
75. (a) 0.88 m; (b) 0.55 m; (c) 0.24 m.
77. 1.5×10^2 N $\leq F \leq 2.2 \times 10^2$ N.
79. 0.051 atm.
81. 0.63 N.
83. 5 km.
85. 5.3×10^{18} kg.
87. 2.6 m.
89. 39 people.
91. 37 N, not float.
93. $d = D[v_0^2/(v_0^2 + 2gy)]^{1/4}$.
95. (a) 3.2 m/s; (b) 19 s.
97. 1.9×10^2 m/s.

CHAPTER 14

1. 0.60 m.
3. 1.15 Hz.
5. (a) 2.4 N/m; (b) 12 Hz.
7. (a) $0.866\,x_{\max}$; (b) $0.500\,x_{\max}$.
9. $0.866\,A$.
11. $[(k_1 + k_2)/m]^{1/2}/2\pi$.
13. (a) $8/7$ s, 0.875 Hz; (b) 3.3 m, -10.4 m/s; (c) $+18$ m/s, -57 m/s^2.
15. 3.6 Hz.
19. (a) $y = -(0.220\text{ m})\sin[(37.1\text{ s}^{-1})t]$;
    (b) maximum extensions at 0.0423 s, 0.211 s, 0.381 s,…; minimum extensions at 0.127 s, 0.296 s, 0.465 s,….
21. $f = (3k/M)^{1/2}/2\pi$.
25. (a) $x = (12.0\text{ cm})\cos[(25.6\text{ s}^{-1})t + 1.89\text{ rad}]$;
    (b) $t_{\max} = 0.294$ s, 0.539 s, 0.784 s,…;
    $t_{\min} = 0.171$ s, 0.416 s, 0.661 s,…;
    (c) -3.77 cm; (d) $+13.1$ N (up);
    (e) 3.07 m/s, 0.110 s.
27. (a) 0.650 m; (b) 1.34 Hz; (c) 29.8 J;
    (d) $K = 25.0$ J, $U = 4.8$ J.
29. 9.37 m/s.
31. $A_1 = 2.24 A_2$.
33. (a) 4.2×10^2 N/m; (b) 3.3 kg.
35. 352.6 m/s.
39. 0.9929 m.
41. (a) 0.248 m; (b) 2.01 s.
43. (a) $-12°$; (b) $+1.9°$; (c) $-13°$.
45. $\tfrac{1}{3}$.
47. 1.08 s.
49. 0.31 g.
51. (a) 1.6 s.
53. 3.5 s.
55. (a) 0.727 s; (b) 0.0755;
    (c) $x = (0.189\text{ m})e^{-(0.108/\text{s})t}\sin[(8.64\text{ s}^{-1})t]$.
57. (a) 8.3×10^{-4}%; (b) 39 periods.
59. (a) 5.03 Hz; (b) 0.0634 s^{-1};
    (c) 110 oscillations.
61. $A_0 k / F_0$

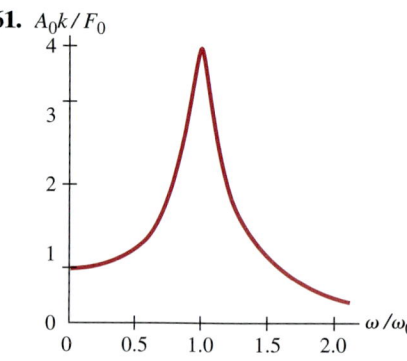

65. (a) 198 s; (b) 8.7×10^{-6} W;
    (c) 8.8×10^{-4} Hz on either side of f_0.
69. (a) 0.63 Hz; (b) 0.65 m/s; (c) 0.077 J.
71. 151 N/m, 20.3 m.
73. 0.11 m.
75. 3.6 Hz.
77. (a) 1.1 Hz; (b) 13 J.
79. (a) 90 N/m; (b) 8.9 cm.
81. $k = \rho_{\text{water}} g A$.
83. Water will oscillate with SHM, $k = 2\rho g A$, the density and the cross section.
85. $T = 2\pi(ma/2k\,\Delta a)^{1/2}$.
87. (a) 1.64 s; (b) 0.67 m.

CHAPTER 15

1. 2.3 m/s.
3. 1.26 m.
5. 0.72 m.
7. 2.7 N.
9. (a) 75 m/s; (b) 7.8×10^3 N.
11. (a) 1.3×10^3 km;
    (b) cannot be determined.
13. (a) 0.25; (b) 0.50.
17. (a) 0.30 W; (b) 0.25 cm.
19. $D = D_M \sin[2\pi(x/\lambda + t/T) + \phi]$.
21. (a) 41 m/s; (b) 6.4×10^4 m/s^2;
    (c) 41 m/s, 8.2×10^3 m/s^2.
23. (a, c)

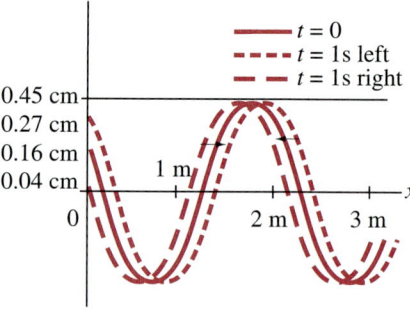

(b) $D = (0.45\text{ m})\cos[(3.0\text{ m}^{-1})x - (6.0\text{ s}^{-1})t + 1.2]$.
(d) $D = (0.45\text{ m})\cos[(3.0\text{ m}^{-1})x + (6.0\text{ s}^{-1})t + 1.2]$.

25. $D = -(0.020\text{ cm})\cos[(8.01\text{ m}^{-1})x - (2.76 \times 10^3\text{ s}^{-1})t]$.
27. The function is a solution.
31. (a) $v_2/v_1 = (\mu_1/\mu_2)^{1/2}$;
    (b) $\lambda_2/\lambda_1 = v_2/v_1 = (\mu_1/\mu_2)^{1/2}$;
    (c) lighter cord.
33. (c) $A_T = [2k_1/(k_2 + k_1)]A = [2v_2/(v_1 + v_2)]A$.
35. (b) $2D_M \cos(\tfrac{1}{2}\phi)$, purely sinusoidal;
    (d) $D = \sqrt{2}\,D_M \sin(kx - \omega t + \pi/4)$.
37. 440 Hz, 880 Hz, 1320 Hz, 1760 Hz.
39. $f_n = n(0.50\text{ Hz})$, $n = 1, 2, 3,\ldots$;
    $T_n = (2.0\text{ s})/n$, $n = 1, 2, 3,\ldots$.
41. 70 Hz.
45. 4.
47. (a) $D_2 = (4.2\text{ cm})\sin[(0.71\text{ cm}^{-1})x + (47\text{ s}^{-1})t + 2.1]$;
    (b) $D_{\text{resultant}} = (8.4\text{ cm})\sin[(0.71\text{ cm}^{-1})x + 2.1]\cos[(47\text{ s}^{-1})t]$.
49. 308 Hz.
51. (a)

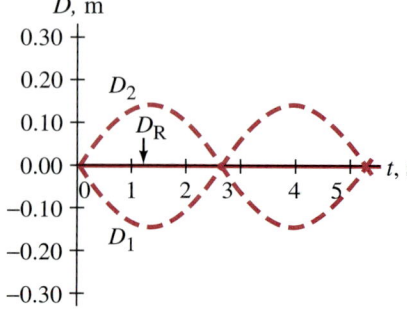

(b)

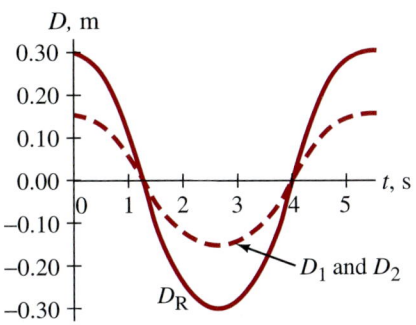

53. 5.4 km/s.

55. 29°.

57. 24°.

59. Speed will be greater in the less dense rod by a factor of $\sqrt{2}$.

61. (a) 0.050 m; (b) 2.3.

63. 0.99 m.

65. (a) solid curves,
(c) dashed curves;

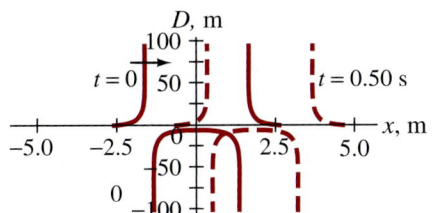

(b) $D = (4.0 \text{ m}^3)/\{[x - (3.0 \text{ m/s})t]^2 - 2.0 \text{ m}^2\}$;

(d)

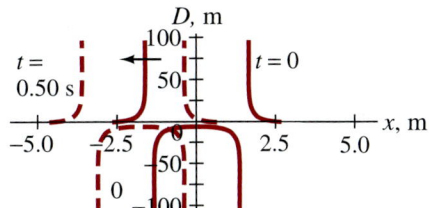

$D = (4.0 \text{ m}^3)/\{[x + (3.0 \text{ m/s})t]^2 - 2.0 \text{ m}^2\}$.

67. (a) 784 Hz, 1176 Hz, 880 Hz, 1320 Hz; (b) 1.26; (c) 1.12; (d) 0.794.

69. $\lambda_n = 4L/(2n - 1), n = 1, 2, 3, \ldots$.

71. $y = (3.5 \text{ cm}) \cos[(1.05 \text{ cm}^{-1})x - (1.39 \text{ s}^{-1})t]$.

73.

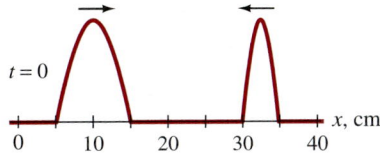

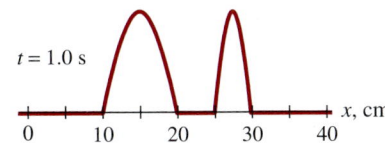

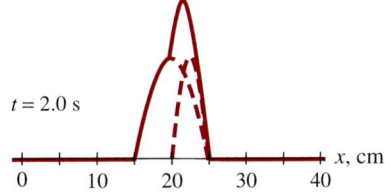

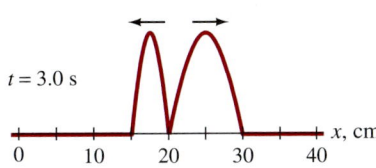

CHAPTER 16

1. 2.6×10^2 m.
3. 5.4×10^2 m.
5. 1200 m, 300 m.
7. (a) 1.1×10^{-8} m; (b) 1.1×10^{-10} m.
9. (a) $\Delta P = (4 \times 10^{-5} \text{ Pa})$
$\sin[(0.949 \text{ m}^{-1})x - (315 \text{ s}^{-1})t]$;
(b) $\Delta P = (4 \times 10^{-3} \text{ Pa})$
$\sin[(94.9 \text{ m}^{-1})x - (3.15 \times 10^4 \text{ s}^{-1})t]$.
11. (a) 49 dB; (b) 3.2×10^{-10} W/m^2.
13. 150 Hz to 20,000 Hz.
15. (a) 9; (b) 9.5 dB.
17. (a) Higher frequency is greater by a factor of 2; (b) 4.
19. (a) 5.0×10^{-13} W; (b) 6.3×10^4 yr.
21. 87 dB.
23. (a) 5.10×10^{-5} m; (b) 29.8 Pa.
25. (a) 1.5×10^3 W; (b) 3.4×10^2 m.
27. (b) 190 dB.
29. (a) 570 Hz; (b) 860 Hz.
31. 8.6 mm $< L <$ 8.6 m.
33. (a) 110 Hz, 330 Hz, 550 Hz, 770 Hz;
(b) 220 Hz, 440 Hz, 660 Hz, 880 Hz.
35. (a) 0.656 m; (b) 262 Hz, 1.31 m;
(c) 1.31 m, 262 Hz.

37. -2.6%.
39. (a) 0.578 m; (b) 869 Hz.
41. 215 m/s.
43. 0.64, 0.20, -2 dB, -7 dB.
45. 28.5 kHz.
47. (a) 130.5 Hz, or 133.5 Hz; (b) $\pm 2.3\%$.
49. (a) 343 Hz; (b) 1030 Hz, 1715 Hz.
53. 346 Hz.
57. (a) 1690 Hz; (b) 1410 Hz.
59. 30,890 Hz.
61. 120 Hz.
63. 91 Hz.
65. 90 beats/min.
67. (a) 570 Hz; (b) 570 Hz; (c) 570 Hz; (d) 570 Hz; (e) 594 Hz; (f) 595 Hz.
71. (a) 120; (b) 0.96°.
73. (a) 37°; (b) 1.7.
75. 0.278 s.
77. 55 m.
79. 410 km/h (255 mi/h).
81. 1, 0.444, 0.198, 0.0878, 0.0389.
83. 18.1 W.
85. 15 W.
87. 2.3 Hz.
89. $\Delta P_M/\Delta P_{M0} = D_M/D_{M0} = 10^6$.
91. 50 dB.
93. 17.5 m/s.
95. 2.3 kHz.
97. 550 Hz.
99. (a) 2.8×10^3 Hz; (b) 1.80 m; (c) 0.12 m.
101. (a) 2.2×10^{-7} m; (b) 5.4×10^{-5} m.

CHAPTER 17

1. 0.548.
3. (a) 20°C; (b) ≈3300°F.
5. 104.0°F.
7. $-40°\text{F} = -40°\text{C}$.
9. $\Delta L_{\text{Invar}} = 2.0 \times 10^{-6}$ m,
$\Delta L_{\text{steel}} = 1.2 \times 10^{-4}$ m,
$\Delta L_{\text{marble}} = 2.5 \times 10^{-5}$ m.
11. $-69°$C.
13. 5.1 mL.
15. 0.06 cm^3.
19. -40 min.
21. -2.8×10^{-3} (0.28%).
23. 3.5×10^7 N/m^2.
25. (a) 27°C; (b) 4.3×10^3 N.
27. $-459.7°$F.
29. 1.07 m^3.
31. 1.43 kg/m^3.

33. (a) 14.8 m³; (b) 1.81 atm.
35. 1.80×10^3 atm.
37. 37°C.
39. 3.43 atm.
41. 0.588 kg/m³, water vapor is not an ideal gas.
43. 2.69×10^{25} molecules/m³.
45. 4.9×10^{22} molecules.
47. 7.7×10^3 N.
49. (a) 71.2 torr; (b) 157°C.
51. (a) 0.19 K; (b) 0.051%.
53. (a) Low; (b) 0.017%.
55. 1/6.
57. 5.1×10^{27} molecules, 8.4×10^3 mol.
59. 11 L, not advisable.
61. (a) 9.3×10^2 kg; (b) 1.0×10^2 kg.
63. 1.1×10^{44} molecules.
65. 3.3×10^{-7} cm.
67. 1.1×10^3 m.
69. 15 h.
71. 0.66×10^3 kg/m³.
73. ± 0.11 C°.
77. 3.6 m.

CHAPTER 18

1. (a) 5.65×10^{-21} J; (b) 7.3×10^3 J.
3. 1.17.
5. (a) 4.5; (b) 5.2.
7. $\sqrt{2}$.
11. (a) 461 m/s; (b) 19 s⁻¹.
13. 1.00429.
17. Vapor.
19. (a) Gas, liquid, vapor; (b) gas, liquid, solid, vapor.
21. 0.69 atm.
23. 11°C.
25. 1.96 atm.
27. 120°C.
29. (a) 5.3×10^6 Pa; (b) 5.7×10^6 Pa.
31. (b) $b = 4.28 \times 10^{-5}$ m³/mol, $a = 0.365$ N·m⁴/mol².
33. (a) 10^{-7} atm; (b) 300 atm.
35. (a) 6.3 cm; (b) 0.58 cm.
37. 2×10^{-7} m.
39. (b) 4.7×10^7 s⁻¹.
43. 7.8 h.
45. (b) 4×10^{-11} mol/s; (c) 0.7 s.
47. 2.6×10^2 m/s, 4×10^{-17} N/m² ≈ 4×10^{-22} atm.
49. (a) 2.9×10^2 m/s; (b) 12 m/s.
51. Reasonable, 70 cm.
53. $mgh = 4.3 \times 10^{-5}(\frac{1}{2}mv_{rms}^2)$, reasonable.
55. $P_2/P_1 = 1.43$, $T_2/T_1 = 1.20$.
57. 1.4×10^5 K.
59. (a) 1.7×10^3 Pa; (b) 7.0×10^2 Pa.
61. 2×10^{13} m.

CHAPTER 19

1. 1.0×10^7 J.
3. 1.8×10^2 J.
5. 2.1×10^2 kg/h.
7. 83 kcal.
9. 4.7×10^6 J.
11. 40 C°.
13. 186°C.
15. 7.1 min.
17. (b) $mc_0[(T_2 - T_1) + a(T_2^2 - T_1^2)/2]$; (c) $c_{mean} = c_0[1 + \frac{1}{2}a(T_2 + T_1)]$.
19. 0.334 kg (0.334 L).
21. $\frac{2}{3}m$ steam and $\frac{4}{3}m$ water at 100°C.
23. 9.4 g.
25. 4.7×10^3 kcal.
27. 1.22×10^4 J/kg.
29. 360 m/s.
31. (a) 0; (b) 5.00×10^3 J.
33.
35. (a) 0; (b) −1300 kJ.
37. (a) 1.6×10^2 J; (b) $+1.6 \times 10^2$ J.
39. $W = 3.46 \times 10^3$ J, $\Delta U = 0$, $Q = +3.46 \times 10^3$ J (into the gas).
41. +129 J.
45. (a) +25 J; (b) +63 J; (c) −95 J; (d) −120 J; (e) −15 J.
47. $W = RT \ln\left(\frac{V_2 - b}{V_1 - b}\right) + a\left(\frac{1}{V_2} - \frac{1}{V_1}\right)$.
49. 22°C/h.
51. 4.98 cal/mol·K, 2.49 kcal/kg·K; 6.97 cal/mol·K, 3.48 kcal/kg·K.
53. 83.7 g/mol, krypton.
55. 46 C°.
57. (a) 2.08×10^3 J; (b) 8.32×10^2 J; (c) 2.91×10^3 J.
59. 0.379 atm, −51°C.
61. 1.33×10^3 J.

63. (a) $T_1 = 317$ K, $T_2 = 153$ K; (b) -1.59×10^4 J; (c) -1.59×10^4 J; (d) $Q = 0$.
65. (a)

(b) 231 K;
(c) $Q_{1 \to 2} = 0$, $\Delta U_{1 \to 2} = -2.01 \times 10^3$ J, $W_{1 \to 2} = +2.01 \times 10^3$ J; $W_{2 \to 3} = -1.31 \times 10^3$ J, $\Delta U_{2 \to 3} = -1.97 \times 10^3$ J, $Q_{2 \to 3} = -3.28 \times 10^3$ J; $W_{3 \to 1} = 0$, $\Delta U_{3 \to 1} = +3.98 \times 10^3$ J, $Q_{3 \to 1} = +3.98 \times 10^3$ J;
(d) $W_{cycle} = +0.70 \times 10^3$ J, $Q_{cycle} = +0.70 \times 10^3$ J, $\Delta U_{cycle} = 0$.
67. (a) 64 W; (b) 22 W.
69. 4.8×10^2 W.
71. 31 h.
73. (a) 1.7×10^{17} W; (b) 278K (5°C).
75. (b) $\Delta Q/\Delta t = A(T_2 - T_1)/\Sigma(\ell_i/k_i)$.
77. 22%.
79. 4×10^{15} J.
81. 2.8 kcal/kg.
83. 30 C°.
85. 682 J.
87. 2.8 C°.
89. 2.58 cm, rod vaporizes.
91. 4.3 kg.
95. (a) 2.3 C°/s; (b) 84°C; (c) convection, conduction, evaporation.
97. (a) $\rho = m/V = (mP/nR)/T$; (b) $\rho = (m/nRT)P$.
99. (a) 1.9×10^5 J; (b) -1.4×10^5 J;
(c)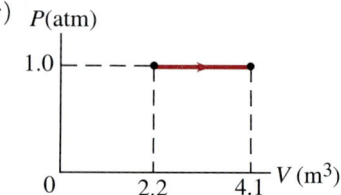

101. 3.2×10^5 s = 3.7 d.
103. 10 C°.

CHAPTER 20

1. 24%.
3. 816 MW.
5. 18%.
7. 13 km³/day, 63 km².
9. 28.0%.
13. 1.2×10^{13} J/h.
15. 1.4×10^3 m/day.
17. 660°C.
19. 3.7×10^8 kg/h.
21. (a) $P_a = 5.15 \times 10^5$ Pa, $P_b = 2.06 \times 10^5$ Pa; (b) $V_c = 30.0$ L, $V_d = 12.0$ L; (c) 2.83×10^3 J; (d) -2.14×10^3 J; (e) 0.69×10^3 J; (f) 24%.
23. 5.7.
25. −21°C.
27. 2.9.
29. (a) 3.9×10^4 J; (b) 3.0 min.
31. 76 L.
33. 0.15 J/K.
35. +11 kcal/K.
37. +0.0104 cal/K·s.
39. 1.7×10^2 J/K.
43. (a) 0.312 kcal/K; (b) > -0.312 kcal/K.
45. (a) Adiabatic process; (b) $\Delta S_i = -nR \ln 2, \Delta S_a = 0$; (c) $\Delta S_{surr,i} = nR \ln 2, \Delta S_{surr,a} = 0$.
47. (a) Entropy is a state function.
51. (a) 5/16; (b) 1/64.
53. (b), (a), (c), (d).
55. 69%.
57. 2.6×10^3 J/K.
59. (a) 17; (b) 5.9×10^7 J/h.
61. (a) 5.3 C°; (b) +77 J/kg·K.
63. $\Delta S = K/T$.
65. (a)

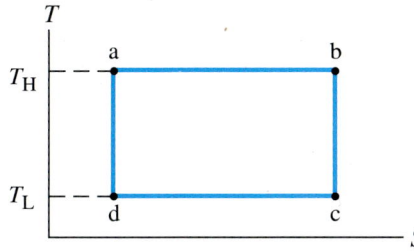

(b) area $= Q_{net} = W_{net}$.
67. $e_{Stirling} = (T_H - T_L) \ln(V_b/V_a)/[T_H \ln(V_b/V_a) + \frac{3}{2}(T_H - T_L)]$, $e_{Stirling} < e_{Carnot}$.
69. 0.091 hp.
71. (a) 1/379; (b) $1/1.59 \times 10^{11}$.

CHAPTER 21

1. 6.3×10^9 N.
3. 2.7×10^{-3} N.
5. 5.5×10^3 N.
7. 8.66 cm.
9. -5.4×10^7 C.
11. 83.8 N away from the center of the triangle.
13. 2.96×10^7 N toward the center of the square.
15. $\mathbf{F}_1 = (kQ^2/\ell^2)[(-2 + 3\sqrt{2}/4)\mathbf{i} + (4 - 3\sqrt{2}/4)\mathbf{j}]$, $\mathbf{F}_2 = (kQ^2/\ell^2)[(2 + 2\sqrt{2})\mathbf{i} + (-6 + 2\sqrt{2})\mathbf{j}]$, $\mathbf{F}_3 = (kQ^2/\ell^2)[(-12 - 3\sqrt{2}/4)\mathbf{i} + (6 + 3\sqrt{2}/4)\mathbf{j}]$, $\mathbf{F}_4 = (kQ^2/\ell^2)[(12 - 2\sqrt{2})\mathbf{i} + (-4 - 2\sqrt{2})\mathbf{j}]$.
17. (a) $Q_1 = Q_2 = \frac{1}{2}Q_T$; (b) Q_1 (or Q_2) = 0.
19. $0.402Q_0, 0.366\ell$ from Q_0.
21. 60.2×10^{-6} C, 29.8×10^{-6} C; -16.8×10^{-6} C, 106.8×10^{-6} C.
23. $\mathbf{F} = -(1.90kQ^2/\ell^2)(\mathbf{i} + \mathbf{j} + \mathbf{k})$.
25. 2.18×10^{-16} N (west).
27. 7.43×10^6 N/C (up).
29. $(1.39 \times 10^2$ N/C$)\mathbf{j}$.
33.

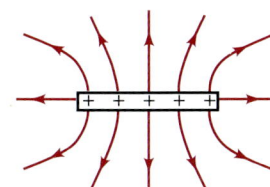

35. 8.26×10^{-10} N/C (south).
37. 4.5×10^6 N/C up, 1.2×10^7 N/C, 56° above the horizontal.
39. 5.61×10^4 N/C away from the opposite corner.
41. $Q_1/Q_2 = \frac{1}{4}$.
43. (a) $2Qy/4\pi\epsilon_0(y^2 + \ell^2)^{3/2} \mathbf{j}$.
45. $\dfrac{Q}{4\pi\epsilon_0} \left[\dfrac{x\mathbf{i} - (2a/\pi)\mathbf{j}}{(x^2 + a^2)^{3/2}} \right]$.
49. $\dfrac{-2\lambda \sin\theta_0}{4\pi\epsilon_0 R} \mathbf{i}$.
51. (a) $\dfrac{\lambda}{4\pi\epsilon_0 x(x^2 + L^2)^{1/2}} \{L\mathbf{i} + [x - (x^2 + L^2)^{1/2}]\mathbf{j}\}$.
53. $(\sigma/2\epsilon_0)\mathbf{k}$.
55. (a) $\mathbf{a} = -(3.5 \times 10^{15}$ m/s²$) \mathbf{i} - (1.41 \times 10^{16}$ m/s²$)\mathbf{j}$; (b) $\theta = -104°$.
57. $\theta = -28°$.
59. (b) $2\pi(4\pi\epsilon_0 mR^3/qQ)^{1/2}$.
61. (a) 3.4×10^{-20} C; (b) No; (c) 8.5×10^{-26} m·N; (d) 2.5×10^{-26} J.
63. (a) $\theta \ll 1$; (b) $(pE/I)^{1/2}/2\pi$.
65. (b) Direction of the dipole.
67. 6.8×10^3 C.
69. 5.7×10^{13} C.
71. $\mathbf{F}_1 = 0.30$ N, 265° from x-axis, $\mathbf{F}_2 = 0.26$ N, 139° from x-axis, $\mathbf{F}_3 = 0.26$ N, 30° from x-axis.
73. 4.2×10^5 N/C up.
75. $0.444Q_0, 0.333\ell$ from Q_0.
77. 5.60 m from the positive charge, and 3.60 m from the negative charge.
79. (a) In the direction of the velocity, to the right; (b) 2.1×10^2 N/C.
81. $\theta_0 = 18°$.
83. $(1.08 \times 10^7$ N/C$)/\{3.00 - \cos[(12.5\text{ s}^{-1})t]\}^2$, up.
85. $E_A = 4.2 \times 10^4$ N/C (right), $E_B = -1.4 \times 10^4$ N/C (left), $E_C = -2.8 \times 10^3$ N/C (left), $E_D = -4.2 \times 10^4$ N/C (left).
87. $d(1 + \sqrt{2})$ from the negative charge, and $d(2 + \sqrt{2})$ from the positive charge.

CHAPTER 22

1. (a) 41 N·m²/C; (b) 29 N·m²/C; (c) 0.
3. $\Phi_{net} = 0$, $\Phi_{x=0} = -(6.50 \times 10^3$ N/C$)\ell^2$, $\Phi_{x=\ell} = +(6.50 \times 10^3$ N/C$)\ell^2$, $\Phi_{all\ others} = 0$.
5. 12.8 nC.
7. −1.2 μC.
9. -3.75×10^{-11} C.
11. (a) -1.0×10^4 N/C (toward wire); (b) -2.5×10^4 N/C (toward wire).
13.

E (10⁷ N/C), 1.9, 0, 7.5, 15, *r* (cm)

15. (a) 5.5×10^7 N/C (away from center); (b) 0.
17. (a) −8.00 μC; (b) +1.00 μC.
19. (a) 0; (b) σ/ϵ_0; (c) unaffected.

21. (a) 0; (b) $\sigma_1 r_1^2/\epsilon_0 r^2$;
(c) $(\sigma_1 r_1^2 + \sigma_2 r_2^2)/\epsilon_0 r^2$;
(d) $\sigma_2/\sigma_1 = -(r_1/r_2)^2$; (e) $\sigma_1 = 0$.

23. (a) $q/4\pi\epsilon_0 r^2$;
(b) $(1/4\pi\epsilon_0)[Q(r^3 - r_1^3) + q(r_0^3 - r_1^3)]/(r_0^3 - r_1^3)r^2$;
(c) $(q + Q)/4\pi\epsilon_0 r^2$.

25. (a) $q/4\pi\epsilon_0 r^2$; (b) $(q + Q)/4\pi\epsilon_0 r^2$;
(c) $E(r < r_0) = Q/4\pi\epsilon_0 r^2$,
$E(r > r_0) = 2Q/4\pi\epsilon_0 r^2$;
(d) $E(r < r_0) = -Q/4\pi\epsilon_0 r^2$,
$E(r > r_0) = 0$.

27. (a) $\sigma R_0/\epsilon_0 r$; (b) 0; (c) same.

29. (a) 0; (b) $Q/2\pi\epsilon_0 Lr$;
(c) 0; (d) $eQ/4\pi\epsilon_0 L$.

31. (a) 0;
(b) -2.3×10^5 N/C (toward the axis);
(c) -1.8×10^4 N/C (toward the axis).

33. (a) $\rho_E r/2\epsilon_0$; (b) $\rho_E R_1^2/2\epsilon_0 r$;
(c) $\rho_E(r^2 + R_1^2 - R_2^2)/2\epsilon_0 r$;
(d) $\rho_E(R_3^2 + R_1^2 - R_2^2)/2\epsilon_0 r$;
(e)

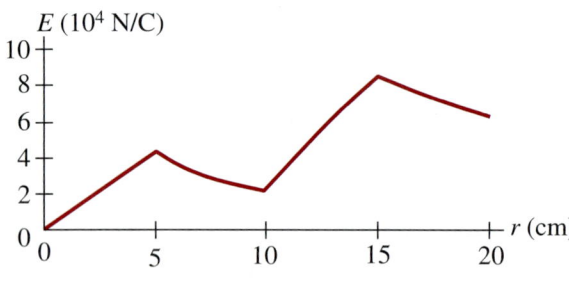

35. $Q/\epsilon_0\sqrt{2}$.

37. $\oint \mathbf{g} \cdot d\mathbf{A} = -4\pi GM$.

39. $Q_{enclosed} = \epsilon_0 b\ell^3$.

41. 3.95×10^2 N·m²/C,
-1.69×10^2 N·m²/C.

43. (a) 0; (b) $Q/25\pi\epsilon_0 r_0^2 \leq E \leq Q/\pi\epsilon_0 r_0^2$;
(c) not perpendicular;
(d) not useful.

45. (a) $0.677e = +1.08 \times 10^{-19}$ C;
(b) 3.5×10^{11} N/C.

47. (a) $\rho_E r_0/6\epsilon_0$ (right);
(b) $-17\rho_E r_0/54\epsilon_0$ (left).

49. (a) 0; (b) 5.65×10^5 N/C (right);
(c) 5.65×10^5 N/C (right);
(d) -5.00×10^{-6} C/m²;
(e) $+5.00 \times 10^{-6}$ C/m².

CHAPTER 23

1. -4.2×10^{-5} J (done by the field).
3. 3.4×10^{-15} J.
5. $V_a - V_b = +72.8$ V.
7. 7.04 V.
9. 0.8 μC.
11. (a) $V_{BA} = 0$; (b) $V_{CB} = -2100$ V;
(c) $V_{CA} = -2100$ V.
13. (a) -9.6×10^8 V;
(b) $V(\infty) = +9.6 \times 10^8$ V.
15. (a) The same;
(b) $Q_2 = r_2 Q/(r_1 + r_2)$.
17. (a) $Q/4\pi\epsilon_0 r$;
(b) $(Q/8\pi\epsilon_0 r_0)[3 - (r^2/r_0^2)]$;
(c)

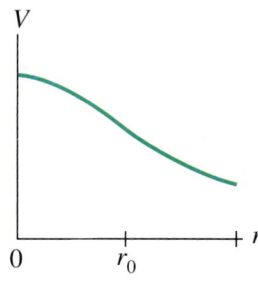

19. (a) $V_0 + (\sigma R_0/\epsilon_0) \ln(R_0/r)$;
(b) $V = V_0$; (c) $V \neq 0$.

21. (a) 29 V;
(b) -29 eV (-4.6×10^{-18} J).

23. $+0.19$ J.

25. 4.2 MV.

27. 2.33×10^7 m/s.

29. $V_{BA} = (1/2\pi\epsilon_0)q(2b - d)/b(d - b)$.

31. $\dfrac{\sigma}{2\epsilon_0}[(x^2 + R_2^2)^{1/2} - (x^2 + R_1^2)^{1/2}]$.

33. $\dfrac{Q}{8\pi\epsilon_0 L} \ln\left(\dfrac{x + L}{x - L}\right)$, $x > L$.

35. $\dfrac{a}{6\epsilon_0}[(x^2 + R^2)^{1/2}(R^2 - 2x^2) + 2x^3]$.

37. 3.2 mm.

39. (a) 8.5×10^{-30} C·m; (b) zero.

41. (a) -0.088 V; (b) 1%.

43. (a) 5.2×10^{-20} C.

47. $\mathbf{E} = 2y(2z - 1)\mathbf{i} - 2(y + x - 2xz)\mathbf{j} + (4xy)\mathbf{k}$.

49. (a) 9.6×10^4 eV; (b) 1.9×10^5 eV.

51. -2.4×10^4 V.

53. (a) $U = (1/4\pi\epsilon_0)(Q_1 Q_2/r_{12} + Q_1 Q_3/r_{13} + Q_1 Q_4/r_{14} + Q_2 Q_3/r_{23} + Q_2 Q_4/r_{24} + Q_3 Q_4/r_{34})$.
(b) $U = (1/4\pi\epsilon_0)(Q_1 Q_2/r_{12} + Q_1 Q_3/r_{13} + Q_1 Q_4/r_{14} + Q_1 Q_5/r_{15} + Q_2 Q_3/r_{23} + Q_2 Q_4/r_{24} + Q_2 Q_5/r_{25} + Q_3 Q_4/r_{34} + Q_3 Q_5/r_{35} + Q_4 Q_5/r_{45})$.

55. (a) 2.0 keV; (b) 42.8.

57. (a) $(-4 + \sqrt{2})Q^2/4\pi\epsilon_0 b$; (b) 0.

59. $3Q^2/20\pi\epsilon_0 r$.

61. 5.4×10^5 V/m.

63. 9×10^2 V.

65. (a) 1.1 MV; (b) 13 kg.

67. 7.2 MV.

69. 1.58×10^{12} electrons.

71. 1.7×10^{-12} V.

73. 1.03×10^6 m/s.

75. $V_a = -3.5\, Q/4\pi\epsilon_0 L$,
$V_b = -5.2\, Q/4\pi\epsilon_0 L$,
$V_c = -6.8\, Q/4\pi\epsilon_0 L$.

77. (a) 5.8×10^5 V; (b) 9.2×10^{-14} J.

79. $V_a - V_b = (\lambda/2\pi\epsilon_0) \ln(R_b/R_a)$.

83. (a) $\rho_E(r_2^3 - r_1^3)/3\epsilon_0 r$;
(b) $(\rho_E/6\epsilon_0)[3r_2^2 - r^2 - (2r_1^3/r)]$;
(c) $(\rho_E/2\epsilon_0)(r_2^2 - r_1^2)$, potential is continuous at r_1 and r_2.

CHAPTER 24

1. 2.6 μF.
3. 6.3 pF.
5. 0.80 μF.
7. 2.0 C.
9. 1.8×10^2 m².
11. 7.1×10^{-4} F.
13. 23 nC.
17. 4.5×10^4 V/m.
19. (a) $\epsilon_0 A/(d - \ell)$; (b) 3.
21. 2880 pF, yes.
23. (a) $(C_1 C_2 + C_1 C_3 + C_2 C_3)/(C_2 + C_3)$;
(b) $Q_1 = 350$ μC, $Q_2 = 117$ μC.
25. (a) 3.71 μF; (b) $V_{ab} = V_1 = 26.0$ V,
$V_2 = 14.9$ V, $V_3 = 11.1$ V.
27. 18 nF (parallel), 1.6 nF (series).
29. (a) $3C/5$; (b) $Q_1 = Q_2 = CV/5$,
$Q_3 = 2CV/5$, $Q_4 = 3CV/5$,
$V_1 = V_2 = V/5$, $V_3 = 2V/5$,
$V_4 = 3V/5$.
31. $Q_1' = C_1 C_2 V_0/(C_1 + C_2)$,
$Q_2' = C_2^2 V_0/(C_1 + C_2)$.

33. (a) $Q_1 = Q_3 = 30 \,\mu$C;
  $Q_2 = Q_4 = 60 \,\mu$C;
  (b) $V_1 = V_2 = V_3 = V_4 = 3.75$ V;
  (c) 7.5 V.
35. 3.0 μF.
37. $C \approx (\epsilon_0 A/d)[1 - \frac{1}{2}(\theta \sqrt{A}/d)]$.
39. 2.0×10^{-3} J.
41. 2.3×10^3 J.
43. 1.65×10^{-7} J.
45. (a) 2.5×10^{-5} J; (b) 6.2×10^{-6} J;
  (c) $Q_{par} = 4.2 \,\mu$C, $Q_{ser} = 1.0 \,\mu$C.
47. (a) 2.2×10^{-4} J; (b) 8.1×10^{-5} J;
  (c) -1.4×10^{-4} J; (d) stored potential energy is not conserved.
51. 1.5×10^{-10} F.
53. 0.46 μC.
55. 3.3×10^2 J.
57. $C = 2\epsilon_0 A K_1 K_2/d(K_1 + K_2)$.
59. (a) $0.40 Q_0, 1.60 Q_0$; (b) $0.40 V_0$.
61. (a) 111 pF; (b) 1.66×10^{-8} C;
  (c) 1.84×10^{-8} C;
  (d) 1.17×10^5 V/m; (e) 3.34×10^4 V/m;
  (f) 150 V;
  (g) 172 pF; (h) 2.58×10^{-8} C.
63. 22%.
65. $Q = 4.41 \times 10^{-7}$ C,
  $Q_{ind} = 3.65 \times 10^{-7}$ C,
  $E_{air} = 2.69 \times 10^4$ V/m,
  $E_{glass} = 4.64 \times 10^3$ V/m;

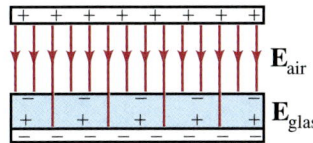

67. 11 μF.
69. (a) $4\times$; (b) $4\times$; (c) $\frac{1}{2}\times$.
71. 10.9 V.
73. (b) 1.5×10^{-10} F/m.
75. $U_2/U_1 = 1/2K$, $E_2/E_1 = 1/K$.
77. (a) 19 J; (b) 0.19 MW.
79. (a) 0.10 MV; (b) voltage will decrease exponentially.
81. 660 pF in parallel.
83. $Q_1 = 11 \,\mu$C, $Q_2 = Q_3 = 13 \,\mu$C,
  $V_1 = 11$ V, $V_2 = 6.5$ V, $V_3 = 4.4$ V.
85. $Q^2 x/2A\epsilon_0$.
87. (a) 7.4 nF, 0.33 μC, 1.5×10^4 V/m, 7.5×10^{-6} J; (b) 27 nF, 1.2 μC, 1.5×10^4 V/m, 2.7×10^{-5} J.
89. (a) 66 pF; (b) 30 μC; (c) 7.0 mm; (d) 450 V.

CHAPTER 25

1. 9.38×10^{18} electron/s.
3. 2.1×10^{-11} A.
5. 7.5×10^2 V.
7. 2.1×10^{21} electron/min.
9. (a) 16 Ω; (b) 6.8×10^3 C.
11. 0.57 mm.
13. $R_{Al} = 1.2 R_{Cu}$.
15. 1/6 the length, 8.3 Ω, 1.7 Ω.
17. 58.3°C.
19. 1.8×10^3 °C.
21. $R_2 = \frac{1}{4} R_1$.
25. $R = (r_2 - r_1)/4\pi\sigma r_1 r_2$.
27. 3.2 W.
29. 37 V.
31. (a) 240 Ω, 0.50 A; (b) 96 Ω, 1.25 A.
33. 0.092 kWh, 22¢/month.
35. 1.1 kWh.
37. 3.
39. 5.3 kW.
41. 0.128 kg/s.
43. 0.094 A.
45. (a) Infinite; (b) 1.9×10^2 Ω.
47. 636 V, 5.66 A.
49. 1.5 kW, 3.0 kW, 0.
51. (a) 7.8×10^{-10} m/s; (b) 10.5 A/m² along the wire; (c) 1.8×10^{-7} V/m.
53. 2.7 A/m² north.
55. 12 h.
57. 6.67×10^{-2} S.
59. (a) 8.6 Ω, 1.1 W; (b) $4\times$.
61. (a) $44; (b) 1.8×10^3 kg/yr.
63. (a) -19.5%; (b) percentage decrease in the power output would be less.
65. (a) 1.44×10^3 W; (b) 17 W; (c) 11 W; (d) 0.8¢/day.
67. (a) 1.5 kW; (b) 12.5 A.
69. 2.
71. 0.303 mm, 28.0 m.
73. (a) 1.2 kW; (b) 100 W.
75. 1.4×10^{12} protons.
77. $j_a = 2.8 \times 10^5$ A/m²,
  $j_b = 1.6 \times 10^5$ A/m².

CHAPTER 26

1. (a) 8.39 V; (b) 8.49 V.
3. 0.060 Ω.
5. 360 Ω, 23 Ω.
7. 25 Ω, 70 Ω, 95 Ω, 18 Ω.
9. Series connection.
11. 4.6 kΩ.
13. 310 Ω, 3.7%.

15. 960 Ω in parallel.
17. 105 Ω.
19. (a) V_1 and V_2 increase; V_3 and V_4 decrease; (b) I_1 ($= I$) and I_2 increase; I_3 and I_4 decrease; (c) increases; (d) $I = I_1 = 0.300$ A, $I_2 = 0, I_3 = I_4 = 0.150$ A; $I = 0.338$ A, $I_2 = I_3 = I_4 = 0.113$ A.
21. 0.4 Ω.
23. 0.41 A.
25. $I_1 = 0.68$ A, $I_2 = -0.40$ A.
27. $I_1 = 0.18$ A right, $I_2 = 0.32$ A left, $I_3 = 0.14$ A up.
29. $I_1 = 0.274$ A, $I_2 = 0.222$ A, $I_3 = 0.266$ A, $I_4 = 0.229$ A, $I_5 = 0.007$ A, $I = 0.496$ A.
31. 52 V, -28 V.
  The negative value means the battery is facing the other direction.
33. 70 V.
35. $I_1 = 0.783$ A.
39. (a) $R(3R + 5R')/8(R + R')$; (b) $R/2$.
41. (a) 3.7 nF; (b) 22 μs.
43. $t = 1.23\tau$.
45. (a) $\tau = R_1 R_2 C/(R_1 + R_2)$;
  (b) $Q_{max} = \mathcal{E} R_2 C/(R_1 + R_2)$.
47. 2.1 μs.
49. 50 μA.
51. (a) 2.9×10^{-4} Ω in parallel;
  (b) 35 kΩ in series.
53. 22 V, 17 V, 14% low.
55. 0.85 mA, 4.3 V.
57. 9.6 V.
61. 3.6×10^{-2} C°.
63. Two resistors in series.
65. 2.2 V, 116 V.
67. 0.19 MΩ.
69. (a) 0.10 A; (b) 0.10 A; (c) 53 mA.
71. 2.5 V.
73. 46.1 V, 0.71 Ω.
75. (a) 72.0 W; (b) 14.2 W; (c) 3.76 W.
77. (a) 40 kΩ; (b) between b and c.
79. 375 cells, 3.8 m $\times$ 0.090 m.
81. (a) 0.50 A; (b) 0.17 A; (c) 3.3 μC; (d) 32 μs.
83. (a) $+ 6.8$ V, 10.2 μC; (b) 28 μs.

CHAPTER 27

1. (a) 6.7 N/m; (b) 4.7 N/m.
3. 2.68 A.
5. 0.243 T.
7. (a) South pole; (b) 3.5×10^2 A; (c) 5.22 N.

9. 5.5×10^3 A.
13. 1.05×10^{-13} N north.
15. (a) Down; (b) in; (c) right.
21. 1.6 m.
23. (a) 2.7 cm; (b) 3.8×10^{-7} s.
25. 1.034×10^8 m/s (west), gravity can be ignored.
27. $(6.4\mathbf{i} - 10.3\mathbf{j} - 0.24\mathbf{k})] \times 10^{-16}$ N.
29. 5.3×10^{-5} m, 3.3×10^{-4} m.
31. (a) 45°; (b) 3.5×10^{-3} m.
33. (a) $2\mu B$; (b) 0.
35. (a) 4.33×10^{-5} m · N; (b) north edge.
39. 29 μA.
41. 1.2×10^5 C/kg.
43. (a) 2.2×10^{-4} V/m; (b) 2.7×10^{-4} m/s; (c) 6.4×10^{28} electrons/m³.
45. (a) Determine polarity of the emf; (b) 0.43 m/s.
47. 1.53 mm, 0.76 mm.
51. 3.0 T up.
53. 1.1×10^{-6} m/s west.
55. 0.17 N, 68° above the horizontal toward the north.
57. 0.20 T, 26.6° from the vertical.
59. (c) 48 MeV.
61. Slower protons will deflect more, and faster protons will deflect less, $\theta = 12°$.
63. 2.0 A, down.
65. 7.3×10^{-3} T.
67. -2.1×10^{-20} J.

CHAPTER 28

1. 1.7×10^{-4} T, 3.1×.
3. 0.18 N attraction.
5.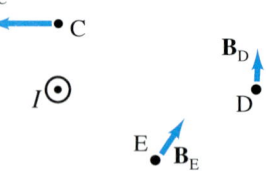
7. 8.9×10^{-5} T, 70° above horizontal.
9. 4.0×10^{-5} T, 15° below the horizontal.
11. (a) $(2.0 \times 10^{-5}$ T/A$)(15$ A $- I)$ up; (b) $(2.0 \times 10^{-5}$ T/A$)(15$ A $+ I)$ down.
13. 21 A down.
15. $[(\mu_0/4\pi)2I(d - 2x)/x(d - x)]\mathbf{j}$.

17. 4.12×10^{-5} T.
19. (b) $(\mu_0/4\pi)(2I/y)$.
21. 0.123 A.
23. (a) 6.4×10^{-3} T; (b) 3.8×10^{-3} T; (c) 2.1×10^{-3} T.
25. (a) 51 cm; (b) 1.3×10^{-2} T.
27. (a) $(\mu_0 I_0/2\pi R_1^2)r$ circular CCW; (b) $\mu_0 I_0/2\pi r$ circular CCW; (c) $(\mu_0 I_0/2\pi r)(R_3^2 - r^2)/(R_3^2 - R_2^2)$ circular CCW; (d) 0.
29. 3.6×10^{-6} T.
31. $\mu_0 I/8R$ out of the page.
33. (a) $\mu_0 I(R_1 + R_2)/4R_1R_2$ into the page; (b) $\frac{1}{2}\pi I(R_1^2 + R_2^2)$ into the page.
35. (a) $\dfrac{Q\omega R^2}{4}\mathbf{i}$;
(b) $\dfrac{\mu_0 Q\omega}{2\pi R^2}\left[\dfrac{R^2 + 2x^2 - 2x\sqrt{R^2 + x^2}}{\sqrt{R^2 + x^2}}\right]\mathbf{i}$;
(c) yes.
37. (b) $B = \mu_0 IL/4\pi y(L^2 + y^2)^{1/2}$ circular.
39. (a) $(\mu_0 I_0/2\pi R)n \tan(\pi/n)$ into the page.
41. $B = \dfrac{\mu_0 I}{4\pi}\Bigg\{\dfrac{(x^2 + y^2)^{1/2}}{xy}$
$+ \dfrac{[(b - x)^2 + y^2]^{1/2}}{y(b - x)}$
$+ \dfrac{[(a - y)^2 + (b - x)^2]^{1/2}}{(a - y)(b - x)}$
$+ \dfrac{[x^2 + (a - y)^2]^{1/2}}{x(a - y)}\Bigg\}$,
out of the page.
43. (a) 26 A · m²; (b) 31 m · N.
45. 30 T.
47. $F_M/L = 5.84 \times 10^{-5}$ up, $F_N/L = 3.37 \times 10^{-5}$ N/m 60° below the line toward P, $F_P/L = 3.37 \times 10^{-5}$ N/m 60° below the line toward N.
49. 0.27 mm, 1.4 cm.
51. $B = \mu_0 jt/2$ parallel to the sheet, perpendicular to the current (opposite directions on the two sides).
53. Between long, thin and short, fat.
55. 3×10^9 A.
57. B will decrease.
59. 2.1×10^{-6} g.
61. $2\mu_0 I/L\pi$ (left).

63. 4×10^{-6} T, about 10% of the Earth's field.

CHAPTER 29

1. -3.8×10^2 V.
3. Counterclockwise.
5. 0.026 V.
7. (a) Counterclockwise; (b) clockwise; (c) zero; (d) counterclockwise.
9. Counterclockwise.
11. (a) Clockwise; (b) 43 mV; (c) 17 mA.
13. 1.1×10^{-5} J.
15. 4.21 C.
17. (a) 5.2×10^{-2} A; (b) 0.32 mW.
19. 1.7×10^{-2} V.
21. $(\mu_0 Ia/2\pi) \ln 2$.
23. (a) 0.15 V; (b) 5.4×10^{-3} A; (c) 4.5×10^{-4} N.
25. (a) Will move at constant speed; (b) $v = v_0 e^{-B^2\ell^2 t/mR}$.
27. (a) $\dfrac{\mu_0 Iv}{2\pi}\ln\left(\dfrac{a + b}{b}\right)$ toward long wire; (b) $\dfrac{\mu_0 Iv}{2\pi}\ln\left(\dfrac{a + b}{b}\right)$ away from long wire.
31. 0.33 kV, 120 rev/s.
33. 100 V.
35. 13 A.
37. 3.54×10^4 turns.
39. 0.18.
41. (a) 5.2 V; (b) step-down transformer.
43. 549 V, 68.6 A.
45. 56.8 kW.
47. 0.188 V/m.
49. (b) Clockwise; (c) $dB/dt > 0$.
51. (a) IR/ℓ (constant); (b) $\dfrac{\mathscr{E}_0}{\ell}e^{-B^2\ell^2 t/mR}$.
53. 31 turns.
55. $v = 0.76$ m/s.
57. 184 kV.
59. 1.5×10^{17}.
61. (a) 23 A; (b) 90 V; (c) 6.9×10^2 W; (d) 75%.
63. (a) 0.85 A; (b) 8.2.
65. $\frac{1}{2}B\omega L^2$ toward the center.
71. $B\omega r$, radially out from the axis.

CHAPTER 30

1. $M = \mu_0 N_1 N_2 A/\ell$.
3. $M/\ell = \mu_0 n_1 n_2 \pi r_2^2$.
5. $M = (\mu_0 w/2\pi)\ln(\ell_2/\ell_1)$.
7. 1.2 H.

9. 2.5×10^{-6} H.
11. $r_1 \geq 2.5$ mm.
13. 3.
15. (a) $L_1 + L_2$; (b) $= L_1L_2/(L_1 + L_2)$.
17. 15.9 J.
19. (a) $u_E = 4.4 \times 10^{-4}$ J/m^3, $u_B = 1.6 \times 10^6$ J/m^3, $u_B \gg u_E$; (b) $E = 6.0 \times 10^8$ V/m.
21. 4.4 J/m^3, 1.6×10^{-14} J/m^3.
23. $(\mu_0 I^2/4\pi) \ln(r_2/r_1)$.
25. $t/\tau = 4.6$.
27. $(dI/dt)_0 = V_0/L$.
29. (a) $(LV^2/2R^2)(1 - 2e^{-t/\tau} + e^{-2t/\tau})$; (b) $t/\tau = 5.3$.
31. (a) 213 pF; (b) 46.5 μH.
35. (a) $Q = Q_0/\sqrt{2}$; (b) $T/8$.
37. $R = 2.30 \, \Omega$.
41. Decrease, 1.15 kΩ.
43. 20 mH, 95 turns.
45. (a) 21 mH; (b) 45 mA; (c) 2.2×10^{-5} J.
47. 3.0×10^3 turns, 95 turns.
51. (b) Positioning one coil perpendicular to the other; (c) $L_1L_2/(L_1 + L_2)$; $(L_1L_2 - M^2)/(L_1 + L_2 \mp 2M)$.
55. (a) $\frac{1}{2}(Q_0^2/C)e^{-Rt/L}$.
57.

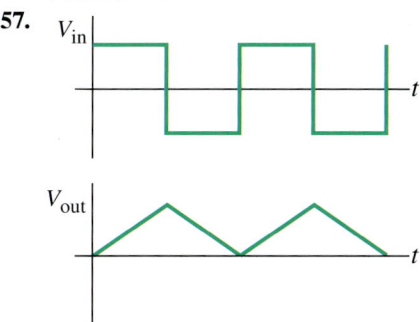

CHAPTER 31

1. (a) $3.7 \times 10^2 \, \Omega$; (b) $2.2 \times 10^{-2} \, \Omega$.
3. 9.90 Hz.
5.

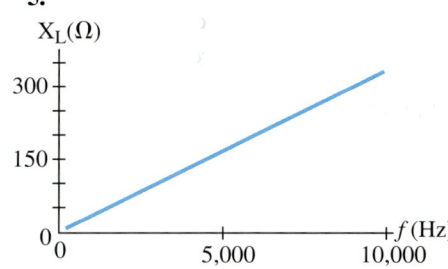

7. 0.13 H.
11. (a) 5.0%; (b) 98%.
13. (a) 9.0 kΩ; (b) 10.2 kΩ.
15. (a) 18 mA; (b) $-29°$; (c) 1.8 W; (d) $V_R = 105$ V, $V_C = 58$ V.
17. (a) 0.38 A; (b) $-89°$; (c) 0.29 W.
19. 332 Ω.
21. (a) 0; (b) $\frac{2}{\pi}V_0$, $\overline{V}_{1/2} = \frac{2\sqrt{2}}{\pi}V_{\text{rms}}$.
23. 8.78 kΩ, $-7.66°$, 91.1 mA.
25. 265 Hz, 324 W.
27. 52.5 mA.
29. (b) $\omega'^2 = [(1/LC) - (R^2/2L^2)]$; (c) $k \leftrightarrow 1/C$, $m \leftrightarrow L$, $b \leftrightarrow R$.
31. (a) $V_0^2 R/[2R^2 + 2(\omega L - 1/\omega C)^2]$; (b) $\omega'^2 = 1/LC$; (c) $\Delta\omega = R/L$.
33. $4 \, \Omega$.
35. 9.76 nF.
37. 27.9 mH.
39. 1.6 kHz.
41. 14 Ω, 75 mH.
43. 2.2×10^3 Hz, 1.1×10^4 Hz.
45. (a) 23.6 kΩ, 10.8°; (b) 1.88×10^{-5} W; (c) 2.8×10^{-5} A, 0.66 V, 4.7×10^{-4} V, 0.126 V.
49. $\{(R_1 + R_2)^2 + [\omega(L_1 + L_2) - (C_1 + C_2)/\omega C_1 C_2]^2\}^{1/2}$.
51. 19 Ω, 62 mH.
53. $I = \left(\dfrac{V_0}{Z}\right) \sin(\omega t + \phi)$,
$I_C = \left(\dfrac{V_0}{X_C}\right)\left[-\left(\dfrac{R}{Z}\right)\cos(\omega t + \phi) + \cos(\omega t)\right]$,
$I_L = \left(\dfrac{V_0}{X_L}\right)\left[\left(\dfrac{R}{Z}\right)\cos(\omega t + \phi) - \cos(\omega t)\right]$,
$Z = \sqrt{R^2 + \left(\dfrac{X_C X_L}{X_L - X_C}\right)^2}$,
$\tan\phi = \dfrac{X_C X_L}{(X_L - X_C)R}$.

CHAPTER 32

1. 9.2×10^4 V/m·s.
3. 7.9×10^{14} V/m·s.
7. $\oint \mathbf{B} \cdot d\mathbf{A} = \mu_0 Q_m$, $\oint \mathbf{E} \cdot d\boldsymbol{\ell} = \mu_0 \, dQ_m/dt - d\Phi_B/dt$.
9. 1.4×10^{-13} T.
11. (a) $B_0 = E_0/c$, $-y$-direction; (b) $-z$-direction.
13. (a) 1.08 cm; (b) 3.0×10^{18} Hz.
15. 314 nm, ultraviolet.
17. (a) 4.3 min; (b) 71 min.
19. 1.77×10^{-6} W/m^2.

21. 7.82×10^{-7} J/h.
23. 4.50×10^{-6} J.
25. 3.8×10^{26} W.
29. $r < 3 \times 10^{-7}$ m.
31. 302 pF.
33. 2.59 nH $\leq L \leq$ 3.89 nH.
35. (a) 441 m; (b) 2.979 m.
37. 5.56 m, 0.372 m.
39. (a) 1.28 s; (b) 4.3 min.
43. 469 V/m.
45. Person at the radio hears the voice 0.14 s sooner.
47. (a) 0.40 W; (b) 12 V/m; (c) 12 V.
49. 1.5×10^{11} W.
51. (a) Parallel; (b) 8.9 pF $\leq C \leq$ 80 pF; (c) 1.05 mH $\leq L \leq$ 1.12 mH.

CHAPTER 33

1. (a) 2.21×10^8 m/s; (b) 1.99×10^8 m/s.
3. 8.33 min.
5. 3 m.
7. 3.4×10^3 rad/s.
9. I_3 is the desired image:

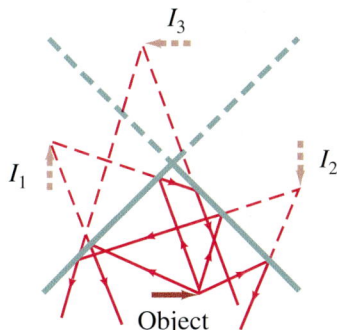

Depending on where you put your eye, two other images may also be visible.
11. 5°.
15. 36.4 cm.
19. 4.5 m.
21. Convex, -20 m.
23. (a) Center of curvature; (b) real; (c) inverted; (d) -1.
29. (a) Convex mirror; (b) 22 cm behind surface; (c) -98 cm; (d) -196 cm.
33. 45.6°.
35. 24.9°.
37. 4.6 m.
43. 3.0%.
45. 0.22°.
47. 61.7°, lucite.
49. 93.5 cm.

51. $n_{\text{liquid}} \geq 1.5$.
55. 17.0 cm below the surface of the glass.
59. (a) 3.0 m, 4.0 m, 7.0 m; (b) toward you, away from you, toward you.
61. -3.80 m.
63. Chose different signs for the magnification; 13.3 cm, 26.7 cm.
65. $\geq 56.1°$.
69. 81 cm (inside the glass).

CHAPTER 34

1. (a)

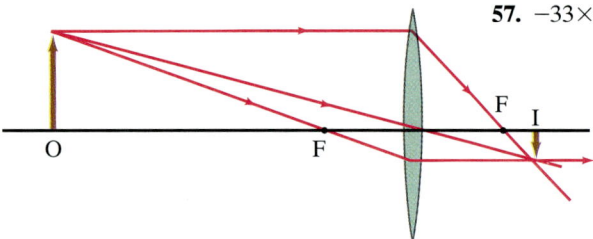

(b) 24.9 cm.
3. (a) 3.64 D, converging; (b) -16.0 cm, diverging.
5. (a) -0.26 mm; (b) -0.47 mm; (c) -1.9 mm.
7. (a) 81 mm; (b) 82 mm; (c) 87 mm; (d) 24 cm.
9. (a) Virtual; (b) converging lens; (c) 7.5 D.
11. (a) -10.5 cm (diverging), virtual; (b) $+203$ cm (converging).
13. 22.9 cm, 53.1 cm.
15. Real and upright.
17. Real, 21.3 cm beyond second lens, -0.708 (inverted).
19. (a) $+7.14$ cm; (b) -0.357 (inverted); (c)

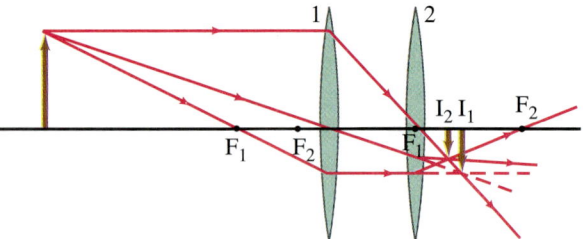

23. 1.54.
25. 8.1 cm.
27. -1.87 m (concave).
29. $+1.15$ D.
31. $f/2.3$.
35. 41 mm.
37. $+2.3$ D.
39. Glasses would be better.
41. (a) -1.33 D; (b) 38 cm.
43. -26.8 cm.
45. 17 cm, 100 cm.
47. 8.3 cm.
49. 6.3 cm from the lens, $3.9\times$.
51. (a) -49.4 cm; (b) $4.7\times$.
53. (a) $7.2\times$; (b) $2.2\times$.
55. 3.2 cm, 83 cm.
57. $-33\times$.
59. $12\times$.
61. $f_o = 4.0$ m, $r = 8.0$ m.
63. $7.5\times$.
65. 1.7 cm.
67. (a) 0.85 cm; (b) $230\times$.
69. (a) 14.4 cm; (b) $137\times$.
71. (a) 15.9 cm; (b) 14.3 cm; (c) 1.6 cm; (d) $r = 0.46$ cm.
73. $6.87\text{ m} \leq d_o \leq \infty$.
75. 100 mm, 200 mm.
77. 79.4 cm, 75.5 cm.
79. 0.101 m, -2.7 m.
81. $0 < -d_o < -f$.
83. (c) $\Delta d = \sqrt{d_T^2 - 4d_T f}$, $m = \left(\dfrac{d_T + \sqrt{d_T^2 - 4d_T f}}{d_T - \sqrt{d_T^2 - 4d_T f}}\right)^2$.
85. $1/f' = [(n/n') - 1][(1/R_1) + (1/R_2)]$; $(1/d_o) + (1/d_i) = 1/f'$, where $1/f' = [(n/n') - 1]/f(n - 1)$; $m = -d_i/d_o$.
87. $+3.6$ D.
89. $2.9\times, 4.1\times$.
91. (a) $-2.5\times$; (b) 5.0 D.
93. $-20\times$.

CHAPTER 35

3. 5.9 μm.
5. 3.9 cm.
7. 0.21 mm.
9. 613 nm.
11. 533 nm.
15. (a) $I_\theta/I_0 = (1 + 4\cos\delta + 4\cos^2\delta)/9$; (b) $\sin\theta_{\max} = m\lambda/d$, $m = 0, \pm 1, \pm 2, \ldots$; $\sin\theta_{\min} = (m + \tfrac{1}{3}k)\lambda/d$; $k = 1, 2$; $m = 0, \pm 1, \pm 2, \ldots$.
17. Orange-red.
19. 179 nm.
21. 9.1 μm.
23. 120 nm, 240 nm.
25. 1.26.
29. 0.47, 0.23.
31. 0.221 mm.
33. 0.289 mm.
35. (a) 17 lm/W; (b) 156.
37. (a) Constructive; (b) destructive.
39. 464 nm.
41. 646 nm.
43. (a) 81.5 nm; (b) 127 nm.
45. 0.5 cm.
47. $\theta = 63.3°$.
49. $\sin\theta_{\max} = (m + \tfrac{1}{2})\lambda/2S$, $m = 0, 1, 2, \ldots$, $\sin\theta_{\min} = m\lambda/2S$, $m = 0, 1, 2, \ldots$.
51. $I/I_0 = \cos^2(2\pi x/\lambda)$.

CHAPTER 36

1. $2.26°$.
3. 2.4 m.
5. 5.8 cm.
7. 4.28 cm.
9. (b) Average intensity is $2\times$.
11. $10.7°$.
13. $d = 4a$.
15. (a) 1.8 cm; (b) 11.0 cm.

17.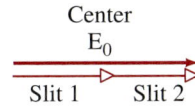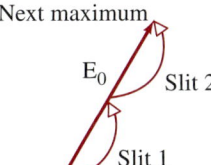

19. 2.4×10^{-7} rad $= (1.4 \times 10^{-5})°$
    $= 0.050''$.
21. 820 lines/mm, 102 lines/mm.
23. 5.61°.
25. Two full orders.
27. 497 nm, 612 nm, 637 nm, 754 nm.
29. 600 nm to 750 nm of second order overlaps with 400 nm to 500 nm of third order.
31. 621 nm, 909 nm.
35. (a) Two orders; (b) 6.44×10^{-5} rad
    $= 13.3'', 7.36 \times 10^{-5}$ rad $= 15.2''$,
    2.52×10^{-4} rad $= 52.0''$.
37. (a) $1.60 \times 10^4, 3.20 \times 10^4$;
    (b) 0.026 nm, 0.013 nm.
39. $\Delta f = f/mN$.
41. (a) 62.0°; (b) 0.21 nm.
43. 0.033.
45. 57.3°.
47. (a) 35°; (b) 63°.
49. 36.9°, 53.1°.
51. $I_0/32$.
55. 31° on either side of the normal.
57. 12,500 lines/cm.
59. $\sin\theta = \sin 20° - (m\lambda/a)$,
    $m = \pm 1, \pm 2, \ldots$.
61. Two orders.
63. 11.7°.
65. (a) 16 km; (b) 0.42'.
67. (a) 0; (b) $0.094 I_0$;
    (c) no light gets transmitted.
69. (a) 30°; (b) 18°; (c) 5.7°.
73. 0.245 nm.

CHAPTER 37

1. (a) 1.00; (b) 0.99995; (c) 0.995;
   (d) 0.436; (e) 0.141; (f) 0.0447.
3. 2.07×10^{-6} s.

5. $0.773c$.
7. $0.141c$.
9. (a) 99.0 yr; (b) 27.7 yr; (c) 26.6 ly;
   (d) $0.960c$.
11. $0.89c = 2.7 \times 10^8$ m/s.
13. (a) (470 m, 20 m, 0);
    (b) (1820 m, 20 m, 0).
15. (a) $0.80c$; (b) $-0.80c$.
17. 60 m/s, 24°.
19. 2.7×10^8 m/s, 43°.
21. (a) $L_0\sqrt{1 - (v/c)^2 \cos^2\theta}$;
    (b) $\tan\theta' = \tan\theta/\sqrt{1 - (v/c)^2}$.
23. Not possible in the boy's frame.
25. $0.866c$.
27. (a) 0.5%; (b) 13%.
29. 5.36×10^{-13} kg.
31. 8.20×10^{-14} J, 0.511 MeV.
33. 9×10^2 kg.
37. (a) 11.2 GeV (1.79×10^{-9} J);
    (b) 6.45×10^{-18} kg·m/s.
39. 7.49×10^{-19} kg·m/s.
41. $0.941c$.
43. $M = 2m/\sqrt{1 - (v^2/c^2)}$,
    $K_{\text{loss}} = 2mc^2\{[1/\sqrt{1 - (v^2/c^2)}] - 1\}$.
45. $0.866c, 4.73 \times 10^{-22}$ kg·m/s.
47. 39 MeV (6.3×10^{-12} J),
    1.5×10^{-19} kg·m/s, $-6\%, -4\%$.
49. $0.804c$.
51. 3.0 T.
57. (a) $0.80c$; (b) $2.0c$.
61. 3.8×10^{-5} s.
63. $\rho = \rho_0/[1 - (v^2/c^2)]$.
65. (a) 1.5 m/s less than c; (b) 30 cm.
67. 1.02 MeV (1.64×10^{-13} J).
69. 2.2 mm.
71. 0.78 MeV.
73. Electron.
75. 5.19×10^{-13}%.
81. (a) $\alpha = \tan^{-1}[(c^2/v^2) - 1]^{1/2}$;
    (c) $\tan\theta = c/v, u = \sqrt{v^2 + c^2}$.

CHAPTER 38

1. 6.59×10^3 K.
3. 5.4×10^{-20} J, 0.34 eV.
5. (a) 114 J; (b) 228 J; (c) 342 J;
   (d) $114n$ J; (e) -456 J.
7. (b) $h = 6.63 \times 10^{-34}$ J·s.
9. 3.67×10^{-7} eV.
11. 2.4×10^{13} Hz, 1.2×10^{-5} m.
13. 400 nm.

15. (a) 2.18 eV; (b) 0.92 V.
17. 3.46 eV.
19. 1.88 eV, 43.3 kcal/mol.
21. (a) 2.43×10^{-12} m;
    (b) 1.32×10^{-15} m.
23. (a) $5.90 \times 10^{-3}, 1.98 \times 10^{-2}$,
    3.89×10^{-2};
    (b) 60.8 eV, 204 eV, 401 eV.
27. 1.82 MeV.
29. 212 MeV, 5.85×10^{-15} m.
31. 3.2×10^{-32} m.
33. 19 V.
35. (a) 0.39 nm; (b) 0.12 nm;
    (c) 0.039 nm.
37. 1.84×10^3.
39. 3.3×10^{-38} m/s, no diffraction.
41. 0.026 nm.
43. 3.4 eV.
45. 122 eV.
49. 52.5 nm.
51.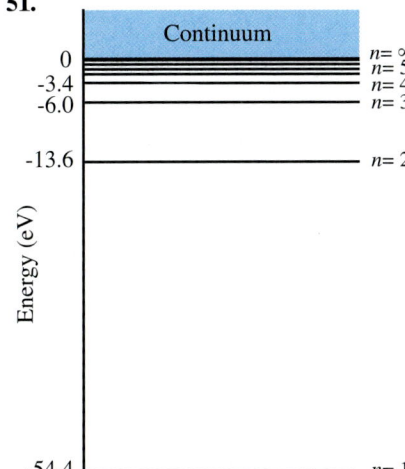

53. $U = -27.2$ eV, $K = +13.6$ eV.
55. Justified.
61. 3.28×10^{15} Hz.
65. 2.78×10^{21} photons/s·m².
67. 8.3×10^6 photons/s.
69. $\theta = 89.4°$.
71. 4.7×10^{-14} m.
73. 10.2 eV.
75. 4.4×10^{-40}, yes.
77. 653 nm, 102 nm, 122 nm.
79. 0.64 V.
83. 5×10^{-12} m.
85. 3.

CHAPTER 39

1. 1.8×10^{-7} m.
3. $\pm 1.3 \times 10^{-11}$ m.
5. 7.2×10^{3} m/s.
7. 2.4×10^{-3} m, 1.4×10^{-32} m.
9. (a) 6.6×10^{-8} eV; (b) 6.5×10^{-9}; (c) 7.9×10^{-7} nm.
13. (a) $\psi = A \sin(3.5 \times 10^{9}\,\mathrm{m}^{-1})x + B \cos(3.5 \times 10^{9}\,\mathrm{m}^{-1})x$;
    (b) $\psi = A \sin(6.3 \times 10^{12}\,\mathrm{m}^{-1})x + B \cos(6.3 \times 10^{12}\,\mathrm{m}^{-1})x$.
17. 3.6×10^{6} m/s.
19. (a) 52 nm; (b) 0.22 nm.
23. $E_1 = 0.094$ eV, $\psi_1 = (1.00\,\mathrm{nm}^{-1/2}) \sin(1.57\,\mathrm{nm}^{-1}\,x)$;
    $E_2 = 0.38$ eV, $\psi_2 = (1.00\,\mathrm{nm}^{-1/2}) \sin(3.14\,\mathrm{nm}^{-1}\,x)$;
    $E_3 = 0.85$ eV, $\psi_3 = (1.00\,\mathrm{nm}^{-1/2}) \sin(4.71\,\mathrm{nm}^{-1}\,x)$;
    $E_4 = 1.51$ eV, $\psi_4 = (1.00\,\mathrm{nm}^{-1/2}) \sin(6.28\,\mathrm{nm}^{-1}\,x)$.
25. (a) 4 GeV; (b) 2 MeV; (c) 2 MeV.
27. (a) 0.18; (b) 0.50; (c) 0.50.
29.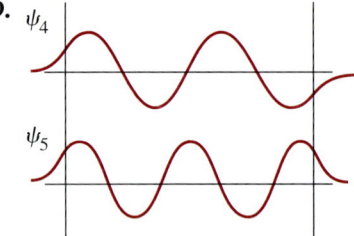
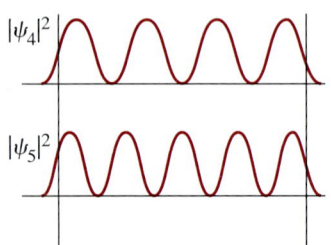
31. 0.03 nm.
33. 9.2 eV.
35. (a) Decreases by 8%; (b) decreases by 5%.
37. (a) 32 MeV; (b) 56 fm; (c) $8.8 \times 10^{20}\,\mathrm{s}^{-1}$, 10^{10} yr.
39. 21 MeV.
41. 3.00×10^{-10} eV/c^2.
43. r_1, the Bohr radius.
45. 0.5 MeV, 5×10^{6} m/s.
47. (b) 4 s.
49. 14% decrease.

CHAPTER 40

1. $\ell = 0, 1, 2, 3, 4, 5$.
3. 32 states, $(4,0,0,-\tfrac{1}{2}), (4,0,0,+\tfrac{1}{2})$,
   $(4,1,-1,-\tfrac{1}{2}), (4,1,-1,+\tfrac{1}{2}), (4,1,0,-\tfrac{1}{2})$,
   $(4,1,0,+\tfrac{1}{2}), (4,1,1,-\tfrac{1}{2}), (4,1,1,+\tfrac{1}{2})$,
   $(4,2,-2,-\tfrac{1}{2}), (4,2,-2,+\tfrac{1}{2})$,
   $(4,2,-1,-\tfrac{1}{2}), (4,2,-1,+\tfrac{1}{2})$,
   $(4,2,0,-\tfrac{1}{2}), (4,2,0,+\tfrac{1}{2}), (4,2,1,-\tfrac{1}{2})$,
   $(4,2,1,+\tfrac{1}{2}), (4,2,2,-\tfrac{1}{2}), (4,2,2,+\tfrac{1}{2})$,
   $(4,3,-3,-\tfrac{1}{2}), (4,3,-3,+\tfrac{1}{2})$,
   $(4,3,-2,-\tfrac{1}{2}), (4,3,-2,+\tfrac{1}{2})$,
   $(4,3,-1,-\tfrac{1}{2}), (4,3,-1,+\tfrac{1}{2}), (4,3,0,-\tfrac{1}{2})$,
   $(4,3,0,+\tfrac{1}{2}), (4,3,1,-\tfrac{1}{2}), (4,3,1,+\tfrac{1}{2})$,
   $(4,3,2,-\tfrac{1}{2}), (4,3,2,+\tfrac{1}{2}), (4,3,3,-\tfrac{1}{2})$,
   $(4,3,3,+\tfrac{1}{2})$.
5. $n \geq 5, m_\ell = -4,-3,-2,-1,0,1,2,3,4$, $m_s = -\tfrac{1}{2}, +\tfrac{1}{2}$.
7. (a) 6; (b) -0.378 eV; (c) $\ell = 4$, $\sqrt{20}\hbar = 4.72 \times 10^{-34}$ kg·m²/s;
   (d) $m_\ell = -4,-3,-2,-1,0,1,2,3,4$.
13. (a) $-[3/(32\pi r_0^3)^{1/2}]\, e^{-5/2}$;
    (b) $(9/32\pi r_0^3)\, e^{-5}$; (c) $(225/8 r_0)\, e^{-5}$.
15. 1.85.
17. 17.3%.
19. (a) $1.34 r_0$; (b) $2.7 r_0$; (c) $4.2 r_0$.
21. $\dfrac{r^4}{24 r_0^5} e^{-r/r_0}$.
27. (a) $\dfrac{4r^2}{27 r_0^3}\left(1 - \dfrac{2r}{3 r_0} + \dfrac{2 r^2}{27 r_0^2}\right)^2 e^{-2r/3 r_0}$;

(b)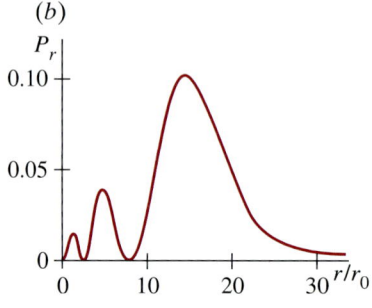

(c) $r = 13 r_0$.
29. (a) $(1,0,0,-\tfrac{1}{2}), (1,0,0,+\tfrac{1}{2})$,
    $(2,0,0,-\tfrac{1}{2}), (2,0,0,+\tfrac{1}{2}), (2,1,-1,-\tfrac{1}{2})$,
    $(2,1,-1,+\tfrac{1}{2})$; (b) $(1,0,0,-\tfrac{1}{2})$,
    $(1,0,0,+\tfrac{1}{2}), (2,0,0,-\tfrac{1}{2}), (2,0,0,+\tfrac{1}{2})$,
    $(2,1,-1,-\tfrac{1}{2}), (2,1,-1,+\tfrac{1}{2}), (2,1,0,-\tfrac{1}{2})$,
    $(2,1,0,+\tfrac{1}{2}), (2,1,-1,-\tfrac{1}{2}), (2,1,-1,+\tfrac{1}{2})$,
    $(3,0,0,-\tfrac{1}{2}), (3,0,0,+\tfrac{1}{2})$.
31. (a) $1s^2 2s^2 2p^6 3s^2 3p^6 3d^{10} 4s^2 4p^4$; (b) $1s^2 2s^2 2p^6 3s^2 3p^6 3d^{10} 4s^2 4p^6 4d^{10} 4f^{14} 5s^2 5p^6 5d^{10} 6s^1$; (c) $1s^2 2s^2 2p^6 3s^2 3p^6 3d^{10} 4s^2 4p^6 4d^{10} 4f^{14} 5s^2 5p^6 5d^{10} 6s^2 6p^6 5f^3 6d^1 7s^2$.
33. 5.8×10^{-13} m, 0.115 MeV.
37. 0.041 nm, 1 nm.
41. 0.19 nm.
43. Chromium.
47. (a) 0.25 mm; (b) 0.13 mm.
49. (a) $\tfrac{1}{2}, \tfrac{3}{2}, \tfrac{1}{2}\sqrt{3}\,\hbar, \tfrac{1}{2}\sqrt{15}\hbar$;
    (b) $\tfrac{5}{2}, \tfrac{7}{2}, \tfrac{1}{2}\sqrt{35}\hbar, \tfrac{1}{2}\sqrt{63}\hbar$;
    (c) $\tfrac{3}{2}, \tfrac{5}{2}, \tfrac{1}{2}\sqrt{15}\hbar, \tfrac{1}{2}\sqrt{35}\hbar$.
51. (a) 0.4 T; (b) 0.4 T.
53. 5.6×10^{-4} rad, 1.7×10^{2} m.
55. 3.7×10^{4} K.
57. (a) 1.56; (b) 1.4×10^{-10} m.
59. (a) $1s^2 2s^2 2p^6 3s^2 3p^6 3d^7 4s^2$;
    (b) $1s^2 2s^2 2p^6 3s^2 3p^6 3d^{10} 4s^2 4p^6$;
    (c) $1s^2 2s^2 2p^6 3s^2 3p^6 3d^{10} 4s^2 4p^6 5s^2$.
61. (a) 2.5×10^{74}; (b) 5.0×10^{74}.
63. $r = 5.24 r_0$.
65. (a) ϕ is unknown; (b) L_x and L_y are unknown.
67. (a) 1.2×10^{-4} eV; (b) 1.1 cm; (c) no difference.
69. (a) $3 \times 10^{-171}, 7 \times 10^{-203}$;
    (b) $1.1 \times 10^{-8}, 6.3 \times 10^{-10}$;
    (c) $6.6 \times 10^{15}, 3.8 \times 10^{14}$;
    (d) 7×10^{23} photons/s, 4×10^{22} photons/s.
71. 182.
73. 2.25.

CHAPTER 41

1. 5.1 eV.
3. 0.7 eV.
7. (a) 13.941 u; (b) 7.0034 u; (c) 0.9801 u.
9. (a) 6.86 u; (b) 1.85×10^{3} N/m.
11. (a) 1.79×10^{-4} eV; (b) 7.16×10^{-4} eV, 1.73 mm.
13. 2.36×10^{-10} m.
15. -7.9 eV.
17. 0.283 nm.
19. (b) -6.9 eV; (c) -10.8 eV; (d) 3%.
21. 1.8×10^{21}.
23. (a) 6.9 eV; (b) 6.8 eV.
25. 6.3%.
27. 3.2 eV, 1.05×10^{6} m/s.
29. (a) 1.79×10^{29} m^{-3}; (b) 3.
33. (a) 0.021, reasonable; (b) 0.979; (c) 0.021.
37. A large energy is required to create a conduction electron by raising an electron from the valence band to the conduction band.
39. 1.1 μm.
41. 5×10^{6}.
43. 1.91 eV.

45. 13 mA.
47. (a) 2.1 mA; (b) 4.3 mA.
49. (a) 9.4 mA (smooth); (b) 6.7 mA (rippled).
51. 4.0 kΩ.
53. 0.43 mA.
55. (a) 3.5×10^4 K; (b) 1.2×10^3 K.
57. (a) 0.9801 u; (b) 482 N/m, 88% of the constant for H_2.
59. States with higher values of L are less likely to be occupied, so less likely to absorb a photon; I will depend on L.
61. 2.8×10^4 J/kg.
63. 1.24 eV.
65. 1.09 μm, could be used.
67. (a) 0.094 eV; (b) 0.63 nm.
69. (a) 145 V ≤ V ≤ 343 V; (b) 3.34 kΩ ≤ R_{load} < ∞.

CHAPTER 42

1. 3729 MeV/c^2.
3. 1.9×10^{-15} m.
5. (a) 2.3×10^{17} kg/m^3; (b) 184 m; (c) 2.6×10^{-10} m.
7. 28 MeV.
9. $^{31}_{15}$P.
11. (a) 1.8×10^3 MeV; (b) 7.3×10^2 MeV.
13. 7.48 MeV.
15. (a) 32.0 MeV, 5.33 MeV; (b) 1636 MeV, 7.87 MeV.
17. 12.4 MeV, 7.0 MeV, neutron is more closely bound in ^{23}Na.
19. (b) Stable.
21. 0.782 MeV.
23. β^+ emitter, 1.82 MeV.
25. $^{228}_{90}$Th, 228.02883 u.
27. (a) $^{32}_{16}$S; (b) 31.97207 u.
29. 0.862 MeV.
31. 5.31 MeV.
33. (b) 0.961 MeV, 0.961 MeV to 0.
35. 3.0 h.
37. 1.2×10^9 decays/s.
39. 1.78×10^{20} nuclei.
41. 7 α particles, 4 β^- particles.
43. (a) 4.8×10^{16} nuclei; (b) 3.2×10^{15} nuclei; (c) 7.2×10^{13} decays/s; (d) 26 min.
45. 1.68×10^{-10} g.
47. 2.6 min.
49. 3.4 mg.
51. 8.6×10^{-7}.
53. $^{211}_{82}$Pb.
55. $N_D = N_0(1 - e^{-\lambda t})$.
57. (a) 0.99946; (b) 1.2×10^{-14}; (c) 2.3×10^{17} kg/m^3, $10^{14}\times$.
59. 28.6 eV.
61. (a) 7.2×10^{55}; (b) 1.2×10^{29} kg; (c) 3.2×10^{11} m/s^2.
63. 6×10^3 yr.
65. 6.64 $T_{1/2}$.
69. Calcium, stored by body in bones, 193 yr, $^{90}_{38}$SR → $^{90}_{39}$Y + $^{0}_{-1}$e + $\bar{\nu}$, $^{90}_{39}$Y is radioactive, $^{90}_{39}$Y → $^{90}_{40}$Zr + $^{0}_{-1}$e + $\bar{\nu}$, $^{90}_{40}$Zr is stable.
73. (a) 0.002603 u, 2.425 MeV/c^2; (b) 0; (c) −0.094909 u, −88.41 MeV/c^2; (d) 0.043924 u, 40.92 MeV/c^2; (e) Δ ≥ 0 for 0 ≤ Z ≤ 8 and Z ≥ 85, Δ < 0 for 9 ≤ Z ≤ 84.
75. 0.083%.
77. $^{228}_{88}$RA, $^{228}_{89}$Ac, $^{228}_{90}$Th, $^{224}_{88}$Ra, $^{220}_{87}$Rn, $^{231}_{90}$Th, $^{231}_{91}$Pa, $^{227}_{89}$Ac, $^{227}_{90}$Th, $^{223}_{88}$Ra.
79. (b) ≈ 10^{17} yr; (c) ≈ 60 yr; (d) 0.4.

CHAPTER 43

1. $^{28}_{13}$Al, β^- emitter, $^{28}_{14}$Si.
3. Possible.
5. 5.701 MeV is released.
7. (a) Can occur; (b) 19.85 MeV.
9. +4.730 MeV.
11. (a) $^{7}_{3}$Li; (b) neutron is stripped from the deuteron; (c) +5.025 MeV, exothermic.
13. (a) $^{31}_{15}$P(p, γ)$^{32}_{16}$S; (b) +8.864 MeV.
15. $\sigma = \pi(R_1 + R_2)^2$.
17. Rate at which incident particles pass through target without scattering.
19. (a) 0.7 μm; (b) 1 mm.
21. 173.2 MeV.
23. 0.116 g.
25. 25.
27. 0.11.
29. 1.3 keV.
33. 6.1×10^{23} MeV/g, 4.9×10^{23} MeV/g, 2.1×10^{24} MeV/g, 5.1×10^{23} MeV/g.
35. Not independent.
37. 3.23×10^9 J, 65×.
39. (a) 4.9×; (b) 1.5×10^9 K.
41. 400 rad.
43. 167 rad.
45. 1.7×10^2 counts/s.
49. 0.225 μg.
51. (a) 0.03 mrem ≈ 0.006% of allowed dose; (b) 0.3 mrem ≈ 0.06% of allowed dose.
53. (a) 1; (b) 1 ≤ m ≤ 2.7.
55. (a) $^{12}_{6}$C; (b) +5.70 MeV.
57. 1.004.
59. 51 mrem/yr.
61. 4.6 m.
63. 18.000953 u.
65. 6.31×10^{14} J/kg, ≈ $10^7 \times$ the heat of combustion of coal.
67. 1×10^{24} neutrinos/yr.
69. (a) 6.8 bn; (b) 3×10^{-14} m.
71. (a) 3.7×10^3 decays/s; (b) 5.2×10^{-4} Sv/yr ≈ 0.15 background.

CHAPTER 44

1. 7.29 GeV.
3. 1.8 T.
5. 13 MHz.
7. $\lambda_\alpha = 2.6 \times 10^{-15}$ m ≈ size of nucleon, $\lambda_p = 5.2 \times 10^{-15}$ m ≈ 2(size of nucleon), α particle is better.
9. 1.4×10^{-18} m.
11. 2.2×10^6 km, 7.5 s.
15. 33.9 MeV.
17. 1.879 GeV.
19. 67.5 MeV.
21. 2.3×10^{-18} m.
23. (b) Uncertainty principle allows energy to not be conserved.
25. 69.3 MeV.
27. 8.6 MeV, 57.4 MeV.
29. 52.3 MeV.
31. 7.5×10^{-21} s.
33. (a) 1.3 keV; (b) 8.9 keV.
35. (a) n = d d u; (b) $\bar{n}$ = $\bar{d}\,\bar{d}\,\bar{u}$; (c) Λ^0 = u d s; (d) Σ^0 = $\bar{u}\,\bar{d}\,\bar{s}$.
37. D^0 = c$\bar{u}$.

39. (a)

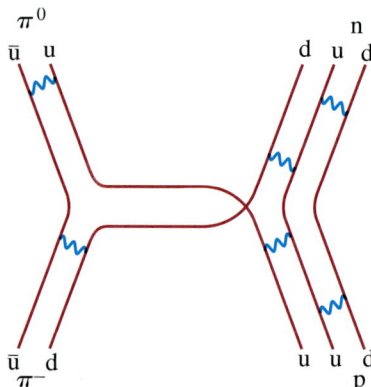

(b)

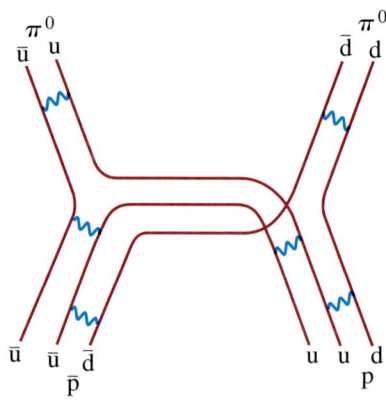

41. 26 GeV, 4.8×10^{-17} m.
43. 5.5 T.
45. (a) Possible, through the strong interaction; (b) possible, through the strong interaction; (c) possible, through the strong interaction; (d) forbidden, charge is not conserved; (e) possible, through the weak interaction; (f) forbidden, charge is not conserved; (g) possible, through the strong interaction; (h) forbidden, strangeness is not conserved; (i) possible, through the weak interaction; (j) possible, through the weak interaction.
49. -135.0 MeV, -140.9 MeV.
51. 64.
53. (b) 10^{27} K.
55. 6.58×10^{-5} m.
57. $\bar{u}\,\bar{u}\,\bar{d} + u\,d\,d \rightarrow \bar{u}\,d + d\,\bar{d}$.

CHAPTER 45

3. 4.8 ly.
5. $0.059''$, 17 pc.
7. Less, $\phi_1/\phi_2 = \tfrac{1}{2}$.
9. 48 W/m².
11. 1.4×10^{-4} kg/m³.
13. 1.8×10^9 kg/m³, 3.3×10^5.
15. -92.2 keV, 7.366 MeV.
17. (a) 9.594 MeV is released; (b) 7.6 MeV; (c) 6×10^{10} K.
19. $d_2/d_1 = 6.5$.
23. 540°.
25. 1.4×10^8 ly.
27. $0.86c$.
29. $0.328c$, 4.6×10^9 ly.
31. 1.1 mm.
33. (a) 10^{-3}; (b) 10^{-10}; (c) 10^{-13}; (d) 10^{-27}.
35. (a) Temperature increases, luminosity is constant, size decreases; (b) temperature is constant, luminosity decreases, size decreases; (c) temperature decreases, luminosity increases, size increases.
37. 8×10^3 rev/s.
39. 7×10^{24} W.
41. (a) 46 eV; (b) 18 eV.
43. $\approx (2 \times 10^{-5})°$.
45. 1.4×10^{16} K, hadron era.
47. Venus is brighter, $\ell_V/\ell_S = 16$.
49. $R \geq GM/c^2$.
51. $R = \dfrac{h^2}{16 m_n^{8/3} GM^{1/3}} \left(\dfrac{18}{\pi^2}\right)^{2/3}$.

Index

Note: The abbreviation *defn* means the page cited gives the definition of the term; *fn* means the reference is in a footnote; *pr* means it is found in a problem or question; *ff* means also the following pages. Page numbers such as A-3 refer to the appendices, after the main text.

A

Aberration:
 chromatic, 856 *fn*, 859
 of lenses, 858–59
 spherical, 817, 829, 858
Absolute luminosity (stars & galaxies), 1145–46
Absolute space, 918, 923
Absolute temperature scale, 447 *ff*, 460, 538
Absolute zero, 455, 538
Absorbed dose, 1102
Absorption lines, 902, 964, 1039, 1042–43
Absorption spectra, 902, 964, 968

Abundancies, natural, 1062
Acceleration, 23–37, 54–55, 64–65, 80–82
 angular, 241–42
 average, 23–24, 54
 centripetal, 64–65, 144 *ff*, 242
 Coriolis, 292–94
 in *g*'s, 36
 of gravity, 31 *ff*, 85, 137–39
 instantaneous, 25, 54
 of the moon, 134
 motion at constant, 26 *ff*, 54–55
 radial, 64–65, 69, 114 *ff*, 242
 related to force, 80–82
 tangential, 121–22, 242
 uniform, 26 *ff*, 55, 243–44
 variable, 36–37
Accelerators, 1115–20
Accelerometer, 94
Acceptor level, 1052
AC circuits, 658 *fn*, 772–83
Accommodation of eye, 851
AC generator, 740–41
Achromatic doublet, 859
Achromatic lens, 859
AC motor, 698
Action at a distance, 147, 554

Action-reaction (Newton's third law), 82–85
Activation energy, 471, 1034
Activity, 1075
Addition of velocities, 66–68
 relativistic, 935–36
Adhesion, 352
Adiabatic process, 495–96, 502–3, 531, 537
Air bags, 30
Air columns, vibrations of, 424–29
Air conditioner, 525–28
Airfoil, 348
Airplane wing, 348
Air resistance, 31–33, 122–23
Airy disc, 896
Alkali metals, 1013
Allowed transition, 1007, 1039, 1042–43
Alpha decay, 1067–70, 1073
 and tunneling, 998, 1069
Alpha particles (or rays), 1068–70
Alternating current (ac), 645–46, 772–83
Alternators, 742
AM, 804
Ammeter, 674–76, 697

Amorphous solids, 1044
Ampère, Andrè, 637, 712, 722, 787
Ampère (unit), 637, 712
   operational definition of, 712
Ampère's law, 712–15, 788–90
Amplifier, 1054–55
Amplitude:
   pressure, 420
   of vibration, 363
   of wave, 390, 396, 419, 422
Amplitude modulation (AM), 804
Analog information, 749
Analyzer (of polarized light), 908
Andromeda, 1143
Aneroid gauge, 338–39
Angle:
   Brewster's, 910
   of dip, 688
   of incidence, 403, 408, 813, 822
   phase, 366, 398, 778
   polarizing, 910
   radian measure of, 240
   of reflection, 403, 813
   of refraction, 408–9, 822
Angstrom (unit), 824 *fn*
Angular acceleration, 241–42
Angular displacement, 246–47
Angular frequency, 366
Angular magnification, 853
Angular momentum, 256–60, 281–89
   in atoms, 965, 1005–7, 1015–18
   conservation, law of, 257–59, 288–89
   of nuclei, 1064
   total, 1017–18
   vector, nature of, 259–60, 281–90
Angular position, 240
Angular quantities, 240–43, 246–47
Angular velocity, 241–44
Annihilation, 957, 1124, 1163
Anode, 605
Antenna, 787, 805, 876–77
Anthropic principle, 1169
Antilock brakes, 119
Antimatter, 1137 *pr*, 1163 *fn*
Antineutrino, 1071
Antineutron, 1124
Antinodes, 405, 427
Antiparticles, 1072, 1124–28, 1163 *fn*
Antiproton, 1124
Antiquark, 1130–31
Apparent brightness (stars and galaxies), 1145
Apparent magnitude, 1172 *pr*

Apparent weightlessness, 142
Approximate calculations, 9–12
Arago, F., 887
Arch, 319–21
Archimedes, 340–41
Archimedes' principle, 340–43
Area:
   formulas for, A-1
   under a curve or graph, 161–64
Aristotle, 2, 78–79
Armature, 698, 744
Asymptotic freedom, 1133
Astigmatism, 851
Astronomical telescope, 854–56
Astrophysics, 1140 *ff*
Atmosphere (unit), 337–40
Atmosphere, scattering of light by, 911
Atmospheric pressure, 336–40
Atomic bomb, 1095
Atomic force microscope, 999
Atomic mass, 446
Atomic mass number, 1062
Atomic mass unit, 7, 446, 1063
Atomic number, 1010, 1012–15, 1062
Atomic spectra, 863–65
Atomic structure:
   Bohr model of, 965–71, 977, 1003–5
   early models of, 862–63
   quantum mechanics of, 1003–24
Atomic weight, 446 *fn*
Atoms, 446–47, 965–73 (*see also* Atomic structure, Kinetic theory)
   angular momentum in, 967
Atoms (*cont.*)
   binding energy in, 968
   complex, 1010–15
   electric charge in, 547
   energy levels in, 967
   hydrogen, 964–71, 1004–10
   ionization energy in, 968, 970
   probability distribution in, 1004, 1008–10
   shells and subshells in, 1012–13
   vector model of, 1028 *pr*
ATP, 1036
Atwood machine, 92, 277 *pr*, 286
Audible range, 418
Aurora borealis, 694
Autoclave, 482 *pr*
Autofocusing camera, 419
Average binding energy per nucleon, 1065
Average lifetime, 1084 *pr*

Average speed, 18
Avogadro, Amedeo, 459
Avogadro's hypothesis, 459
Avogadro's number, 459
Axis, instantaneous, 246
Axis of lens, 837
Axis of rotation (*defn*), 240

B

Back, forces in, 331 *pr*
Back emf, 742–43
Background radiation, cosmic microwave, 1159–61, 1164
Balance, 308
Ballistic galvanometer, 755 *pr*
Ballistic pendulum, 218
Balmer, J. J., 964
Balmer formula, 964, 969
Balmer series, 964, 968–69
Band gap, 1049–50
Band spectra, 1038
Band theory of solids, 1048–50
Banking of curves, 118–20
Bar (unit), 337
Bar-code reader, 1022
Barn (unit), 1089
Barometer, 339
Barrier penetration, 996–99, 1069–70, 1155 *fn*, 1168
Baryon (*defn*), 1127
Baryon number, 1124–26, 1132, 1162
Base (of transistor), 1054
Base and derived quantities and units, 8
Battery, 635–36, 659, 668
Beam splitter, 881
Beat frequency, 431
Beats, 429–32
Becquerel, H., 1067
Becquerel (unit), 1101
Bel, 421
Bell, Alexander Graham, 421
Bernoulli, Daniel, 345–46
Bernoulli's equation, 345–47
Bernoulli's principle, 345–49
Beta decay, 1067, 1070–73, 1125
Beta particle, 1067, 1070
Betatron, 754 *pr*
Biasing and bias voltage, 1052, 1054–55
Big Bang, 1136, 1159–64
Big crunch, 1165

Bimetallic-strip thermometer, 448
Binary system, 1146, 1155
Binding energy:
  in atoms, 968
  in molecules, 1031, 1033, 1036
  in nuclei, 1064–66
  in solids, 1044
  total, 1064
Binding energy per nucleon, 1065
Binoculars, 827, 856
Binomial expansion, A-1
Biological damage by radiation, 1101–2
Biot, Jean Baptist, 719
Biot-Savart law, 719–21
Blackbody, 950
Blackbody radiation, 950–51, 1160
Black dwarf, 1149
Black hole, 149, 1144, 1151, 1155, 1158, 1166, 1168
Blood flow, 344–45
Blueshift, 1156
Blue sky, 911
Bohr, Niels, 958, 965, 969, 985
Bohr magneton, 1016, 1064
Bohr model of the atom, 965–71, 977, 1003–5
Bohr radius, 590 pr, 966, 1004
Boiling, 475–76, 490–93 (see also Phase, changes of)
Boiling point, 475–76
Boltzmann, Ludwig, 535–36
Boltzmann's constant, 459
Boltzmann distribution, 1020
Boltzmann factor, 1020, 1046
Bomb calorimeter, 490
Bond:
  covalent, 1030–33, 1044
  dipole, 1036–37
  hydrogen, 1036
  ionic, 1031–33, 1044
  metallic, 1044
  molecular, 1030–33
  partially covalent/ionic, 1032–33, 1044
  strong (defn), 1036
  van der Waals, 1036–37
  weak, 1036–37, 1045
Bond energy, 1031, 1036
Bonding:
  in molecules, 1030–33
  in solids, 1044–45
Bone density, 957
Bose, S. N., 1011 fn

Bose-Einstein statistics, 1046 fn
Bosons, 1011 fn, 1046 fn, 1126–27, 1132–34
Bottomness and bottom quark, 1130–31
Boundary conditions, 990, 995
Bound charge, 625
Bound state, 994
Bow wave, 436
Boyle, Robert, 455
Boyle's law, 455–56, 467
Bragg, W. H., 906
Bragg, W. L., 906
Bragg equation, 906
Brahe, Tycho, 143
Brake, hydraulic, 338
Braking distance of car, 29, 167
Brayton cycle, 544 pr
Breakdown voltage, 596
Break-even (fusion), 1099
Breaking point, 309
Breeder reactor, 1094
Bremsstrahlung, 1014–15
Brewster, D., 910
Brewster's angle and law, 910
Bridges, 315–18
Brightness, apparent, 1145
British system of units, 8
Broglie, Louis de (see de Broglie)
Bronchoscope, 827
Brown, Robert, 446
Brown dwarfs, 1166
Brownian movement, 446
Brushes, 698, 740
BSCCO, 650
Btu, 486
Bubble chamber, 1120, 1124
Bulk modulus, 310, 312
Buoyancy, 340–43
Buoyant force, 340–43
Burglar alarm, 954

C

Calculator errors, 5
Caloric theory, 486
Calorie (unit), 486–87
  related to the joule, 486
Calorimeter, 489–90
Calorimetry, 489–90
Camera, 848–50
  autofocusing, 419
  gamma, 1104

Camera flash unit, 621
Cancer, 1101, 1104
Candela (unit), 882
Cantilever, 304–5
Capacitance, 614–21
Capacitance bridge, 630 pr
Capacitive reactance, 775
Capacitor discharge, 672
Capacitor microphone, 677
Capacitors, 613–21
  charging of, 789–90
  in circuits, 669–73, 764–65, 774 ff
  as filter, 776
  reactance of, 775
  in series and parallel, 617–20
  uses of, 776
Capacity (see Capacitance)
Capillaries, 352
Capillarity, 351–52
Car, stopping distance, 29, 167
Carbon cycle, 1097, 1112 pr, 1148
Carbon dating, 1061, 1078–79
Carburetor of car, 348
Carnot, S., 520
Carnot cycle, 520–25
Carnot efficiency, 522–23
Carnot engine, 520–25
Carnot's theorem, 522–23
Carrier frequency, 804
Carrier of forces, 1121 ff
Cassegrainian focus, 856
Cathedrals, 319
Cathode, 605
Cathode rays, 605, 699 (see also Electron)
Cathode ray tube (CRT), 605–6, 700, 805
CAT scan, 1105–6
Causality, 146
Cavendish, Henry, 135, 137, 148
CD player, 1022, 1054
Cell, electric, 635
Cellular telephone, 806
Celsius temperature scale, 448–49
Center of buoyancy, 357 pr
Center of gravity (CG), 223
Center of mass (CM), 221–24
  for human body, 221–22
  translational motion and, 225–27, 262–65
Centigrade (see Celsius temperature scale)
Centipoise (unit), 350
Centrifugal (pseudo) force, 115, 291

Centrifugal pump, 353
Centripetal acceleration, 64–65, 114 *ff*, 242
Centripetal force, 114
Cepheid variables, 1170 *pr*
Čerenkov radiation, 886 *pr*
CERN, 1117–18, 1123
CGS system of units, 8
Chadwick, J., 1062
Chain reaction, 1091–95
Chandrasekhar limit, 1149
Change of phase (or state), 473–76, 490–93
Characteristic expansion time, 1158
Characteristic X-rays, 1013
Charge (*see* Electric charge)
Charge density, 582
Charging a battery, 659 *fn*, 668
Charles, Jacques, 455
Charles' law, 455
Charm, 1130–31
Chemical bonds, 1030–37
Chemical reaction, rate of, 471
Chernobyl, 1094
Chip, 1030, 1052–55
Christmas tree lights, 658
Chromatic aberration, 856 *fn*, 859
Circle of confusion, 849
Circle of least confusion, 858
Circuit (*see* Electric circuits)
Circuit breaker, 644
Circular motion, 63–65, 114–22
 nonuniform, 121–22
 uniform, 63–65, 114 *ff*, 371
Circulating pump, 353
Classical physics (*defn*), 1, 916, 978
Clausius, R. J. E., 517, 533
Clausius equation of state, 477
Clausius statement of second law of thermodynamics, 517, 526, 533
Clock paradox (*see* Twin paradox)
Closed system, 493
Cloud chamber, 1120
Coating of lenses, optical, 880–81
Coaxial cable, 632 *pr*, 715, 760–61, 800
Coefficient (*see* name of)
Coefficient of performance, 526
Coherent light, 873–74, 877, 1019
Coherent sources of light, 873–74, 877
Cohesion, 352
Coil (*see* Inductor)
Collector, 1054

Colliding beams, 1118–19
Collimated beam, 1105
Collision, 211–20
 elastic, 214–17
 inelastic, 214, 217–18
Collision frequency, 483 *pr*
Colonoscope, 827
Color:
 of light related to frequency and wavelength, 872, 878
 of quarks, 1132–33
 of sky, 911
 of stars, 950, 1146
Color charge, 1133–34
Color force, 1133–34
Coma, 859
Commutative property, 160
Commutator, 698
Compass, magnetic, 688, 710
Complementarity, principle of, 958–59
Completely inelastic collision, 217–18
Complex atoms, 1010–15
Complex quantities, 980 *fn*, 986 *fn*, 989
Components of vector, 48–52
Composite wave, 401
Compound lenses, 859, 899
Compound microscope, 856–58
Compound nucleus, 1090
Compression (longitudinal wave), 391
Compressive strength, 313
Compressive stress, 311
Compton, A. H., 955–57, 1092
Compton effect, 955–57
Compton shift, 956
Compton wavelength, 956, 975 *pr*
Computer:
 chip, 1030
 digital information and, 749
 disks, 749
 hard drive, 243–44
 keyboards, 616
 monitor, 606
Computerized axial tomography, 1105–6
Computerized tomography, 1105–6
Concave mirror, 816, 820
Concentration gradient, 480
Concrete, prestressed and reinforced, 313
Condensation, 475
Condensed-matter, physics, 1044–50

Conductance, 656 *pr*
Conduction, of heat, 503–5
Conduction band, 1049–50
Conduction current (*defn*), 791
Conductivity:
 electrical, 640, 649
 thermal, 503–5
Conductors:
 electric, 547–48, 562–63, 640
 heat, 504
 quantum theory of, 1048–50
Confinement (fusion), 1098–99
Conical pendulum, 116
Conservation laws:
 of angular momentum, 257–59, 288–89
 of baryon number, 1124–26, 1135, 1162
 of electric charge, 546–47
 of energy, 182–92, 493–95
 of lepton number, 1124–26, 1162
 of linear momentum, 208–11
 of nucleon number, 1073, 1124–26
 in nuclear and particle physics, 1073, 1124–26
 of strangeness, 1129
Conservative force, 177–78, 749
Conserved quantity, 183
Constants, values of, inside front cover
Constant-volume gas thermometer, 449, 460
Constructive interference, 404, 430, 871, 1031
Contact force, 86, 148
Contact lenses, 852
Contact potential, 677 *fn*
Continental drift, 342
Continuity, equation of, 344
Continuous laser, 1022
Continuous spectrum, 902
Control rods, 1093
Convection, 505
Conventional current (*defn*), 637
Conventions, sign (geometric optics), 819, 822
Converging lens, 836 *ff*
Conversion factors, 8, inside front cover
Convex mirror, 816
Conveyor belt, 228
Coordinate systems, 17
Copenhagen interpretation of quantum mechanics, 985

Copier, electrostatic, 555
Core, of reactor, 1093
Coriolis force, 292–94
Cornea, 850
Corrective lenses, 850–52
Correspondence principle, 943, 971, 976 pr, 978
Cosmic microwave background radiation, 1159–61, 1164
Cosmic rays, 1102, 1114
Cosmological principle, 1157 ff
  "perfect," 1159 fn
Cosmology, 1136, 1140–69
Coulomb, Charles, 549–50
Coulomb (unit), 550, 712
  operational definition of, 712
Coulomb barrier, 998, 1098, 1148
Coulomb's law, 549–53, 575–86, 966
  vector form of, 553
Coulomb potential, 1034
Counter emf, 742–43
Counter torque, 743
Covalent bond, 1030–33, 1044
Crane, 304
Creativity in science, 3
Crick, F.H.C., 906
Critical damping, 376
Critical density, of universe, 1165–68
Critical mass, 1092–95
Critical point, 473
Critical temperature, 473
Crossed Polaroids, 908
Cross product, 279–80
Cross section, 1088–90
CRT, 605–6, 700, 805
Crystal lattice, 446–47, 1044
Crystallography, 906
CT scan, 1105–6
Curie, M. and P., 726, 1067
Curie (unit), 1101
Curie temperature, 722
Curie's law, 726
Current (see Electric current)
Current density, 647–49
Current gain, 1055
Current sensitivity, 674
Curvature of field, 859
Curvature of universe (space-time), 149, 1151–55, 1165
Cycle (defn), 363
Cyclic universe, 1168
Cyclotron, 707 pr, 1116–17
Cyclotron frequency, 694, 1116

D

Damped harmonic motion, 374–77
Dark matter, 1166
Dating, radioactive, 1078–79
Daughter nucleus (defn), 1068
Davisson, C. J., 960
dB (unit), 421
DC circuits, 658–77
DC generator, 741
DC motor, 698
de Broglie, Louis, 959–61, 971, 977–78
de Broglie wavelength, 959, 971–72
Debye's law, 514 pr
Decay:
  of elementary particles, 1124–34
  radioactive (see Radioactivity)
Decay constant, 1073–76
Decay series, 1077–78
Deceleration (defn), 24
Deceleration parameter, 1168
Decibel (db), 421
Declination, magnetic, 688
Decommissioning nuclear plant, 1094
Dee, 1116
Defects of the eye, 851
Defibrillator, 632 pr, 651 fn
Degrees of freedom, 500–501
Del, 603 fn
Delayed neutron, 1093
Delta particle, 1128
Demagnetization, 725
Density, 332–33
  and floating, 342
Density of occupied states, 1047
Density of states, 1045–48
Depth of field, 849
Derivatives, 21–22, A-4–A-5
  partial, 399
Derived quantities, 8
Descartes, R., 146
Destructive interference, 404, 430, 871, 1031
Detection of radiation, 1080
Determinism, 147, 984–85
Deuterium, 1086, 1092, 1095–1100
Deuteron, 1086
Dew point, 476
Diamagnetism, 725–26
Diamond, 826–27
Dielectric constant, 621–22
Dielectrics, 621–26
  molecular description of, 624–26

Dielectric strength, 622
Diesel engine, 540 pr
Differential cross section, 1089
Diffraction, 867, 887–911, 1020
  by circular opening, 896–97
  as distinguished from interference, 896
  in double slit experiment, 896
  of electrons, 960
  Fraunhofer, 888 fn
  Fresnel, 888 fn
  of light, 867, 887–911
  as limit to resolution, 896–97
  by single slit, 888–93
  of water waves, 410
  X-ray, 905–6
Diffraction grating, 900–905
  resolving power of, 903–5
Diffraction limit of lens resolution, 896–97
Diffraction patterns, 888 ff
  of circular opening, 896–97
  of single slit, 888–93
  X-ray, 905–6
Diffraction spot or disc, 896
Diffuse reflection, 813
Diffusion, 479–80
  Fick's law of, 480
Diffusion constant, 480
Diffusion equation, 480
Digital information, 749
Digital voltmeter, 675
Dimensional analysis, 12–13
Dimensions, 12–13
Diodes, 1052–54
Diopter, 838
Dip, angle of, 688
Dipole antenna, 793
Dipole bonds, 1036–37
Dipoles and dipole moments:
  of atoms, 1015–18
  electric, 565–67, 578, 601–3
  magnetic, 695–97, 721
Dirac, P. A. M., 1005, 1046 fn
Direct current (dc) (see Electric current)
Discharge tube, 963
"Discovery" in science, 700
Disintegration energy (defn), 1068–69
Disorder and order, 533–34
Dispersion, 402, 824–25
Displacement, 17–18, 46–47
  angular, 246–47

Index **A-37**

Displacement (*continued*)
  vector, 46–47, 53
  in vibrational motion, 363
  of wave, 396
Displacement current, 791
Dissipative force, 189–92
Dissociation energy, 1031
Distance:
  image, 814, 819
  object, 814, 819
Distortion, by lenses, 859
Distributive property, 160
Diverging lens, 836 *ff*
DNA, 573 *pr*, 906, 1036–37
Domains, magnetic, 722
Dome, 319–21
Donor level, 1052
Doorbell, 724
Door opener, automatic, 954
Doping of semiconductors, 1051 *ff*
Doppler, J.C., 432 *fn*
Doppler effect:
  for light, 435, 942–43, 1156–57
  for sound, 432–35
Dose, 1101–3
Dosimetry, 1101–3
Dot product, 159–61
Double-slit experiment (light), 870–73
  for electrons, 978–79
  intensity in pattern, 874–77, 895
Drag force, 122–23
Drift velocity, 647–49, 701
Dry cell, 636
Duality, wave-particle, 958–59, 972
Dulong and Petit value, 501
Dynamic lift, 348
Dynamics, 16, 77 *ff*
  of rotational motion, 247 *ff*
Dynamo (*see* Electric generator)
Dyne (unit), 81
Dynode, 1080

E

$E = mc^2$, 938–42
Ear, response of, 422
Earth, rocks and earliest life, 1079
Earthquake waves, 394, 396, 409
ECG, 606
Eddy currents, 744
Edison, Thomas, 605
Effective cross section, 1089

Effective dose, 1102
Efficiency of heat engine, 519, 522–23
Einstein, A., 148, 169, 446, 916, 922–23, 940–41, 965, 977, 985, 1095, 1134, 1151–55
Einstein cross, 1153
EKG, 606
Elastic collision, 214–17
Elasticity, 309–12
Elastic limit, 309
Elastic moduli, 309–12
  and speed of sound waves, 393
Elastic potential energy, 181, 369
Elastic region, 309
Elastic scattering, 1089
Elastic spring, 181
Electric battery, 635–36, 659, 668
Electric car, 592, 657 *pr*
Electric cell, 635
Electric charge, 545 *ff*
  in atom, 547
  bound, 625
  charge density, 582
  conservation of, 546–47
  of electron, 550, 722–23
  elementary, 550
  free, 625
  induced, 548–49, 625
  motion of, in electric field, 564–65
  motion of, in magnetic field, 692–95
  point (*defn*), 551
  quantization of, 550
  "test," 554
Electric circuits, 636, 658–77, 772–83
  ac, 658 *fn*, 772–83
  containing capacitors, 669–73, 764–67, 774 *ff*
  dc, 658–77
  grounding, 651–52
  household, 644–45
  impedance matching of, 781–82
  integrated, 1054–55
  and Kirchhoff's rules, 664–69
  LC, 764–65
  LR, 762–63
  LRC, 766–67, 776–79
  RC, 669–73
  rectifier, 1053–54
  resonant, 780
  time constants of, 670, 762
Electric current, 634–52 (*see also* Electric circuits)
  ac, 645–46, 772–83

  conduction (*defn*), 791
  conventional, 637
  dc (*defn*), 645
  density, 647–49
  displacement, 791
  hazards of, 651–52
  induced, 735
  leakage, 652
  magnetic force on, 689–92
  microscopic view, 647–49
  and Ohm's law, 638–39
  peak, 645
  produced by magnetic field, 747–49
  produces magnetic field, 689, 719–21
  rms, 646
Electric dipole, 565–67, 578, 601–3
Electric energy, 603–5, 620–21
  stored in capacitor, 620–21
  stored in electric field, 321
Electric field, 545 *ff*, 595–97, 602–3
  calculation of, 558–61, 595, 602–3
  and conductors, 562–63
  in dielectric, 622–25
  of dipole, 565–67, 603
  in EM wave, 792–803
  energy stored in, 621
  motion of charged particle in, 564–65, 694
  produced by changing magnetic field, 747–49
  produced by dipole, 566
  produces magnetic field, 788–91
  relation to electric potential, 595–97, 602–3
Electric field lines, 561–62, 576
Electric flux, 575–78, 789
Electric force, 545 *ff*, 694
  Coulomb's law for, 549–53
Electric generator, 740–42
Electric hazards, 651–52
Electric heater, 643
Electricity, static, 545
Electric motor, 698, 742–43
  counter emf in, 742–43
Electric potential, 591–604 (*see also* Potential difference)
  of dipole, 601–2
  due to any charge distribution, 599
  relation to electric field, 595–97, 602–3
  of single point charge, 597–98
Electric potential energy, 591–94, 603–5

Electric power, 642–44
  in ac circuits, 778
  generation, 741
  in household circuits, 644–45
  and impedance matching, 781–82
  transmission of, 746–47
Electric shocks, 651–52
Electrocardiogram (ECG, EKG), 606
Electrode, 635
Electrolyte, 635
Electromagnet, 723–24
Electromagnetic force, 1121–22, 1126–27, 1134–36
Electromagnetic induction, 734 *ff*
Electromagnetic oscillations, 764–65
Electromagnetic pumping, 703 *pr*
Electromagnetic spectrum, 798–800, 874–75
Electromagnetic (EM) waves, 792–806, 866 (*see also* Light)
Electrometer, 548
Electromotive force (emf), 659–60, 734–35, 739–40, 742–43
  back, 742–43
  counter, 742–43
  of generator, 740–42
  Hall, 701
Electromotive force (emf) *(cont.)*
  induced, 734–35, 739–40
  motional, 739
  series and parallel, 668
  sources of, 659–60, 734–35
Electron:
  as beta particle, 1067, 1070
  charge on, 547, 550, 722–23
  discovery of, 699–700
  in double-slit experiment, 978–79
  as elementary particle, 1126
  free, 548
  mass of, 1063
  in pair production, 957–58
  wavelength of, 959–60
Electron band theory, 1048–50
Electron capture, 1072
Electron cloud, 1004, 1030–31
Electron configuration, 925–26
Electron gun, 606
Electronically steered phase array, 877
Electron microscope, 899, 961, 999
Electron sharing, 1030
Electron spin, 723, 1005–6, 1017–18
Electron spin resonance, 1029 *pr*
Electron volt (unit), 603–5, 1063
Electroscope, 548–49, 635 *fn*

Electrostatic copier, 555
Electrostatic potential energy, 603–5
Electrostatics, 551 *ff*
Electrostatic unit, 550 *fn*
Electroweak force, 148, 545 *fn*
Electroweak theory, 1132–34
Elementary charge, 550
Elementary particle physics, 1114–36
Elements:
  origin of in universe, 1149–50
  periodic table of, 1011–13, 1062 *fn*, inside rear cover
  transmutation of, 1068, 1070, 1085–88
Elevator and counterweight, 92
Ellipse, 143
EMF (*see* Electromotive force)
Emission spectra, 902, 949–51, 963–65, 968–69
Emission tomography, 1006–7
Emissivity, 506
Emitter, 1054
EM waves (*see* Electromagnetic waves)
Endoergic reaction (*defn*), 1087
Endoscope, 827
Endothermic reaction, 1086–87
Energy, 155, 165–69, 176–97, 260–65, 487–88, 493–94
  activation, 471, 1034
  binding, 968, 1031, 1033, 1036, 1064–66
  chemical, 190
  conservation of, 182–92, 493–95, 1086
  degradation of, 535
  disintegration, 1068
  distinguished from heat and temperature, 487
  electric (*see* Electric energy)
  in EM waves, 800–802
  equipartition of, 500–501
  equivalence to mass, 938–42
  and first law of thermodynamics, 493–94
  internal, 189–90, 487–88
  kinetic, 164–69, 182 *ff*, 260–65
  mechanical, 182–89
  molecular rotational and vibrational, 500–501
  nuclear, 990–1016
  potential (*see* Potential energy)
  quantization of, 951, 965–71, 991–92

  reaction, 1086
  relation to work, 164–69, 178–82, 190–91, 195, 260–61
  rest, 939, 1063
  rotational, 260–65
  in simple harmonic motion, 369–70
  thermal, 189–90, 487–88
  threshold, 1088, 1113 *pr*
  total mechanical energy (*defn*), 183
  transformation of, 189–90
  unavailability of, 534–35
  and uncertainty principle, 982–83, 996
  units of, 167
  vibrational, 369–70
  of waves, 395–96
  zero point, 992, 1041
Energy bands, 1048–50
Energy density:
  in electric field, 621
  in magnetic field, 761
Energy gap, 1049–50
Energy levels:
  in atoms, 967
  for fluorescence, 1019
  for lasers, 1019–22
Energy states, in atoms, 967
Energy transfer, heat as, 189–90, 485 *ff*
Engine, heat (*see* Heat engine)
Enriched uranium, 1092
Entropy, 528–37
  in life processes, 534
  statistics and, 535–37
Environmental pollution, 525
Enzyme, 1037
Equally-tempered chromatic scale, 424
Equation of continuity, 344
Equation of motion, oscillations, 364, 375, 377
Equation of state, 454, 457, 477–78
  Clausius, 477
  ideal gas, 457
  van der Waals, 477–78
Equilibrium, 197–98, 300–303, 308
  conditions for, 301–3
  stable, unstable, neutral, 197–98, 308
  thermal, 449–50
Equilibrium position (vibrational motion), 363
Equilibrium state (*defn*), 454
Equipartition of energy, 500–501

Equipotential lines, 600–601
Equipotential surface, 600–601
Equivalence, principle of, 148, 1151–52
Erg (unit), 156
Escape velocity, 192–94
Estimated uncertainty, 4
Estimating, 9–12
Ether, 919–22
Euclidean space, 1153
Evaporation, 474–75, 493
Event horizon, 1155
Everest, Mt., 138
Evolution and entropy, 534
Exact differential, 494 fn
Exchange coupling, 723
Excited state, of atom, 957 ff
Exclusion principle, 1010–12, 1046, 1132–33, 1150
Exoergic reaction (defn), 1086
Exothermic reaction, 1086
Expansion, thermal (see Thermal expansion)
Expansion of universe, 1156–59
Expansions, in waves, 391
Expansions, mathematical, A-1
Exponential curves, 1074–75
Exponential decay, 1074–75
Extragalactic (defn), 1143
Eye:
  accommodation, 851
  defects of, 851
  far and near points of, 851
  normal (defn), 851
  resolution of, 899
  structure and function of, 850–52
Eyeglass lenses, 851–52
Eyepiece, 854

F

Fahrenheit temperature scale, 448–49
Falling bodies, 31–36
Fallout, radioactive, 1095
Farad (unit of capacitance), 614
Faraday, Michael, 147, 554, 622, 734, 736, 787
Faraday's law, 736–38, 747–48, 756, 792
Far point of eye, 851
Farsighted eye, 851–52
Fermat's principle, 835 pr

Fermi, E., 11, 958, 978, 1011 fn, 1046 fn, 1071, 1088, 1090–95, 1128
Fermi-Dirac statistics, 1046–48
Fermi energy, 1046–48
Fermi factor, 1046–47
Fermi gas, 1046
Fermilab, 1114, 1117
Fermi level, 1046–48
Fermions, 1011 fn, 1046, 1132–33
Fermi speed, 1047
Fermi temperature, 1060 pr
Ferris wheel, 124 pr
Ferromagnetism and ferromagnetic materials, 687, 722–23
Feynman, R., 1121
Feynman diagram, 1121–22
Fiber optics, 826–27
Fick's law of diffusion, 480
Field, 147 (see also Electric field, Gravitational field, Magnetic field)
  conservative and nonconservative, 748–49
  in elementary particles, 1121
Film badge, 1103
Filter circuit, 776, 785 pr
Fine structure, 977, 1003, 1006, 1018
Fine structure constant, 1018
First law of motion, 78 ff
First law of thermodynamics, 493–98
Fission, nuclear, 1090–95
Fission bomb, 1095
Fission fragments, 1090–95
Fitzgerald, G. F., 922
Flashlight battery, 636
Flashlight bulb, 639
Flavor (of elementary particles), 1132–33
Floating, 342
Flow of fluids, 343–45
  laminar, 343–45
  streamline, 343–45
  in tubes, 351
  turbulent, 343
Flow rate, 343–45
Fluids, 332–61 (see also Gases, Liquids)
Fluorescence, 1019
Fluorescent lightbulbs, 1019
Flux:
  electric, 575–78, 789
  magnetic, 736, 791
Flying buttress, 319
FM radio, 804

Focal length:
  of lens, 838, 840–41
  of spherical mirror, 817, 821
Focal plane, 838
Focal point, 817, 821, 837–38
Focus, 817
Focusing, of camera, 849
Foot-candle (defn), 882 fn
Foot-pound (unit), 156, 167
Forbidden transition, 1007, 1020 fn, 1041 fn
Force, in general, 77–85, 147–48, 177–78 (see also Electric force, Gravitational force, Magnetic force)
  buoyant, 340–43
  centrifugal, 115
  centripetal, 114
  color, 1133–34
  conservative and nonconservative, 177–78
  contact, 86, 148
  Coriolis, 292–94
  damping, 374–77 (see also Drag force)
  dissipative, 189–92
  drag, 122–23
  elastic, 162
  electromagnetic, 1121–22, 1126–27, 1134–36
  electroweak, 148
  in equilibrium, 300–303
  exerted by inanimate object, 83–84
  fictitious, 291
  of friction, 78, 106–14, 118–20, 178, 190–92
  of gravity, 85–86, 133 ff, 1123, 1150–55
  impulsive, 212
  inertial, 291–92
  long and short range, 1066
  measurement of, 78
  in muscles and joints, 306
  net, 80, 88
  in Newton's laws, 78–85, 207–9
  nonconservative, 178
  normal, 85–87, 107
  nuclear, 1064–66, 1072, 1121–36
  pseudoforce, 291
  relation of momentum to, 206–8
  resistive, 122–23
  restoring, 162, 363
  strong, 1064–66, 1121–36

types of, in nature, 147–48, 545 *fn*, 1123, 1136
units of, 81
van der Waals, 1036–37
velocity-dependent, 122–23
viscous, 344, 350–51
weak, 1066, 1072, 1121–36
work done by, 156 *ff*
Forced convection, 505
Force diagram (*see* Free-body diagram)
Forced vibrations, 378–79
Force pump, 353
Fossil-fuel power plants, 741
Foucault, J., 868
Four-dimensional space-time, 932
Fourier integral, 401
Fourier's theorem, 401
Fovea, 850, 899
Fractal behavior, 446
Fracture, 312–15
Frames of reference, 17, 79, 291
Franklin, Benjamin, 546, 586
Fraunhofer diffraction, 888 *fn*
Free-body diagram, 88–96, 303
Free charge, 625
Free electrons, 548
Free-electron theory of metals, 1045–48
Free expansion, 498, 531, 537
Free fall, 31–33
Free particle, and Schrödinger equation, 989–94
Freezing (*see* Phase, changes of)
Frequency, 65, 242
  angular, 366
  of audible sound, 422
  beat, 429–32
  carrier, 804
  of circular motion, 65
  cyclotron, 694
  fundamental, 406, 424–26
  infrasonic, 419
  of light, 799, 869
  natural, 366, 378, 405
  resonant, 378–79, 405, 780
  of rotation, 242–43
  ultrasonic, 418, 437
  of vibration, 363, 366
  of wave, 390
Frequency modulation (FM), 804
Fresnel, A., 887
Fresnel diffraction, 888 *fn*
Friction, 78, 106–14, 118–20, 178, 184
  coefficients of, 107, 112
  kinetic, 106 *ff*
  in rolling, 106, 263–64, 267
  static, 107 *ff*
Fringes, interference, 871
Frisch, O., 1090
f-stop, 848
Full-wave rectifier, 1053
Fundamental frequency, 406, 424–26
Fuse, 644–45
Fusion, heat of, 490–93
Fusion, nuclear, 1095–1100
  in stars, 1148–49, 1163
Fusion reactor, 1097

G

g-factor, 1017
Galaxies, 1141–45, 1155 *ff*
Galaxy clusters, 1143
Galilean-Newton relativity, 917–19, 932–36
Galilean telescope, 856
Galilean transformation, 932–36
Galileo, 2–3, 16, 31–32, 56–57, 78–79, 146, 339, 448, 811, 854–56, 854 *fn*, 917–18, 932, 1141, 1150
Galvani, Luigi, 635
Galvanometer, 674, 697, 755 *pr*
Gamma camera, 1104
Gamma decay, 1067, 1072–73
Gamma rays, 957, 1067, 1072–73
Gamow, G., 1160
Gas constant, 457
Gases, 332, 445–46, 454–81
  adiabatic expansion of, 502–3
  change of phase, 473–76, 490–93
  heat capacities of, 498–99
  ideal, 456–60, 466–72
  real, 473–74
  work done by, 496–98
Gas laws, 454–59
Gauge bosons, 1026–27, 1132, 1134
Gauge pressure, 337
Gauges, pressure, 338–40
Gauge theory, 1134
Gauss, C. F., 1154
Gauss (unit), 691
Gauss's law, 575–86, 713
  for magnetism, 791
Gay-Lussac, Joseph, 456
Gay-Lussac's law, 456
Geiger counter, 612 *pr*, 1080
Gell-Mann, M., 1130
General motion, 262–68, 283–84
General theory of relativity, 148, 149, 169, 922, 1140–41, 1151–55
Generator, electric, 740–42
Genetic damage, 1101
Geodesic, 1153–54
Geological dating, 1079
Geometric optics, 811 *ff*
Geophone, 759–60
Geosynchronous satellite, 140, 145
Germer, L. H., 960
Glaser, D., 1120
Glashow, S., 1134
Glasses, eye, 851–52
Glueballs, 1133 *fn*
Gluino, 1136
Gluons, 1123, 1133
Gradient:
  concentration, 480
  electric potential, 602
  pressure, 351
  velocity, 350
Gradient operator, 603 *fn*
Gram (unit), 7
Grand unified era, 1162
Grand unified theories, 148, 1134–36
Grating, diffraction (*see* Diffraction grating)
Gravitation, universal law of, 133 *ff*
Gravitational collapse, 1155
Gravitational constant (*G*), 135
Gravitational field, 146–47, 562, 1151–55
Gravitational force, 85–86, 133 *ff*
  due to spherical mass distribution, A-6–A-8
Gravitational mass, 148, 1152
Gravitational potential energy, 178–80, 192–94
Gravitino, 1164 *fn*
Graviton, 1122–23
Gravity, 32–33, 85, 133 *ff*, 1123, 1150–55
  acceleration of, 32–33, 85, 137–39
  center of, 223
  free fall under, 31–36
  specific, 333
Gravity anomalies, 138
Gray (unit), 1102
Greek alphabet, inside front cover
Grimaldi, F., 867, 872
Grounding, electrical, 651–52

Ground state, of atom, 967 ff
GUT, 148, 1134–36
Gyromagnetic ratio, 1017
Gyroscope, 289

H

h-bar (ℏ), 1013
Hadron era, 1162
Hadrons, 1126–36, 1162
Hahn, O., 1090
Hair dryer, 646
Halflife, 1073–76
Half-wave rectification, 1053
Hall, E. A., 701
Hall effect, Hall emf, Hall field, Hall probe, 701
Halogen, 1013
Hard drive, 243–44
Harmonic motion:
  damped, 374–77
  forced, 378–79
  simple, 364–71, 390
Harmonics, 406, 424, 426–27
Harmonic wave, 398
Hazards of electricity, 651–52
Headlights, 643
Hearing, threshold of, 422
Heart, 353
  defibrillator, 632 pr, 651 fn
  pacemaker, 673, 758
Heat, 485–544
  compared to work, 493
  conduction, convection, radiation, 503–8
  distinguished from internal energy and temperature, 487
  as energy transfer, 189–90, 485 ff
  in first law of thermodynamics, 493–98
  as flow of energy, 485–86
  of fusion, 490–93
  latent, 490–93
  mechanical equivalent of, 486
  specific, 488 ff, 498–99
  of vaporization, 491
Heat capacity, 515 pr (see also Specific heat)
Heat conduction, 503–5
Heat death, 534–35
Heat engine, 517–25, 1093
  Carnot, 520–25
  efficiency of, 519, 522–23
  internal combustion, 518–19
  steam, 518–19
Heating element, 642
Heat of fusion, 490–93
Heat pump, 527–28
Heat reservoir, 495
Heat transfer, 503–8
Heat of vaporization, 491
Heavy water, 1092
Heisenberg, W., 949, 978, 981–84
Heisenberg uncertainty principle, 981–84, 996, 1031
Helium:
  liquid, 474
  primordial production, 1164
Helium-neon laser, 1021
Helmholtz coils, 733 pr
Henry, Joseph, 734
Henry (unit), 757
Hertz, Heinrich, 798
Hertz (unit of frequency), 243, 363, 418
Hertzsprung-Russell diagram, 1146–49
Higgs boson, 1134
Higgs field, 1134
High energy accelerators, 1115–20
High energy physics, 1114–36
Highway curves, banked and unbanked, 118–20
Holes (in semiconductors), 1050–52
Holograms, 1023–24
Hooke, Robert, 309, 878 fn
Hooke's law, 162, 181, 309
Horsepower (unit), 195
H-R diagram, 1146–49
Hubble, E., 943, 1143, 1156–57
Hubble age, 1158
Hubble's law, 1156–58
Hubble parameter, 1156–58
Hubble Space Telescope, 897, 1140
Humidity, 476
Huygens, C., 867
Huygens' principle, 867–69
Hydraulic brakes, 338
Hydraulic lift, 338
Hydrodynamics, 343 ff
Hydroelectric power, 741
Hydrogen atoms, 604–5
  Bohr theory of, 965–71
  magnetic moment of, 697
  quantum mechanics of, 1004–10
  spectrum of, 968–70
Hydrogen bomb, 1098

Hydrogen bond, 1036
Hydrogen-like atoms, 966 fn
Hydrogen molecule, 1030–31, 1033, 1040–42
Hydrometer, 342
Hyperopia, 851
Hysteresis, 724–25
Hysteresis loop, 725

I

Ideal gas, 456–60, 466 ff
  internal energy of, 487–88
Ideal gas law, 456–60, 466 ff
Ideal gas temperature scale, 460
Ignition (fusion), 1099–1100
Illuminance, 882
Image:
  as tiny diffraction pattern, 896–97
  CAT scan, 1105–6
  fiber optic, 826–27
  formed by lens, 836 ff
  formed by plane mirror, 812–15
  formed by spherical mirror, 816–22
  NMR, 1064, 1107–8
  PET and SPET, 1106–7
  real, 814, 818, 839
  ultrasound, 437
  virtual, 814, 821, 840
Image distance, 814, 819
Imaging, medical, 437, 1105–9
Imbalance, rotational, 287–88
Impedance, 759, 774–75, 781–82
Impedance matching, 781–82
Impulse, 211–13
Impulsive forces, 212
Incidence, angle of, 403, 408, 813, 822
Inclines, motion on, 94–95, 262–64
Incoherent source of light, 873–74, 877
Indeterminacy principle (see Heisenberg uncertainty principle)
Index of refraction, 811–12
  in Snell's law, 822–24
Induced electric charge, 548–49, 625
Induced emf, 734–35, 739–40
  counter, 742–43
  in electric generator, 740–42
  in transformer, 744–47
Inductance, 756–60
  in ac circuits, 758–63, 772–83
  mutual, 756–58
  self-, 758–60

Induction, electromagnetic, 734 *ff*
Induction, Faraday's law, 736–38, 792
Inductive reactance, 774
Inductor, 758–59
  in circuits, 758–63, 772–83
  energy stored in, 760–61
  reactance of, 774
Inelastic collision, 214, 217–18
Inelastic scattering, 1089
Inertia, 79
  law of, 78–79
  moment of, 249–56, 374
  rotational (*defn*), 249–50
Inertial confinement, 1098–99
Inertial forces, 291–92
Inertial mass, 148, 1152
Inertial reference frame, 79, 291, 917 *ff*
  equivalence of all, 918–19, 922
  transformations, 932–36
Inflationary universe, 1164
Infrared radiation (IR), 506, 799, 825
Infrasonic waves, 419
Initial conditions, 365
Instantaneous axis, 246
Instruments, wind and stringed, 424–29
Insulators:
  electrical, 547–48, 640, 1049–50
  thermal, 504, 1049–50
Integrals and integration, 36–37, 161 *ff*, A-4–A-5, B-5
Integrated circuits, 1054–55
Intensity:
  of coherent and incoherent sources, 877
  in interference and diffraction patterns, 874–77, 890–93
  of light, 877, 882
  of sound, 420–23
  of waves, 395–96, 420–23
Intensity level (*see* Sound level)
Interference, 404–5
  constructive, 404, 430, 871, 1031
  destructive, 404, 430, 871, 1031
  as distinguished from diffraction, 895
  of light waves, 870–81, 894–96, 900
  of sound waves, 429–32
  by thin films, 877–81
  of water waves, 404–5
  of waves on a string, 404
Interference fringes, 871

Interference pattern:
  double slit, 870–73, 978–79
  double slit, including diffraction, 893–95
  multiple slit, 900–905
Interferometers, 881
Internal combustion engine, 518–19
Internal conversion, 1073
Internal energy, 189–90, 487–88
Internal reflection, total, 415 *pr*, 826–27
Internal resistance, 659
Intrinsic semiconductor (*defn*), 1049
Inverted population, 1020
Ion, 547
Ionic bonds, 1031–33, 1044
Ionic cohesive energy, 1044
Ionization energy, 968, 970
Ionizing radiation (*defn*), 1100
IR radiation, 506, 799, 825
Irreversible process, 520–25
Isobaric process (*defn*), 496–97
Isochoric process (*defn*), 496–97
Isolated system, 209, 493
Isomer, 1073
Isotherm, 495
Isothermal process (*defn*), 495
Isotopes, 702, 1062
  table of, A-9–A-13
Isotropic material, 867

J

J (total angular momentum), 1017–18
J/ψ particle, 1131
Jeweler's loupe, 854
Joint, 315
Joule, James Prescott, 486
Joule (unit), 156, 167, 248 *fn*
  relation to calorie, 486
Jump start, 668–69
Junction diode, 1054–55
Junction transistor, 1054–55
Jupiter, moons of, 152, 854

K

Kant, Immanuel, 1143
Kaon, 1127
K-capture, 1072
K lines, 1013

Kelvin (unit), 455
Kelvin-Planck statement of second law of thermodynamics, 520, 523, 533
Kelvin temperature scale, 455, 460, 538
Kepler, Johannes, 143, 854 *fn*, 1150
Keplerian telescope, 854
Kepler's laws, 143–46, 288–89
Keyboard, computer, 616
Kilocalorie (unit), 486–87
Kilogram (unit), 7, 79
Kilowatt-hour (unit), 643
Kinematics, 16–67, 240–44 (*see also* Motion)
  for rotational motion, 240–44
  for translational motion, 16–67
  for uniform circular motion, 63–65
  vector kinematics, 53–55
Kinetic energy, 164–69, 182 *ff*, 260–65
  in collisions, 214–17, 219
  molecular, relation to temperature, 469
  relativistic, 938–42
  rotational, 260–65
  total, 262–65
  translational, 165–69
Kinetic friction, 106 *ff*
Kinetic theory, 445, 466 *ff*
Kirchhoff, G. R., 665
Kirchhoff's rules, 664–69

L

Ladder, 306–7
Laminar flow, 343–45
Land, Edwin, 907
Large Magellanic Cloud, 1150
Laser light, 873
Lasers, 873, 887, 1019–22, 1100
Latent heats, 490–93
Lateral magnification, 819, 841
Lattice structure, 446–47, 1044, 1051–52
Laue, Max von, 905
Lawrence, E. O., 1116
Laws, 3–4 (*see also* specific name of law)
Lawson, J. D., 1099
Lawson criterion, 1099–1100
LC circuit, 764–65
LCD, 909

LC oscillation, 766–67
Leakage current, 652
LED, 1054
Length, standard of, 6, 881
Length contraction, 930–31
Lens, 836–60
 achromatic, 859
 camera, 848–50
 coating of, 80–81
 color corrected, 859
 compound, 859, 899
 contact, 852
 converging, 836 *ff*
 corrective, 851–52
 cylindrical, 851
 diverging, 836 *ff*
 of eye, 850–51
 eyeglass, 851–52
 eyepiece, 854
 focal length of, 838, 840–41
 magnetic, 961
 magnification of, 841–43
 normal, 850
 objective, 854–55, 898–99
 ocular, 856
 positive and negative, 841
 power of (diopters), 838
 resolution of, 896–99
 telephoto, 850
 thin (*defn*), 837
 wide angle, 850
 zoom, 850
Lens aberrations, 858–59
Lens elements, 859
Lens equation, 840–43, 862 *pr*
Lensmaker's equation, 845–47
Lenz's law, 736–38, 758
LEP collider, 1118–19
Lepton era, 1162–63
Leptons and Lepton number, 1121, 1124–27, 1130–32, 1134, 1162
Level range formula, 60
Lever, 303
Lever arm, 247–48
LHC collider, 1118
Lifetime, 1075, 1127 (*see also* Halflife)
Lift, dynamic, 348
Light, 810–912 (*see also* Diffraction, Intensity, Interference, Interference pattern, Reflection, Refraction)
 coherent sources of, 873–74, 877
 color of, and wavelength, 872
 dispersion of, 824–25
 as electromagnetic wave, 798–800, 810
 frequencies of, 798, 869
 gravitational deflection of, 1152
 incoherent sources of, 873–74, 877
 infrared (IR), 506, 799, 825
 intensity of, 877, 882
 monochromatic (*defn*), 870
 photon (particle) theory of, 952–55
 polarized, 907–10
 ray model of, 811 *ff*
 reflection of, 813–22
 refraction of, 822–24, 867–69
 scattering, 911
 from sky, 911
 speed of, 6, 797, 811–12, 867–68, 919–23, 938
 ultraviolet (UV), 799, 825
 unpolarized (*defn*), 907
 velocity of, 6, 797, 811–12, 867–68, 919–23, 938
 visible, 811, 867 *ff*
 wavelengths of, 799, 868–69, 871–73
 wave-particle duality of, 958–59
 wave theory of, 866–911
 white, 825
Light bulb, 634, 642, 664, 882, 954, 1019
Light-emitting diode, 1054
Light-gathering power, 855
Light meter, 954
Lightning, 591, 643
Light pipe, 827
Light-year (unit), 14 *pr*, 1141
Linac, 1118
Linear accelerator, 1118
Linear expansion, coefficient of, 450
Linear momentum, 206 *ff*
Lines of force, 562
Line spectrum, 902, 963 *ff*, 977
Line voltage, 646, 646 *fn*
Liquefaction, 473
Liquid crystal display (LCD), 909
Liquid drop model, 1090
Liquids, 332 *ff*, 446–47 (*see also* Phase, changes of)
Liquid scintillation, 1080
Lloyd's mirror, 886 *pr*
Logarithms, A-3
Longitudinal magnification, 833 *pr*
Longitudinal wave, 391 *ff*, 419–20 (*see also* Sound waves)

Long-range force, 1066
Lorentz, H. A., 922
Lorentz equation, 694
Lorentz transformation, 932–36
Los Alamos laboratory, 1095
Loudness, 418, 420–23 (*see also* Intensity)
Loudness level, 422
Loudspeakers, 367, 421, 698, 776, 954
LRC circuit, 766–67, 776–79
LR circuit, 762–63
Lumen (unit), 882
Luminosity (stars and galaxies), 1145–46
Luminous efficiency, 885 *pr*
Luminous flux, 882
Luminous intensity, 882
Lyman series, 964, 968–69

M

Mach, E., 435 *fn*
Mach number, 435
MACHOS, 1166
Macroscopic description of system, 445
Macrostate, 535–36
Madelung constant, 1044
Magnet, 686–88
 domains of, 722
 electro-, 723–24
 permanent, 722
 superconducting, 650, 723
Magnetic confinement, 1098–99
Magnetic damping, 751 *pr*
Magnetic declination, 688
Magnetic dipoles and dipole moments, 695–97, 721
Magnetic domains, 722
Magnetic field, 686–702, 709–26
 of circular loop, 720–21
 definition of, 687–88
 determination of, 691
 direction of, 689–92
 of Earth, 688
 energy stored in, 760–61
 induces emf, 736–38
 motion charged particle in, 692–95
 produced by changing electric field, 788–91
 produced by electric current, 689, 719–21 (*see also* Ampère's law)

produces electric field and current, 747–49
of solenoid, 716–18
sources of, 709–26
of straight wire, 710, 720
of toroid, 708
Magnetic field lines, 687
Magnetic flux, 736, 791
Magnetic flux density (*see* Magnetic field)
Magnetic force, 686, 689–95
on electric current, 689–92
on moving electric charges, 692–95
Magnetic induction (*see* Magnetic field)
Magnetic lens, 961
Magnetic moment, 695–97, 1015–18, 1064
Magnetic permeability, 710, 724–25
Magnetic poles, 686–87
of Earth, 688
Magnetic quantum number, 1005–6, 1016
Magnetic resonance imaging, 1064, 1108–9
Magnetic tape and discs, 749
Magnetism, 686–726
Magnetization (vector), 726
Magnification:
angular, 853
of electron microscope, 961
lateral, 819, 841
of lens, 841–43
of lens combination, 843–45
longitudinal, 833 *pr*
of magnifying glass, 853–54
of microscope, 856–58, 898–99
of mirror, 819
sign conventions for, 819, 841
of telescope, 855, 898–99
useful, 899, 961
Magnifier, simple, 853–54
Magnifying glass, 836, 853–54
Magnifying power, 853, 855 (*see also* Magnification)
Magnitude (stars and galaxies), 1172 *pr*
Main-sequence, stars, 1146–49
Manhattan Project, 1095
Marconi, Guglielmo, 803
Manometer, 338
Mass, 7, 79, 82
atomic, 446
center of, 221–24

critical, 1092–95
gravitational vs. inertial, 148, 1152
inertial, 148, 1152
molecular, 446
in relativity theory, 936–38
rest, 938–40, 1063
standard of, 7
units of, 7, 79
variable, 227–29
Mass-energy transformation, 938–42
Mass excess, 1084 *pr*
Mass increase in relativity, 936–38
Mass spectrometer (spectrograph), 702
Matter, states of, 332, 446–47
Matter-dominated universe, 1161, 1164
Maxwell, James Clerk, 470, 787, 792, 797–98, 918
Maxwell distribution of molecular speeds, 470–72
Maxwell's equations, 787, 792–806, 918–19, 934
Mean free path, 478–79
Mean life, 1075, 1084 *pr*
Mean speed, 469
Measurement, 4–5, 981
Mechanical advantage, 93, 303
Mechanical energy, 182–89
Mechanical equivalent of heat, 486
Mechanical waves, 388–411
Mechanics (*defn*), 16
Medical imaging, 437, 1105–9
Meitner, L., 978, 1090
Melting point, 491 (*see also* Phase, changes of)
Mendeleev, D., 1012
Meson exchange, 1122
Mesons, 1122–36
Metallic bond, 1044
Metals, free electron theory of, 1045–48
Metastable state, 1019–20, 1073
Meter (unit), 6, 812 *fn*, 881
Meters, 674–76, 697
correction for resistance of, 676
Metric system, 6–8
MeV (million electron volts) (*see* Electron volt)
Mho (unit), 656 *pr*
Michelson, A., 811–12, 881, 919–22
Michelson interferometer, 881, 919–22

Michelson-Morley experiment, 919–23, 923 *fn*
Micrometer, 10
Microphones:
capacitor, 677
magnetic, 749
ribbon, 749
Microscope:
compound, 856–58
electron, 899, 961, 999
magnification of, 856–58, 898–99
resolving power of, 898–99
useful magnification, 899
Microscopic description of a system, 445
Microstate, 535–36
Microwaves, 798
Milky Way, 854, 1141
Millikan, R. A., 700, 953
Millikan oil-drop experiment, 700
Mirage, 869
Mirror equation, 819–22
Mirrors:
concave and convex, 819 *ff*
focal length of, 817, 821
plane, 812–15
spherical, 816–22
used in telescope, 856
Missing mass, 1166–68
MKS units (*see* SI units)
mm-Hg (unit), 339
Models, 3
Moderator, 1092–94
Modern physics (*defn*), 1, 916
Modulation, 804
Moduli of elasticity, 309–12
Molar specific heat, 498–99
Mole (*defn*), 456
volume of, for ideal gas, 457
Molecular mass, 446
Molecular rotation, 1038–40
Molecular spectra, 1037–43
Molecular speeds, 466–72
Molecular vibrations, 501, 1040–42
Molecules, 1030–43
bonding in, 1030–33
polar, 1032–33
P E diagrams for, 1033–36
spectra of, 1037–43
Moment arm, 247
Moment of a force, 247
Moment of inertia, 249–56, 374
Momentum, 206 *ff*, 936–38, 941
angular, 256–60, 281–89, 965

Momentum (*continued*)
 in collisions, 208–20
  conservation of angular, 257–59, 288–89
  conservation of linear, 208–11
  linear, 206 *ff*
  relation of force to, 206–8
  relativistic, 936–38, 941
  total, of systems of particles, 222–23
Monochromatic aberration, 859
Monochromatic light (*defn*), 870
Moon, 134, 1141
Morley, E. W., 919–22
Moseley, H. G. J., 1014
Moseley plot, 1014
Motion, 16–299, 916–44
 of charged particle in electric field, 564–65
 at constant acceleration, 26 *ff*
 damped harmonic, 374–77
 description of (kinematics), 16–67
 dynamics of, 77 *ff*
 general, 262–68, 283–84
 under gravity (free fall), 31–36
 harmonic, 364 *ff*
 Kepler's laws of planetary, 143–46, 288–89
 linear, 16–37
 Newton's laws of, 78–85 and *ff*, 207, 209, 225, 226, 250, 256, 282, 283–85, 916, 938, 978, 985–86
 nonuniform circular, 121–22
 oscillatory, 362–80
 periodic (*defn*), 362
 projectile, 55–63
 rectilinear, 16–37
Motion (*cont.*)
 relative, 66–68, 916–44
 rolling, 244–45, 262–68
 rotational, 239–94
 simple harmonic, 364 *ff*
 translational, 16–238, 262–65
 uniform circular, 63–65, 114 *ff*, 371
 uniformly accelerated, 26 *ff*, 54–55
 uniform rotational, 243–44
 vibrational, 362–80
 of waves, 388 *ff*
Motional emf, 739
Motor, electric, 698, 742–43
 counter emf in, 742–43
Movie projector, 954
MRI, 1064, 1108–9
Multimeter, 675
Multiplication factor, 1093

Muon, 928, 1114, 1125–27
Muscles and joints, forces in, 306
Musical instruments, 424–29
Mutation, 1101
Mutual inductance, 756–58
Myopic eye, 851

N

n-type semiconductor, 1051 *ff*
Natural abundance, 1062
Natural convection, 505
Natural frequency, 366, 378, 405 (*see also* Resonant frequency)
Natural radioactive background, 1102
Near field, 793
Near point, of eye, 851
Nearsighted eye, 851–52
Nebulae, 1143
Negative, 848 *fn*
Negative electric charge, 546, 593
Negative lens, 841
Neon tubes, 1003
Neptune, 146
Neptunium, 1088
Net force, 80, 88
Neutral equilibrium, 198, 308
Neutralino, 1166
Neutrino, 1071, 1126–28, 1121, 1163, 1166–68
 mass of, 1166–67
Neutron, 547, 1062–63
 delayed, 1093
 in fission, 1090–95
 thermal, 1089
Neutron cross section, 1089
Neutron number, 1062
Neutron star, 1144, 1150
Newton, Isaac, 16, 79, 133–34, 146–48, 207, 856 *fn*, 917, 1151, 1154 *fn*
Newton (unit), 81
Newtonian focus, 862
Newtonian mechanics, 79–146
Newton's law of cooling, 515 *pr*
Newton's law of universal gravitation, 133–37, 1151–52
Newton's laws of motion, 78–85 and *ff*, 207, 209, 225, 226, 916, 938, 978, 985–86
 for rotational motion, 250, 256, 282, 283–85
Newton's rings, 878

Newton's synthesis, 143–46
NMR, 1064, 1107–8
Noble gases, 1012–13, 1045
Nodes, 405, 427
Noise, 429
Nonconductors, 547–48, 640
Nonconservative field, 748–49
Nonconservative force, 178
Non-Euclidean space, 1153
Noninductive winding, 758
Noninertial reference frame, 79, 291
Nonlinear device, 1054
Nonohmic device, 638
Nonreflecting glass, 880
Nonuniform circular motion, 121–22
Normal eye, 851
Normal force, 85–87, 107
Normalization, 987, 990 *fn*, 993
Normal lens, 850
North pole, 687
Novae, 1144
NOVA laser, 1100
Nuclear angular momentum, 1064
Nuclear binding energy, 1064–66
Nuclear collision, 217, 220
Nuclear fission, 1090–95
Nuclear forces, 1064–66, 1072, 1121–36
Nuclear fusion, 1095–1100, 1148–51, 1163
Nuclear magnetic resonance, 1064, 1107–8
Nuclear magneton, 1064
Nuclear medicine, 1104
Nuclear physics, 1061–1113
Nuclear power, 741, 1093–94
Nuclear reactions, 1085–88
Nuclear reactors, 1090–95
Nuclear spin, 1064
Nuclear weapons testing, 1095
Nucleon (*defn*), 1062
Nucleon number, 1073
Nucleosynthesis, 1140, 1149–51, 1163
Nucleus, 1061 *ff*
 compound, 1090
 daughter and parent (*defn*), 1068
Nuclide (*defn*), 1062
Null result, 922
Numerical integration, 36–37

O

Object distance, 814, 819, 828–29
Objective lens, 854

Occhialini, G., 1122
Ocular lens (*see* Eyepiece)
Oersted, H. C., 688–89, 709, 787–88
Off-axis astigmatism, 859
Ohm, G. S., 638
Ohm (unit), 638
Ohmmeter, 675, 697
Ohm's law, 638–39
Oil-drop experiment, 700
Onnes, H. K., 650
Open system, 493
Operating temperatures, 517
Operational definitions, 8, 712
Oppenheimer, J. R., 1095
Optical coating, 880–81
Optical illusion, 869
Optical instruments, 827, 836–60
Optical pumping, 1021
Optical tweezers, 803
Optics, geometric, 811 *ff*
Orbital angular momentum, in atoms, 1005–7, 1015–18
Orbital quantum number, 1005–6
Order and disorder, 533–34
Order of interference or diffraction pattern, 871
Order of magnitude and rapid estimating, 9, 96
Organ pipe, 427–28
Orion, 1143
Oscillations (*see* Vibrations)
Oscillator, sawtooth, 685 *pr*
Oscilloscope, 605–6, 782
Osteoporosis, 957
Otto cycle, 524
Overdamping, 376
Overexposure, 848
Overtone, 406, 426–27

P

P waves, 394
p-type semiconductor, 1051 *ff*
p-n junction, 1052–54, 1080
Pacemaker, 673, 758
Pair production, 957–58
Parabola (projectile), 63
Parabolic reflector, 817
Parallax, 1144–45
Parallel axis theorem, 255–56
Parallelogram method of adding vectors, 47
Paramagnetism, 725–26

Paraxial rays (*defn*), 817
Parent nucleus, 1068
Parsec (unit), 1144
Partial ionic character, 1032–33
Partial pressure, 476
Particle (*defn*), 16
Particle acclerators, 1115–20
Particle exchange, 1121–23
Particle interactions, 1124–26
Particle physics, 1114–36
Particle resonance, 1127–28
Pascal, Blaise, 333, 337–38
Pascal (unit of pressure), 333
Pascal's principle, 337–38
Paschen series, 964, 968–69
Pauli, W., 978, 1011, 1071
Pauli exclusion principle, 1010–12, 1046, 1132–33, 1150
Peak current, 645
Peak voltage, 645
Pendulum:
  ballistic, 218
  conical, 116
  physical, 373–74
  simple, 13, 188, 371–73
  torsion, 374
Pentium chip, 1030
Penzias, A., 1160
Percent uncertainty, 4
Perfect cosmological principle, 1159 *fn*
Performance, coefficient of, 526
Perfume atomizer, 348
Period, 65, 243
  of circular motion, 65
  of pendulums, 13, 372, 374
  of planets, 143
  of rotation, 243
  of vibration, 363, 366
Periodic motion, 362 *ff*
Periodic table, 1011–13, 1062 *fn*, inside rear cover
Periodic wave, 389–90
Permeability, magnetic, 710, 724–25
Permittivity, 551, 622
Perpendicular-axis theorem, 255–56
Perturbation, 145
PET, 1106–7
Phase:
  in ac circuit, 772–80
  changes of, 473–76, 490–93
  of matter, 332, 446–47
  of waves, 397
Phase angle, 366, 398, 778

Phase diagram, 474
Phase-echo technique, 437
Phase transition, 473–76, 490–93
Phase velocity, 397
Phasor diagram:
  ac circuits, 777–78
  interference and diffraction of light, 874–75, 903–4
Phons (unit), 422
Phosphor, 1080
Phosphorescence, 1019
Photino, 1136, 1164 *fn*
Photocathode, 1080
Photocell, 952
Photodiode, 954, 1054
Photoelectric effect, 952–55, 957
Photographic emulsion, 1120
Photographic film, 848
Photomultiplier tube, 1080
Photon, 952–57, 978–79, 1121, 1126–28, 1162–64 (*see also* Gamma rays, X-rays)
Photon exchange, 1121–23
Photon interactions, 957–58
Photon theory of light, 952–55
Photosynthesis, 955
Physical pendulum, 373–74
Piano tuner example, 12
"Pick-up" nuclear reaction, 1111 *pr*
Piezoelectric effect, 677
Pin, structural, 305, 315
Pion, 940, 1122–27
Pipe, vibrating air columns in, 427–28
Pitch of a sound, 418
Pixel, 1106
Planck, Max, 949, 951, 965
Planck's constant, 951
Planck's quantum hypothesis, 949–51
Plane geometry, A-2
Plane-polarized light, 907–10
Planetary motion, 143–46
Plane waves, 403, 793–94, 989–90
Plasma, 1085, 1098–99
Plastic region, 309
Plate tectonics, 342
Plum-pudding model of atom, 962
Pluto, 146, 1141
Plutonium, 1088, 1094
Point charge (*defn*), 551
  potential, 597–98
Poise (unit), 350–51
Poiseuille, J. L., 351
Poiseuille's equation, 351
Poisson, S., 887

Polarization of light, 907–10
  plane, 907–10
  by reflection, 910, 911
  of skylight, 911
Polarized light (see Polarization of light)
Polarizer, 908
Polarizing angle, 910
Polar molecules, 547, 565, 624–25, 1032–33
Polaroid, 907–9
Poles, magnetic, 686–87
  of Earth, 688
Pole vault, 176, 185–86
Pollution, 525
  thermal, 525
Position vector, 53, 55, 240, 243–44
Positive electric charge, 546
Positive lens, 841
Positron, 1072, 1106, 1121
Positron emission tomography, 1106–7
Post-and-beam construction, 319
Potential difference, electric, 591–94
  (see also Electric potential)
Potential energy, 178–82, 192–93, 197–98
  diagrams, 197–98
  elastic, 181, 369
  electric, 591–94, 603–5
  gravitational, 178–80, 192–94
  for molecular bonds, 1033–36, 1041
  for square well and barriers, 994–99
Potential well, 994–96, 1009 fn
Potentiometer, 685 pr
Pound (unit), 81
Powell, C.F., 1122
Power, 195–97, 642–44, 778 (see also Electric power)
Power factor (ac circuit), 778
Power generation, 741
Power of a lens, 838
Power plants, 741, 1094
Power reactor, 1093
Powers of ten, 5
Power transmission, 746–47, 782
Poynting, J. H., 801 fn
Poynting vector, 800–802
Precession, 290
Presbyopia, 851
Pressure, 333–40
  absolute, 337
  atmospheric, 336–40
  in fluids, 333–36
  in a gas, 337, 454–56, 468
  gauge, 337
  hydraulic, 338
  measurement of, 338–40
  partial, 476
  radiation, 802–3
  units for and conversions, 333, 337
  vapor, 475–76
Pressure amplitude, 420
Pressure cooker, 476
Pressure gradient, 351
Pressure head, 335
Pressure transducers, 677
Pressure waves, 417 ff
Prestressed concrete, 313
Principal axis, 817
Principle of complementarity, 958–59
Principle of correspondence, 943, 971, 976 pr
Principle of equivalence, 148, 1151–52
Principle of superposition, 400–402
Principle quantum number, 1005–6
Principles (see names of)
Prism, 825
Prism binoculars, 856
Probability, in quantum mechanics, 980, 984–85
Probability and entropy, 536–37
Probability density, 980, 989, 992, 995, 1004, 1008–10, 1031–32
Problem solving techniques, 28, 89, 96, 159
Process, reversible and irreversible (defn), 520–25
Projectile, horizontal range of, 60–61
Projectile motion, 55–63
Proper length, 930
Proper time, 928, 1139 pr
Proportional limit, 309
Proton, 1062–63, 1127
  decay (?) of, 1135
  electric charge on, 547
Proton-proton collision, 220
Proton-proton cycle, 999–1100, 1148
Protostar, 1148
Proxima Centauri, 1141
Pseudoforce, 291
Pseudovector, 246 fn
psi (unit), 333
PT diagram, 474
Pulley, 93
Pulsar, 1150
Pulsating universe, 1168
Pulsed laser, 1021–22
Pulse-echo technique, 437
Pumps, 353
  heart as, 353
  heat, 527–28
Pupil, 850, 899
PV diagram, 473
P waves, 394
Pythagorean theorem, A-2

Q

Q-value, 380, 771 pr, 785 pr, 1068–69, 1086
QCD, 1123, 1133–35
QED, 1121
Quadratic formula, 35, A-1
Quadrupole, 573 pr
Quality factor, 1102
Quality of a sound, 429
Quality value (Q-value) of a resonant system, 380, 771 pr, 785 pr, 1068–69, 1086
Quantities, base and derived, 8
Quantization:
  of angular momentum, 965
  of electric charge, 550
  of energy, 951, 965–71, 991–92
Quantum chromodynamics, 1123, 1133–35
Quantum condition, Bohr's, 965, 972
Quantum electrodynamics, 1121
Quantum hypothesis, Planck's, 949–51
Quantum mechanics, 972, 977–1060
Quantum numbers, 965 ff, 992, 1004–7
  in H atom, 1004–7
  for complex atoms, 1010–11
  for molecules, 1038–43
Quantum theory, 916, 949–72
  of atoms, 965–71
  of blackbody radiation, 951
  of light, 949–58
Quarks, 550 fn, 1121, 1123, 1130–32, 1134, 1162
Quasars, 1144, 1157–58
Quasistatic process (defn), 495

R

Rad (unit), 1102
Radar, 435, 876–77

Radial acceleration, 64–65, 69, 114 ff, 240
Radial probability density, 1008–10
Radian measure for angles, 240
Radiant flux, 882
Radiation:
  blackbody, 950–51, 1160
  Čerenkov, 886 pr
  cosmic, 1102, 1114
  detection of, 1080, 1102
  from hot objects, 506–8, 950–51
  infrared (IR), 506, 799, 825
  ionizing (defn), 1100
  measurement of, 1101–3
  microwave, 798
  nuclear, 1067–80, 1100–1103
  synchrotron, 1117
  thermal, 506–8
  ultraviolet (UV), 799, 825
  X (see X-rays)
Radiation damage, 1100–1101
Radiation-dominated universe, 1161, 1163
Radiation dose, 1101–3
Radiation era, 1162–63
Radiation field, 793
Radiation film badge, 1103
Radiation pressure, 802–3
Radiation sickness, 1103–4
Radiation therapy, 1104
Radio, 632, 745, 780, 803–6
Radioactive dating, 1078–79
Radioactive decay (see Radioactivity)
Radioactive decay constant, 1073–76
Radioactive decay law, 1074
Radioactive decay series, 1077–78
Radioactive fallout, 1095
Radioactive tracers, 1104
Radioactive waste, 1093–94
Radioactivity, 998, 1067–80
  alpha, 1067–70, 1073
  artificial, 1067
  beta, 1067, 1070–73
  dosage of, 1101–3
  gamma, 1037, 1072–73
  natural, 1067, 1102
  probabilistic nature of, 1073
  rate of decay, 1073–78
Radionuclide (defn), 1101
Radio waves, 787, 798
Radius of gyration, 273 pr
Radius of nuclei, 1062
Radon, 1102

Rainbow, 825
Random access memory (RAM), 613
Range of projectile, 60–61
Rapid estimating, 9–12
Rarefaction, in wave, 391
Ray, 403, 811
  paraxial (defn), 817
Ray diagramming, 837 ff
Rayleigh, Lord, 897, 951
Rayleigh criterion, 897
Rayleigh-Jeans theory, 951
Ray model of light, 811 ff
RBE, 1102
RC circuit, 669–73
Reactance, 759, 774–77 (see also Impedance)
Reaction energy, 1086
Reactions, nuclear, 1085–88
Reactors, nuclear, 1090–95
Real image, 814, 818, 839
Rearview mirror, 822
Receivers, radio and television, 805
Recombination, 1170 pr
Rectifiers, 1053–54
Red giant star, 1144, 1146, 1148
Redshift, 435, 943, 1145, 1156 ff
Reduced mass, 1039
Reference frames, 17, 917 ff
  accelerating, 79, 148, 291
  inertial, 79, 291, 917 ff
  noninertial, 79, 291
  rotating, 291
  transformations between, 932–36
Reflecting telescope, 856
Reflection:
  angle of, 403, 813
  diffuse, 813
  law of, 403, 813
  of light, 813–22
  phase changes during, 877–81
  polarization by, 910
  specular, 813
  from thin films, 877–81
  total internal, 415 pr, 826–27
  of water waves, 402–3
  of waves on a string, 402
Reflection coefficient, 997
Reflection grating, 900
Refracting telescope, 854
Refraction, 408–9, 822–24
  angle of, 408–9, 822
  index of, 811–12
  law of, 409, 823–24, 867–69
  of light, 822–24, 867–69

  and Snell's law, 822–24, 868
  at spherical surface, 828–30
  by thin lenses, 837–40
  of water waves, 408–9
Refrigerator, 525–26
Reinforced concrete, 313
Relative biological effectiveness (RBE), 1102
Relative humidity, 476
Relative motion, 66–68, 916–41
Relative permeability, 725
Relative velocity, 66–68
Relativistic mass, 936–38
Relativistic momentum, 936–38, 941
Relativity, Galilean-Newtonian, 917–19, 932–36
Relativity, general theory of, 148, 149, 169, 922, 1140–41, 1151–55
Relativity, special theory of, 916–44
  and appearance of object, 931
  constancy of speed of light, 923
  four-dimensional space-time, 932
  and length, 930–31
  and Lorentz transformation, 932–36
  and mass, 936–38
  mass-energy relation in, 938–42
  postulates of, 922–23
  simultaneity in, 924–26
  and time, 926–29, 932
Relativity principle, 917–19, 923 ff
Relay, 727
Rem (unit), 1102–3
Research reactor, 1093
Resistance and resistors, 638–39
  in ac circuit, 773, 776–79
  with capacitor, 669–73, 776–79
  color code, 639
  electric currents and, 634–52
  with inductor, 762–63, 766–67, 776–79
  internal, 659
  in series and parallel, 660–64
  shunt, 674
Resistance thermometer, 641, 676
Resistive force, 122–23
Resistivity, 640–42
  temperature coefficient of, 641–42
Resolution:
  of diffraction grating, 903–5
  of electron microscope, 961
  of eye, 899
  of lens, 896–99
  of light microscope, 898–99

Resolution (*continued*)
  of telescope, 898–99
  of vectors, 48–52
Resolving power, 898–99, 903–5 (*see also* Resolution)
Resonance, 378–79, 405–8, 780
  in ac circuit, 780
  particle, 1127–28
Resonant frequency, 378–79, 405, 780
Resources, energy, 741
Rest energy, 939, 1063
Restitution, coefficient of, 234 *pr*
Rest mass, 938–40
Restoring force, 162, 363
Resultant vector, 47, 50
Retentivity (magnetic), 725
Retina, 850
Reversible process, 520–25
Reynolds' number, 359 *pr*
Right-hand rule, 246, 280, 689, 690, 692, 711
Rigid body (*defn*), 239
  rotational motion of, 239–68, 285–90
Rigid box, particle in, 990–94
Ripple voltage, 1060 *pr*
rms (root-mean-square):
  current, 646
  velocity, 469
  voltage, 646
Rockets, 83, 210, 227–29
Roemer, Olaf, 811
Roentgen, W. C., 905
Roentgen (unit), 1102
Rolling friction, 106, 263–64, 267
Rolling motion, 244–45, 262–68
Root-mean-square (rms) current, 646
Root-mean-square (rms) speed, 469
Root-mean-square (rms) voltage, 646
Rotation, frequency of, 242–43
Rotational angular momentum quantum number, 1038–39, 1042–43
Rotational imbalance, 287–88
Rotational inertia (*defn*), 249–50
Rotational kinetic energy, 260–65
  molecular, 500–501
Rotational transitions, 1038–40
Rotation and rotational motion, 239–94
Rotor, 698, 742
Ruby laser, 1021
Russell, Bertrand, 960

Rutherford, Ernest, 962, 1061, 1067, 1085
Rutherford's model of the atom, 962–63
R-value, 505
Rydberg constant, 964, 969
Rydberg states, 1029 *pr*

S

S waves, 394
Safety factor, 312
Sailboat and Bernoulli's principle, 348
Salam, A., 1134
Satellites, 139–43, 145
Saturated vapor pressure, 475
Saturation (magnetic), 724
Savart, Felix, 719
Sawtooth oscillator, 685 *pr*
Sawtooth voltage, 673
Scalar (*defn*), 45
Scalar product, 159–61
Scales, musical, 424
Scanning electron microscope, 961
Scanning tunneling electron microscope, 999
Scattering of light, 911
Schrödinger, Erwin, 869, 978
Schrödinger equation, 978, 985–95, 1004–7
Schwarzschild radius, 1155
Scientific notation, 5
Scintillation counter, 1080
Scintillator, 1080, 1120
Search coil, 754 *pr*
Second (unit), 7
Second law of motion, 80 *ff*, 250 *ff*
Second law of thermodynamics, 516–37
  Clausius statement of, 517, 526, 533
  entropy and, 529–33
  general statements of, 532–33
  Kelvin-Planck statement of, 520, 523, 533
  statistical interpretation of, 535–37
Seismograph, 749–50
Selection rules, 1007, 1042
Self-inductance, 758–60
Self-sustaining chain reaction, 1091–95
Semiconductor detector, 1080, 1120

Semiconductor doping, 1051 *ff*
Semiconductors, 548, 640, 954, 1049–55
  n and *p* types, 1051 *ff*
  resistivity of, 640
Sensitivity of meters, 675–76
Separation of variables, 988
Shear modulus, 312
Shear strength, 313
Shear stress, 311–12
Shells and subshells, 1012–13
SHM (*see* Simple harmonic motion)
Shock waves, 435–36
Short circuit, 644
Short-range forces, 1066
Shunt resistor, 674
Shutter speed, 848, 850
Sievert (unit), 1102
Sign conventions (geometric optics), 819, 822
Significant figures, 4–5
Silicon, 1051 *ff*
Simple harmonic motion (SHM), 364–71, 390
  applied to pendulum, 371–73
Simple harmonic oscillator (SHO), 364–71
  molecular vibration as, 1040–42
Simple pendulum, 13, 188, 371–73
Simultaneity, 924–26
Single photon emission tomography, 1106–7
Single slit diffraction, 888–93
Singularity, 1155
Sinusoidal curve, 364 *ff*
Siphon, 354 *pr*
SI units (Système Internationale), 6–8, 81
Sky, color of, 911
SLAC, 1118
Slepton, 1136
Slope, of a curve, 21
Slow neutron reaction, 1087
Slug (unit), 81
Smoke detector, 954, 1070
SN 1987a, 1150, 1167
Snell, W., 823
Snell's law, 822–24, 826
Soap bubble, 877, 880
Soaps, 352
Sodium chloride bond, 628 *pr*, 1031–32, 1034–35, 1044
Solar cell, 1054
Solar constant, 507

Solenoid, 709, 716–18, 723–24, 757, 759
Solids, 309 ff, 332, 446–47, 474 (see also Phase, changes of)
Solid state physics, 1044–50
Somatic damage, 1101
Sonar, 437
Sonic boom, 435–36
Sound "barrier," 435
Sounding board, 425
Sounding box, 425
Sound level, 421–22
Sound spectrum, 429
Sound track, 954
Sound waves, 417–37
  Doppler shift of, 432–35
  infrasonic, 419
  intensity of (and dB), 420–23
  interference of, 429–32
  quality of, 429
  sources of, 424–29
  speed of, 418
  supersonic, 418
  ultrasonic, 418
Source activity, 1101
Source of emf, 659–60, 734–35
South pole, 687
Space:
  absolute, 918, 923
  relativity of, 930–36
Space quantization, 1005, 1016–17
Space-time, 932
  curvature of, 149, 1151–55, 1165
Special theory of relativity (see Relativity, special theory of)
Specific gravity, 333
Specific heat, 488 ff, 498–99
Spectrograph, mass, 702
Spectrometer:
  light, 901–2
  mass, 702
Spectrophotometer, 901 fn
Spectroscope and spectroscopy, 901–2
Spectroscopic notation, 1018
Spectrum, 901
  absorption, 902, 964, 968
  atomic emission, 902, 963–65
  band, 1038
  continuous, 902
  electromagnetic, 798–800, 874–75
  emitted by hot object, 902, 949–51
  line, 902, 963 ff
  molecular, 1037–43
  visible light, 799, 824–25
  X-ray, 1013–15

Specular reflection, 813
Speed, 18–22 (see also Velocity)
  average, 18
  of EM waves, 794–97
  mean (molecular), 469
  molecular, 466–72
  most probable, 471–72
  rms, 469
Speed of light, 6, 797, 811–12, 867–68, 919–23, 938
  constancy of, 923
  measurement of, 811–12
  as ultimate speed, 938
Speed of sound, 418
  supersonic, 418, 435–36
  ultrasonic, 418
SPET, 1106–7
Spherical aberration, 817, 829, 858
Spherical mirrors, image formed by, 816–22
Spherical waves, 395
Spin, electron, 723, 1005–6, 1017–18
Spin-echo technique, 1109
Spinning top, 290
Spin-orbit interaction, 1018
Spin quantum number, 1005
Spiral galaxy, 1143
Spring:
  potential energy of, 181, 187, 369
  vibration of, 363 ff
Spring constant, 162, 363
Spring equation, 162
Spyglass, 856
Squark, 1136
Stable and unstable equilibrium, 197–98, 308
Standard conditions (STP), 457
Standard length, 6, 881
Standard mass, 7, 79
Standard model:
  cosmology, 1161–64
  elementary particles, 1132–34
Standards and units, 6–8
Standard temperature and pressure (STP), 457
Standing waves, 405–8, 424–28, 971
Star clusters, 1143
Stars, 1141–51
Starter, automobile, 724
State:
  changes of, 473–76, 490–93
  energy (see Energy states)
  equation of, 454, 457, 459, 477–78
  of matter, 332, 445–47

  as physical condition of system, 445
State variable, 445, 494, 529
Static electricity, 545
Static equilibrium, 300–331
Static friction, 107 ff
Statics, 300–321
Stationary states in atom, 965 (see also Energy states)
Statistics and entropy, 535–37
Stator, 742
Steady state model of universe, 1159
Steam engine, 518–19
Steam power plant, 1094
Stefan-Boltzmann constant, 506
Stefan-Boltzmann law (or equation), 506
Stellar evolution, 1145–51, 1159
Step-down transformer, 745
Step-up transformer, 745
Stern-Gerlach experiment, 1016–18
Stereo broadcasting, FM, 805 fn
Stimulated emission, 1019–22
Stirling cycle, 543 pr
Stopping distance for car, 29, 167
Stopping potential, 952
Storage rings, 1118–19
STP, 457
Strain, 309–12
Strain factor, 683 pr
Strain gauge, 677
Strangeness and strange particles, 1129
Strassmann, F., 1090
Streamline (defn), 343
Streamline flow, 343–45
Strength of materials, 309, 312–15
Stress, 309–12
  compressive, 311
  shear, 311–12
  tensile, 311
  thermal, 454
Stringed instruments, 424–25
Strings, vibrating, 405–8, 424–25
String theory, 1136
"Stripping" nuclear reaction, 1111 pr
Strong bonds, 1036
Strongly-interacting particles (defn), 1126
Strong nuclear force, 545 fn, 1066, 1121–36
Sublimation, 474
Subshell, atomic, 1012–13
Suction, 340

Sun, energy source of, 1096–98
Sunglasses, polarized, 908–10
Sunsets, 911
Supercluster, 1143
Superconducting magnets, 650, 723
Superconducting supercollider, 1118
Superconductivity, 650, 723
Superfluidity, 474
Supernovae, 1144, 1150, 1167
Superposition, principle of, 400, 401–2, 552, 555
Supersaturated air, 476
Supersonic speed, 418, 435–36
Superstring theory, 1136
Supersymmetry, 1136
Surface tension, 351–52
Surface waves, 394
Surfactants, 352
Surgery, laser, 1022
Survival equation, 483 pr
Suspension bridge, 318
S waves, 394
Symmetry, 10, 35, 93, 557–58, 714, 715, 716
Symmetry breaking, 1134, 1162, 1164
Synchrocyclotron, 1117
Synchrotron, 1117
Synchrotron radiation, 1117
Système Internationale (SI), 6–8, 81
Systems, 445, 493
　closed, 493
　isolated, 209, 493
　open, 493
　as set of objects, 445
　of units, 6–8

T

Tangential acceleration, 121–22, 242
Tape-recorder head, 749
Tau lepton, 1126–27, 1131
Telephone, cellular, 806
Telephoto lens, 850
Telescopes, 854–56
　astronomical, 854–56
　field-lens, 856
　Galilean, 856
　Keplerian, 854
　magnification of, 855, 899
　reflecting, 856
　refracting, 854
　resolution of, 898–99
　terrestrial, 856

Television, 595, 605–6, 803–6
Temperature, 447 ff, 460, 537–38
　absolute, 455, 460, 538
　Celsius (or centigrade), 448–49
　critical, 473
　Curie, 722
　distinguished from heat and internal energy, 487
　Fahrenheit, 448–49
　Fermi, 1060 pr
　human body, 449
　Kelvin, 455, 460, 538
　molecular interpretation of, 466–70
　operating (of heat engine), 517
　relation to molecular velocities, 470–72
　scales of, 448, 455, 460, 538
　transition of, 650
Temperature coefficient of resistivity, 641–42
Tensile strength, 313
Tensile stress, 311
Tension, 90, 311
Terminal, 635
Terminal speed or velocity, 32 fn, 122–23
Terminal voltage, 659–60
Terrestrial telescope, 856
Tesla (unit), 691
Test charge, 554
Tevatron, 1117
Theories (general), 3
Theories of everything, 1136
Theory of relativity (see Relativity)
Thermal conductivity, 503–5
Thermal contact, 449
Thermal energy, 189–90, 487–88 (see also Internal energy)
　distinguished from heat and temperature, 487
Thermal equilibrium, 449–50
Thermal expansion, 450–53
　coefficients of, 450–51
　of water, 453
Thermal neutron, 1089
Thermal pollution, 525
Thermal radiation, 506–8
Thermal stress, 454
Thermionic emission, 605–6
Thermistor, 641, 676
Thermochemical calorie, 486–87
Thermocouple, 677
Thermodynamic probability, 536–37
Thermodynamics, 445, 485–544

　first law of, 493–98
　second law of, 516–37
　third law of, 529 fn, 537–38
　zeroth law of, 449–50
Thermodynamic temperature scale, 537–38
Thermoelectric effect, 677
Thermography, 507–8
Thermometers, 447–49
　constant-volume gas, 449
　resistance, 641, 676
Thermonuclear device, 1098
Thermostat, 461 pr
Thin-film interference, 877–81
Thin lenses, 837–40
Third law of motion, 82 ff
Third law of thermodynamics, 529 fn, 537–38
Thomson, G. P., 960
Thomson, J. J., 699–700, 960, 962
Three Mile Island, 1094
Threshold energy, 1088, 1113 pr
Thrust, 228
TIA, 349
Timbre, 429
Time:
　absolute, 918
　proper, 928
　relativity of, 924–29, 931–36
　standard of, 7
Time constant, 670, 762
Time dilation, 926–29
Time's arrow, 533
Tire pressure, 458
Tire pressure gauge, 338
Tokamak, 1085, 1099
Tokamak Fusion Test Reactor, 1099
Tomography, 1105–6
Tone color, 429
Top quark and topness, 1130–31
Toroid, 718
Torque, 247–50 and ff, 266–67, 280–81 and ff, 302 ff
　counter, 743
　on current loop, 695–97
Torr (unit), 339
Torricelli, Evangelista, 339, 347
Torricelli's theorem, 347
Torsion pendulum, 374
Total angular momentum, 1017–18
Total binding energy, 1064
Total cross section, 1089
Total internal reflection, 415 pr, 826–27

Total mechanical energy, 183
Total reaction cross section, 1089
Townsend, J. S., 700
Tracers, 1104
Transducers, 676–77
Transformations:
   Galilean, 932–36
   Lorentz, 932–36
Transformer, 744–47, 758
Transformer equation, 745
Transient ischemic attack (TIA), 349
Transistors, 1054–55
Transition elements, 1013
Transition temperature, 650
Translational motion, 16–238, 262–65
Transmission coefficient, 997
Transmission electron microscope, 961
Transmission grating, 900 *ff*
Transmission lines, 746–47, 782
Transmission of electricity, 746–47, 782
Transmutation of elements, 1068, 1070, 1085–88
Transverse waves, 391 *ff*, 794, 907
Trap, sink, 349
Triangulation, 11, 1144 *fn*, 1153
Trigonometric functions and identities, 49, A-2–A-3, inside back cover
Triple point, 460, 474
Tritium, 1084 *pr*, 1097–99
Trusses, 315–18
Tube, vibrating column of air in, 425–28
Tubes, flow in, 351
Tunnel diode, 998–99
Tunneling through a barrier, 996–99, 1069–70, 1155 *fn*, 1168
Turbine, 741
Turbulent flow, 343
Turning points, 197
Turn signal, automobile, 673
Tweeter, 776
Twin paradox, 926–29
Tycho Brahe, 143

U

UHF, 804–5
Ultimate speed, 938
Ultimate strength, 309, 312–13
Ultrasonic waves, 418, 437

Ultrasound, 437
Ultrasound imaging, 437
Ultraviolet (UV) light, 799, 825
Unavailability of energy, 534–35
Uncertainty (in measurements), 4, 981
Uncertainty principle, 981–84, 996, 1031
Underdamping, 376
Underexposure, 848
Unification scale, 1134
Unified atomic mass unit, 7, 446, 1063
Uniform circular motion, 63–65, 114 *ff*, 371
   dynamics of, 114–17
   kinematics of, 63–65
Uniformly accelerated motion, 26 *ff*, 54–55
Uniformly accelerated rotational motion, 243–44
Unit conversion, 8–9, inside front cover
Units of measurement, 6–8
Unit vectors, 52–53
Universal gas constant, 457
Universal law of gravitation, 133–37 and *ff*
Universe (*see also* Cosmology):
   age of, 1158–59
   Big Bang theory of, 1136, 1159–64
   curvature of, 149, 1151–55, 1165
   expanding, 1156–59
   future of, 1165–69
   inflation scenario of, 1164
   matter-dominater, 1161, 1164
   open or closed, 1151, 1165–69
   pulsating, 1168
   radiation-dominated, 1161, 1163
   standard model of, 1161–64
   steady state model of, 1159
Unpolarized light (*defn*), 907
Unstable equilibrium, 198, 308
Uranium, 1077, 1079, 1090–95
Uranus, 146
Useful magnification, 899, 961
UV light, 799, 825

V

Vacuum pump, 353
Vacuum tube, 803
Valence, 1013
Valence band, 1049–50

Van de Graaff generator and accelerator, 612 *pr*, 1115
van der Waals, J. D., 477
van der Waals bonds and forces, 1036–37
van der Waals equation of state, 477–78
Vapor, 473 (*see also* Gases)
Vaporization, latent heat of, 491
Vapor pressure, 474–76
Variable-mass systems, 227–29
Vector displacement, 46–47, 53
Vector model (atoms) 1028 *pr*
Vector product, 279–80
Vectors, 17, 46–53, 159–61, 279–80, A-3
   addition of, 46–52
   components of, 48–52
   cross product, 279–80
   multiplication of, 48, 159–61, 279–80
   multiplication by a scalar, 48
   position, 53, 55, 240, 243–44
   resolution of, 48–52
   resultant, 47, 50
   scalar (dot) product, 159–61
   subtraction of, 48
   sum, 46–52
   unit, 52–53
   vector (cross) product, 279–80
Velocity, 18–22 and *ff*, 36–37, 45, 54–55
   addition of, 66–68
   angular, 241–44
   average, 18–19, 21, 54
   drift, 647–49, 701
   of EM waves, 794–97
   escape, 192–94
   gradient, 350
   instantaneous, 20–22, 26–27, 54–55
   of light, 6, 797, 811–12, 867–68, 919–23, 938
   molecular, and relation to temperature, 466–72
   phase, 397
   relative, 66–68
   rms, 469
   of sound, 418
   supersonic, 418, 435–36
   terminal, 32 *fn*, 122–23
   of waves, 390–94, 397, 418
Velocity selector, 695
Venturi meter, 348
Venturi tube, 348

Index **A-53**

Venus, phases of, 854
Vibrational energy, 369–70
  molecular, 500–501, 1040–42
Vibrational motion, 362–80
Vibrational quantum number, 1041
Vibrational transition, 1040–42
Vibrations, 362–80, 405
  of air columns, 425–28
  of atoms and molecules, 500–501
  forced, 378–79
  as source of wave, 388–90, 417, 424–28
  of springs, 363 *ff*
  of strings, 405–8, 424–25
Virtual image, 814, 821, 840
Virtual particles, 1121
Virtual photon, 1121
Viscosity, 344, 350–51
  coefficient of, 350
Viscous force, 344, 350–51
Visible light, wavelengths of, 799, 824–25, 867 *ff*
Visible spectrum, 824–25
Volt (unit), 592
Volta, Alessandro, 592, 634–35
Voltage, 592–93 (*see also* Electric potential)
  bias, 1054–55
  breakdown, 596
  line, 646, 646 *fn*
  peak, 645
  ripple, 1060 *pr*
  rms, 646
  sawtooth, 673
  terminal, 659–60
Voltage drop, 665
Voltage gain, 1055
Voltmeter, 674–76, 697
  digital, 675
Volume, formulas for, A-1
Volume expansion, coefficient of, 452

W

W particle, 1123, 1126–27, 1133
Walking, 83
Water:
  anomalous behavior below 4°, 453
  cohesion of, 352
  dipole moment of, 566
  dissolving power of, 628 *pr*
  expansion of, 453
  heavy, 1092
  latent heats of, 490–93
  molecule, 1032–34
  polar nature of, 547, 565
  saturated vapor pressure, 475
  thermal expansion of, 453
  triple point of, 460, 474
Water equivalent, 490
Watson, J.D., 906
Watt, James, 195 *fn*
Watt (unit), 195, 642
Wave(s), 388–411 (*see also* Light, Sound waves)
  amplitude of, 390, 396
  composite, 401
  continuous (*defn*), 389–90
  diffraction of, 410
  displacement of, 396 *ff*
  earthquake, 394, 396, 409
  electromagnetic, 792–806, 866
  energy in, 395–96
  frequency, 390
  harmonic (*defn*), 398
  infrasonic, 419
  intensity, 395–96, 420–23
  interference of, 404–5
  light, 866–911
  longitudinal, 391 *ff*, 419–20
  mathematical representation of, 396–401, 419–20
  mechanical, 388–411
  P-, 394
  periodic (*defn*), 389–90
  phase, 397
  plane (*defn*), 403, 793–94, 989–90
  pressure, 419 *ff*
  pulse, 389
  radio, 787, 798
  reflection of, 402–3
  refraction of, 408–9
  S-, 394
  shock, 435–36
  sound, 417–37
  source is vibration, 388–90, 417, 424–28
  spherical, 395
  standing, 405–8, 424–28, 971
  surface, 394
  transmission, 402–3
  transverse, 391 *ff*, 794, 907
  ultrasonic, 418, 437
  velocity of, 390–94, 397, 418
  water, 388 *ff*
Wave equation, 399–401, 797
  Schrödinger, 978, 985–95, 1004–7
Waveform, 429
Wave front (*defn*), 403
Wave function, 978–79, 985–95, 1003–4, 1007–10
  for H atom, 1003–4, 1007–10
  for square well, 994–96
  measurement of, 901–15
  of visible light, 798
Wave intensity, 395–96, 420–23 (*see also* Light)
Wavelength (*defn*), 390
  Compton, 956, 975 *pr*
  de Broglie, 959, 971–72
  depending on index of refraction, 869
  as limit to resolution, 898–99
  of material particles, 959
Wave motion, 388 *ff*
Wave nature of matter, 959 *ff*
Wave number (*defn*), 397
Wave packet, 989–90
Wave-particle duality:
  of light, 958–59
  of matter, 959, 972, 978, 981–82
Wave pulse, 389
Wave theory of light, 866–911
Wave velocity, 390–94, 397, 418
Weak bonds, 1036–37, 1045
Weak charge, 1134
Weakly interacting massive particles (WIMPS), 1166
Weak nuclear force, 1066, 1072, 1123–36
Weber (unit), 736
Weight, 79, 85–87, 137–39
Weightlessness, 139–42
Weinberg, S., 1134
Wheatstone bridge, 677
Whirlpool galaxy, 1143
White dwarf, 1144, 1146, 1148–49
White-light holograms, 1024
Whole-body dose, 1103
Wide-angle lens, 850
Wien's displacement law, 950, 1146
Wien's radiation theory, 950
Wilson, R., 1160
WIMPS, 1166
Wind instruments, 425–27
Windshield wipers, 673
Wing of an airplane, lift on, 348
Wire drift chamber, 1120
Woofer, 776
Work, 155–69, 260–61
  compared to heat, 493
  defined, 156

done by a gas, 496–98
in first law of thermodynamics, 494 *ff*
from heat engines, 517 *ff*
relation to energy, 164–69, 177–83, 190–91, 193, 195, 260–61
units of, 156
Work-energy principle, 165–68, 182, 190–91, 261
Work function, 953, 1048
Working substance (*defn*), 519
Wright, Thomas, 1141

X

X-ray crystallography, 906
X-ray diffraction, 905–6
X-rays, 799, 905–6, 1013–15, 1073, 1105–9
and atomic number, 1013–15
in electromagnetic spectrum, 799
spectra, 1013–15

Y

Young, Thomas, 870
Young's double-slit experiment, 870–73
Young's modulus, 309–10
Yukawa, H., 1121–22

Z

Z^0 particle, 1123, 1126–28, 1133
Zeeman effect, 708 *pr*, 1005, 1016, 1018, 1107
Zener diode, 1053
Zero, absolute, 455, 538
Zero-point energy, 992, 1041
Zeroth law of thermodynamics, 449–50
Zoom lens, 850
Zweig, G., 1130

Photo Credits

CO-1 NOAA/Phil Degginger/Color-Pic, Inc. **1-1a** Philip H. Coblentz/Tony Stone Images **1-1b** Richard Berenholtz/The Stock Market **1-1c** Antranig M. Ouzoonian, P.E./Weidlinger Associates, Inc. **1-2** Mary Teresa Giancoli **1-3a** Oliver Meckes/E.O.S./MPI-Tubingen/Photo Researchers, Inc. **1-3b** Douglas C. Giancoli **1-4** International Bureau of Weights and Measures, Sevres, France **1-5** Douglas C. Giancoli **1-6** Doug Martin/Photo Researchers, Inc. **1-9** David Parker/Science Photo Library/Photo Researchers, Inc. **1-10** The Image Works **CO-2** Joe Brake Photography/Crest Communications, Inc. **2-22** Justus Sustermans, painting of Galileo Galilei/The Granger Collection **2-23** Photograph by Dr. Harold E. Edgerton, © The Harold E. Edgerton 1992 Trust. Courtesy Palm Press, Inc. **CO-3** Michel Hans/Vandystadt Allsport Photography (USA), Inc **3-19** Berenice Abbott/Commerce Graphics Ltd., Inc. **3-21** Richard Megna/Fundamental Photographs **3-30a** Don Farrall/PhotoDisc, Inc. **3-30b** Robert Frerck/Tony Stone Images **3-30c** Richard Megna/Fundamental Photographs **CO-4** Mark Wagner/Tony Stone Images **4-1** AP/Wide World Photos **4-3** Central Scientific Company **4-5** Sir Godfrey Kneller, Sir Isaac Newton, 1702. Oil on canvass. The Granger Collection **4-6** Gerard Vandystadt/Agence Vandystadt/Photo Researchers, Inc. **4-8** David Jones/Photo Researchers, Inc. **4-11** Tsado/NASA/Tom Stack & Associates **4-31** Lars Ternblad/The Image Bank **4-34** Kathleen Schiaparelli **4-36** Jeff Greenberg/Photo Researchers, Inc. **CO-5a** Jess Stock/Tony Stone Images **CO-5b** Werner H. Muller/Peter Arnold, Inc. **5-12** Jay Brousseau/The Image Bank **5-17** Guido Alberto Rossi/The Image Bank **5-33** C. Grzimek/Okapia/Photo Researchers, Inc. **5-34** Photofest **5-38** Cedar Point Photo by Dan Feicht **CO-6** Earth Imaging/Tony Stone Images **6-9** NASA/Johnson Space Center **6-13** I. M. House/Tony Stone Images **6-14L** Jon Feingersh/The Stock Market **6-14M** © Johan Elbers 1995 **6-14R** Peter Grumann/The Image Bank **CO-7** Eric Miller/Reuters/Corbis **7-20a** Stanford Linear Accelerator Center/Science Photo Library/Photo Researchers, Inc. **7-20b** Account Phototake/Phototake NYC **7-24** Official U.S. Navy Photo **CO-8 and 8-10** Photograph by Dr. Harold E. Edgerton, © The Harold E. Edgerton 1992 Trust. Courtesy Palm Press **8-11** David Madison/David Madison Photography **8-16** AP/Wide World Photos **8-23** M. C. Escher's "Waterfall" © 1996 Cordon Art-Baarn-Holland. All rights reserved. **CO-9** Richard Megna/Fundamental Photographs **9-9** Photograph by Dr. Harold E. Edgerton, © The Harold E. Edgerton 1992 Trust. Courtesy Palm Press **9-16** D.J. Johnson **9-20** Courtesy Brookhaven National Laboratory **9-22** Berenice Abbott/Photo Researchers, Inc. **CO-10** Ch. Russeil/Kipa/Sygma Photo News **10-10b** Mary Teresa Giancoli **10-14a** Richard Megna/Fundamental Photographs **10-14b** Photoquest, Inc. **10-47** Jens Hartmann/AP/Wide World Photos **10-48** Focus on Sports, Inc. **10-49** Karl Weatherly/PhotoDisc, Inc **CO-11** AP/Wide World Photos **11-19c** NOAA/Phil Degginger/Color-Pic, Inc. **11-19d** NASA/TSADO/Tom Stack & Associates **11-34** Michael Kevin Daly/The Stock Market **CO-12** Steve Vidler/Leo de Wys, Inc. **12-2** AP/Wide World Photos **12-7** T. Kitchin/Tom Stack & Associates **12-10** The Stock Market **12-21** Douglas C. Giancoli **12-23** Mary Teresa Giancoli **12-27** Fabricius & Taylor/Liaison Agency, Inc. **12-36** Henryk T. Kaiser/Leo de Wys, Inc. **12-38** Douglas C. Giancoli **12-39** Galen Rowell/Mountain Light Photography, Inc. **12-41** Douglas C. Giancoli **12-43** Giovanni Paolo Panini (Roman, 1691-1765), Interior of the Pantheon, Rome c. 1734. Oil on canvas. 1.283 × .991 (50 1/2 × 39); framed: 1.441 × 1.143 (56 3/4 × 45). © 1995 Board of Trustees, National Gallery of Art, Washington. Samuel H. Kress Collection. Photo by Richard Carafelli. **12-44** Robert Holmes/Corbis **12-45** Italian Government Tourist Board **CO-13** Steven Frink/Tony Stone Images **13-12** Corbis **13-20** Department of Mechanical and Aerospace Engineering, Princeton University **13-31** Rod Planck/Thomas Stack & Associates **13-33** Alan Blank/Bruce Coleman Inc. **13-42** Douglas C. Giancoli **13-45** Galen Rowell/Mountain Light Photography, Inc. **13-51** NASA/Goddard Space Flight Center/Science Source/Photo Researchers, Inc. **CO-14** Richard Megna/Fundamental Photographs **14-4** Mark E. Gibson/Visuals Unlimited **14-14** Fundamental Photographs **14-15** Douglas C. Giancoli **14-22** Taylor Devices, Inc. **14-25** Martin Bough/Fundamental Photographs **14-26a** AP/Wide World Photos **14-26b** Paul X. Scott/Sygma Photo News **CO-15** Richard Megna/Fundamental Photographs **15-1** Douglas C. Giancoli **15-24** Douglas C. Giancoli **15-30** Martin G. Miller/Visuals Unlimited **15-32** Richard Megna/Fundamental Photographs **CO-16** Photographic Archives, Teatro alla Scala, Milan, Italy **16-6** Yoav Levy/Phototake NYC **16-9a** David Pollack/The Stock Market **16-9b** Andrea Brizzi/The Stock Market **16-10** Bob Daemmrich/The Image Works **16-13** Bildarchiv Foto Marburg/Art Resource, N.Y. **16-24a** Norman Owen Tomalin/Bruce Coleman Inc. **16-25b** Sandia National Laboratories, New Mexico **16-28a** P. Saada/Eurelios/Science Photo Library/Photo Researchers, Inc. **16-28b** Howard Sochurek/Medical Images Inc. **CO-17** Le Matin de Lausanne/Sygma Photo News **17-3** Bob Daemmrich/Stock Boston **17-5** Leonard Lessin/Peter Arnold, Inc. **17-14** Leonard Lessin/Peter Arnold, Inc. **17-17** Brian Yarvin/Photo Researchers, Inc. **CO-18** Tom Till/DRK Photo **18-9** Paul Silverman/Fundamental Photographs **18-14** Mary Teresa Giancoli **CO-19** Tom Bean/DRK Photo **19-25** Science Photo Library/Photo Researchers, Inc. **CO-20a** David Woodfall/Tony Stone Images **CO-20b** AP/Wide World Photos **20-7a** Sandia National Laboratories, New Mexico **20-7b** Martin Bond/Science Photo Library/Photo Researchers, Inc **20-7c** Lionel Delevingne/Stock Boston **20-13** Leonard Lessin/Peter Arnold, Inc. **CO-21** Fundamental Photographs **21-38** Michael J. Lutch/Boston Museum of Science **CO-23** Gene Moore/Phototake NYC **23-21** Jon Feingersh/The Stock Market **CO-24** Paul Silverman/Fundamental Photographs **CO-25** Mahaux Photography/The Image Bank **25-1** J.-L. Charmet/Science Photo Library/Photo Researchers, Inc. **25-10** T.J. Florian/Rainbow **25-13** Richard Megna/Fundamental Photographs **25-16** Barbara Filet/Tony Stone Images **25-26** Takeshi Takahara/Photo Researchers, Inc. **25-33** Liaison Agency, Inc. **CO-26a** Steve Weinrebe/Stock Boston **CO-26b** Sony Electronics, Inc. **26-22** Paul Silverman/Fundamental Photographs **26-26** Paul Silverman/Fundamental Photographs **CO-27** Richard Megna/Fundamental Photographs **27-4** Richard Megna/Fundamental Photographs **27-6** Mary Teresa Giancoli **27-8** Richard Megna/Fundamental Photographs **27-19** Pekka Parviainen/Science Photo Library/Photo Researchers, Inc. **CO-28** Manfred Kage/Peter Arnold, Inc. **28-21** Richard Megna/Fundamental Photographs **CO-29** Richard Megna/Fundamental Photographs **29-10** Werner H. Muller/Peter Arnold, Inc.

29-15 Tomas D.W. Friedmann/Photo Researchers, Inc. **29-20** Jon Feingersh/Comstock **29-28b** National Earthquake Information Center, U.S.G.S. **CO-30** Adam Hart-Davis/Science Photo Library/Photo Researchers, Inc. **CO-31v1** Albert J. Copley/Visuals Unlimited **CO-31v2** Mary Teresa Giancoli **CO-32** Richard Megna/Fundamental Photographs **32-1** AIP Emilio Segre' Visual Archives **32-14** The Image Works **CO-33** Douglas C. Giancoli **33-6** Douglas C. Giancoli **33-11a** Mary Teresa Giancoli and Suzanne Saylor **33-11b** Mary Teresa Giancoli **33-21** Mary Teresa Giancoli **33-25** David Parker/Science Photo Library/Photo Researchers, Inc. **33-28b** Michael Giannechini/Photo Researchers, Inc. **33-34b** S. Elleringmann/Bilderberg/Aurora & Quanta Productions **33-40** Douglas C. Giancoli **33-42** Mary Teresa Giancoli **CO-34** Mary Teresa Giancoli **34-1c** Douglas C. Giancoli **34-1d** Douglas C. Giancoli **34-2** Douglas C. Giancoli and Howard Shugat **34-4** Douglas C. Giancoli **34-7a** Douglas C. Giancoli **34-7b** Douglas C. Giancoli **34-18** Mary Teresa Giancoli **34-19a** Mary Teresa Giancoli **34-19b** Mary Teresa Giancoli **34-26** Mary Teresa Giancoli **34-30a** Franca Principe/Istituto e Museo di Storia della Scienza, Florence, Italy **34-32** Yerkes Observatory, University of Chicago **34-33c** Palomar Observatory/California Institute of Technology **34-33d** Joe McNally/Joe McNally Photography **34-35b** Olympus America Inc. **CO-35** Larry Mulvehill/Photo Researchers, Inc. **35-4a** John M. Dunay IV/Fundamental Photographs **35-9a** Bausch & Lomb Incorporated **35-16a** Paul Silverman/Fundamental Photographs **35-16b** Richard Megna/Fundamental Photographs **35-16c** Yoav Levy/Phototake NYC **35-18b** Ken Kay/Fundamental Photographs **35-20b** Bausch & Lomb Incorporated **35-20c** Bausch & Lomb Incorporated **35-22** Kristen Brochmann/Fundamental Photographs **CO-36** Richard Megna/Fundamental Photographs **36-2a** Reprinted with permission from P.M. Rinard, American Journal of Physics, Vol. 44, #1, 1976, p. 70. Copyright 1976 American Association of Physics Teachers. **36-2b** Ken Kay/Fundamental Photographs **36-2c** Ken Kay/Fundamental Photographs **36-11a** Richard Megna/Fundamental Photographs **36-11b** Richard Megna/Fundamental Photographs **36-12a and b** Reproduced by permission from M. Cagnet, M. Francon, and J. Thrier, The Atlas of Optical Phenomena. Berlin: Springer-Verlag, 1962. **36-15** Space Telescope Science Institute **36-16** The Arecibo Observatory is part of the National Astronomy and Ionosphere Center which is operated by Cornell University under a cooperative agreement with the National Science Foundation. **36-21** Wabash Instrument Corp./Fundamental Photographs **36-26** Photo by W. Friedrich/Max von Laue. Burndy Library, Dibner Institute for the History of Science and Technology, Cambridge, Massachusetts. **36-29b** Bausch & Lomb Incorporated **36-36** Diane Schiumo/Fundamental Photographs **36-39a** Douglas C. Giancoli **36-39b** Douglas C. Giancoli **CO-37** Image of Albert Einstein licensed by Einstein Archives, Hebrew University, Jerusalem, represented by Roger Richman Agency, Beverly Hills, California **37-1** AIP Emilio Segre Visual Archives **CO-38** Wabash Instrument Corp./Fundamental Photographs **38-10** Photo by S.A. Goudsmit, AIP Emilio Segre' Visual Archives **38-11** Education Development Center, Inc. **38-20b** Richard Megna/Fundamental Photographs **38-21abc** Wabash Instrument Corp./Fundamental Photographs **CO-39** Institut International de Physique/American Institute of Physics/Emilio Segre Visual Archives **39-01** American Institute of Physics/Emilio Segre Visual Archives **39-02** F.D. Rosetti/American Institute of Physics/Emilio Segre Visual Archives **39-04** Advanced Research Laboratory/Hitachi Metals America, Ltd. **39-18** Driscoll, Youngquist, and Baldeschwieler, Caltech/Science Photo Library/Photo Researchers, Inc. **CO-40** Patricia Peticolas/Fundamental Photographs **40-16** Paul Silverman/Fundamental Photographs **40-22** Yoav Levy/Phototake NYC **40-24b** Paul Silverman/Fundamental Photographs **CO-41** Charles O'Rear/Corbis **CO-42** Reuters Newmedia Inc/Corbis **42-03** Chemical Heritage Foundation **42-07** University of Chicago, Courtesy of AIP Emilio Segre Visual Archives **CO-43** AP/Wide World Photos **43-07** Gary Sheahan "Birth of the Atomic Age" Chicago (Illinois); 1957. Chicago Historical Society. **43-10** Liaison Agency, Inc. **43-11** LeRoy N. Sanchez/Los Alamos National Laboratory **43-12** Corbis **43-16a** Lawrence Livermore National Laboratory/Science Source/Photo Researchers, Inc. **43-16b** Gary Stone/Lawrence Livermore National Laboratory **43-22a** Martin M. Rotker/Martin M. Rotker **43-22b** Simon Fraser/Science Photo Library/Photo Researchers, Inc. **43-26b** Southern Illinois University/Peter Arnold, Inc. **43-28** Mehau Kulyk/Science Photo Library/Photo Researchers, Inc. **CO-44** Fermilab Visual Media Services **44-06** CERN/Science Photo Library/Photo Researchers, Inc. **44-08** Fermilab Visual Media Services **CO-45** Jeff Hester and Paul Scowen, Arizona State University, and NASA **45-01** NASA Headquarters **45-02c** NASA/Johnson Space Center **45-03** U.S. Naval Observatory Photo/NASA Headquarters **45-04** National Optical Astronomy Observatories **45-05a** R.J. Dufour, Rice University **45-05b** U.S. Naval Observatory **45-05c** National Optical Astronomy Observatories **45-10** Space Telescope Science Institute/NASA Headquarters **45-11** National Optical Astronomy Observatories **45-15** European Space Agency/NASA Headquarters **45-22** Courtesy of Lucent Technologies/Bell Laboratories

Table of Contents Photos p. v (left) NOAA/Phil Degginger/Color-Pic, Inc. **p. v** (right) Mark Wagner/Tony Stone Images **p. vi** (left) Jess Stock/Tony Stone Images **p. vi** (right) Photograph by Dr. Harold E. Edgerton, © The Harold E. Edgerton 1992 Trust. Courtesy Palm Press **p. vii** (left) Ch. Russeil/Kipa/Sygma Photo News **p.vii** (right) Tibor Bognar/The Stock Market **p. viii** (left) Richard Megna/Fundamental Photographs **p. viii** (right) Douglas C. Giancoli **p. ix** (top) S. Feval/Le Matin de Lausanne/Sygma Photo News **p. ix** (bottom) Richard A. Cooke III/Tony Stone Images (right) David Woodfall/Tony Stone Images and AP/ Wide World Photos **p. x** (left) Fundamental Photographs **p. x** (right) Richard Kaylin/Tony Stone Images **p. xi** (left) Mahaux Photography/The Image Bank **p. xi** (right) Manfred Cage/Peter Arnold, Inc. **p. xii** Werner H. Muller/Peter Arnold, Inc. **p. xiii** (left) Mary Teresa Giancoli. **p. xiii** (right) Larry Mulvehill/Photo Researchers, Inc. **p. xiv** Image of Albert Einstein licensed by Einstein Archives, Hebrew University, Jerusalem, represented by Roger Richman Agency, Beverly Hills, California **p. xiv** (right) Donna McWilliam/AP/Wide World Photos **p. xv** (left) Charles O'Rear/Corbis (right) Fermilab **p. xvi** (left) Jeff Hester and Paul Scowen, Arizona State University, and NASA (right) R. J. Dufour, Rice University.

Periodic Table of the Elements§

Transition Elements

Symbol — Cl 17 — Atomic Number
Atomic Mass§ — 35.4527
3p⁵ — Electron Configuration (outer shells only)

Group I	Group II											Group III	Group IV	Group V	Group VI	Group VII	Group VIII
H 1 1.00794 $1s^1$																	He 2 4.002602 $1s^2$
Li 3 6.941 $2s^1$	Be 4 9.012182 $2s^2$											B 5 10.811 $2p^1$	C 6 12.0107 $2p^2$	N 7 14.00674 $2p^3$	O 8 15.9994 $2p^4$	F 9 18.9984032 $2p^5$	Ne 10 20.1797 $2p^6$
Na 11 22.989770 $3s^1$	Mg 12 24.3050 $3s^2$											Al 13 26.981538 $3p^1$	Si 14 28.0855 $3p^2$	P 15 30.973761 $3p^3$	S 16 32.066 $3p^4$	Cl 17 35.4527 $3p^5$	Ar 18 39.948 $3p^6$
K 19 39.0983 $4s^1$	Ca 20 40.078 $4s^2$	Sc 21 44.955910 $3d^14s^2$	Ti 22 47.867 $3d^24s^2$	V 23 50.9415 $3d^34s^2$	Cr 24 51.9961 $3d^54s^1$	Mn 25 54.938049 $3d^54s^2$	Fe 26 55.845 $3d^64s^2$	Co 27 58.933200 $3d^74s^2$	Ni 28 58.6934 $3d^84s^2$	Cu 29 63.546 $3d^{10}4s^1$	Zn 30 65.39 $3d^{10}4s^2$	Ga 31 69.723 $4p^1$	Ge 32 72.61 $4p^2$	As 33 74.92160 $4p^3$	Se 34 78.96 $4p^4$	Br 35 79.904 $4p^5$	Kr 36 83.80 $4p^6$
Rb 37 85.4678 $5s^1$	Sr 38 87.62 $5s^2$	Y 39 88.90585 $4d^15s^2$	Zr 40 91.224 $4d^25s^2$	Nb 41 92.90638 $4d^45s^1$	Mo 42 95.94 $4d^55s^1$	Tc 43 (98) $4d^55s^2$	Ru 44 101.07 $4d^75s^1$	Rh 45 102.90550 $4d^85s^1$	Pd 46 106.42 $4d^{10}5s^0$	Ag 47 107.8682 $4d^{10}5s^1$	Cd 48 112.411 $4d^{10}5s^2$	In 49 114.818 $5p^1$	Sn 50 118.710 $5p^2$	Sb 51 121.760 $5p^3$	Te 52 127.60 $5p^4$	I 53 126.90447 $5p^5$	Xe 54 131.29 $5p^6$
Cs 55 132.90545 $6s^1$	Ba 56 137.327 $6s^2$	57–71†	Hf 72 178.49 $5d^26s^2$	Ta 73 180.9479 $5d^36s^2$	W 74 183.84 $5d^46s^2$	Re 75 186.207 $5d^56s^2$	Os 76 190.23 $5d^66s^2$	Ir 77 192.217 $5d^76s^2$	Pt 78 195.078 $5d^96s^1$	Au 79 196.96655 $5d^{10}6s^1$	Hg 80 200.59 $5d^{10}6s^2$	Tl 81 204.3833 $6p^1$	Pb 82 207.2 $6p^2$	Bi 83 208.98038 $6p^3$	Po 84 (209) $6p^4$	At 85 (210) $6p^5$	Rn 86 (222) $6p^6$
Fr 87 (223) $7s^1$	Ra 88 (226) $7s^2$	89–103‡	Rf 104 (261) $6d^27s^2$	Db 105 (262) $6d^37s^2$	Sg 106 (266) $6d^47s^2$	Bh 107 (264) $6d^57s^2$	Hs 108 (269)	Mt 109 (268)	110 (271)	111 (272)	112 (277)		114 (289)		116 (289)		118 (293)

†Lanthanide Series

La 57 138.9055 $5d^16s^2$	Ce 58 140.115 $4f^15d^16s^2$	Pr 59 140.90765 $4f^36s^2$	Nd 60 144.24 $4f^45d^06s^2$	Pm 61 (145) $4f^55d^06s^2$	Sm 62 150.36 $4f^65d^06s^2$	Eu 63 151.964 $4f^75d^06s^2$	Gd 64 157.25 $4f^75d^16s^2$	Tb 65 158.92534 $4f^95d^06s^2$	Dy 66 162.50 $4f^{10}5d^06s^2$	Ho 67 164.93032 $4f^{11}5d^06s^2$	Er 68 167.26 $4f^{12}5d^06s^2$	Tm 69 168.93421 $4f^{13}5d^06s^2$	Yb 70 173.04 $4f^{14}5d^06s^2$	Lu 71 174.967 $4f^{14}5d^16s^2$

‡Actinide Series

Ac 89 (227.02775) $6d^17s^2$	Th 90 232.0381 $6d^27s^2$	Pa 91 231.03588 $5f^26d^17s^2$	U 92 238.0289 $5f^36d^17s^2$	Np 93 (237) $5f^46d^17s^2$	Pu 94 (244) $5f^66d^07s^2$	Am 95 (243) $5f^76d^07s^2$	Cm 96 (247) $5f^76d^17s^2$	Bk 97 (247) $5f^96d^07s^2$	Cf 98 (251) $5f^{10}6d^07s^2$	Es 99 (252) $5f^{11}6d^07s^2$	Fm 100 (257) $5f^{12}6d^07s^2$	Md 101 (258) $5f^{13}6d^07s^2$	No 102 (259) $5f^{14}6d^07s^2$	Lr 103 (262) $5f^{14}6d^17s^2$

§ Atomic mass values averaged over isotopes in percentages they occur on Earth's surface. For many unstable elements, mass of the longest-lived known isotope is given in parentheses. 1999 revisions. (See also Appendix D.)